Lecture Notes on Data Engineering and Communications Technologies

Volume 140

Series Editor

Fatos Xhafa, Technical University of Catalonia, Barcelona, Spain

The aim of the book series is to present cutting edge engineering approaches to data technologies and communications. It will publish latest advances on the engineering task of building and deploying distributed, scalable and reliable data infrastructures and communication systems.

The series will have a prominent applied focus on data technologies and communications with aim to promote the bridging from fundamental research on data science and networking to data engineering and communications that lead to industry products, business knowledge and standardisation.

Indexed by SCOPUS, INSPEC, EI Compendex.

All books published in the series are submitted for consideration in Web of Science.

Joong Hoon Kim · Kusum Deep ·
Zong Woo Geem · Ali Sadollah · Anupam Yadav
Editors

Proceedings of 7th International Conference on Harmony Search, Soft Computing and Applications

ICHSA 2022

Springer

Editors
Joong Hoon Kim
School of Civil, Environmental and Architectural Engineering
Korea University
Seoul, Korea (Republic of)

Zong Woo Geem
Department of Energy IT
Gachon University
Seongnam, Korea (Republic of)

Anupam Yadav
Department of Mathematics
Dr. B. R. Ambedkar National Institute of Technology
Jalandhar, India

Kusum Deep
Department of Mathematics
Indian Institute of Technology Roorkee
Roorkee, India

Ali Sadollah
School of Mechanical Engineering
University of Science and Culture
Tehran, Iran

ISSN 2367-4512 ISSN 2367-4520 (electronic)
Lecture Notes on Data Engineering and Communications Technologies
ISBN 978-981-19-2947-2 ISBN 978-981-19-2948-9 (eBook)
https://doi.org/10.1007/978-981-19-2948-9

This Springer imprint is published by the registered company Springer Nature Singapore Pte Ltd.
The registered company address is: 152 Beach Road, #21-01/04 Gateway East, Singapore 189721, Singapore

Preface

The International Conference on Harmony Search, Soft Computing and Applications (ICHSA) is the flagship conference and forum of harmony search and soft computing for researchers and practitioners in the highly active fields of theoretical fundamentals and large-scale applications. It has a glorious history; the earlier events of the conference were held in South Korea, Spain, India, China, and Turkey. After the successful 6th ICHSA held in Istanbul, Turkey, in July 2020, it was decided to host it back at Korea University in South Korea online during February 23–24, 2022.

The 7th ICHSA (ICHSA 2022) consists of eight themes which include from advances in harmony search and soft computing to artificial intelligence and various engineering applications. This book is the outcome of the research papers presented at ICHSA 2022. It contains a varied range of chapters, and these chapters throw light on the state-of-the-art status of research in the areas of harmony search, soft computing technique, and their applications in engineering and science.

Seoul, Korea (Republic of)	Joong Hoon Kim
Roorkee, India	Kusum Deep
Seongnam, Korea (Republic of)	Zong Woo Geem
Tehran, Iran	Ali Sadollah
Jalandhar, India	Anupam Yadav

Contents

About the Editors

Prof. Joong Hoon Kim dean of Engineering College of Korea University, obtained his Ph.D. degree from the University of Texas at Austin in 1992 with the thesis title "Optimal replacement/rehabilitation model for water distribution systems". Professor Kim's major areas of interest include: optimal design and management of water distribution systems, application of optimization techniques to various engineering problems, and development and application of evolutionary algorithms. He has been the faculty of School of Civil, Environmental and Architectural Engineering at Korea University since 1993 and is now serving as the dean of Engineering College. He has hosted international conferences including APHW 2013, ICHSA 2014 and 2015, and HIC 2016 and has given keynote speeches at many international conferences including AOGS 2013, GCIS 2013, SocPros 2014 and 2015, SWGIC 2017, and RTORS 2017. He is a member of National Academy of Engineering of Korea since 2017.

Dr. Kusum Deep is a full Professor (HAG), with the Department of Mathematics as well as Joint Faculty at the Centre for Artificial Intelligence and Data Science at the Indian Institute of Technology Roorkee, India. Also, she is a visiting professor, Liverpool Hope University, UK, University of Technology Sydney, Australia, and University of Wollongong, Australia. With B.Sc. Hons and M.Sc. Hons. School from Centre for Advanced Studies, Panjab University, Chandigarh, she is an M.Phil. Gold Medalist. She earned her Ph.D. from UOR (now IIT Roorkee) in 1988. She has been a national scholarship holder and a postdoctoral from Loughborough University, UK, assisted by International Bursary funded by Commission of European Communities, Brussels. She has won numerous awards like Khosla Research Award, UGC Career Award, Starred Performer of IITR Faculty, best paper awards by Railway Bulletin of Indian Railways, special facilitation in memory of late Prof. M. C. Puri, AIAP Excellence Award. Recently, she was one of the four women from IIT Roorkee to feature in the ebook "Women in STEM-2021" celebrating the contributions made by 50 Indian women in STEM published by Confederation of Indian Industries. She has authored two books, supervised 20 Ph.Ds. and published 125 research papers. She is a Senior Member of ORSI, CSI, IMS, and ISIM. She is the executive editor of

International Journal of Swarm Intelligence, Inderscience. She is associate editor of *Swarm and Evolutionary Algorithms*, Elsevier, and is on the editorial board of many journals. She is the founder president of Soft Computing Research Society, India. She is the general chair of series of International Conference on Soft Computing for Problems Solving (SocProS). She has a vast teaching experience in Mathematics, Operations Research, Numerical and Analytical Optimization, Parallel Computing, Computer Programming, Numerical Methods, etc. Her research interests are nature-inspired optimization techniques, particularly Evolutionary Algorithms, and Swarm Intelligence Techniques and their applications to solve real-life problems.

Zong Woo Geem is a Faculty of IT Convergence College at Gachon University, South Korea. He has obtained B.Eng. in Chung-Ang University, Ph.D. in Korea University, and M.Sc. in Johns Hopkins University and researched at Virginia Tech, University of Maryland-College Park, and Johns Hopkins University. He invented a music-inspired optimization algorithm, harmony search, which has been applied to various scientific and engineering problems. His research interest includes phenomenon-mimicking algorithms and their applications to energy, environment, and water fields. He has served as an editor/editorial board for various journals including *Engineering Optimization, Swarm Evolutionary Computation, International Journal of Bio-Inspired Computation, Journal of Applied Mathematics, Applied Sciences, Complexity, and Sustainability.*

Dr. Ali Sadollah received his B.S. degree in mechanical engineering, solid states, from Azad University, Semnan Branch, Iran in 2007. He received his M.S. degree in mechanical engineering, applied mechanics, from Semnan University, Semnan, Iran in 2010. He obtained his Ph.D. at the faculty of mechanical engineering, University of Malaya, Kuala Lumpur, Malaysia in 2013. Also, since 2014, he served as a postdoctoral research fellow for more than 2 years at Korea University, Seoul, South Korea. In 2016 for a year, he served as a research staff at Nanyang Technological University, Singapore. Afterwards, he was a guest assistant professor at the Sharif University of Technology, Tehran, Iran. Currently, he is an assistant professor at the department of mechanical engineering at the University of Science and Culture, Tehran, Iran. His research interests include algorithm development, engineering optimization, metaheuristics, applications of soft computing methods in engineering, artificial neural networks, and computational solid mechanics.

Dr. Anupam Yadav assistant professor, Department of Mathematics, Dr. B. R. Ambedkar National Institute of Technology Jalandhar, India. His research area includes numerical optimization, soft computing, and artificial intelligence, and he has more than 10 years of research experience in the areas of soft computing and optimization. Dr. Yadav has done a Ph.D. in soft computing from the Indian Institute of Technology Roorkee, and he had worked as a research professor at Korea University. He has published more than 25 research articles in journals of international repute and has published more than fifteen research articles in conference proceedings. Dr. Yadav has authored a textbook entitled *An Introduction to Neural Network*

Methods for Differential Equations. He has edited three books which are published by AISC, Springer Series. Dr. Yadav was the general chair, convener, and member of the steering committee of several international conferences. He is a member of various research societies.

COVID-19 Chest X-rays Classification Through the Fusion of Deep Transfer Learning and Machine Learning Methods

Nour Eldeen M. Khalifa, Mohamed Hamed N. Taha, Ripon K. Chakrabortty, and Mohamed Loey

Abstract One of the most challenging issues that humans face in the last decade is in the health sector, and it is threatening his existence. The COVID-19 is one of those health threats as declared by the World Health Organization (WHO). This spread of COVID-19 forced WHO to declare this virus as a pandemic in 2019. In this paper, COVID-19 chest X-rays classification through the fusion of deep transfer learning and machine learning methods will be presented. The dataset "DLAI3 Hackathon Phase3 COVID-19 CXR Challenge" is used in this research for investigation. The dataset consists of three classes of X-rays images. The classes are COVID-19, Thorax Disease, and No Finding. The proposed model is made up of two main parts. The first part for feature extraction, which is accomplished using three deep transfer learning algorithms: AlexNet, VGG19, and InceptionV3. The second part is the classification using three machine learning methods: K-nearest neighbor, support vector machine, and decision trees. The results of the experiments show that the proposed model using VGG19 as a feature extractor and support vector machine. It reached the highest conceivable testing accuracy with 97.4%. Moreover, the proposed model achieves a superior testing accuracy than VGG19, InceptionV3, and other related works. The obtained results are supported by performance criteria such as precision, recall, and F1 score.

N. E. M. Khalifa · M. H. N. Taha
Department of Information Technology, Faculty of Computers and Artificial Intelligence, Cairo University, Cairo 12613, Egypt
e-mail: nourmahmoud@cu.edu.eg

M. H. N. Taha
e-mail: mnasrtaha@cu.edu.eg

R. K. Chakrabortty
School of Engineering and IT, UNSW Canberra at ADFA, Canberra, Australia
e-mail: r.chakrabortty@adfa.edu.au

M. Loey (✉)
Department of Computer Science, Faculty of Computers and Artificial Intelligence, Benha University, Benha 13518, Egypt
e-mail: mloey@fci.bu.edu.eg

J. H. Kim et al. (eds.), *Proceedings of 7th International Conference on Harmony Search, Soft Computing and Applications*, Lecture Notes on Data Engineering and Communications Technologies 140, https://doi.org/10.1007/978-981-19-2948-9_1

Keywords Coronavirus · COVID-19 · Deep transfer learning · Machine learning · Image classification

1 Introduction

The novel coronavirus disease of 2019 (COVID-19) has had a global effect with over 3,000,000 deaths and over 160 million infected cases so far [1]. The epidemic was affirmed a Public Health Emergency of International Concern in January 2020 [2]. In February 2020, the 2019 latest coronavirus named COVID-19 by World Health Organization (WHO) [3, 4]. A COVID-19 patient may acquire multiple displays and signs of contagion once infected, including high temperature, cough, and respiratory disease [5]. Coronaviruses are classified into four subgroups: α, β, δ, and γ. New SARS (2019) was identified as a coronavirus organ of the β community [6]. β-coronaviruses have induced illness in humans who have come into contact with wild creatures, most often bats or rodents [7].

Deep learning (DL) is now increasingly becoming a key technology in the classification and detection of image/video. Transfer learning (TL) is an alternate method of training deep learning models, whereby a deep learning network is pre-weighted from a different domain with the output of a previous training model [8]. This approach is widely used as a basis for the initialization of deep learning models that are then fine-tuned using the restricted medical sample data available [9–11].

In this research, DL models were presented to automatically classify the digital images of COVID-19 chest X-rays scans. The medical chest X-rays scans were enhanced using classical data augmentation enlarge the dataset number of images. Following that, a machine learning classifier is used to combine the classification outcomes' class outputs. This analysis is novel according to the following points:

(i) The presented DL models have an end-to-end configuration and are constructed using conventional function extraction.
(ii) Chest X-rays scans are one of the most effective methods for COVID-19 classification.
(iii) In the small COVID-19 dataset, the DL models with machine learning have been shown to be extremely accurate.

The remainder of the research is structured appropriately. Section 2 includes a summary of the related work. Section 3 introduced the material of the work. Section 4 shows the presented model. Section 5, preliminary outcomes are evaluated and measurement, and the underlying assumptions and future research are introduced in Sect. 6.

2 Related Works

In [12], the authors proposed “CoroNet,” an Xception model to detect coronavirus contagion from CXRs images. The trial outcomes showed that the proposed model achieved an overall accuracy of 98.6% with a precision of 93% and a recall of 98.2% for the four classes. Abbas et al. [13] introduced Transfer, and Compose (DeTraC) with DL to diagnose COVID-19 CXRs images. The authors used five transfer learning models (ResNet, AlexNet, GoogleNet, VGG19, and SqueezeNet) with DeTraC. The outcomes demonstrated DeTraC’s capacity to identify COVID-19 events. The introduced model achieved an accuracy of 93.1% in the classification of COVID-19 CRXs images from normal and infected cases.

Rajaraman et al. [14] proposed a custom deep learning model with iteratively pruned ensembles to classify CXRs images of coronavirus. The proposed model reached a recognition rate of 99.01% with a sensitivity and precision of 99.01% on test data from a small coronavirus dataset of 27 COVID-19 images. In [15], authors introduced a DL model to diagnose COVID-19 based on CXRs images. The research measures the impact of a VGG16 DTL model for the recognition of coronavirus using radiographs. Comprehensive study outcomes have shown that the proposed model classifies coronavirus with a classification rate of 85%. Narayan Das et al. [16] proposed a TL-based (Xception) model for COVID-19 detection. The authors model achieves 97.4% accuracy with a sensitivity and specificity of 97.09, 97.29% on test data from a small coronavirus dataset.

Agrawal et al. [17] proposed “FocusCovid,” an advanced COVID-19 identification system based on deep learning and chest X-rays pictures. To test the proposed model, two distinct datasets are used. The authors’ model achieves 99.2% and 95.2 accuracy for binary and multi-class recognition with a sensitivity of 99.2%, 95.2% and precision of 99.2% and 95.6% on test data. Gaur et al. [18] proposed a deep convolution neural network to detect COVID-19 from chest X-rays. TL is used to test three pre-trained CNN models (VGG16, Inception, and EfficientNet) that are ideal for mobile applications. The findings indicate that the proposed solution provided a model of high quality, with an overall accuracy of 92.9% and a sensitivity of 94.8%.

3 Dataset Description

The dataset utilized in this study is an online collection titled “DLAI3 Hackathon Phase3 COVID-19 CXR Challenge” [19]. The dataset is created for Phase3 of the third deep learning and AI summer/winter school (DLAI3). The dataset author collected its COVID-19 chest X-rays images from online public various sources. It is divided into three classes (COVID-19, Thorax Disease, and No Finding). It contains 363 images for the “COVID-19” class, 3736 images for the “Thorax Disease” class, and 1408 images for the “No Finding” class with total 5507 images. The source of

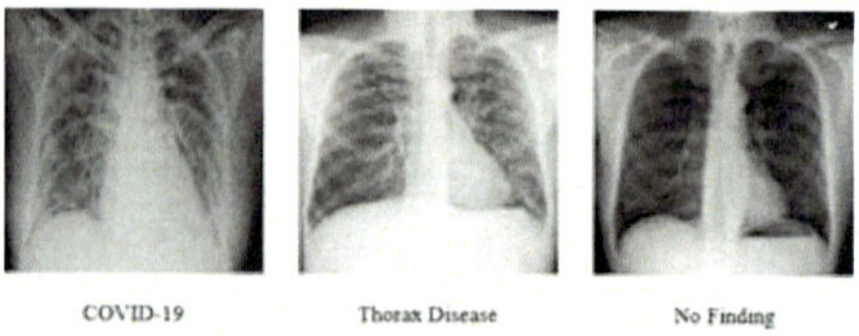

Fig. 1 Image samples for various classes in the dataset

all images used in this dataset is described in detail in [19]. A sample image for every class is presented in Fig. 1.

4 The Proposed Model Architecture

The proposed model is based on the fusion of deep transfer learning architectures with classical machine learning methods. It consists of two parts. The first part is dedicated to the feature extractions using different deep transfer learning architectures. The second part is for the classification process using different classical machine learning methods. The selected deep transfer learning architectures are AlexNet [20], VGG19 [21], and InceptionV3 [22]. The selected machine learning methods in the proposed model for the second part are K-nearest neighbor (KNN) [23], support vector machine (SVM) [24], and decision trees (DT) [25]. Figure 2 illustrates the proposed model with different deep transfer learning architecture and machine learning methods.

4.1 Deep Transfer Learning

DL design is often used to implement image analysis and medical machine vision in other regions, in addition to their application for medical X-rays detection [26]. AlexNet [20] consists of eight layers in total, of which five are convolutional layers and three are fully connected layers. In VGG19 [21], the architecture's input dimensions are fixed at 244×244. As part of a preprocessing step, the average RGB value of each pixel in an image is subtracted. Many CNNs based on TL models were used as feature extractor like AlexNet, VGG19. The TL models outcome a feature vector from an input image before the classification layers.

4.2 K-Nearest Neighbor

KNNs work by calculating the distances between an unlabeled object and all data instances. The KNN technique makes the assumption that the new case/data and previous cases are comparable and assigns the new case to the category that is most

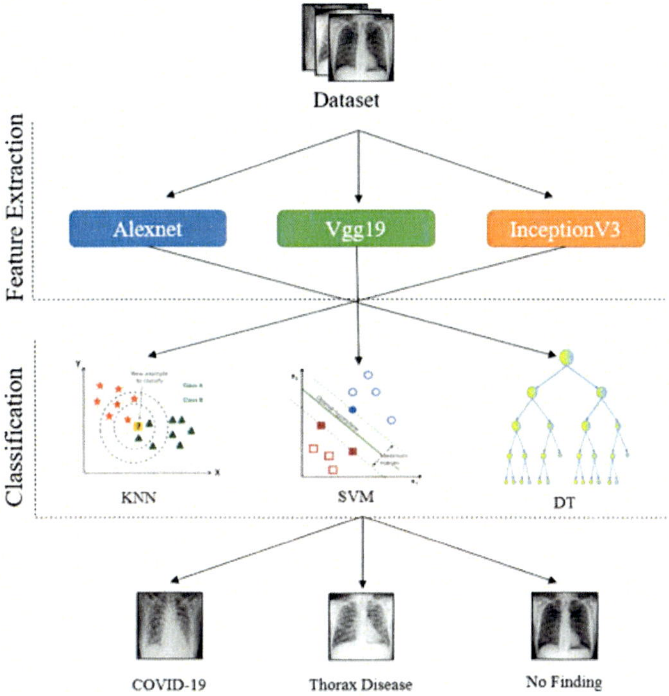

Fig. 2 Suggested model architecture

similar to the existing categories [27]. The KNN algorithm employs distance functions, often the standard Euclidean distance $d(x, y)$ between two points x and y (1):

$$D(x, y) = \sqrt{\sum_{i=1}^{n}(x_i - y_i)^2} \tag{1}$$

4.3 Support Vector Machine

The support vector machine (SVM) algorithm's objective is to determine the optimal line or decision boundary for categorizing n-dimensional space into classes in such a way that more data points may be readily classified in the future. The optimal choice boundary is referred to as a hyperplane [24]. Equation (2) is a popular implementation of the SVM algorithm, where k is a numeric value between 0 and 1, $w.s - t$ is the result, w and t are coefficients of linear categorization, and s is a vector-based input. Equation (3) will force a reduction in the loss function [28].

$$\mathrm{SVM}_{\mathrm{a_b}} = \max(0,\ 1 - k_\mathrm{b}(w.s_\mathrm{b} - t)) \tag{2}$$

$$\mathrm{SVM}_{\mathrm{loss}} = \frac{1}{d}\sum_{g=1}^{d}\max\left(0,\ p_g\right) \tag{3}$$

4.4 *Decision Tree*

In a decision tree, the tree is divided into subtrees based on the answer (yes/no) [25]. The decision tree relies on entropy and creation of information. Entropy measures the degree of data instability as shown in Eq. (4), where O is the data, c denotes the label result, and $p(x)$ denotes the percentage of u label. The knowledge acquisition (KA) is calculated by determine the entropy difference between outcomes, as given in Eq. (5). q denotes a subset of the dataset [29].

$$E(O) = \sum_{j=1}^{u} -p\left(c_j\right).\log\left(p\left(c_j\right)\right) \tag{4}$$

$$\mathrm{KA} = E(O) - \sum_{\mathrm{q}\in\mathrm{O}} p(q)E(q) \tag{5}$$

5 Experimental Results

All tests were conducted on a computer server equipped with 96 GB of RAM and a Xeon CPU from Intel (2 GHz). The experiments were created using the MATLAB software package and were CPU-specific. During the experiments, the following specifications were chosen.

- Three deep transfer learning models for feature extractions (AlexNet, VGG19, and InceptionV3). Three different classifiers are evaluated (K-nearest neighbor, SVM, and decision trees).
- The dataset was partitioned into two halves (80% of the data is used in the training phase, and 20% is used in the testing process). Data augmentation was applied to the training data.
- Testing accuracy, precision (P), recall (R), and F1 score are selected as performance metrics and presented from Eqs. (6)–(9) along with the consumed time during the training process.

$$\text{Testing Accuracy} = \frac{\text{TPos} + \text{TNeg}}{(\text{TPos} + \text{FPos}) + (\text{TNeg} + \text{FNeg})} \tag{6}$$

$$P = \frac{\text{TPos}}{(\text{TPos} + \text{FPos})} \tag{7}$$

$$R = \frac{\text{TPos}}{(\text{TPos} + \text{FNeg})} \tag{8}$$

$$\text{Score} = 2 * \frac{P * R}{(P + R)} \tag{9}$$

where TPos indicates the number of samples with a True Positive result, TNeg indicates the number of samples with True Negative values, FPos indicates the number of samples with False Positive results, and FNeg when it comes to calculating the False Negative rate, it measures the number of False Negatives from a confusion matrix.

5.1 Deep Transfer Learning as Features Extractor

Table 1 presents the testing accuracy for AlexNet as a feature extractor with different machine learning methods. The SVM classifier achieved the highest testing accuracy for every class with 96.7, 95.1, and 97.2% for COVID-19, Thorax Disease, and No Finding class consequently. The overall testing accuracy for AlexNet with SVM is 96.6%. Table 2 shows the testing accuracy for VGG19 as a feature extractor with different machine learning methods. Also, for every class, and for all tests, the SVM

Table 1 Testing accuracy for AlexNet

AlexNet	KNN (%)	SVM (%)	DT (%)
"COVID-19" class	80.3	96.7	67.9
"Thorax Disease" class	95.8	97.2	91.2
"No Finding" class	87.3	95.1	80.7
Overall testing accuracy	92.6	96.6	87.5

Table 2 Testing accuracy for VGG19

VGG19	KNN (%)	SVM (%)	DT (%)
"COVID-19" class	78.9	95.5	63.2
"Thorax Disease" class	97.1	99.0	90.4
"No Finding" class	91.2	93.5	79.3
Overall testing accuracy	94.3	97.4	85.9

Table 3 Testing accuracy for InceptionV3

InceptionV3	KNN (%)	SVM (%)	DT (%)
"COVID-19" class	75.0	88.1	65.5
"Thorax Disease" class	74.4	88.2	64.9
"No Finding" class	96.0	96.9	88.6
Overall testing accuracy	88.2	94.1	80.7

Table 4 Testing accuracy for VGG19 as feature extractor with different machine learning methods

InceptionV3	Recall (%)	Precision (%)	F1 score (%)	Testing accuracy (%)
KNN	90.2	89.1	89.6	94.3
SVM	94.1	96.0	95.1	97.4
DT	76.34	77.63	76.98	85.9

classifier obtained the best results of accuracy 97.4%. Table 3 illustrates the testing accuracy for InceptionV3 as a feature extractor with different machine learning methods. Also, the InceptionV3 classifier achieved the highest testing accuracy for every class and with overall testing accuracy of 94.1%.

Tables 1, 2, and 3 illustrate that the SVM classifier achieved the highest testing accuracies with the different deep transfer learning methods. The highest testing accuracy is achieved when the VGG19 and SVM classifier is selected with 97.4%. It seems that the features of images are vanishing when deeper neural networks were used such as InceptionV3 as reflected in the testing accuracy throughout the different deep transfer learning models.

5.2 *Proposed Model Performance Metrics*

Performance metrics such as precision, recall, and F1 score indicate the performance of any proposed model. Section 5.1 concludes that VGG19 with SVM classifier is the optimal model as it achieved 97.4%. The performance metrics for VGG19 are presented in Table 4. The proposed model with VGG19 and SVM classifier achieved a 94.1% in recall, 96% in precision, and 95.1% in F1 score. The performance metrics strengthen the obtained results about the selection of VGG19 with the SVM classifier as an optimal model.

5.3 *Results and Discussion*

In this section, a comparison between the proposed model against deep transfer learning as a feature extractor and a classifier is presented. Figure 3 illustrates

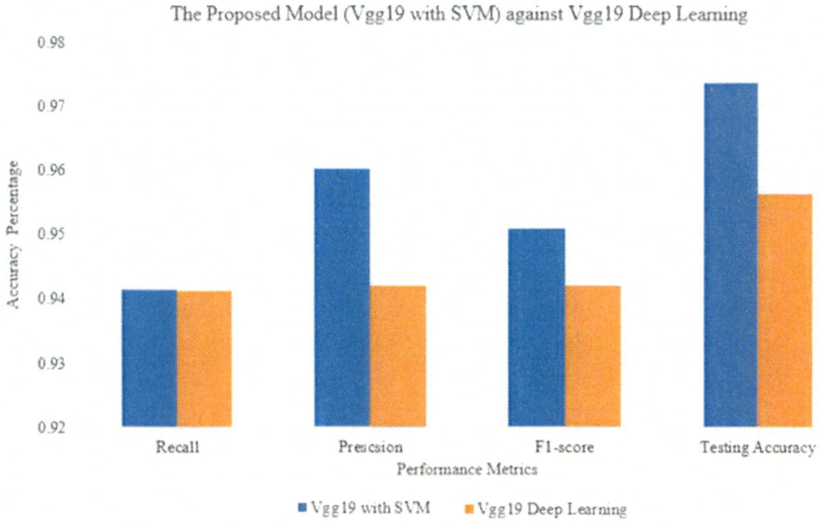

Fig. 3 Proposed model performance metrics versus deep learning

the proposed model (VGG19 with SVM) testing accuracy along with performance metrics against VGG19 as features extractor and as classifier as well. Figure 3 illustrates that the proposed model achieved better accuracies in performance metrics and testing accuracy over the deep learning model. The proposed model achieved 96.1% in the precision metric, while the deep learning model achieved 94.2%. In the F1 score, the proposed model achieved 95.1%, while the deep learning model achieved 94.2%. In the overall testing accuracy, the proposed model achieved 97.4%, while the deep learning model achieved 95.6%. In the recall metric, both models achieved a competitive accuracy with 94.1%.

In the related work, we noted a small number of new studies that have used the TL model for COVID-19 CXRs diagnosis. Our goal is to select a suitable TL model with machine learning through relative testing dataset. The related work is also used to address the limited data samples of coronavirus CXRs images. The experimental outcome shows that the proposed model using VGG19 as a feature extractor with SVM as classifier achieved the highest testing accuracy possible with 97.4%. Moreover, the proposed model achieves a higher testing accuracy than VGG19 as a deep learning model.

6 Conclusion and Future Works

The coronavirus pandemic is trying to jeopardize global healthcare systems. With advances in computer algorithms, especially artificial intelligence, early detection of this kind of virus will help in the patients' rapid recovery. The categorization of

COVID-19 chest X-rays using a combination of deep transfer learning and machine learning techniques was described in this article. The dataset “DLAI3 Hackathon Phase3 COVID-19 CXR Challenge” was used in this research for investigation. It consisted of three classes of X-rays images. The classes are COVID-19, Thorax Disease, and No Finding. The proposed model was made up of two parts. The first part was feature extraction, which was accomplished through the use of three deep transfer learning algorithms: AlexNet, VGG19, and InceptionV3. The second part was the classification using three machine learning methods, and they were K-nearest neighbor, support vector machine, and decision trees. The experimental results showed that the proposed model using VGG19 as a feature extractor and support vector machine as classifier achieved the highest testing accuracy possible with 97.4%.

References

1. Coronavirus (COVID-19) (2021) Google news. https://news.google.com/covid19/map?hl=en-US&gl=US&ceid=US:en. Accessed 18 May 2021
2. Li J et al (2020) Game consumption and the 2019 novel coronavirus. Lancet Infect Dis 20(3):275–276. https://doi.org/10.1016/S1473-3099(20)30063-3
3. Loey M, Manogaran G, Taha MHN, Khalifa NEM (2021) A hybrid deep transfer learning model with machine learning methods for face mask detection in the era of the COVID-19 pandemic. Measurement 167:108288. https://doi.org/10.1016/j.measurement.2020.108288
4. Loey M, Smarandache F, Khalifa NEM (2020) Within the lack of chest COVID-19 x-ray dataset: a novel detection model based on GAN and deep transfer learning. Symmetry 12(4), 4. https://doi.org/10.3390/sym12040651
5. Mahase E (2020) Coronavirus: covid-19 has killed more people than SARS and MERS combined, despite lower case fatality rate. BMJ 368:m641. https://doi.org/10.1136/bmj.m641
6. Decaro N, Lorusso A (2020) Novel human coronavirus (SARS-CoV-2): a lesson from animal coronaviruses. Vet Microbiol 244:108693. https://doi.org/10.1016/j.vetmic.2020.108693
7. Chang L, Yan Y, Wang L (2020) Coronavirus disease 2019: coronaviruses and blood safety. Transfus Med Rev 34(2):75–80. https://doi.org/10.1016/j.tmrv.2020.02.003
8. Ribani R, Marengoni M (2019) A survey of transfer learning for convolutional neural networks. In: 2019 32nd SIBGRAPI conference on graphics, patterns and images tutorials (SIBGRAPI-T). pp 47–57. https://doi.org/10.1109/SIBGRAPI-T.2019.00010
9. Loey M, ElSawy A, Afify M (2020) Deep learning in plant diseases detection for agricultural crops: a survey. Int J Serv Sci Manage Eng Technol (IJSSMET) www.igi-global.com/article/deep-learning-in-plant-diseases-detection-for-agricultural-crops/248499. Accessed 11 Apr 2020
10. Loey M, Naman MR, Zayed HH (2020) A survey on blood image diseases detection using deep learning. Int J Serv Sci Manage Eng Technol (IJSSMET) www.igi-global.com/article/a-survey-on-blood-image-diseases-detection-using-deep-learning/256653. Accessed 17 June 2020
11. Khalifa N, Loey M, Taha M, Mohamed H (2019) Deep transfer learning models for medical diabetic retinopathy detection. Acta Inform Medica 27(5):327. https://doi.org/10.5455/aim.2019.27.327-332
12. Khan AI, Shah JL, Bhat MM (2020) CoroNet: a deep neural network for detection and diagnosis of COVID-19 from chest x-ray images. Comput Methods Programs Biomed 196:105581. https://doi.org/10.1016/j.cmpb.2020.105581

13. Abbas A, Abdelsamea MM, Gaber MM (2020) Classification of COVID-19 in chest X-ray images using DeTraC deep convolutional neural network. Appl Intell. https://doi.org/10.1007/s10489-020-01829-7
14. Rajaraman S, Siegelman J, Alderson PO, Folio LS, Folio LR, Antani SK (2020) Iteratively pruned deep learning ensembles for COVID-19 detection in chest X-rays. IEEE Access 8:115041–115050. https://doi.org/10.1109/ACCESS.2020.3003810
15. Civit-Masot J, Luna-Perejón F, Domínguez Morales M, Civit A (2020) Deep learning system for COVID-19 diagnosis aid using X-ray pulmonary images. Appl Sci 10(13), 13. https://doi.org/10.3390/app10134640
16. Narayan Das N, Kumar N, Kaur M, Kumar V, Singh D (2020) Automated deep transfer learning-based approach for detection of COVID-19 infection in chest X-rays. IRBM. https://doi.org/10.1016/j.irbm.2020.07.001
17. Agrawal T, Choudhary P (2021) FocusCovid: automated COVID-19 detection using deep learning with chest X-ray images. Evol Syst. https://doi.org/10.1007/s12530-021-09385-2
18. Gaur L, Bhatia U, Jhanjhi NZ, Muhammad G, Masud M (2021) Medical image-based detection of COVID-19 using deep convolution neural networks. Multimed Syst. https://doi.org/10.1007/s00530-021-00794-6
19. Jonathan HC (2020) DLAI3 Hackathon phase3 COVID-19 CXR challenge. https://www.kaggle.com/jonathanchan/dlai3-hackathon-phase3-covid19-cxr-challenge. Accessed 26 Sep 2020
20. Krizhevsky A, Sutskever I, Hinton GE (2012) Imagenet classification with deep convolutional neural networks. In: Advances in neural information processing systems. pp 1097–1105
21. Liu S, Deng W (2015) Very deep convolutional neural network based image classification using small training sample size. In: 2015 3rd IAPR Asian conference on pattern recognition (ACPR). pp 730–734. https://doi.org/10.1109/ACPR.2015.7486599
22. Szegedy C, Vanhoucke V, Ioffe S, Shlens J, Wojna Z (2016) Rethinking the inception architecture for computer vision. In: Proceedings of the IEEE conference on computer vision and pattern recognition. pp 2818–2826
23. Jiang S, Pang G, Wu M, Kuang L (2012) An improved K-nearest-neighbor algorithm for text categorization. Expert Syst Appl 39(1):1503–1509
24. Noble WS (2006) What is a support vector machine? Nat Biotechnol 24(12):1565–1567
25. Quinlan JR (1986) Induction of decision trees. Mach Learn 1(1):81–106
26. Khalifa NEM, Loey M, Taha MHN (2020) Insect pests recognition based on deep transfer learning models. J Theor Appl Inf Technol 98(1)
27. Taunk K, De S, Verma S, Swetapadma A (2019) A brief review of nearest neighbor algorithm for learning and classification. In: 2019 International conference on intelligent computing and control systems (ICCS). pp 1255–1260. https://doi.org/10.1109/ICCS45141.2019.9065747
28. Çayir A, Yenidoğan I, Dağ H (2018) Feature extraction based on deep learning for some traditional machine learning methods. In: 2018 3rd International conference on computer science and engineering (UBMK). pp 494–497. https://doi.org/10.1109/UBMK.2018.8566383
29. Navada A, Ansari AN, Patil S, Sonkamble BA (2011) Overview of use of decision tree algorithms in machine learning. In: 2011 IEEE control and system graduate research colloquium. pp 37–42. https://doi.org/10.1109/ICSGRC.2011.5991826

Quantitative and Qualitative Analysis of Harmony Search Algorithm in Geomechanics and Its Applications

Sina Shaffiee Haghshenas, Nicola Careddu, Saeid Jafarzadeh Ghoushchi, Reza Mikaeil, Tae-Hyung Kim, and Zong Woo Geem

Abstract The harmony search (HS) algorithm is one of the meta-heuristic algorithms that was inspired by the concept of the musical process with the aim of harmony, to achieve the best solution. In comparison with other meta-heuristic algorithms, one of the most significant characteristics of this algorithm that has increased the flexibility of the algorithm in search of solution spaces is the use of all the solutions in its memory. The literature review shows that, according to the high efficiency of the harmony search algorithm, it has been widely used in various sciences in recent years. Hence, the main purpose of this study is to review the applications of the harmony search algorithm in geomechanics which is a significant research topic in the engineering and academic sectors. For this purpose, articles of geomechanics including the two main disciplines, namely soil mechanics and rock mechanics, are evaluated from 2011 to 2021. Also, two qualitative and quantitative investigations are applied to review articles based on the Web of Science (WOS) platform. This study indicates that the harmony search algorithm can be applied as a powerful tool for modeling some problems involved in geomechanics.

S. Shaffiee Haghshenas
Department of Civil Engineering, University of Calabria, 87036 Rende, Italy

N. Careddu
Department of Civil, Environmental Engineering and Architecture (DICAAr), University of Cagliari, Institute of Environmental Geology and Geoengineering, IGAG, CNR, via Marengo 2, 09123 Cagliari, Italy

S. Jafarzadeh Ghoushchi
Faculty of Industrial Engineering, Urmia University of Technology, Urmia, Iran

R. Mikaeil
Department of Mining and Engineering, Faculty of Environment, Urmia University of Technology, Urmia, Iran

T.-H. Kim
Department of Civil Engineering, Korea Maritime and Ocean University, Pusan 49112, Korea

Z. W. Geem (✉)
College of IT Convergence, Gachon University, Seongnam 13120, Korea
e-mail: zwgeem@gmail.com

J. H. Kim et al. (eds.), *Proceedings of 7th International Conference on Harmony Search, Soft Computing and Applications*, Lecture Notes on Data Engineering and Communications Technologies 140, https://doi.org/10.1007/978-981-19-2948-9_2

Keywords Harmony search algorithm · Geomechanics · Rock mechanics · Soil mechanics · Literature review · WOS

1 Introduction

Geomechanics is the science of studying the mechanical behavior of soil and rock. Soil mechanics and rock mechanics are the two main branches of geomechanics that are used in a wide range of projects of theoretical and practical including engineering geology, mining, petroleum, and civil engineering. In many construction and operation projects, the geomechanical study of projects plays a key role in the success of a project; therefore, insufficient attention to this issue can lead to irreparable loss of life, the environment, and finance. Extensive studies have been conducted in various branches of geomechanics such as rock mechanics [1–11], geotechnics [12–17], and tunneling and underground spaces [18–24]. Li et al. [25] introduced a forecast process of the uniaxial compressive strength by developing the group method of data handling algorithm. Their results approved that the proposed GMDH model can be considered a robust prediction system [25]. Ching et al. [26] evaluated the deformation modulus of the rock mass. They used the four methods to determine such modulus. Based on their results, the hierarchical Bayesian model was a reliable model in comparison with other models [26]. Bagheri and Rezania evaluated the geological and geotechnical properties of London clay from the Isle of Sheppey. Their study integrates the earlier and recent studies with the aim to develop the current knowledge of the geotechnical properties of this stiff clay from the east of the London basin [27]. The unconfined compressive strength of clay material mixed with recycled additive was predicted by Al-Bared et al. [28]. They designed a series of laboratory tests, and then, they used two hybrid intelligence systems. The results obtained demonstrated the effectiveness of both hybrid predictive models [28]. Faradonbeh et al. [10] investigated rockburst hazards using two robust clustering techniques. They provided some suggestions to reduce the effect of rockburst by optimizing the diameter and shape of the underground openings [29]. Geological conditions and clogging of tunneling were investigated by Bai et al. [30]. They used machine learning-based approaches, and the results obtained indicated that there was suitable compliance between measurements and the application of machine learning-based approaches [30]. Pham et al. [31] have focused on presenting new models for classification of soils based on laboratory tests. They applied the three artificial intelligence techniques, namely Adaboost, Tree, and ANN. In addition, 440 samples of the real project were applied to develop and present the proposed methodology, and this data included clay content, moisture content, specific gravity, void ratio, plastic, and liquid limit parameters. Finally, their proposed methodology could achieve highly acceptable degrees of accuracy, and also, it could decrease the cost of projects [31].

A review of the previous literature reveals that various approaches and techniques have been used in the study, evaluation, and analysis of research related to geomechanical problems. Also, in recent years, the use of artificial intelligence methods has been relatively widely used in this research. Therefore, the purpose of this study is to review and analyze the qualities and quantities of the articles related to the application of harmony search algorithm as one of the practical and robust algorithms in the field of artificial intelligence in the branches of geomechanics during the last decade. It should be noted that all documents are reviewed from the Science Citation Index (SCI), which is achieved by subscription from the ISI, Web of Science, Philadelphia, PA, USA.

2 Methodology

Literature review plays a key role in assessing the body of knowledge to identify potential research gaps and highlight their boundaries [32, 33]. Therefore, qualitative and quantitative analysis is presented by reviewing all documents and articles related to the branches of geomechanics and harmony search algorithm based on the Web of Science (WOS) platform in this study. For this purpose, firstly a review of the application of the harmony search algorithm is performed in all disciplines, and then, a review of articles and other documents related to the application of the harmony search algorithm in geomechanics is reviewed. It should be noted that the harmony search algorithm was introduced in 2001, and we wanted to show that after a decade of introducing the algorithm, how much is the participation of this algorithm in scientific issues; therefore, this study will focus on published documents from the January of 2011 to the July of 2021.

2.1 Quantitative Analysis

The SCI is systematically investigated during the quantitative analysis from 2011 to the July of 2021. As mentioned before, first, a quantitative review of the application of the HS algorithm in all disciplines is done. Then, a quantitative review of the application of the HS algorithm in branches of geomechanics is conducted. For searching in the second quantitative review, the set of keyword combinations are determined, such as "HS algorithm AND Rock Mechanics," "HS algorithm AND Rock Breakage," "HS algorithm AND Ornamental Stone," "HS algorithm AND Rock Drilling," "HS algorithm AND Land-slide," "HS algorithm AND Geotechnics," "HS algorithm AND Soil Mechanics," "HS algorithm AND Foundations," "HS algorithm AND Flyrock," "HS algorithm AND Tunneling," and "HS algorithm AND Slope Stability." It should be noted that this search was selected only in the title to focus more on articles. The parameters that were analyzed included authorship, patterns of international collaboration, number of times cited, and reprint author's address. Citation analysis was conducted according to the impact factor which is determined by the Journal Citation Reports (JCR) and on Citations per Publications (CPP), which are applied to evaluate the impact of a journal relative to all fields.

2.2 *Qualitative Analysis*

The historical method is applied for qualitative analysis that is rich and complex. In this type of analysis, it's helpful to look into the places, times, and contexts in which events happen to understand them better. In this study, a qualitative study of the application of the HS algorithm in branches of geomechanics in the last ten years is performed. Based on the results, a summary of the possibility of using this algorithm in future work is given.

3 Harmony Search (HS) Algorithm

Meta-heuristic algorithms are one of the most powerful tools for solving optimization problems and also increase the ability to find high-precision solutions to difficult optimization problems [34–42]. The harmony search algorithm is one of the most practical meta-heuristic algorithms, inspired by the process of making music by a composer to harmonize a piece of music [43, 44]. The harmony search algorithm was introduced in 2001, and it has been used successfully in a wide range of engineering and scientific issues. Search harmony algorithm has unique features that have made this algorithm one of the most widely used optimization algorithms in recent years in various problems, including the flexibility of the algorithm in search of better solutions for discrete and continuous optimization problems, low mathematical calculations, and low parameters [45, 46].

4 Review of Application of HS

4.1 *Quantitative Evaluation of Publications and Citations*

According to the data obtained from the Web of Science for a period of ten years from January 2011 to August 2021, 1313 documents were determined. The distribution of the number of publications per year is shown in Fig. 1. According to the distribution of publications in Fig. 1, it is clear that most of the number of publications in 2016 was 169 documents. Also, harmony search in the title in 2021 has had 51 documents in 7 months since the beginning of the year. Based on Fig. 1, it can be seen that there is an approximate decrease in the number of documents published since 2016, which one of the most important reasons is the introduction of a wide range of meta-heuristic optimization algorithms and the willingness of researchers to use the new titles in their researches during this period. Although the number of published documents has fluctuated over the past 10 years, in most of these years more than 100 articles have been published each year that harmony search was in their titles, indicating the algorithm's place in the field of scientific publishing.

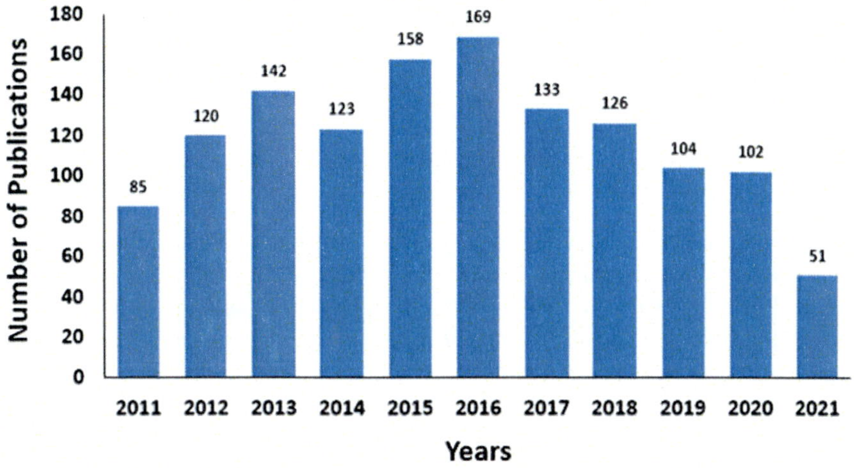

Fig. 1 Number of published documents per each year from January 2011 to August 2021

4.2 Qualitative Analysis

The top authors and the number of published papers in the top ten countries in the field of harmony search algorithm are shown in Table 1. The largest number of documents published in this period belonged to Geem with 52 documents. Also, Al-Betar, Del Ser, Gao, Khader, and Bekdas were authors with over 20 papers on the top authors' list. Among the countries, China ranks first with the release of 336 documents, and India and Iran were the next countries in producing content based on the harmony search algorithm with 230 and 186 documents, respectively. At the end of the top ten countries list, Jordan and Mexico with 32 print documents have an equal share

Table 1 Top ten authors and countries from January 2011 to August 2021

Top ten countries			Top ten authors		
% of 1313	Record count	Countries	% of 1313	Record count	Authors
25.57	336	CHINA	3.95	52	Geem ZW
17.5	230	INDIA	1.9	25	Al-betar MA
14.15	186	IRAN	1.59	21	Del Ser J
7.68	101	SOUTH KOREA	1.59	21	Gao LQ
7	92	MALAYSIA	1.59	21	Khader AT
5.02	66	TURKEY	1.52	20	Bekdas G
4.94	65	USA	1.44	19	Awadallah MA
2.81	37	AUSTRALIA	1.44	19	Kim JH
2.43	32	JORDAN	1.44	19	Li S
2.43	32	MEXICO	1.44	19	Pan QK

with each other in printing the documents which were related to harmony search algorithm.

5 Review of Application of HS in Geomechanics

In the previous sections, a review of the position of the harmony search algorithm in various disciplines over the past decade from January 2011 to August 2021 was conducted that the results showed the appropriate process of citations as well as the ability and efficiency of the algorithm in solving various problems in different countries of the world. Although the harmony search algorithm had a significant share in many disciplines, in some fields, especially geomechanics, it had a much smaller share that one of the reasons could be the desire of more geomechanical researchers to solve related problems in this field using classical methods or relatively older algorithms. However, this review has shown that in recent years, the innovative use of this algorithm in various branches of geomechanics has led to articles with suitable citations. Therefore, on the one hand, given the wide variety of applications of geomechanics in many fields such as tunneling, dam construction, drilling, civil engineering structures, and petroleum engineering, and, on the other hand, appropriate trend of increasing citations in documents related to the harmony search algorithm, it can be a good opportunity for researchers in the field of geomechanics to not only use a new and innovative method in their research but also increase the chances of growing citations. Some of which will be discussed in the next section. It should be noted that in this review to focus more on the application of the harmony search algorithm in various subjects, the harmony search algorithm with other related words was searched only in the title of documents.

5.1 Review the Top-Cited Papers in Rock Mechanics

In the field of tunneling as a subset of geomechanics, the study of Mikaeil et al. [47] gained the highest citations. In their study, the harmony search algorithm was applied for optimizing the K-means clustering algorithm to investigate the geological hazards in a tunneling project. As a result, a comparison was made between the obtained results from data analysis using hybridization of K-means and harmony search algorithm with the observed results, and this comparison showed that harmony search algorithm had the high efficiency and performance to optimize K-means algorithm for modeling some tunneling problems [47]. In another study, Mikaeil et al. [48] used the harmony search algorithm for modeling a problem involved in rock mechanics engineering. They used the hybridization of K-means and harmony search algorithm to evaluate the performance of diamond wire saw by rock characteristics and controlled parameters related to characteristics of the cutting machine and operational parameters. Their results obtained showed that hybridization of K-means and

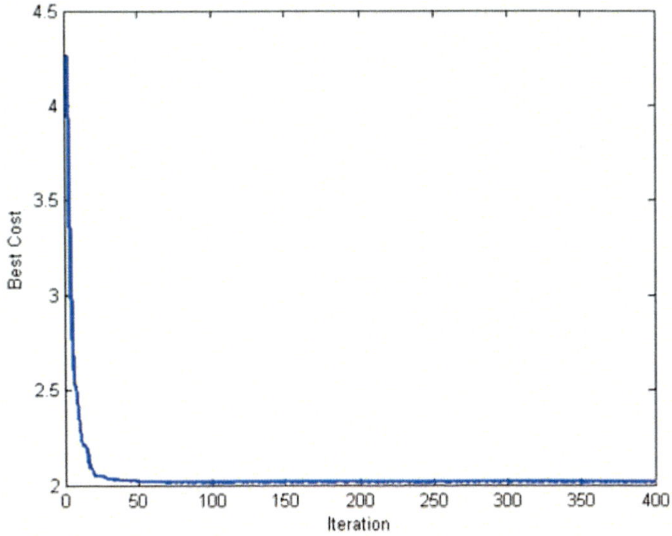

Fig. 2 Minimum cost per iteration in HSA

harmony search algorithm was a reliable system modeling technique. Figure 2 shows the minimum cost per iteration in process of HAS- K-means optimization. This study could achieve 16 citations according to the WOS platform [48].

In another problem in the field of rock mechanics, Hasanipanah et al. [49] published in the journal of engineering with computer that used the dynamical harmony search algorithm in the training of artificial neural networks. This article with 15 citations on the WOS platform is one of the most cited articles in this field. In fact, they used the ANN-adaptive dynamical harmony search algorithm to predicted undesirable and inevitable effects of blasting in rock excavation, namely flyrock. The results of this novel hybrid ANN-adaptive dynamical harmony search algorithm had higher degrees of accuracy and robustness compared to other used methods [49].

5.2 *Review the Top-Cited Papers in Soil Mechanics*

Cheng et al. [50] carried out an investigation into using two improved harmony search methods in geotechnical problems that were published in the Journal of Mechanics from Oxford University Press. The efficiency of the proposed methods was tested with three difficult examples, and the results showed the high degree of accuracy of these methods in solving geotechnical problems [50]. In another study, a coupled particle swarm and harmony search optimization algorithm was applied for difficult geotechnical problems by Cheng et al. [51]. They found out that due to the high degree of efficiency and robustness of the proposed method in solving a very complicated hydropower problem, this proposed method can be used in many complex geotechnical problems [51]. Bekdas et al. [52] carried out investigations for the optimal design of cantilever soldier pile retaining walls embedded in frictional soils. They applied the harmony search algorithm to conduct parametrical analyses. Consequently, a comparison was made between the results of optimization and finite element solutions and indicated the effectiveness of the usage of optimization algorithms in soil mechanics problems. This study was published in the Applied Sciences from Multidisciplinary Digital Publishing Institute (MDPI), and it could receive seven citations during fifteen months [52].

6 Conclusions

In this paper, the application of the harmony search algorithm in geomechanics problems from 2011 to 2021 was reviewed. For this purpose, first, the application of the harmony search algorithm in all disciplines in this time period and based on the WOS platform was reviewed and evaluated. The reviews indicated that the amount of citations in articles related to this algorithm has an increasing trend. All evaluations and analyses showed the efficiency and acceptability of the harmony search algorithm among researchers in a wide range of disciplines. In the next step, a set of keywords related to the two main branches in geomechanics was defined and all documents were reviewed and evaluated on the WOS platform. The results of this review showed that the presented papers in different branches of geomechanics were innovative due to their connection with the harmony search algorithm, and the documents that used this algorithm in their analysis received significant citations. It is important to note that the contribution of these documents is small compared to other disciplines that have used harmony search algorithm. As a result, it is recommended that researchers use the harmony search algorithm as a powerful and robust tool in solving various problems in geomechanics according to the high efficiency of this algorithm, the increasing trend of citations to works related to this algorithm, and its significant contribution among the reputable publishers.

References

1. Careddu N, Siotto G (2010) Geo-mechanical characterization for the underground marble mine "Su Cuccumiau" (Sardinia). In: Ersoy M, Çelik MY, Yeşilkaya L (ed) Proceedings of 7th international marble and natural stones congress of Turkey (Mersem VII), 14–15 Oct 2010. Afyonkarahisar, Turkey, pp 25–32. ISBN: 978-605-01-0023-5
2. Ozcelik Y, Careddu N, Yilmazkaya E (2012) The effects of freeze–thaw cycles on the gloss values of polished stone surfaces. Cold Reg Sci Technol 82:49–55
3. Tumac D, Shaterpour-Mamaghani A (2018) Estimating the sawability of large diameter circular saws based on classification of natural stone types according to the geological origin. Int J Rock Mech Min Sci 101:18–32

4. Mikaeil R, Haghshenas SS, Haghshenas SS, Ataei M (2018) Performance prediction of circular saw machine using imperialist competitive algorithm and fuzzy clustering technique. Neural Comput Appl 29(6):283–292
5. Mikaeil R, Haghshenas SS, Hoseinie SH (2018) Rock penetrability classification using artificial bee colony (ABC) algorithm and self-organizing map. Geotech Geol Eng 36(2):1309–1318
6. Mikaeil R, Haghshenas SS, Ozcelik Y, Gharehgheshlagh HH (2018) Performance evaluation of adaptive neuro-fuzzy inference system and group method of data handling-type neural network for estimating wear rate of diamond wire saw. Geotech Geol Eng 36(6):3779–3791
7. Aryafar A, Mikaeil R, Haghshenas SS, Haghshenas SS (2018) Application of metaheuristic algorithms to optimal clustering of sawing machine vibration. Meas 124:20–31
8. Haghshenas SS, Faradonbeh RS, Mikaeil R, Haghshenas SS, Taheri A, Saghatforoush A, Dormishi A (2019) A new conventional criterion for the performance evaluation of gang saw machines. Meas 146:159–170
9. Dormishi A, Ataei M, Mikaeil R, Khalokakaei R, Haghshenas SS (2019) Evaluation of gang saws' performance in the carbonate rock cutting process using feasibility of intelligent approaches. Int J Eng Sci Technol 22(3):990–1000
10. Faradonbeh RS, Taheri A, e Sousa LR, Karakus M (2020) Rockburst assessment in deep geotechnical conditions using true-triaxial tests and data-driven approaches. Int J Rock Mech Min Sci 128:104279
11. Li Y, Bahrani N (2020) A numerical study on the failure process and strength of heterogeneous rocks and highly interlocked jointed pillars. In: 54th US rock mechanics/geomechanics symposium. OnePetro
12. Moayedi H, Mosallanezhad M, Nazir R (2017) Evaluation of maintained load test (MLT) and pile driving analyzer (PDA) in measuring bearing capacity of driven reinforced concrete piles. Soil Mech Found Eng 54(3):150–154
13. Bai Z, Xu H, Geng H, Qi L (2019) A study of uniaxial compressive strength of shale based on homogenization method. Geotech Geol Eng 37(6):5485–5497
14. Rahardjo H, Kim Y, Satyanaga A (2019) Role of unsaturated soil mechanics in geotechnical engineering. Int J Geotech Eng 10(1):1–23
15. Sarro R, María Mateos R, Reichenbach P, Aguilera H, Riquelme A, Hernández-Gutiérrez LE, Martín A, Barra A, Solari L, Monserrat O, Alvioli M, Fernández-Merodo JA, López-Vinielles J, Herrera G (2020) Geotechnics for rockfall assessment in the volcanic island of Gran Canaria (Canary Islands, Spain). J Maps 16(2):605–613
16. Gu X, Zhang J, Huang X (2020) DEM analysis of monotonic and cyclic behaviors of sand based on critical state soil mechanics framework. Comput Geotech 128:103787
17. Shastri A, Sánchez M, Gai X, Lee MY, Dewers T (2021) Mechanical behavior of frozen soils: experimental investigation and numerical modeling. Comput Geotech 138:104361
18. Noori AM, Mikaeil R, Mokhtarian M, Haghshenas SS, Foroughi M (2020) Feasibility of intelligent models for prediction of utilization factor of TBM. Geotech Geol Eng 38(3):3125–3143
19. Haghshenas SS, Haghshenas SS, Mikaeil R, Sirati Moghadam P, Haghshenas AS (2017) A new model for evaluating the geological risk based on geomechanical properties—case study: the second part of Emamzade Hashem tunnel. Electron J Geotech Eng 22(01):309–320
20. Salemi A, Mikaeil R, Haghshenas SS (2018) Integration of finite difference method and genetic algorithm to seismic analysis of circular shallow tunnels (Case study: Tabriz urban railway tunnels). KSCE J Civ Eng 22(5):1978–1990
21. Mikaeil R, Beigmohammadi M, Bakhtavar E, Haghshenas SS (2019) Assessment of risks of tunneling project in Iran using artificial bee colony algorithm. SN Appl Sci 1(12):1–9
22. Mikaeil R, Haghshenas SS, Sedaghati Z (2019) Geotechnical risk evaluation of tunneling projects using optimization techniques (Case study: the second part of Emamzade Hashem tunnel). Nat Hazards 97(3):1099–1113
23. Mikaeil R, Bakhshinezhad H, Haghshenas SS, Ataei M (2019) Stability analysis of tunnel support systems using numerical and intelligent simulations (Case study: Kouhin tunnel of Qazvin-Rasht railway). RGN zbornik. 34(2):1–10

24. Li S, Tan Z, Yang Y (2020) Mechanical analyses and controlling measures for large deformations of inclined and laminar stratum during tunnelling. Geotech Geol Eng 38(3):3095–3112
25. Li D, Armaghani DJ, Zhou J, Lai SH, Hasanipanah M (2020) A GMDH predictive model to predict rock material strength using three non-destructive tests. J Nondestruct Eval 39(4):1–14
26. Ching J, Phoon KK, Ho YH, Weng MC (2021) Quasi-site-specific prediction for deformation modulus of rock mass. Can Geotech J 58(7):936–951
27. Bagheri M, Rezania M (2021) Geological and geotechnical characteristics of London clay from the isle of Sheppey. Geotech Geol Eng 39(2):1701–1713
28. Al-Bared MAM, Mustaffa Z, Armaghani DJ, Marto A, Yunus NZM, Hasanipanah M (2021) Application of hybrid intelligent systems in predicting the unconfined compressive strength of clay material mixed with recycled additive. Transp Geotech 100627
29. Faradonbeh RS, Haghshenas SS, Taheri A, Mikaeil R (2020) Application of self-organizing map and fuzzy c-mean techniques for rockburst clustering in deep underground projects. Neural Comput Appl 32(12):8545–8559
30. Bai XD, Cheng WC, Ong DE, Li G (2021) Evaluation of geological conditions and clogging of tunneling using machine learning. Geomech Eng 25(1):59–73
31. Pham BT, Nguyen MD, Nguyen-Thoi T, Ho LS, Koopialipoor M, Quoc NK, Armaghani DJ, Van Le H (2021) A novel approach for classification of soils based on laboratory tests using adaboost, tree and ANN modeling. Transp Geotech 27:100508
32. Jafarzadeh-Ghoushchi S (2018) Qualitative and quantitative analysis of green supply chain management (GSCM) literature from 2000 to 2015. Int J Supply Chain Manag 7(1):77–86
33. Maghami MR, esmaeil Rezadad M, Ebrahim NA, Gomes C (2015) Qualitative and quantitative analysis of solar hydrogen generation literature from 2001 to 2014. Scientometrics 105(2):759–771
34. Mirjalili S, Saremi S, Mirjalili SM, Coelho LDS (2016) Multi-objective grey wolf optimizer: a novel algorithm for multi-criterion optimization. Expert Syst Appl 47:106–119
35. Hasanipanah M, Armaghani DJ, Amnieh HB, Abd Majid MZ, Tahir MM (2017) Application of PSO to develop a powerful equation for prediction of flyrock due to blasting. Neural Comput Appl 28(1):1043–1050
36. Choopan Y, Emami S (2019) Optimal operation of dam reservoir using gray wolf optimizer algorithm (Case study: Urmia Shaharchay dam in Iran). J Soft Comput Civ Eng 3(3):47–61
37. Hajihassani M, Armaghani DJ, Marto A, Mohamad ET (2015) Ground vibration prediction in quarry blasting through an artificial neural network optimized by imperialist competitive algorithm. Bull Eng Geol Env 74(3):873–886
38. Golafshani EM, Behnood A, Arashpour M (2020) Predicting the compressive strength of normal and high-performance concretes using ANN and ANFIS hybridized with grey wolf optimizer. Constr Build Mater 232:117266
39. Guido G, Haghshenas SS, Haghshenas SS, Vitale A, Astarita V, Haghshenas AS (2020) Feasibility of stochastic models for evaluation of potential factors for safety: a case study in Southern Italy. Sustainability 12(18):7541
40. Keshtegar B, Correia JA, Trung NT (2020) Optimisation of nanocomposite pipes under internal fluid reinforced by FRP and CNTs under seismic load. IJHM 3(3):213–227
41. Golafshani EM, Behnood A, Hosseinikebria SS, Arashpour M (2021) Novel metaheuristic-based type-2 fuzzy inference system for predicting the compressive strength of recycled aggregate concrete. J Clean Prod 128771
42. Mikaeil R, Mokhtarian M, Haghshenas SS, Careddu N, Alipour A (2021) Assessing the system vibration of circular sawing machine in carbonate rock sawing process using experimental study and machine learning. Geotech Geol Eng 1–17
43. Geem ZW, Kim JH, Loganathan GV (2001) A new heuristic optimization algorithm: harmony search. Simulation 76(2):60–68
44. Geem ZW (ed) (2009) Music-inspired harmony search algorithm: theory and applications, vol 191. Springer
45. Lee KS, Geem ZW (2005) A new meta-heuristic algorithm for continuous engineering optimization: harmony search theory and practice. Comput Methods Appl Mech Eng 194(36–38):3902–3933

46. Geem ZW, Lee KS, Park Y (2005) Application of harmony search to vehicle routing. Am J Appl Sci 2(12):1552–1557
47. Mikaeil R, Haghshenas SS, Shirvand Y, Hasanluy MV, Roshanaei V (2016) Risk assessment of geological hazards in a tunneling project using harmony search algorithm (Case study: Ardabil-Mianeh railway tunnel). Civ Eng J 2(10):546–554
48. Mikaeil R, Ozcelik Y, Ataei M, Shaffiee Haghshenas S (2019) Application of harmony search algorithm to evaluate performance of diamond wire saw. JEM 10(1):27–36
49. Hasanipanah M, Keshtegar B, Thai DK, Troung NT (2020) An ANN-adaptive dynamical harmony search algorithm to approximate the flyrock resulting from blasting. Eng Comput 1–13
50. Cheng YM, Li L, Fang SS (2011) Improved harmony search methods to replace variational principle in geotechnical problems. J Mech 27(1):107–119
51. Cheng YM, Li L, Sun YJ, Au SK (2012) A coupled particle swarm and harmony search optimization algorithm for difficult geotechnical problems. Struct Multidiscipl Optim 45(4):489–501
52. Bekdaş G, Arama ZA, Kayabekir AE, Geem ZW (2020) Optimal design of cantilever soldier pile retaining walls embedded in frictional soils with harmony search algorithm. Appl Sci 10(9):3232

The Robustness of Tuned Liquid Dampers Optimized via Metaheuristic Methods

Ayla Ocak, Sinan Melih Nigdeli, and Gebrail Bekdaş

Abstract In the optimization made with metaheuristic algorithms, each fixed value and design constraint entered into the system play an important role in determining the optimum parameters. Looking at the general logic of passive control devices used for structure control, it is seen that damping is aimed with the help of a spring and mass. For damping devices, these constants are usually values such as structure mass, stiffness, and damping coefficient. Building mass is one of the most important parameters used by dampers to provide structure control. It is known that the mass affects the properties of the structure through frequency and damping coefficient. In the optimization for the structure control, since the behavior of the building mass under the effect of variable loads is not known exactly, the optimization process was carried out with a fixed mass value. This study aims to investigate the effect of the design parameters such as tank radius, height, period, and damping ratio optimized for a constant mass value, on the displacement and total acceleration values of the structure for different building mass values for a tuned liquid damper (TLD) device in which water is used as a liquid. The optimum TLD parameters are also acceptable and effective when the mass of the structure is different from the taken in the optimization process.

Keywords Tuned liquid dampers · Structural control · Optimization · Variable loads · Metaheuristics

A. Ocak (✉) · S. M. Nigdeli · G. Bekdaş
Department of Civil Engineering, Istanbul University-Cerrahpaşa, 34320 Avcılar, Istanbul, Turkey
e-mail: aylaocak@outlook.com

S. M. Nigdeli
e-mail: melihnig@iuc.edu.tr

G. Bekdaş
e-mail: bekdas@iuc.edu.tr

J. H. Kim et al. (eds.), *Proceedings of 7th International Conference on Harmony Search, Soft Computing and Applications*, Lecture Notes on Data Engineering and Communications Technologies 140, https://doi.org/10.1007/978-981-19-2948-9_3

1 Introduction

Metaheuristic algorithms are algorithms created by different mathematical models inspired by natural events and instinctive living behaviors. It has an open-source to produce new algorithms, in line with the inspiration for its development. There are types inspired by different living behaviors in nature, such as the Genetic Algorithm (GA), Simulating Annealing (SA), Ant Colony Optimization (ACO), Particle Swarm Optimization (PSO), Harmony Search Algorithm (HS), Flower Pollination Algorithm (FPA), Jaya Algorithm, Artificial Bee Colony (ABC), Teaching-Learning-Based Optimization (TLBO), and Bat Algorithm (BA) [1–10]. These algorithms, which are frequently used for the solution of various engineering problems, are effective in obtaining the optimum design parameters for the problem. There are many studies on the use of metaheuristic algorithms [11–19].

The application of the optimization process in the design of tuned liquid damper devices dates to the 90s. When the studies conducted since this date are examined, parametric optimization applications are generally preferred, and it is aimed to obtain optimum results in structure control by applying various dynamic loads to the structure [20–26].

In this study, the performance of the design parameters of the tuned liquid damper device, which was optimized with the TLBO algorithm and in which water is used as a liquid, was analyzed for 5 different structure masses that are different from the used in the optimization process, and then, the effect on the structure displacement and total acceleration values was investigated.

2 Methodology

2.1 *Optimum Design of Tuned Liquid Dampers via TLBO Algorithm*

TLBO algorithm is a 2-phase metaheuristic algorithm developed by Rao et al. [9]. For optimization, all kinds of design constants are introduced to the system at the beginning. Equation 1 shows the design constraint.

$$X_j^{\min} \leq X_j \leq X_j^{\max} \quad j = 1, 2, 3, \ldots, \text{Tn} \tag{1}$$

X_j represents the variables in the design; the value of j is the number of variables from 1 to Tn. After assigning random values for the design, it is turned into a matrix (Eq. 2).

$$A = \begin{bmatrix} X_{1,1} & \cdots & X_{1,T} \\ \vdots & \ddots & \vdots \\ X_{Pn,1} & \cdots & X_{Pn,T} \end{bmatrix} \tag{2}$$

The results obtained by using the values in the A matrix in the objective function are converted into a vector (Eq. 3).

$$f(X) = \begin{bmatrix} f(X_1) \\ f(X_2) \\ \vdots \\ \vdots \\ \vdots \\ f(X_{Pn}) \end{bmatrix} \tag{3}$$

The average of the population is taken and written into M_T as in Eq. 4.

$$M_T = [m_1, m_2, \ldots, m_T] \tag{4}$$

The best solution is defined as $X_{teacher}$ as in Eq. 5. This value is the minimum value defined in the vector $f(x)$.

$$X_{teacher} = X_{\min f(x)} \tag{5}$$

The value will be written as in Eq. 6 and will serve as a tool in producing new solutions.

$$M_{New,T} = X_{teacher,T} \tag{6}$$

The difference between the old mean and the new best solution is shown in Eq. (7). The teaching factor (T_f) determined according to the average to be changed is randomly decided to be 1 or 2 according to Eq. (8). Here, r is a random number in the range [0, 1].

$$\text{Difference}_T = r\left(M_{New,T} - T_f M_T\right) \tag{7}$$

$$T_f = \text{round}[1 + \text{rand}(0,\ 1)\{2 - 1\}] \tag{8}$$

The solutions are updated with the difference value obtained in Eq. 7, and new solutions are obtained as in Eq. 9.

$$X_{New,T} = X_{Old,T} + \text{Difference}_T \tag{9}$$

After this stage, the 2nd phase, in which the students interact with themselves, takes place. 2 students are randomly selected from the population (X_p and X_t), and according to the better solution, one of the formulas in Eqs. 10 and 11 is chosen and new solutions are obtained, and iterations are repeated.

$$r \to [0,\ 1]$$

$$X_{\text{New}} = X_{\text{P}} + r(X_{\text{P}} - X_{\text{T}}) \tag{10}$$

$$X_{\text{New}} = X_{\text{P}} + r(X_{\text{T}} - X_{\text{P}}) \tag{11}$$

In the design of tuned liquid dampers, the mass of sloshing liquid m_s and the mass of empty tank + non-sloshing liquid m_d are given in Eqs. 12 and 13, respectively.

$$m_s = m_{st} \times R \times \frac{\tan h\left(\frac{1.84\,h}{R}\right)}{2.2\,h} \tag{12}$$

$$m_d = m_{tld} - m_s \tag{13}$$

Here, m_{TLD} denotes total TLD mass, and h and R represent TLD tank height and radius, respectively. The stiffness and damping coefficients of TLD and sloshing are as in Eqs. 14, 15, 16, and 17.

$$k_d = m_d \times \left(\frac{2\pi}{T_d}\right)^2 \tag{14}$$

$$k_s = m_{st} \times \frac{g\left\{\tan \text{h}\left(\frac{1.84\,h}{R}\right)\right\}^2}{1.19\,h} \tag{15}$$

$$c_d = 2 \times \zeta_d \times \sqrt{m_d \times k_d} \tag{16}$$

$$c_s = \zeta_s \times 2\sqrt{m_s K_s} \tag{17}$$

The damping ratio calculation of the system with TLD is shown in Eq. 18.

$$\zeta_d = \frac{c_d}{2m_d\sqrt{\frac{k_d}{m_d}}} \tag{18}$$

Mass, stiffness, and damping coefficient matrices of the system are given in Eqs. 19, 20, and 21, respectively. Expressions with subscript "*s*" denote the sloshing liquid; expressions with "*d*" denote the properties of the TLD.

$$\begin{bmatrix} m & 0 & 0 \\ 0 & m_{\mathrm{d}} & 0 \\ 0 & 0 & m_{\mathrm{s}} \end{bmatrix} \tag{19}$$

$$\begin{bmatrix} K+K_{\mathrm{d}} & -K_{\mathrm{d}} & 0 \\ -K_{\mathrm{d}} & K_{\mathrm{d}}+K_{\mathrm{s}} & -K_{\mathrm{s}} \\ 0 & -K_{\mathrm{s}} & K_{\mathrm{s}} \end{bmatrix} \tag{20}$$

$$\begin{bmatrix} C+C_{\mathrm{d}} & -C_{\mathrm{d}} & 0 \\ -C_{\mathrm{d}} & C_{\mathrm{d}}+C_{\mathrm{s}} & -C_{\mathrm{s}} \\ 0 & -C_{\mathrm{s}} & C_{\mathrm{s}} \end{bmatrix} \tag{21}$$

In its simplest form, the equation of motion is given in Eq. 22.

$$[M]\{\ddot{X}\} + [C]\{\dot{X}\} + [K]\{X\} = -[M]1\{\ddot{X}_g\} \tag{22}$$

The coupled equation of motion is solved on MATLAB with the help of Simulink [27]. Earthquake ground acceleration of far fault records in FEMA: Quantification of Building Seismic Performance Factors [28] was taken as the earthquake dataset.

3 The Numerical Example

In this research, a tuned liquid damper (TLD) device with a cylindrical tank was placed in a single-story structure [29]. By creating a 3DOF structure + TLD system shown in Fig. 1, seismic excitations consisting of 22 earthquake records were directed to the structure [28]. The building mass was initially taken as 100 t, and the stiffness

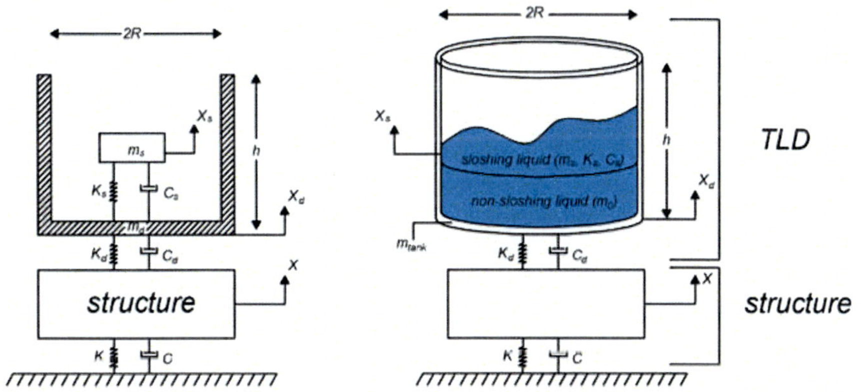

Fig. 1 Structure + TLD system model

Table 1 Optimum results for TLD-water optimized with TLBO [29]

Variables	Optimized values
$\boldsymbol{T}_{\rm d}$(s)	0.8902
ζ	0 0253
$\boldsymbol{R}$(m)	0.5575
$\boldsymbol{h}$(m)	3.2575

and damping coefficient values were calculated according to the mass, and respectively, 3.95 MN/m and 0.06 MNs/m were obtained. For TLD, a mass ratio of 5% was chosen and a 5 t mass was used. The period of the structure was determined as 1 s, and the TLD period was optimized to be between 0.5 and 1.5 times the period of the structure.

Minimum and maximum design constraints for TLD tank radius and height were chosen as 0.1 and 10 m. The ratio between tank diameter and liquid height in the tank Fujino et al. According to the study conducted by the system, the design constraint was defined as $\frac{h}{2R} > 0.15$ [30]. For optimization, the pn value is 10 and the $\mathrm{st}_{\max}$ value is 2, and the processes are repeated for 1000 iterations. The optimum parameters obtained are given in Table 1 [29].

Since the optimized values are tested for a fixed building mass, the efficiency of the optimum parameters is also tested for variable building masses. By taking the building mass of 80, 90, 100, 110, and 120 t, the structural stiffness and damping coefficient values obtained for the 100 t building mass were used as constants. From the values obtained under earthquake excitations, the displacement and total acceleration values for the critical earthquake are shown in Table 2. The displacement and total acceleration graphs of the critical earthquake for the building mass values in the 80–120 t range are given, respectively, in Figs. 2, 3, 4, 5, and 6.

Table 2 TLD-water critical earthquake analysis results for values taken in the range of 80–120 t for the structure mass

Structure mass (ton)	Without TLD structure		TLD-water	
	Displacement (m)	Total acceleration (m/s^2)	Displacement (m)	Total acceleration (m/s^2)
80	0.2610085	12.9673247	0.2102628	10.0832103
90	0.2797414	12.3429855	0.2246641	9.4989821
100	0.2873419	11.4077630	0.2314739	8.7552966
110	0.2874282	10.3619123	0.2440711	8.3517830
120	0.2844103	9.3895882	0.2519835	7.9295577

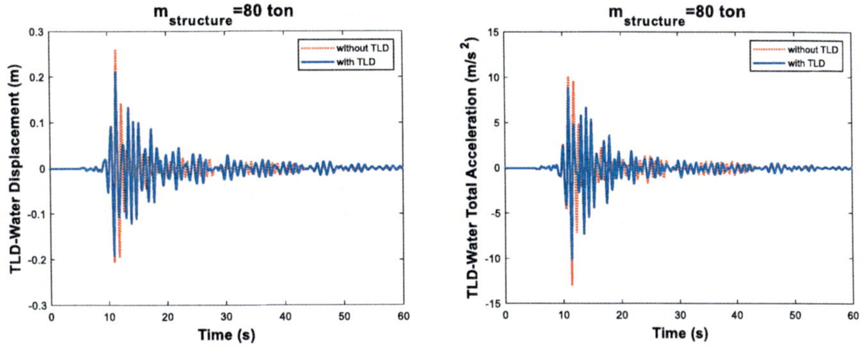

Fig. 2 Displacement and total acceleration graphs of critical earthquake for structural mass 80 t

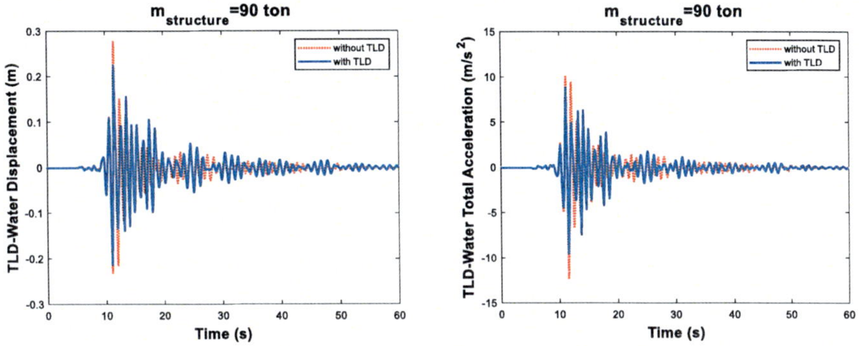

Fig. 3 Displacement and total acceleration graphs of critical earthquake for structural mass 90 t

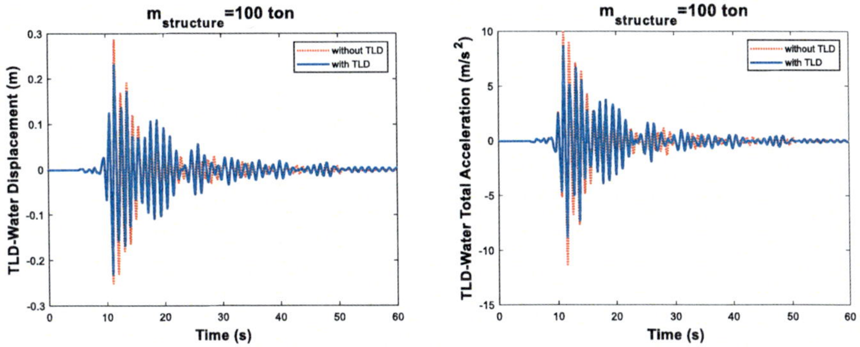

Fig. 4 Displacement and total acceleration graphs of critical earthquake for structural mass 100 t

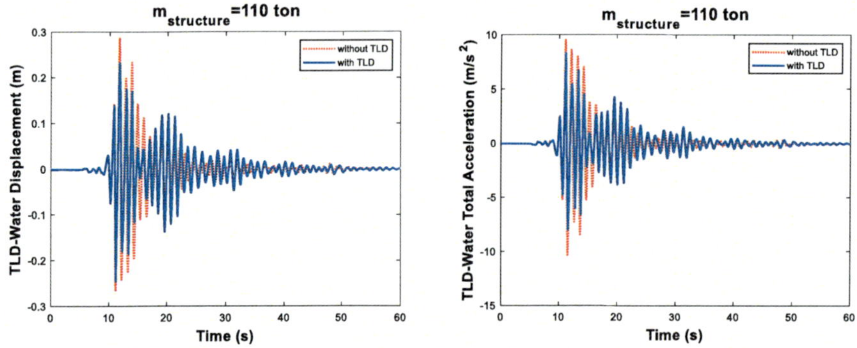

Fig. 5 Displacement and total acceleration graphs of critical earthquake for structural mass 110 t

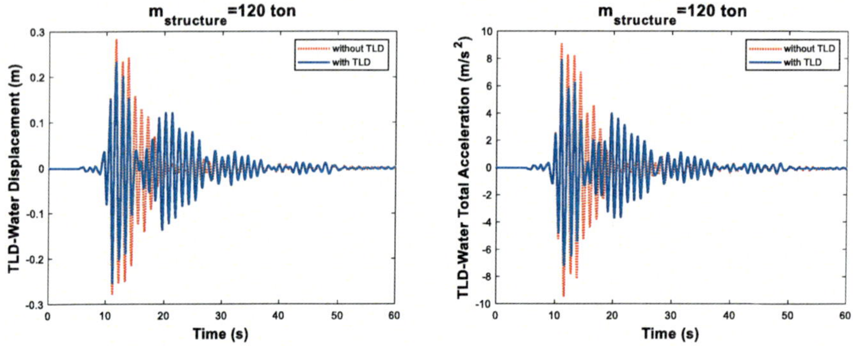

Fig. 6 Displacement and total acceleration graphs of critical earthquake for structural mass 120 t

4 Conclusions

It is known that remarkable results are obtained in the optimization process with metaheuristic algorithms. In optimization, design factors of metaheuristic algorithm change according to the type of metaheuristic algorithm, and limit values affect the result of this process. In this study, TLD parameters were optimized with the TLBO algorithm to show the minimum displacement under earthquake excitations. A study was conducted to observe the effect of building mass, one of the design constants used for TLBO, on the durability of TLD. By using TLD parameters optimized with TLBO, TLD was analyzed for structural masses varying between 80 and 120 t, with the same structural stiffness and damping coefficient values.

The results obtained from the analysis are given below.

- For 80 t, 90 t, 100 t, 110 t, and 120 t structure masses, the structural displacement reduction percentages for the critical earthquake were 19.44%, 19.49%, 19.44%, 15.09%, and 11.40%, respectively. Considering these values, it was determined that the displacement reduction ratio decreased rapidly for the mass values taken after 100 t.

- The total acceleration reduction percentages of the same structure mass values are 22.24%, 23.04%, 23.25%, 19.40%, and 15.55%, respectively. According to these percentages, the increase in the structure mass caused a faster decrease compared to the other mass results obtained after 100 t, as in the displacement. It has been understood that this has a remarkable effect on reducing the total acceleration values of TLD by optimizing the TLD.
- The fact that the values in the displacement drops obtained for 80 and 100 t are very close to each other showed that optimum results can give ideal results for more than one mass value.

Considering these results, it can be said that the selected building mass for optimization is better than the other masses. It is understood that the correct choice of the building mass as a design variable affects the design.

References

1. Holland JH (1975) Adaptation in natural and artificial systems. University of Michigan Press, Ann Arbor, MI
2. Kirkpatrick S, Gelatt CD, Vecchi MP (1983) Optimization by simulated annealing. Science 220(4598):671–680
3. Dorigo M, Maniezzo V, Colorni A (1996) The ant system: an autocatalytic optimizing process. IEEE Trans Syst Man Cybernet B 26:29–41
4. Kennedy J, Eberhart RC (1995) Particle swarm optimization. In: Proceedings of IEEE international conference on neural networks, 27 Nov–1 Dec, no 4. IEEE Service Center, Piscataway, Perth, NJ, pp 1942–1948
5. Geem ZW, Kim JH, Loganathan GV (2001) A new heuristic optimization algorithm: harmony search. SIMULATION 76(2):60–68
6. Yang XS (2012) Flower pollination algorithm for global optimization. In: Durand-Lose J, Jonoska N (eds) Lecture notes in computer science, vol 7445. 27, Springer, London, pp 240–249
7. Rao R (2016) Jaya: a simple and new optimization algorithm for solving constrained and unconstrained optimization problems. Int J Ind Eng Comput 7(1):19–34
8. Karaboğa D (2005) An idea based on honey bee swarm for numerical optimization, vol 200. Technical report-tr06, Erciyes University, Engineering Faculty, Computer Engineering Department, pp 1–10
9. Rao RV, Savsani VJ, Vakharia DP (2011) Teaching-learning-based optimization: a novel method for constrained mechanical design optimization problems. Comput Aided Des 43:303–315
10. Yang XS (2010) A new metaheuristic bat-inspired algorithm. In: Nature inspired cooperative strategies for optimization (NICSO 2010). Springer, Berlin, pp 65–74
11. Atmaca B (2021) Determination of proper post-tensioning cable force of cable-stayed footbridge with TLBO algorithm. Steel Compos Struct 40(6):805–816
12. Bekdaş G, Niğdeli SM, Aydın A (2017) Optimization of tuned mass damper for multi-storey structures by using impulsive motions. In: 2nd International conference on civil and environmental engineering (ICOCEE 2017). Cappadocia, Turkey
13. Nigdeli SM, Bekdaş G, Alhan C (2014) Optimization of seismic isolation systems via harmony search. Eng Optim 46(11):1553–1569

14. Öztürk HT (2018) Modeling of concrete compressive strength with Jaya and teaching-learning based optimization (TLBO) algorithms. J Invest Eng Technol 1(2):24–29
15. Yang Y, Li C (2017) Performance of tuned tandem mass dampers for structures under the ground acceleration. Struct Control Health Monit 24(10):e1974
16. Kaveh A, Hosseini SM, Zaerreza A (2021) Improved Shuffled Jaya algorithm for sizing optimization of skeletal structures with discrete variables. In: Structures, vol 29. Elsevier, pp 107–128
17. Zhang HY, Zhang LJ (2017) Tuned mass damper system of high-rise intake towers optimized by improved harmony search algorithm. Eng Struct 138:270–282
18. Syrimi PG, Sapountzakis EJ, Tsiatas GC, Antoniadis IA (2017) Parameter optimization of the KDamper concept in seismic isolation of bridges using harmony search algorithm. In: Proceedings of the 6th COMPDYN
19. Talatahari S, Hosseini A, Mirghaderi SR, Rezazadeh F (2014) Optimum performance-based seismic design using a hybrid optimization algorithm. Math Probl Eng
20. Gao H, Kwok KCS, Samali B (1997) Optimization of tuned liquid column dampers. Eng Struct 19(6):476–486
21. Xue SD, Ko JM, Xu YL (2000) Optimum parameters of tuned liquid column damper for suppressing pitching vibration of an undamped structure. J Sound Vib 235(4):639–653
22. Taflanidis AA, Angelides DC, MAnos GC (2005) Optimal design and performance of liquid column mass dampers for rotational vibration control of structures under white noise excitation. Eng Struct 27(4):524–534
23. Shum KM (2009) Closed form optimal solution of a tuned liquid column damper for suppressing harmonic vibration of structures. Eng Struct 31(1):84–92
24. Debbarma R, Chakraborty S, Ghosh SK (2010) Optimum design of tuned liquid column dampers under stochastic earthquake load considering uncertain bounded system parameters. Int J Mech Sci 52(10):1385–1393
25. Tanveer M, Usman M, Khan IU, Farooq SH, Kasım (2020) Material optimization of tuned liquid column ball damper (TLCBD) for the vibration control of multi-storey structure using various liquid and ball densities
26. Mehrkian B, Altay O (2020) Mathematical modelling and optimization scheme for omnidirectional tuned liquid column dampers. J Sound Vib 484:115523
27. Matlab R2018a (2018) The mathworks. Natick, MA
28. FEMA P-695 (2009) Quantification of building seismic performance factors. Washington
29. Ocak A (2021) Optimization of Tuned liquid dampers for structures, master's thesis. Istanbul University-Cerrahpasa Institute of Graduate Studies, Istanbul
30. Fujino Y, Sun LM (1993) Vibration control by multiple tuned liquid dampers (MTLDs). J Struct Eng 119(12):3482–3502

Hybrid Generalized Normal Distribution Optimization with Sine Cosine Algorithm for Global Optimization

Jingwei Too, Ali Safaa Sadiq, Hesam Akbari, Guo Ren Mong, and Seyedali Mirjalili

Abstract This paper proposes two hybrid versions of the generalized normal distribution optimization (GNDO) and sine cosine algorithm (SCA) for global optimization. The proposed hybrid methods combine the excellent characteristics of the GNDO and SCA algorithms to enhance the exploration and exploitation behaviors. Moreover, an additional weight parameter is introduced to further improve the search ability of the hybrid methods. The proposed methods are tested with 23 mathematical optimization problems. Our results reveal that the proposed hybrid method was very competitive compared to the other metaheuristic algorithms.

Keywords Optimization · Generalized normal distribution optimization · Sine cosine algorithm

J. Too
Faculty of Electrical Engineering, Universiti Teknikal Malaysia Melaka, Hang Tuah Jaya, Durian Tunggal, 76100 Melaka, Malaysia

A. S. Sadiq
Department of Computer Science, Nottingham Trent University, Clifton Lane, Nottingham NG11 8NS, UK

H. Akbari
Department of Biomedical Engineering, South Tehran Branch, Islamic Azad University, Tehran, Iran

G. R. Mong
School of Energy and Chemical Engineering, Xiamen University Malaysia, Jalan Sunsuria, Bandar Sunsuria, 43900 Sepang, Selangor, Malaysia

S. Mirjalili (✉)
Center for Artificial Intelligence Research and Optimization, Torrens University Australia, Fortitude Valley, Brisbane, QLD 4006, Australia
e-mail: ali.mirjalili@gmail.com

Yonsei Frontier Lab, Yonsei University, Seoul, South Korea

J. H. Kim et al. (eds.), *Proceedings of 7th International Conference on Harmony Search, Soft Computing and Applications*, Lecture Notes on Data Engineering and Communications Technologies 140, https://doi.org/10.1007/978-981-19-2948-9_4

1 Introduction

Many real-world problems in engineering, business, and transportation applications can be transformed into optimization problems. Initially, the mathematical or numerical methods are used to find the optimal solution by attaining the zero-derivative point. Nevertheless, these methods are having high difficulty in solving the complex problem with many variables and constraints, mainly due to the increment of the search space [1, 2]. Thus, metaheuristic algorithms have been widely used to tackle the complex optimization problems.

Metaheuristic algorithms are inspired by nature, and they can be categorized into evolution-based algorithm, swarm-based algorithm, physic-based algorithm, and human-based algorithm. Evolution-based algorithm adopts the mechanisms inspired by biological concepts including mutation, selection, and reproduction strategies. Example of evolution-based algorithm is genetic algorithm (GA) [3, 4] and differential evolution (DE) [5, 6]. Swarm-based algorithms such as particle swarm optimization (PSO) [7, 8] have been received a lot of attentions since last decade. Some of the recent ones are: African vulture optimization algorithm (AVOA) [9], artificial gorilla troops optimizer (AGTO) [10], and artificial hummingbird algorithm (AHA) [11]. Such algorithms are flexible, and they are using the collective behavior of a group of solutions to find the global optimal. Physic-based algorithms simulate the physical laws such as simulated annealing (SA) [12], while human-based algorithms mimic human-related activities (e.g., harmony search [13] and teaching–learning-based optimization [14]).

Generalized normal distribution optimization (GNDO) [15] is a recent metaheuristic algorithm. It has shown competitive performance against many optimizers when solving the optimization problems. However, GNDO has the limitation of local optima avoidance, which reduces the effectiveness of GNDO in finding the best solution. To enhance the search ability of GNDO, we hybrid the GNDO with another stochastic optimizer called sine cosine algorithm (SCA) [16, 17]. Hybridizing metaheuristics the benefit from their advantages and alleviate their drawbacks have been widely used in the literature [18–22], which motivated our attempts to hybridize GNDO and SCA.

In this study, we present two hybrid versions of GNDOSCA algorithms (GNDOSCA and GNDOSCA2) by combining the excellent characteristics of the GNDO and SCA algorithms. In the proposed models, the mechanism of SCA is associated and incorporated into the GNDO to construct the hybrid algorithms, which aims to improve the exploitation and exploration behaviors. The proposed hybrid algorithms are tested by using several unimodal and multimodal function to verify its effectiveness in solving the global optimization problems.

2 Proposed Approaches

This section presents the details of the proposed hybrid GNDOSCA algorithms.

2.1 Hybrid GNDOSCA Approach 1

Figure 1 depicts the detail of the GNDOSCA approach with 10 solutions. The main operations of GNDOSCA are originated from the global strategy (from GNDO) and SCA mechanism to find the optimal solution. Initially, the solutions are sorted according to fitness values. The first half of solutions is considered as best solutions, and the SCA mechanism is implemented to perform the local exploitation. The rest half of the solutions is known as worst solutions. These worst solutions should be discarded, and the global exploration strategy from GNDO is applied to generate new global solutions.

In the GNDOSCA, the new solutions are generated as follows:

$$v_i = \begin{cases} \begin{cases} w_1 x_i(t) + r_1 \times \sin(r_2) \times |r_3 x_{\text{best}} - x_i(t)| & r_4 < 0.5 \\ w_1 x_i(t) + r_1 \times \cos(r_2) \times |r_3 x_{\text{best}} - x_i(t)| & r_4 \geq 0.5 \end{cases} & \text{if } i \leq \frac{N}{2} \\ x_i(t) + \beta \times (|\lambda_3| \times v_1) + (1 - \beta) \times (|\lambda_4| \times v_2) & \text{otherwise} \end{cases} \quad (1)$$

$$v_1 = \begin{cases} x_i(t) - x_{p1}(t), & \text{if } F(x_i(t)) < F\big(x_{p1}(t)\big) \\ x_{p1}(t) - x_i(t), & \text{otherwise} \end{cases} \quad (2)$$

$$v_2 = \begin{cases} x_{p2}(t) - x_{p3}(t), & \text{if } F\big(x_{p2}(t)\big) < F\big(x_{p3}(t)\big) \\ x_{p3}(t) - x_{p2}(t), & \text{otherwise} \end{cases} \quad (3)$$

where v is the new solutions, w_1 is a weight parameter randomly distributed between 0 and 0.5 in each iteration, x_{best} is the best solution, $r_1, r_2, r_3, r_4, \beta$ are random number

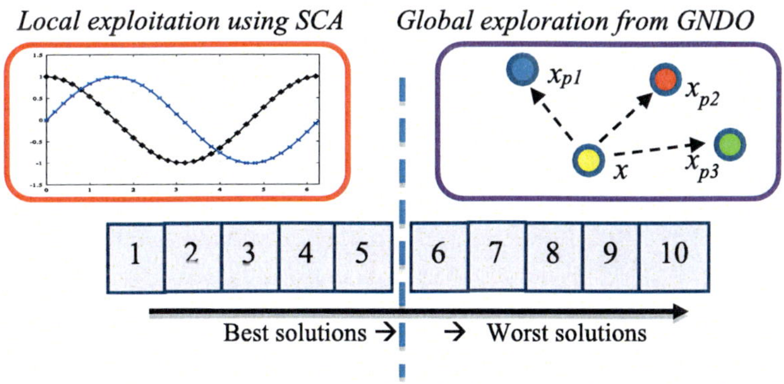

Fig. 1 Concept of hybrid GNDOSCA algorithm

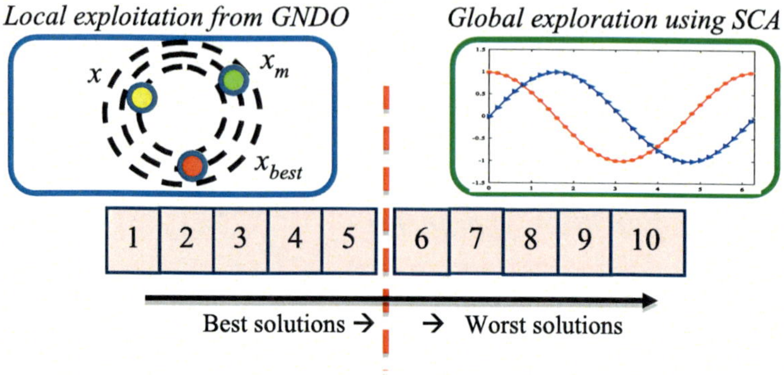

Fig. 2 Concept of hybrid GNDOSCA2 algorithm

in [0, 1], λ_3 and λ_4 are two random vectors generated from normal distribution, $p1$, $p2$, and $p3$ are three random integers between 1 and N (note that $p1 \neq p2 \neq p3$). As given in Eq. (1), an additional weight (w_1) is added to limit the history of self-experience as well as to enhance the history of best experience. In this way, it is believed that the exploitative behavior of the algorithm can be improved.

2.2 *Hybrid GNDOSCA Approach 2*

This subsection describes the second version of the proposed hybrid model. The concept of the GNDOSCA2 approach is illustrated in Fig. 2. Unlike GNDOSCA, the GNDOSCA2 utilizes the local exploitation strategy from GNDO to handle the exploitation process. On the contrary, the SCA mechanism is adopted in the exploration process. In the first step, the solutions are sorted from best to worst. Afterward, the exploitation and exploration strategies, respectively, are used to construct the new solutions from the first half and second half of the population.

$$v_i = \begin{cases} \mu_i + \delta_i \times \eta & \text{if } i \leq N/2 \\ \begin{cases} w_2 x_i(t) + r_1 \times \sin(r_2) \times |r_3 x_{\text{best}} - x_i(t)| & r_4 < 0.5 \\ w_2 x_i(t) + r_1 \times \cos(r_2) \times |r_3 x_{\text{best}} - x_i(t)| & r_4 \geq 0.5 \end{cases} & \text{otherwise} \end{cases} \quad (4)$$

$$u_i = \frac{1}{3}(x_i(t) + x_{\text{best}} + x_m) \quad (5)$$

$$\delta_i = \sqrt{\frac{1}{3}\left[(x_i(t) - \mu)^2 + (x_{\text{best}} - \mu)^2 + (x_m - \mu)^2\right]} \quad (6)$$

$$\eta = \begin{cases} \sqrt{-\log(\lambda_1)} \times \cos(2\pi\lambda_2), & \text{if } a \leq b \\ \sqrt{-\log(\lambda_1)} \times \cos(2\pi\lambda_2 + \pi), & \text{otherwise} \end{cases} \quad (7)$$

where a, b, λ_1, and λ_2 are random numbers in [0, 1], x_m is the mean of current population, w_2 is a weight parameter randomly distributed between 0.5 and 1 in each iteration. As given in Eq. (4), a higher value of weight (w_2) is assigned to restrict the global best experience for local optimal avoidance.

3 Experiment Results

This section discusses the finding of the experiment. The efficiencies of the proposed hybrid algorithms are tested on 23 standard mathematical optimization problems which consist of 7 unimodal (F1–F7), 6 multimodal (F8–F13), and 10 fixed-dimension multimodal (F14–F23) benchmark functions. The detail of the F1–F23 can be found in [23].

Table 1 displays the results of the function F1–F13. The GNDO, SCA, and CSA [24] are used in performance comparison. In this research, the population size is set at 30, while the maximum iteration is set to 500. All algorithms are repeated with 30 runs, and the mean and standard deviation are reported. Based on the results obtained, the proposed GNDOSCA and GNDOSCA2 are showing good performance. For unimodal and multimodal functions, the proposed GNDOSCA has overtook its competitors in at least 9 cases. Unlike GNDO and SCA, the hybrid GNDOSCA is more capable of escaping the local optima and searching for the global best solution. Our results affirm the superiority of the GNDOSCA in exploitation.

Table 2 presents the results of function F14—F23. Among the optimizers, the proposed GNDOSCA has scored the best results, followed by CSA and GNDOSCA2. Based on the findings in Tables 1 and 2, it can be inferred that the best algorithm was GNDOSCA. As compared to GNDOSCA2, GNDO, SCA, and CSA, the GNDOSCA can effectively explore and exploit the search space, which enables it to obtain the most promising solution.

4 Conclusion

In this research, the hybrid versions of the GNDO and SCA have been proposed for global optimization. The proposed methods combined the excellent characteristics of both algorithms to enhance the exploration and exploitation behaviors. An extensive study was performed on 23 benchmark functions to verify the efficacy of the proposed methods. According to findings, the proposed GNDOSCA was found to be enough competitive when compared to GNDO, SCA, and CSA algorithms. In the future, the proposed GNDOSCA can be applied to optimize the machine learning model, feature selection, and parameter optimization for photovoltaic model.

Table 1 Results of function F1–F13

		GNDOSCA	GNDOSCA2	GNDO	SCA	CSA
F1	Mean	**0**	4.00E−40	330.5667	14.85479	9.202114
	STD	0	2.04E−39	210.8284	20.84788	3.900283
F2	Mean	**3.71E−245**	1.37E−23	5.45171	0.013089	4.435599
	STD	0	5.20E−23	2.377451	0.017935	1.266218
F3	Mean	**0**	1.61E−13	1275.671	7903.172	511.4868
	STD	0	5.61E−13	618.3774	5848.88	134.4408
F4	Mean	**2.77E−229**	1.64E−10	18.2126	35.68591	9.210046
	STD	0	4.46E−10	3.319087	12.29331	1.992767
F5	Mean	28.52174	**28.15631**	34,563.7	53,121.4	557.1902
	STD	0.2591	0.276545	38,874.85	92,441.31	304.2184
F6	Mean	3.236369	**2.307143**	361.6245	14.78492	9.743103
	STD	0.55595	0.350175	274.311	13.22647	5.467811
F7	Mean	**0.000348**	0.000873	0.38196	0.111201	0.056915
	STD	0.000275	0.000469	0.228928	0.161681	0.022118
F8	Mean	−4931.51	−3586.71	−5753.35	−3734.54	**−6358.17**
	STD	683.294	283.6949	1297.715	281.8864	667.3067
F9	Mean	**0**	0	29.67206	50.84872	33.06867
	STD	0	0	9.157078	48.63632	10.08792
F10	Mean	**8.88E−16**	1.10E−13	9.369656	14.22922	4.817434
	STD	0	5.76E−13	1.249003	8.666451	1.142132
F11	Mean	**0**	0.000551	3.912507	0.91439	1.083082
	STD	0	0.001714	1.944045	0.428465	0.035206
F12	Mean	0.255241	**0.246914**	90.3632	40,327.77	6.261803
	STD	0.078958	0.054334	454.2597	146,702.7	1.964754
F13	Mean	**1.626686**	1.933632	12,246.88	669,443.4	26.61534
	STD	0.210659	0.214316	58,652.81	2,564,850	18.63015

Table 2 Results of function F14–F23

		GNDOSCA	GNDOSCA2	GNDO	SCA	CSA
F14	Mean	**0.998**	1.5611	1.0641	2.2512	1.2298
	STD	0	0.6738	0.3622	1.8878	0.5005
F15	Mean	**0.0005**	**0.0005**	0.0016	0.001	0.0007
	STD	0.0003	0.0001	0.0051	0.0004	0.0004
F16	Mean	**−1.0316**	**−1.0316**	**−1.0316**	**−1.0316**	**−1.0316**
	STD	0	0	0	0.0001	0
F17	Mean	**0.3979**	0.3981	**0.3979**	0.4001	**0.3979**

(continued)

Table 2 (continued)

		GNDOSCA	GNDOSCA2	GNDO	SCA	CSA
	STD	0	0.0005	0	0.0023	0
F18	Mean	**3**	**3**	**3**	3.0001	**3**
	STD	0	0	0	0.0001	0
F19	Mean	**−3.8628**	**−3.8628**	−3.837	−3.8542	**−3.8628**
	STD	0	0	0.1411	0.0033	0
F20	Mean	−3.2742	−3.2706	−3.2824	−2.8823	**−3.2975**
	STD	0.0528	0.0658	0.057	0.326	0.0559
F21	Mean	**−9.7023**	−8.1934	−6.667	−2.1516	−8.4133
	STD	1.2934	1.3324	3.7906	1.872	3.0821
F22	Mean	**−10.1648**	−8.1922	−7.8864	−2.7098	−9.3842
	STD	0.9848	1.9449	3.6238	1.9244	2.6417
F23	Mean	**−10.3229**	−9.6776	−8.5566	−4.1564	−8.9932
	STD	0.6345	1.5292	3.3464	1.5765	3.0824

References

1. Bogar E, Beyhan S (2020) Adolescent Identity Search Algorithm (AISA): a novel metaheuristic approach for solving optimization problems. Appl Soft Comput 95:106503
2. Zervoudakis K, Tsafarakis S (2020) A mayfly optimization algorithm. Comput Ind Eng 145:106559
3. Holland JH (1992) Genetic algorithms. Sci Am 267(1):66–73
4. Mirjalili S (2019) Genetic algorithm. Evolutionary algorithms and neural networks. Springer, pp 43–55
5. He Y, Zhang F, Mirjalili S, Zhang T (2022) Novel binary differential evolution algorithm based on Taper-shaped transfer functions for binary optimization problems. Swarm Evol Comput 69:101022
6. Price KV (2013) Differential evolution. Handbook of optimization. Springer, pp 187–214
7. Kennedy J, Eberhart R (1995) Particle swarm optimization. In: Proceedings of ICNN'95-international conference on neural networks, vol 4. IEEE, pp 1942–1948
8. Shami TM, El-Saleh AA, Alswaitti M, Al-Tashi Q, Summakieh MA, Mirjalili S (2022) Particle Swarm optimization: a comprehensive survey. IEEE Access
9. Abdollahzadeh B, Gharehchopogh FS, Mirjalili S (2021) African vultures optimization algorithm: a new nature-inspired metaheuristic algorithm for global optimization problems. Comput Ind Eng 158:107408
10. Abdollahzadeh B, Soleimanian Gharehchopogh F, Mirjalili S (2021) Artificial gorilla troops optimizer: a new nature-inspired metaheuristic algorithm for global optimization problems. Int J Intell Syst 36(10):5887–5958
11. Zhao W, Wang L, Mirjalili S (2022) Artificial hummingbird algorithm: a new bio-inspired optimizer with its engineering applications. Comput Methods Appl Mech Eng 388:114194
12. Van Laarhoven PJ, Aarts EH (1987) Simulated annealing. Simulated annealing: theory and applications. Springer, pp 7–15
13. Geem ZW, Kim JH, Loganathan GV (2001) A new heuristic optimization algorithm: harmony search. Simulation 76(2), 60–68
14. Rao RV, Savsani VJ, Vakharia D (2011) Teaching–learning-based optimization: a novel method for constrained mechanical design optimization problems. Comput Aided Des 43(3):303–315

15. Zhang Y, Jin Z, Mirjalili S (2020) Generalized normal distribution optimization and its applications in parameter extraction of photovoltaic models. Energy Convers Manage 224:113301
16. Mirjalili S (2016) SCA: a sine cosine algorithm for solving optimization problems. Knowl-Based Syst 96:120–133
17. Taghian S, Nadimi-Shahraki MH (2019) A binary metaheuristic algorithm for wrapper feature selection. Int J Comput Sci Eng (IJCSE) 8(5):168–172
18. Althobiani F, Khatir S, Brahim B, Ghandourah E, Mirjalili S, Wahab MA (2021) A hybrid PSO and Grey Wolf optimization algorithm for static and dynamic Crack identification. Theor Appl Fract Mech, 103213
19. Talbi E-G (2002) A taxonomy of hybrid metaheuristics. J Heuristics 8(5):541–564
20. Blum C, Puchinger J, Raidl GR, Roli A (2011) Hybrid metaheuristics in combinatorial optimization: a survey. Appl Soft Comput 11(6):4135–4151
21. Blum C, Roli A, Sampels M (2008) Hybrid metaheuristics: an emerging approach to optimization. Springer
22. Blum C, Raidl GR (2016) Hybrid metaheuristics: powerful tools for optimization. Springer
23. Kaveh A, Talatahari S, Khodadadi N (2020) Stochastic paint optimizer: theory and application in civil engineering. Eng Comput 1–32
24. Askarzadeh A (2016) A novel metaheuristic method for solving constrained engineering optimization problems: crow search algorithm. Comput Struct 169:1–12

Sensitivity Analysis on Structural Optimization Using Jaya Algorithm

Mehmet Berat Bilgin, Sinan Melih Nigdeli, and Gebrail Bekdaş

Abstract In this study, three different structural engineering benchmark design problems were optimized using the Jaya algorithm, one of the metaheuristic algorithms. The optimization process was repeated many times for different iteration numbers, and the results were evaluated. Thus, the accuracy of the optimum result reached by the algorithm has been checked. It is found that Jaya algorithm is a robust one if the required number of iterations is done, but it must be noted that several failures can be also observed in several runs when the maximum iteration number is low although the problems are small in numbers of design variables.

Keywords Jaya algorithm · Structural engineering · Optimum design · Optimization · Robustness

1 Introduction

Heuristic optimization processes of living things in nature attracted the attention of researchers working on basic sciences, and they produced various algorithms that express these processes mathematically. Although there are many types of algorithms called metaheuristics, one of the easiest to implement in the literature is the Jaya algorithm due to having a single phase of iterative optimization. This algorithm, whose main goal is to both approach the best result and get away from the worst result, uses the following equation to reach the new value in each iteration [1]:

M. B. Bilgin (✉) · S. M. Nigdeli · G. Bekdaş
Department of Civil Engineering, Istanbul University - Cerrahpaşa, 34320 Avcılar, Istanbul, Turkey
e-mail: mehmetberatbilgin@gmail.com

S. M. Nigdeli
e-mail: melihnig@iuc.edu.tr

G. Bekdaş
e-mail: bekdas@iuc.edu.tr

J. H. Kim et al. (eds.), *Proceedings of 7th International Conference on Harmony Search, Soft Computing and Applications*, Lecture Notes on Data Engineering and Communications Technologies 140, https://doi.org/10.1007/978-981-19-2948-9_5

$$X_{i,new} = X_{i,j} + r_1\left(X_{i,best} - \left|X_{i,j}\right|\right) - r_2\left(X_{i,worst} - \left|X_{i,j}\right|\right) \quad (1)$$

where r_1 and r_2 are random numbers between 0 and 1. As can be seen, the Jaya algorithm uses a single equation to obtain the new solution for all variables. In this phase, a new ith variable $(X_{i,new})$ is generated by using existing solution of jth population $(X_{i,j})$ together with best $(X_{i,best})$ and worst $(X_{i,worst})$ existing solutions. Although it is known that algorithms are successful in reaching the right result in practice, they are not without the possibility of making mistakes. A sensitivity analysis is presented on three benchmark structural engineering problems by using Jaya algorithm that was previously applied to several structural engineering problems including steel grillage structures [2], truss structures [3, 4], braced dome structures [5], structural damage identification problem [6], plate-like structures [7], crack identification in plates [8], reinforced concrete frames [9], tuned mass dampers considering soil-structure interaction [10], and tuning of PID controllers for seismic structures [11].

2 The Methodology

In this study, three different civil engineering design problems are discussed. Optimization of these problems has been achieved by using the Jaya algorithm with the codes written in MatLab. Then, in order to test the reliability of the algorithm, the optimization process was run 30 and 50 times with different iteration numbers for three different problems, and the results are saved. Among the results, the best and worst objective function values and their standard deviations are tabulated.

2.1 *Optimization Problem 1*

The first optimization problem involves cost optimization of a tubular section column under pressure load [12]. In this problem, section center diameter "*d*" and thickness "*t*" parameters are sought, which will enable the column to carry the load on it safely and also to reveal the cheapest cost and construction cost.

In Fig. 1, "*P* (kN)" represents the axial compressive force, and "*l* (m)" represents the column length. Together with the modulus of elasticity "*E*" and the yield stress "σ_y," these four parameters can be considered as design constants. And with this information, the objective function that expresses the total cost minimization is calculated as in Eq. 2 depending on *d* and *t*.

$$f(d,\ t) = 9,\ 8\ dt + 2d \quad (2)$$

Apart from this, there are two design constraints related to the axial load capacity of the column (g_1) and the buckling limit of the column (g_2). These constraints can

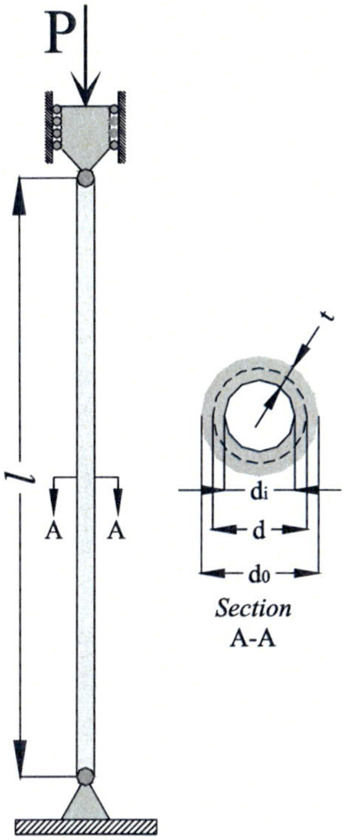

Fig. 1 Tubular column and section detail

be expressed as

$$g_1 = \frac{P}{\pi d t \sigma_y} - 1 \leq 0 \tag{3}$$

$$g_2 = \frac{8Pl^2}{\pi^3 E d t \left(d^2 + t^2\right)} - 1 \leq 0 \tag{4}$$

By using this objective function and design constraints, the optimization problem was performed 30 and 50 times with 100, 250, 500, and 1000 iteration values by Jaya algorithm for predefined design constants. The results obtained are shown in Tables 1 and 2.

2.2 *Optimization Problem 2*

The second optimization problem is related to the minimization of the vertical displacement in I section beam under the applied loads [13]. As seen in Fig. 2, the design variables are "*h*" (beam height), "*b*" (flange width), "t_w" (web thickness),

Table 1 Results for optimization problem 1 with 30 runs

Number of runs of optimization problem: 30	Number of ıterations			
	100	250	500	1000
The best obj. func. value	26.499523	26.499497	26.499497	26.499497
The worst obj. func. value	30.276016	27.797218	30.111089	29.884612
Standard deviation	0.722645	0.275019	0.787221	0.672087

Table 2 Results for optimization problem 1 with 50 runs

Number of runs of optimization problem: 50	Number of ıterations			
	100	250	500	1000
The best obj. func. value	26.499545	26.499497	26.499497	26.499497
The worst obj. func. value	32.527837	27.667700	27.702154	30.232555
Standard deviation	1.144700	0.174401	0.226670	0.717758

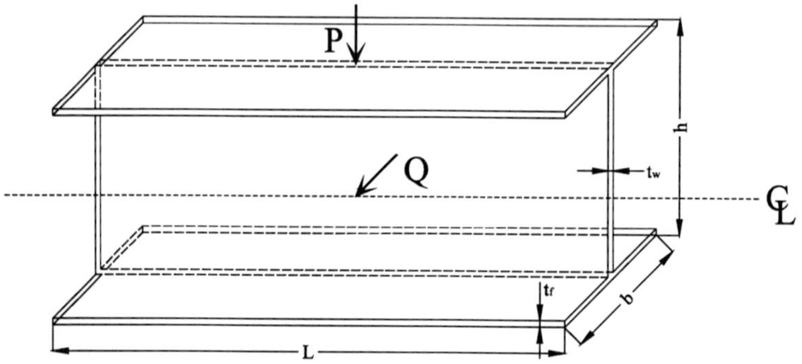

Fig. 2 I-Beam section properties and design loads

"t_f" (flange thickness), and the design constants are vertical load "*P*", and lateral load "*Q*," modulus of elasticity "*E*" and beam length "*L*."

And the displacement that will occur in the beam due to the application of loads can be expressed by Eq. 5.

$$f(x) = \frac{PL^3}{48EI} \tag{5}$$

Here, *I* represents the moment of inertia. By substituting the moment of inertia in Eq. 5 together with the parameters determining the beam section, the objective function determined as vertical displacement minimization can be written as follows:

$$\min f(h, b, t_w, t_f) = \frac{5000}{\left(\frac{t_w(h-2t_f)^3}{12} + \frac{bt_f^3}{6} + 2bt_f\left(\frac{h-t_f}{2}\right)^2\right)} \tag{6}$$

Table 3 Results for optimization problem 2 with 30 runs

Number of runs of optimization problem: 30	Number of iterations			
	100	250	500	1000
The best obj. func. value	0.013074	0.013074	0.013074	0.013074
The worst obj. func. value	0.020759	0.033693	0.023758	0.014212
Standard deviation	0.001921	0.003933	0.002876	0.000208

Table 4 Results for optimization problem 2 with 50 runs

Number of runs of optimization problem: 50	Number of iterations			
	100	250	500	1000
The best obj. func. value	0.013074	0.013074	0.013074	0.013074
The worst obj. func. value	0.022869	0.024868	0.031044	0.020279
Standard deviation	0.002124	0.002109	0.003293	0.001019

There are two design constraints for this optimization problem. The first of these states that the beam section cannot be larger than 300 cm^2, while the other states that the allowable moment stress cannot be more than 6 kN/cm^2.

$$g_1 = 2bt_f + t_w(h - 2t_f) \leq 300 \tag{7}$$

$$g_2 = \frac{18000\,h}{t_w(h - 2t_f)^3 + 2bt_f\left(4t_f^2 + 3h(h - 2t_f)\right)} + \frac{15000\,b}{t_w^3(h - 2t_f) + 2t_w b^3} \leq 6 \tag{8}$$

By using this objective function and design constraints, the optimization problem was performed 30 and 50 times with 100, 250, 500, and 1000 iteration values for predefined design constants. The results obtained are shown in Tables 3 and 4.

2.3 Optimization Problem 3

In optimization problem 3, the beam model to be used in the weight optimization application of the cantilever beam is divided into five separate sections and has a hollow section as can be seen in Fig. 3 [14].

The "l_j ($j = 1, 2, 3, 4, 5$)" values represent the length of each beam segment; the "P" represents the vertical load, and "x_j ($j = 1, 2, 3, 4, 5$)" shows sizes of the hollow section. Also, section thickness is taken 2/3 cm. For the optimization of the beam weight, the relevant objective function was calculated as in Eq. 9.

$$\min f\left(x_j\right) = 0,0624(x_1 + x_2 + x_3 + x_4 + x_5) \tag{9}$$

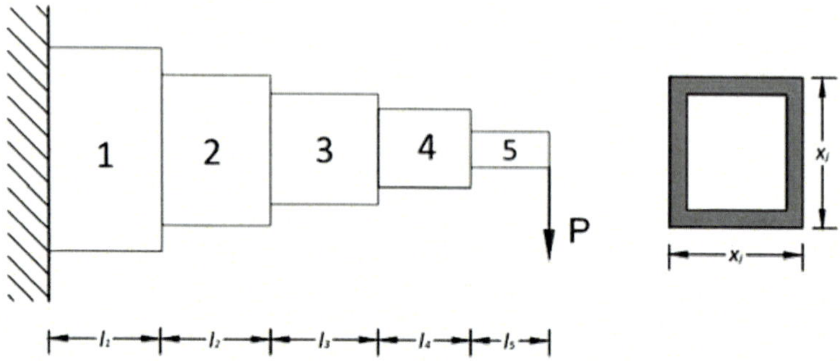

Fig. 3 Cantilever beam model and hollow section sizes

On the other hand, there is only one design constraint for this optimization problem that expressed in Eq. 10:

$$g_1 = \frac{61}{x_1^3} + \frac{37}{x_2^3} + \frac{19}{x_3^3} + \frac{7}{x_4^3} + \frac{1}{x_5^3} - 1 \leq 0 \tag{10}$$

By using this objective function and design constraints, the optimization problem was performed 30 and 50 times with 500, 1000, 2500, 5000, and 10,000 iteration values for predefined design constants. The results obtained are shown in Tables 5 and 6.

Table 5 Results for optimization problem 3 with 30 runs

Number of runs of optimization problem: 30	Number of iterations				
	500	1000	2500	5000	10,000
The best obj. func. value	1.339961	1.339959	1.339957	1.339956	1.339956
The worst obj. func. value	1.340077	1.339981	1.339959	1.339957	1.339957
Standard deviation	0.000027	0.000006	0.000001	0.000000	0.000000

Table 6 Results for optimization problem 3 with 50 runs

Number of runs of optimization problem: 50	Number of iterations				
	500	1000	2500	5000	10,000
The best obj. func. value	1.339964	1.339958	1.339957	1.339956	1.339956
The worst obj. func. value	1.340040	1.339980	1.339960	1.339958	1.339957
Standard deviation	0.000021	0.000005	0.000001	0.000000	0.000000

3 Conclusion

Firstly, when the first optimization problem is compared in the case of 100 iterations for different number of runs, it is seen that the standard deviation is higher at 50 runs than at 30 runs. The reason for this is that the relatively low number of 100 iterations makes it difficult to reach the correct result, while running the problem 50 times increases the probability of getting the wrong result.

Another issue is that regardless of the number of iterations, the best objective function values are the same for different iteration and the number of runs for all three problems. In addition, standard deviations are also low. Therefore, it can be said that the optimization process is very successful.

Finally, the iteration numbers in optimization problem 3 are higher than the iteration numbers in the other two problems due to the number of design variables. As a result of this situation, it is observed that the standard deviations are almost 0 for all cases. Therefore, a large number of iterations has a positive effect on the optimization process reaching the correct result as expected. In the future, the evaluation done in this study can be conducted for real-size structural engineering problems like optimization of a whole structural system including trusses and frames.

References

1. Rao R (2016) Jaya: a simple and new optimization algorithm for solving constrained and unconstrained optimization problems. Int J Ind Eng Comput 7(1):19–34
2. Dede T (2018) Jaya algorithm to solve single objective size optimization problem for steel grillage structures. Steel Compos Struct 26(2):163–170
3. Degertekin SO, Lamberti L, Ugur IB (2018) Sizing, layout and topology design optimization of truss structures using the Jaya algorithm. Appl Soft Comput 70:903–928
4. Bekdaş G, Yucel M, Nigdeli SM (2021) Evaluation of metaheuristic-based methods for optimization of truss structures via various algorithms and Lèvy flight modification. Buildings 11(2):49
5. Grzywinski M, Dede T, Ozdemir YI (2019) Optimization of the braced dome structures by using Jaya algorithm with frequency constraints. Steel Compos Struct 30(1):47–55
6. Du DC, Vinh HH, Trung VD, Hong Quyen NT, Trung NT (2018) Efficiency of Jaya algorithm for solving the optimization-based structural damage identification problem based on a hybrid objective function. Eng Optim 50(8):1233–1251
7. Dinh-Cong D, Vo-Duy T, Ho-Huu V, Nguyen-Thoi T (2019) Damage assessment in plate-like structures using a two-stage method based on modal strain energy change and Jaya algorithm. Inverse Prob Sci Eng 27(2):166–189
8. Khatir S, Boutchicha D, Le Thanh C, Tran-Ngoc H, Nguyen TN, Abdel-Wahab M (2020) Improved ANN technique combined with Jaya algorithm for crack identification in plates using XIGA and experimental analysis. Theoret Appl Fract Mech 107:102554
9. Rakıcı E, Bekdaş G, Nigdeli SM (2020) Optimal cost design of single-story reinforced concrete frames using Jaya algorithm. In: International conference on harmony search algorithm. Springer, Singapore, pp 179–186
10. Bekdaş G, Kayabekir AE, Nigdeli SM, Toklu YC (2019) Tranfer function amplitude minimization for structures with tuned mass dampers considering soil-structure interaction. Soil Dyn Earthq Eng 116:552–562

11. Ulusoy S, Nigdeli SM, Bekdaş G (2021) Novel metaheuristic-based tuning of PID controllers for seismic structures and verification of robustness. J Build Eng 33:101647
12. Rao SS (2019) Engineering optimization: theory and practice. Wiley
13. Gold S, Krishnamurty S (1997) Trade-offs in robust engineering design. In: Proceedings of the ASME design engineering technical conferences
14. Fleury C, Braibant V (1986) Structural optimization: a new dual method using mixed variables. Int J Numer Meth Eng 23:409–428

Upgrading Urban Drainage Systems for Extreme Rainfall Events Using Multi-objective Optimization: Case Study of Tan Hoa-Lo Gom Drainage Catchment, HCMC, Vietnam

Check for updates

Hoa Van Ho, Truong-Huy Nguyen, Loc Huu Ho, Quang Nguyen Xuan Chau, Linh Ngoc Trinh, and Joong Hoon Kim

Abstract Ho Chi Minh City (HCMC) is the largest urban area in Vietnam where the urban inundation has increasingly become severe recently due to a rapid expansion of its urbanization and an increase of extreme rainfalls in terms of both intensity and frequency. The proper designs of the new or upgraded drainage (canal) systems based on the recent extreme rainfall data can reduce the flooding impacts. Hence, a new extreme rainfall intensity-duration-frequency (IDF) relation is first constructed for HCMC based on the most up-to-date observed data from 1982 to 2019. This rainfall IDF relation is then used as an input to investigate and select a suitable drainage routine, shape, and size for upgrading the urban drainage systems in HCMC. These optimal cross-sections as well as routine selections were determined based on an optimization model through a large number of numerical simulations. In this paper, an interfacing Visual C++ programming tool with an SWMM DLL for rainfall-runoff computation in a drainage network completes the optimization. The multi-objective genetic algorithm (MOGA) is used to find the optimum solutions by minimizing both the total flooding overflow and the installing cost as the objective functions. The developed model was applied to the case study of Tan Hoa-Lo Gom drainage

H. Van Ho (✉) · Q. N. X. Chau · L. N. Trinh
Department of Hydrology and Water Resources, Institute for Environment and Resources, Vietnam National University - Ho Chi Minh City (VNU-HCM), Ho Chi Minh City, 700000, Vietnam
e-mail: harryhoa@gmail.com

T.-H. Nguyen (✉)
McGill University, Montreal, Canada
e-mail: huytruongnguyen.fwr@gmail.com

University of Science and Technology—University of Danang, Danang, Vietnam

L. H. Ho
Water Engineering and Management, School of Engineering and Technology, Asian Institute of Technology, Pathumthani, Thailand

J. H. Kim
School of Civil, Environmental and Architectural Engineering, Korea University, Seoul, South Korea 02841

J. H. Kim et al. (eds.), *Proceedings of 7th International Conference on Harmony Search, Soft Computing and Applications*, Lecture Notes on Data Engineering and Communications Technologies 140, https://doi.org/10.1007/978-981-19-2948-9_6

basin in HCMC. Model performance was evaluated to propose the optimum solution to the drainage system upgrade. By considering the most recent extreme rainfalls and urbanization conditions, with the optimal drainage routine and cross-section identified by the model, loss flooding in Lo Gom canal after upgrading can be reduced by 57.7%. This optimal drainage model is, therefore, premise for further studies in selecting the best option of flooding prevention in urban areas.

Keywords Reduce urban flood · IDF Ho Chi Minh City · MOGA

1 Introduction

Ho Chi Minh City is the largest and most developed city in Vietnam with the population increasing rapidly and is approximately 9.1 million people [1]. The city has master plans for urban drainage systems up to 2020 [2] with the orientation of expanding the service scope of the drainage system in order to cover 70% of the urban areas. Under the impacts of urbanization and climate change, many different factors affect the drainage systems in a negative direction, such as the changes in land use (i.e., increase of the residential land areas, decrease of the water room areas), increase of extreme storm intensity and frequency, increase of tide level, and so on. In addition, the drainage system has been built for a long time; many sections are no longer suitable for use. Many technical infrastructures that work in general are in a state of deterioration, and investment in new construction is still slow and cannot keep up with the ever-increasing demand. As a result, flood situation in HCMC has become increasingly severe in the last few decades. This reduces the rainwater drainage capacity of the system and causes severe flooding events, which affect the residents' livelihoods and properties in the area. Although the city municipality has spent billions of dollars (USD) to invest in the urban water infrastructure, the situation is far from indefinitely solved [3].

Optimization in the design of drainage system is a new research direction in Vietnam. It has been applied to select the parameters of the design network to achieve a feasible low cost with an optimal solution [4]. It has been used in the management of operating procedures and upgrading the system or specific work in the system such as reservoirs, pump types, improvement of hydraulic pipeline cross-sections [5]. In addition, the study of Dang Thanh Lam (2015) has succeeded in applying the hydraulic model to solve urban hydraulic problems in tidal areas and also provide solutions to increase the size of canals and pipes as effective solutions to reduce flooding for the THLG basin, but the optimal channel size has not yet been determined in terms of cost and reduces flooding for the channel (sewer) that needs upgrading [6].

Some researchers on genetic algorithm (GA) have found that if prior knowledge exists or can be generated at a low computational cost, seeding GAs with good initial estimates may generate better solutions with faster convergence [7]. Vast computation costs can be saved by providing the GA execution with a high-quality initial population. In the light of their findings, the authors proposed a novel hybrid sewer design method, named as cellular automata and genetic algorithm for sewers in network optimization (CA-GASiNO), by combining CASiNO and a multi-objective GA, specifically a constrained non-dominated sorting genetic algorithm (NSGAII) [8, 9].

Besides, the rainfall IDF curve often used for the design of urban drainage system is outdated in term of both the historical data and the representative probability model. Several papers used data from 1980 to 2016 to construct IDF curves [10, 11]. However, a recent record-breaking extreme rainfall storm of 407.2 mm occurred on 11/25–26/2018. In addition, the GEV probability distribution model has been recently advocated by many research to represent extreme rainfall events.

Therefore, the objective of this study is to propose a method to set up a scheme to optimize the cost of upgrading the main drainage axes by applying a multi-target genetic algorithm combined with hydraulic simulation software EPA-SWMM. The model automatically calculates the hydraulics with new parameters of the size and shape of the path to be improved and, at the same time, outputs the results of the flood reduction efficiency of each simulated option. In addition, the unit price calculated in this study applies the correct method of cost estimation in Vietnam. The selected option will have the lowest investment cost and inundation level and satisfy the given constraints on cost and route size in the study basin.

2 Material and Methods

2.1 *Study Area*

THLG canal basin is in the southwest of the inner HCMC bordering with the suburbs Fig. 1a. The canal flows from the northeast to the southwest area and flows through 5 districts: Tan Binh district, Districts 11, 6, 8, and Tan Phu district, then terminates at Tau Hu canal. The total area is 2498 ha (3.8% of the city), and length is 7.84 km. This is the 12th of 21 city's basin. The THLG canal area is spontaneous, so the technical infrastructure is very weak and does not respond to urban standards. The city's development has extended the center to almost the entire canal basin, so the role of the canal basin has become increasingly important to the city. As a result, the residents are constantly confronted with floods during heavy storms and/or high tides. The problem of upgrading canals is necessary for the urban area which has been developed for a long time, to create an effective space for floodwaters due to rain and tides.

The model is corrected so that the water depth in the upstream area is consistent with the observed data measured on September 15, 2010. The total rain volume was 80.3 mm at Tan Son Hoa station and tide data of the same day at Phu An station. The parameters of the model are corrected, including percent (%) of impermeable surface area, sub-basin surface roughness factor, depth of depression in the basins,

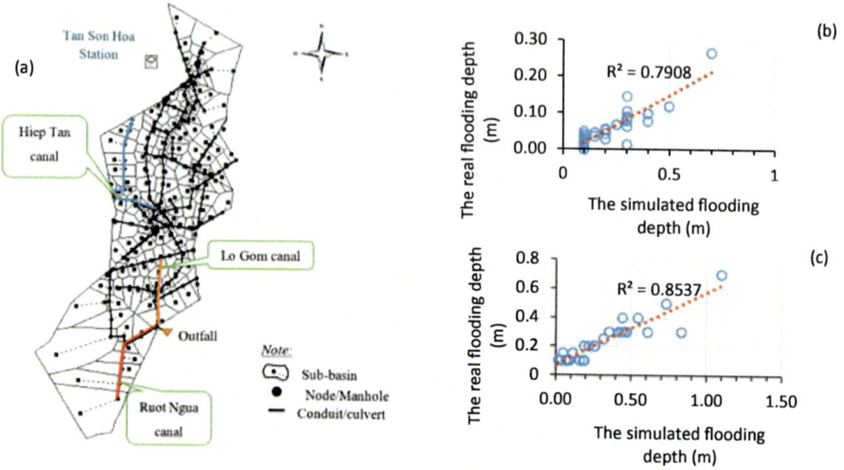

Fig. 1 **a** Study area THLG and three main canals selected for the optimal study; **b** and **c** correlation coefficient between the simulated flooding depth and the real flooding depth for the calibration and validation, respectively

roughness factor in a transmission system. The model's ability in predicting the flood areas has been pre-calibrated with the correlation coefficient R^2 of 0.79 for flooding depths and Nash–Sutcliffe coefficient of 0.85. In addition, the model has also been validated using the rain data on November 25–26, 2018, with the total rain volume of 403.7 mm at Tan Son Hoa station combined with the tidal data at Phu An station. The correlation coefficient R^2 reached 0.92 for the flooding depth; Nash–Sutcliffe coefficient reached 0.86 for flood depth (Fig. 1b, c). The results of calibration and validation are satisfactory; hence, the model can be used to simulate the situations in the THLG canal basin.

2.2 *IDF Rainfall Design for HCM City*

The extreme rainfall intensity-duration-frequency (IDF) curves represent the relationships between rainfall intensities of different rainfall durations (usually, a few minutes to a day) and of different frequencies (usually, two to a hundred years). The IDF curves for HCMC are out of date in terms of both data and methodology. Hence, in this study, a new rainfall IDF curve is constructed for HCMC using the most up-to-date observed data (1982–2019) from the Tan Son Hoa site. The annual maximum series corresponding to nine different rainfall durations ($D = 15$, 30, 45 min, and 1, 2, 3, 6, 12, 24 h) was extracted from the observed full-time series using the moving windows.

The generalized extreme values (GEV) distribution was used to fit to the extreme series. The L-moment method was used to estimate the three parameters of the

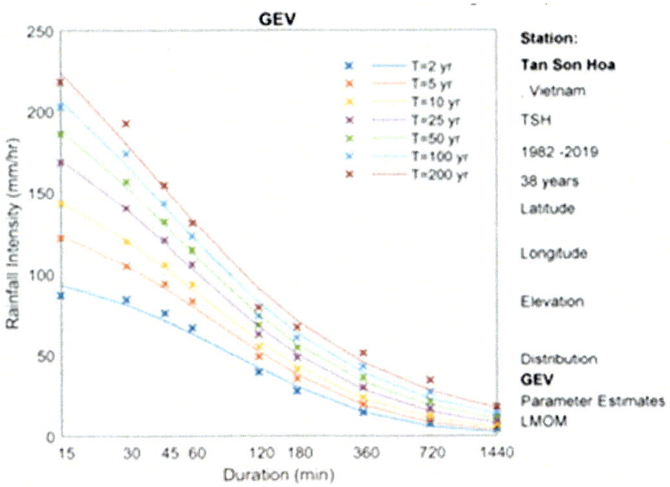

Fig. 2 IDF curve at Tan Son Hoa station with rainfall data from 1982 to 2019

GEV distribution for each rainfall duration series. The extreme rainfall quantiles of different return periods ($T = 2$–200 years) were then estimated based on the GEV quantile function and the known parameters obtained above. The three-coefficient regression model, $I = a/(D + c)^b$ (where a, b, c are coefficients; I and D are rainfall intensity and duration, respectively), was used to fit these theoretical rainfall quantiles. All this work was done using the SMExRain software developed at McGill University, Canada [12]. The new rainfall IDF curves constructed for the study area are shown in Fig. 2. For the 10-year return period, this value is slightly higher (0.5%) than the one. But for the 100-year return period, it is much significant (8.8%) than the one [11]. In this study, we use a frequency design of 10 years and a rainfall duration of 3 h to simulate the optimization cases, which was a proposal by the Ministry of Construction and was published by MOST Vietnam (2008) [13].

2.3 *Hydrology-Runoff EPA. SWMM and Optimization Model*

The EPA Storm Water Management Model (EPA-SWMM) is a dynamic rainfall-runoff simulation model used for single event or long-term (continuous) simulation of runoff quantity and quality from primarily urban areas. The conservation of mass and momentum for unsteady free-surface flow through a channel or pipe is known as the Saint Venant equations.

GAs, as well as evolutionary algorithms, are based on a concept that is considered an axiom consistent with objective reality. The process of evolution shows the optimality in that the next generation is always better than the previous one. The strength of the genetic optimization algorithm is that it can find the optimal solution in a faster time because for iterative problems that combine all billion cases to find the most valuable solution, and it can be used to solve multivariable nonlinear problems with high speed and accurate solutions with high values. Lim et al. (2014) successfully used a multi-objective genetic algorithm to optimize the drainage system by optimizing the location of the underground reservoir and the volume, cost, and effectiveness of flood reduction for Seoul urban area, South Korea [14].

The above advantages and results are the foundation for the author to apply the genetic optimization algorithm to optimize the parameters of the urban drainage system. The algorithm diagram for solving an urban drainage optimization problem using a genetic algorithm is shown in Fig. 3. An efficient methodology and computer

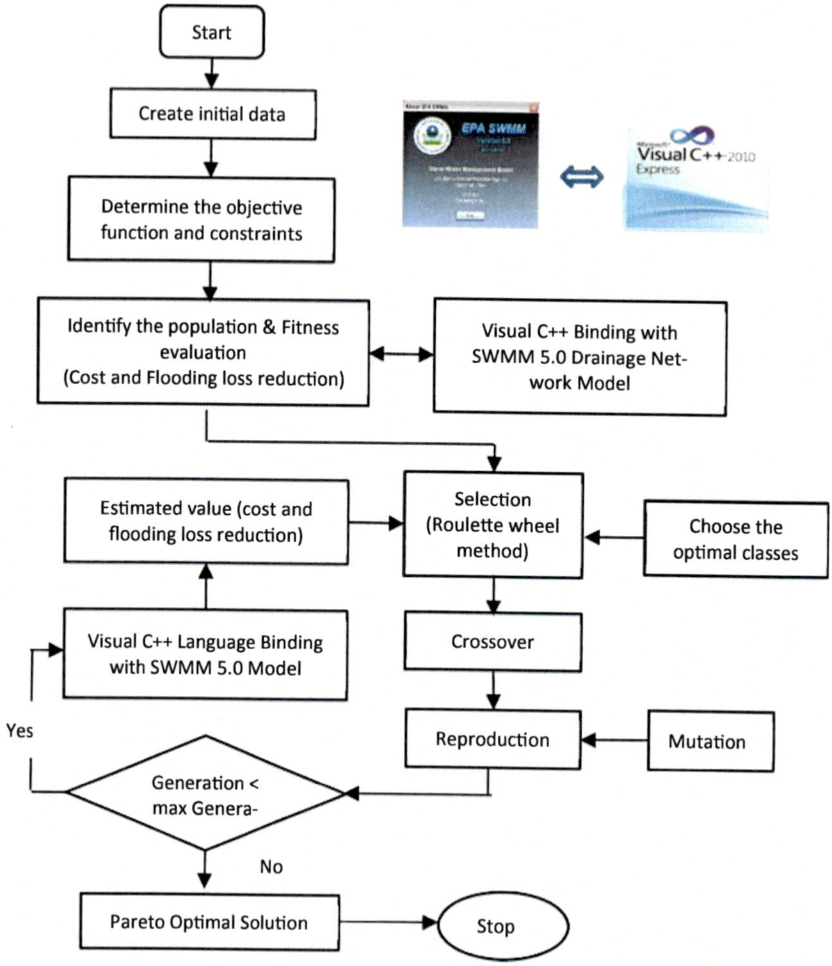

Fig. 3 Flowchart of optimization using MOGAs linkage with EPA. SWMM

software model are developed to optimize urban drainage systems. This software is written in C++ language, linking with SWMM 5.1 libraries which are freely available. The software can change the parameters in the drainage model, then automatically let the SWMM model run the hydraulic calculations and read the hydraulic results of the drainage model parameters set in the SWMM. From there, take the parameters of the inundation, the flow flowing through the manholes to calculate, evaluate and select the options to change the channel cross-section well.

Details of the optimization process for each canals are presented below:

1. Initialize route position, gland shape, and gland size by gene sequences (0;1); (0;2);... and fed into the EPA-SWMM model.
2. Establishment of family gene sequence (initial set of gene sequences).
3. Implement the program to run the hydraulic calculation EPA-SWMM, taking the results of the flood reduction efficiency, the overflow through the manhole as the evaluation value for the converted gene sequence.
4. Selecting N gene sequences based on evaluated value (flood reduction efficiency) by Roulette method and sorted in ascending order.
5. Then, perform the method of cross-breeding or mutating the gene sequences to produce offspring from the good gene sequences.
6. Repeat Steps (3) and (5) until the program stopping criteria are satisfied; if the criteria are met, the Pareto-optimal solutions are obtained.

Flood damage is classified into two typical criteria [15]. The first criterion is based on the quantitative nature of the distinction between tangible and intangible damages. Tangible damages are impacts on things that can be quantified in terms of money including property damage, loss of profits due to business interruption [16].

1. *Construction cost function*: Minimum cost of drainage upgrade: Minimize C_{total}

$$\text{Minimize} \sum_{j=1}^{m} C_j \tag{1}$$

   C_j is the formula to calculate the cost of the upgrade option C_j depending on the volume of the construction.

   With $C_j = 932.3 \times \text{Vj}$ ($US). The cost of construction and installation is calculated according to the standard 1 m^3 of concrete = $1 \times 1 \times 1$ m which is dug into and buried in the work, above there are items of earth filling and grass planting (calculated according to preliminary estimate of unit price of Ho Chi Minh City, 2017).

   V_j: Volume of the *j*th upgrade option.

   For rectangular channel cross-section: $Cj_{\text{rectangular}} = Cj*0.75$;

   For trapezoidal channel cross-section: $Cj_{\text{trapezoid}} = Cj*0.95$;

   For circular section: $Cj_{\text{circle}} = Cj*0.785$.
2. *Flood reduction objective*: Maximum flood reduction effect.

$$\text{Maximize}\left(W_{\text{total flooding status}} - \sum_{1}^{n} W_{i\ \text{flood}}\right) \tag{2}$$

where W_i: flooding loss (m^3) at number of node i flooded in the drainage.

3. *Contrast conditions.*

Based on the city finance, the total value of investment is upgrading < [50 million $US] and number of km of upgraded route <= [5 km].

3 Results and Discussion

Advantages: The optimal program only needs to run the calculation 300 times, and within an appropriate 20 h, the optimal results are obtained as shown in the graphs Figs. 4 and 5 for Lo Gom canal. A similar approach was followed for Hiep Tan canal and Ruot Ngua canal. The rainfall obtained from the new IDF design from Fig. 2 is 122.3 mm in 3 h and, together with the peak tide, is 1.68 m for 10 years frequency.

From Figs. 4 and 5 and Table 1, it can be seen that the optimal solution for Lo Gom canal of effective flooding reduction is 57.7%, and upgrade cost of 10.71 million $US is the most effective option and the most expensive compared to other options. Figure 5a shows the optimal cross-sections with the near join node then need the big cross-section to store water, and the final cross-section in the outlet is small because this basin is flooded by tide, so the inlet of tide needs to be small to less water

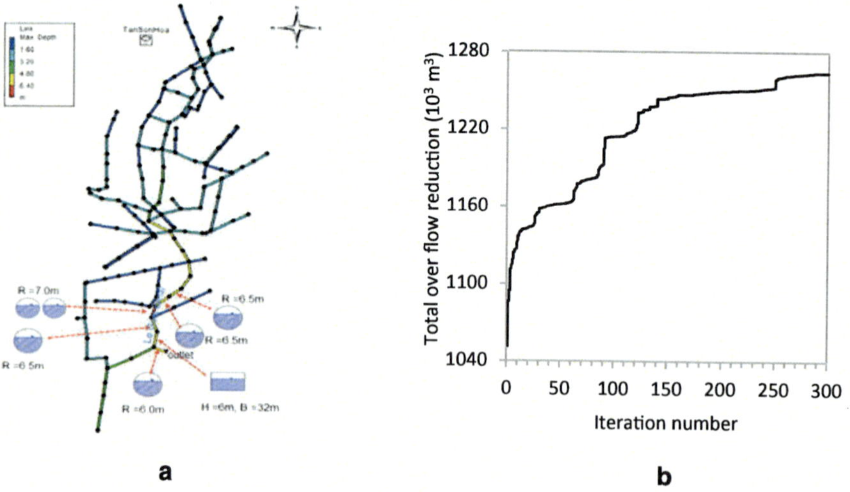

Fig. 4 **a** Optimal routine in rainfall 122.36 mm for Lo Gom canal; tide 1.68 mm; **b** GA process according to iterations

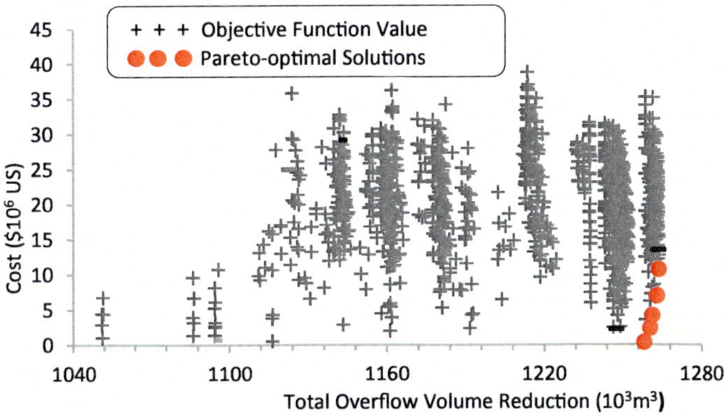

Fig. 5 Best Pareto-optimal solutions of Lo Gom canal with duration $T = 10$ years in THLG basin

Table 1 Benefit values of the optimal upgraded solution of three selected canals

No.	Name of the canal	Total overflow volume reduction (10^3 m^3)	Effective (%)	Cost (10^6 $US)
1	Tan Hiep	442.1	20.2	9.49
2	Ruot Ngua	1096.2	50.7	13.07
3	Lo Gom	1263.7	57.7	10.71

to come from the sea (consider as width gate control). That is the important thing when designing urban drainage networks for areas under tidal impacts. In essence, additional spaces to store pluvial water are required alongside sluice gates to prevent fluvial water.

In addition, from Table 1, it is shown that the optimal solution for the Ruot Ngua canal is the one with the total efficiency in reducing the amount of flooding 50.7% reducing the total amount of flooding compared to the current situation, and the total investment cost of 13.07 million $US is the most effective option and the most expensive. Hiep Tan canal can only be upgraded within the permitted land area. Although, it has a over midling investment rate is 9.49 million $US, but the flood reduction effect of Hiep Tan canal route is quite small. The total cost amount of upgrading of the Lo Gom canal option is 1.1 times greater, but the total amount of flood reduction is 2.85 times greater compared to Hiep Tan canal, so it led to making the Lo Gom canal area almost no longer flooded than the option of upgrading Ruot Ngua canal and Hiep Tan canal (Table 1). This can be concluded that priority upgrading the Lo Gom canal is the first, then Ruot Ngua canal, and finally Hiep Tan for the solution of reducing flooding for Tan Hoa-Lo Gom basin.

4 Conclusion

Through the optimization process, the author found that the genetic optimization algorithm (GA) is a very effective algorithm in supporting the search for a fast optimal solution without having to evaluate all thousands of cases (e.g., combination) of the proposed options as shown in the resulting figures above.

The above results show that expanding the cross-section of Lo Gom, Tan Hiep, and Ruot Ngua canals has contributed to reducing flooding in the THLG basin through an increase in storage volume and faster flow clearance. Between the three routes of Lo Gom, Hiep Tan, and Ruot Ngua canal, after applying the optimization search options in terms of cross-section, the results show that the effect of the Lo Gom canal in reducing flooding after upgrading the canal is much larger times (57.7%) and reduces flooding compared to 50.7% of Ruot Ngua canal and 20.2% of Hiep Tan canal, although the cost is more than less 1.1 times. And, the solution still has not gone beyond the bounds of economic conditions. Therefore, through this article, it is possible to recommend that the upgrading of the Lo Gom canal will contribute to reducing flooding of the THLG basin due to tides and rain first. This study shows the important thing when we design drainage network urban in tide area. More specifically, additional spaces to store pluvial water are required alongside sluice gates to prevent fluvial water. The new IDF rainfall helped for the found solution more precise in drainage system design of study area to applied reality upgrade.

GA combined with SWMM using Visual C++ language helps to automatically calculate and read data and optimal analysis results, to lead to the optimal solution to reduce flooding as well as save construction costs. This has contributed to supporting decision-making options in the design of urban drainage canal upgrades. Optimization is important against the backdrop of the ever-increasing demand for land resources.

Future studies are strongly recommended to search for the optimal solution for the entire HCM City's network while integrating other aspects, such as environmental and social factors. The optimal program can combine selected canals in the simulated process that can reduce the computed runtime needed to upgrade in the future.

Acknowledgements The authors acknowledge the financial support from the research project under contract no. 62/2019/HD-QPTKHCN dated on September 06, 2019, funded by the Department of Science and Technology in Ho Chi Minh City.

References

1. General Statistics Office, Vietnam, Population and labour (2019)
2. 752/QD-TTg, Decision on approving the master plan on the drainage system of Ho Chi Minh City until 2020, Hanoi, 19 June 2001 (2001)

3. Ho Chi Minh City Steering center of urban flood control, Official Dispatch Provide general information about flood situation in the city. Official No. 157/TTCN-QLTN, dated 09/02/2015, Dispatch (2015)
4. Quan LV, On TV (2015) Research to establish the problem of optimizing investment costs for urban drainage system construction: a case study for the western basin of Hanoi. J Sci Technol Irrig Environ 03(48):92–98
5. Can PV (2013) Optimizing water supply and drainage systems and the environment. Construction Publishing Company, Ha Noi
6. Dang Thanh Lam (2015) Thesis of building an appropriate model for the calculation of urban water drainage system in the tidal affected area
7. Keedwell E, Khu ST (2005) Using cellular automata to seed genetic algorithms for water distribution network design problems, 18(4), 461–472
8. Deb K, Agrawal S, Pratap A, Meyarivan T (2000) A fast elitist non-dominated sorting genetic algorithm for multi-objective optimization: NSGA-II. Lecture Notes in Computer Science (including Subseries, Lecture Notes in Artificial Intelligence and Lecture Notes in Bioinformatics), vol 1917, pp 849–858. https://doi.org/10.1007/3-540-45356-3_83
9. Deb K (2001) Constrained test problems for multi-objective evolutionary optimization. In: Proceedings of 1st international conference on evolutionary multi-criterion optimization, pp 284–298
10. Quan T, Nhi PTT, Khoi DN (2017) Building an extreme rain IDF curve for Tan Son Hoa station in the period 1980–2015. J Sci Technol Dev 20(M2) 2017
11. Binh LTH, Agilan V, Umamahesh NV, Rathnam EV (2019) Modelling spatial variation of extreme precipitation over Ho Chi Minh City under nonstationary condition. AcGeo 67(3):849–861
12. Huy NT, Van NVT (2020) Linking climate change to urban storm drainage system design: an innovative approach to modeling of extreme rainfall processes over different spatial and temporal scales. J Hydro-Environ Res 29:80–95
13. Most V (2008) Drainage and sewerage-External networks and facilities-Design Standard
14. Lim S, Ho VH, Lee SY, Yoo DG, Kim JH (2014) Determination of optimal location and capacity of detention facilities. Procedia Eng 70:1037–1045
15. Hammond MJ, Chen AS, Djordjević S, Butler D, Mark O (2015) Urban flood impact assessment: a state-of-the-art review. Urban Water J 12(1):14–29
16. Ingle Smith D (1999) Floods: physical processes and human impacts by K. Smith and R. Ward. Wiley, Chichester, p 382, 1998. Earth Surf Process Land 24(13), 1261

Comparative Study on Optimization of Cantilever Retaining Walls via Several Metaheuristics

Sena Aral, Gebrail Bekdaş, and Sinan Melih Nigdeli

Abstract In this study, using metaheuristic algorithms, reinforced concrete retaining walls under static and dynamic loads were dimensioned and designed with optimum cost. Reinforced concrete design was made following the rules of the Turkey Building Earthquake Code (TBDY 2018). The objective function of the optimization problem is defined depending on the reinforced concrete section dimensions and the minimization of the reinforcement area. In the studies developed to find the lowest cost, four different metaheuristic algorithms (Jaya algorithm (JA), teaching-learning-based optimization algorithm (TLBO), flower pollination algorithm (FPA), and harmony search algorithm (HS)) were used. When the optimization results using metaheuristic algorithms were evaluated, it was determined that the methods were successful in finding low-cost reinforced concrete designs. In addition, it has been seen that all algorithms used in retaining wall optimization give close and effective results with each other. Thus, with the developed method, it is possible to minimize the cost in structural design and the targeted results can be achieved in a shorter time.

Keywords Reinforced concrete · Structural engineering · Optimum design · Optimization · Retaining wall

1 Introduction

Metaheuristic algorithms are used to solve the optimization problems performed to find the best-targeted design. For the best design under the current constraints and conditions, minimization or maximization of the objective function is made according

S. Aral (✉) · G. Bekdaş · S. M. Nigdeli
Department of Civil Engineering, Istanbul University-Cerrahpaşa, 34320 Avcılar, Istanbul, Turkey
e-mail: senaaral@ogr.iu.edu.tr

G. Bekdaş
e-mail: bekdas@iuc.edu.tr

S. M. Nigdeli
e-mail: melihnig@iuc.edu.tr

J. H. Kim et al. (eds.), *Proceedings of 7th International Conference on Harmony Search, Soft Computing and Applications*, Lecture Notes on Data Engineering and Communications Technologies 140, https://doi.org/10.1007/978-981-19-2948-9_7

to the type of problem. Metaheuristics aim to use the search space more effectively by further developing nature-inspired heuristics. They do not always guarantee to find the best, but they are preferred because they can solve large-scale and complex problems. At the same time, these methods have been highly developed today and have proven to be effective in finding near-best results in studies in the literature. Civil engineering structural design problems are multifaceted and do not have a linear nature. For this reason, metaheuristic algorithms are preferred in the field of structural design.

Retaining walls are retaining structures built to prevent soil from slipping on sloping lands. Retaining structures are divided into three groups as rigid, semirigid, and flexible according to TS 7994 [1]. There are many recent studies on the optimization of retaining walls in the literature. Kaveh and Farhoud [2] used dolphin detection optimization that is the method inspired by dolphins detecting their prey via biosonar systems in the design of cantilever retaining walls. They determined that the optimization method gave better results compared to the harmony search algorithm. Bekdas et al. [3] provided cost and cross-sectional optimization of piled cantilever retaining structures that adapt to frictional soils by utilizing the harmony search (HS) algorithm. The results showed that the algorithm used is effective in terms of both cost and dimensional balance. Kayabekir et al. [4] made the optimum reinforced concrete design of a cantilever retaining wall under static and dynamic loads by developing a method using the teaching–learning-based optimization (TLBO) algorithm. The results showed that the algorithm used was successful in retaining wall design. As seen from these examples, the performance varies according to the problem. Aral et al. employed Jaya algorithm for toe projection restricted reinforced concrete retaining walls [5].

In this study, reinforced concrete retaining walls were examined. Four different metaheuristic algorithms (Jaya algorithm (JA) [6], teaching–learning-based optimization algorithm (TLBO) [7], flower pollination algorithm (FPA) [8], and harmony search algorithm (HS) [9] were used for the structural design. By using optimization methods, it is aimed to obtain the most optimum solution among all solutions. These solutions are created by checking their compliance with the specifications and regulation rules. The objective function of the defined optimization problem is cost minimization.

2 The Design Methodology

2.1 Design Variables

In the retaining wall optimization problem, four geometric design variables were used: top wall thickness (b_1), lower wall thickness (b_2), back part length (b_3), and base plate thickness (h) (Fig. 1). Regarding the reinforcement design, five design

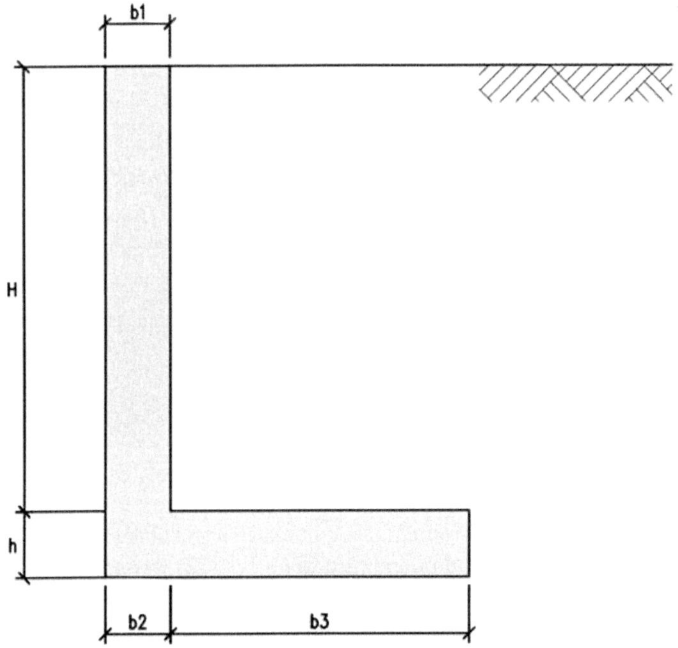

Fig. 1 A cantilever retaining wall cross section

variables, A_{s1}, A_{s2}, A_{s3}, A_{s4}, and A_{s5} were used. All these design variables are included in Table 1.

Table 1 Design variables

Variables	Description	Design variable
Geometry-related variables	Top wall thickness	b_1
	Bottom wall thickness	b_2
	Back part length	b_3
	Base plate thickness	h
Variables related to reinforcement design	Wall reinforcement area (Bottom)	A_{S1}
	Wall reinforcement area (Middle)	A_{S2}
	Wall reinforcement area (Top)	A_{S3}
	Base plate reinforcement area (Back)	A_{S4}
	Base plate reinforcement area (Front)	A_{S5}

Table 2 Design constraints

Explanation	Constraints
Overturning safety	$g_1(X)$: $SF_{O,design} \geq SF_O$
Sliping safety	$g_2(X)$: $SF_{S,design} \geq SF_S$
Maximum bearing compression	$g_3(X)$: $SF_{B,design} \geq SF_B$
Minimum bearing tension	$g_4(X)$: $q_{min} \geq 0$
Minimum reinforcement area, As_{min}	$g_{5-9}(X)$: $A_{s,design} \geq As_{min}$
Maximum reinforcement area, As_{max}	$g_{10-14}(X)$: $A_{s,design} \leq As_{max}$
Maximum shear capacity, V_{max}	$g_{15-16}(X)$: $V_{max} \leq V_{cr}$

2.2 *Design Constraints*

Design constraints were calculated according to the Turkish Building Earthquake Code (TBDY 2018) [10]. In this context, overturning, slip and bearing safety controls and the suitability of the reinforcement area calculated according to the internal forces were checked under the maximum and minimum boundary conditions specified in the regulation. In the analysis, it is aimed to reach the best solution by penalizing the situations that do not comply with the boundary conditions. Definition of constraints and their formulations are given in Table 2.

The optimization problem is designed to meet the design constraints stated in Table 2. The SF_O overturn safety coefficient in Table 2 is 1.3, and the SF_S shear safety coefficient is 1.1 according to TBDY 2018. SF_B must be 1. V_{cr} shear force symbolizes the capacity of the section.

3 Numerical Examples

In this part, a cantilever reinforced concrete retaining wall design example made by Celep [11] has been optimized by using four different algorithms, namely Jaya algorithm, teaching–learning-based optimization algorithm, flower pollination algorithm, and harmony search algorithm. The stopping criterion for optimization (maximum number of iterations) was defined as 100,000, and the number of population (solution set) was defined as 20. Calculations were made for the 1 m length of the cantilever retaining wall. The parameters used in the design are given in Table 3. In the optimization problem, S420 steel class was used, and the unit price of the steel was 1400 TL/ton. C30 was used as the concrete class, and the unit price of the concrete was defined as 111 TL/m^3. These unit costs were taken as in the study conducted by Aral et al. [5].

While designing reinforced concrete, the characteristic compressive strength of the concrete and the characteristic yield strength of the steel were divided by the coefficients of 1.5 and 1.15, respectively. In the example of the cantilever retaining

Table 3 Design parameters

Design parameters	Values
Material class	C30/S420
Concrete compressive strength	30 MPa
Steel yield strength	420 MPa
Ground acceleration coefficient, S_{DS}	0.5
Ground safety tension	350 kN/m^2
Coefficient of friction	0.6
Internal friction angle of infill soil	30°
Infill slope	0°
Unit volume weight of infill soil	18 kN/m^3
Concrete unit weight	25 kN/m^3
Steel unit volume weight	7.85 ton/m^3
Surcharge load	5–10–15–20 kN/m^2
Wall height	5.6 m
R_a, factor of safety	1.5

wall made by Celep [11], the unit volume weight of the infilled soil is 18 kN/m^3, the surcharge load is 10 kN/m^2, and the height of the cantilever wall is 5.6 m. The lower and upper limit ranges for the design variables are defined as in Table 4 so that they remain within the TS 7994 boundary conditions.

The reinforced concrete cantilever retaining wall example was numerically solved by Celep without optimization and is named Solution 1. The optimized solution using Jaya algorithm is named Solution 2. Table 5 shows the comparison of the results of these two conditions. The construction cost obtained in the design made by Celep was calculated as approximately 693.90 TL/m according to the unit costs used in this study. The cost of the design obtained using the Jaya algorithm was calculated as 565.71 TL/m. When the results obtained with the optimization are compared with the design made by Celep, lower construction costs of 128.19 TL/m can be achieved. According to these results, it was determined that 19% lower cost design was obtained with optimization.

Table 4 Design limits

Design variable	Unite	Lower limit	Upper limit
b_1	m	0.2	3
b_2	m	0.2	3
b_3	m	0.2	10
h	m	0.2	10

Table 5 Comparisons of geometric dimension, reinforced concrete design, and total cost

Variables	Design variable	Solution 1 [11]	Solution 2
Geometry-related variables	b_1 (m)	0.40	0.20
	b_2 (m)	0.40	0.48
	b_3 (m)	4.10	3.30
	h (m)	0.40	0.36
Variables related to reinforced concrete design	A_{S1} (mm^2)	3197	2558
	A_{S2} (mm^2)	926	1050
	A_{S3} (mm^2)	119	234
	A_{S4} (mm^2)	4026	3979
	A_{S5} (mm^2)	1741	1518
Total cost	$f(x)$ (TL/m)	693.90	565.71

4 Discussion and Conclusion

In this section, the performances of the algorithms used in the optimization problem of the retaining wall are discussed. In this context, analyses were carried out under four different surcharge loading (5–10–15–20 kN/m^2). The analyses were repeated 20 times for four different algorithms and loading conditions. 100,000 iterations were performed in each analysis. After the analyses were carried out to show the stability, performance, and reproducibility of the algorithms, the closeness of the solutions with the best results was tested, and the ability of each individual to converge to the optimum value was examined. In the analysis groups performed for the algorithms, independent random populations were repeated 20 times in each optimization analysis. Accordingly, the results obtained in the analyses using the Jaya algorithm, the teaching–learning-based optimization algorithm, the flower pollination algorithm, and the harmony search algorithm are given in Tables 6, 7, 8 and 9.

According to the results, similar optimum values (528, 565, 606, and 651 TL/m) were obtained in all algorithms in optimization studies performed under four different surcharge loadings. According to the analysis results for $Q = 10$ kN/m^2 loading with four different algorithms, the standard deviation values among the best solutions were obtained as 0.0000–7.3838–0.0000–0.0035 TL/m, respectively. Therefore, it has been determined that all algorithms are successful in finding optimum or near-optimal values and can be used in optimization. In other words, the global minimum has been reached in the analyses made in the cost optimization. However, when the standard deviation values between the independent analyses are examined, it is understood that the Jaya algorithm is superior to other algorithms. Although the values obtained with the TLBO algorithm are relatively acceptable, due to the high standard deviation value and the high transaction volume, it cannot be superior to the Jaya algorithm. As a result of the comparison of the reliability and performance of the metaheuristic algorithms used in the retaining wall optimization problem, it

Table 6 Optimization results with Jaya algorithm

Design variable	Q:5 kN/m²		Q:10 kN/m²		Q:15 kN/m²		Q:20 kN/m²	
	Average	Standard deviation	Average	Standard deviation	Average	Standard deviation	Average	Standard deviation
h (m)	0.35	0	0.36	0	0.39	0	0.4	0
*b*1 (m)	0.2	0	0.2	0	0.2	0	0.2	0
*b*2 (m)	0.45	0	0.48	0	0.44	0	0.46	0
*b*3 (m)	3.21	0	3.3	0	3.52	0	3.73	0
A_{S1} (mm²)	2444.82	0.0287	2558.44	0.0272	3220.54	0.0035	3350.52	0.0043
A_{S2} (mm²)	945.04	0.0087	1050.47	0.0089	1331.6	0.0011	1441.03	0.0014
A_{S3} (mm²)	186.8	0.0011	234.38	0.0013	307.42	0.0002	354.82	0.0002
A_{S4} (mm²)	3741.14	0.0391	3979.32	0.0475	4154.52	0.0047	4328.39	0.0044
A_{S5} (mm²)	1436.75	0.0118	1518.01	0.0129	1643.65	0.0015	1718.01	0.0014
$f(x)$ (TL/m)	528.15	0	565.71	0	606.46	0	651.39	0

Table 7 Optimization results with TLBO algorithm

Design variable	Q:5 kN/m²		Q:10 kN/m²		Q:15 kN/m²		Q:20 kN/m²	
	Average	Standard deviation	Average	Standard deviation	Average	Standard deviation	Average	Standard deviation
h (m)	0.35	0	0.36	0	0.39	0	0.4	0
*b*1 (m)	0.2	0	0.2	0	0.2	0	0.2	0
*b*2 (m)	0.45	0	0.48	0	0.44	0	0.46	0
*b*3 (m)	3.21	0	3.3	0	3.52	0	3.73	0
A_{S1} (mm²)	2444.82	0.0457	2558.44	0.0329	3220.54	0.0015	3350.51	0.0024
A_{S2} (mm²)	945.04	0.0139	1050.48	0.0108	1331.6	0.0005	1441.03	0.0008
A_{S3} (mm²)	186.8	0.0018	234.38	0.0016	307.42	0.0001	354.81	0.0001
A_{S4} (mm²)	3741.15	0.05	3990.92	265.1621	4154.52	0.002	4328.39	0.0021
A_{S5} (mm²)	1436.75	0.0124	1587.34	92.0094	1643.65	0.0005	1718.01	0.0005

was determined that the Jaya algorithm is more suitable than the other algorithms. In the analyses to be made after this section, the Jaya algorithm, which is determined to have the best performance in optimization, will be used.

Table 8 Optimization results with FPA algorithm

Design variable	Q:5 kN/m²		Q:10 kN/m²		Q:15 kN/m²		Q:20 kN/m²	
	Average	Standard deviation	Average	Standard deviation	Average	Standard deviation	Average	Standard deviation
h (m)	0.35	0.0000	0.36	0.0000	0.39	0.0000	0.40	0.0000
b_1 (m)	0.20	0.0000	0.20	0.0000	0.20	0.0000	0.20	0.0000
b_2 (m)	0.45	0.0000	0.48	0.0000	0.44	0.0000	0.46	0.0000
b_3 (m)	3.21	0.0000	3.30	0.0000	3.52	0.0000	3.73	0.0000
A_{S1} (mm²)	2444.68	0.1659	2558.52	0.1598	3220.54	0.0042	3350.33	0.1210
A_{S2} (mm²)	945.00	0.0503	1050.50	0.0524	1331.60	0.0013	1440.97	0.0404
A_{S3} (mm²)	186.80	0.0065	234.38	0.0078	307.42	0.0002	354.81	0.0064
A_{S4} (mm²)	3741.17	0.2136	3979.23	0.1522	4154.49	0.0261	4328.79	0.2927
A_{S5} (mm²)	1436.74	0.0687	1518.00	0.0459	1643.64	0.0079	1718.11	0.0775
$f(x)$ (TL/m)	528.15	0.0000	565.71	0.0000	606.46	0.0000	651.39	0.0000

Table 9 Optimization results with HS algorithm

Design variable	Q:5 kN/m²		Q:10 kN/m²		Q:15 kN/m²		Q:20 kN/m²	
	Average	Standard deviation	Average	Standard deviation	Average	Standard deviation	Average	Standard deviation
h (m)	0.35	0.0006	0.36	0.0006	0.38	0.0009	0.40	0.0009
b_1 (m)	0.20	0.0000	0.20	0.0000	0.20	0.0000	0.20	0.0000
b_2 (m)	0.45	0.0021	0.48	0.0018	0.44	0.0014	0.46	0.0016
b_3 (m)	3.21	0.0034	3.30	0.0030	3.52	0.0004	3.73	0.0002
A_{S1} (mm²)	2446.42	15.2841	2556.55	12.6302	3221.93	14.7124	3340.62	16.3107
A_{S2} (mm²)	945.52	4.6353	1049.85	4.1448	1332.03	4.6404	1437.71	5.4484
A_{S3} (mm²)	186.86	0.5971	234.29	0.6149	307.48	0.6806	354.29	0.8676
A_{S4} (mm²)	3739.91	11.9344	3990.62	11.1566	4164.45	16.1900	4327.85	16.4938
A_{S5} (mm²)	1436.58	4.3484	1521.00	3.1032	1646.73	4.8600	1716.98	5.8585
$f(x)$ (TL/m)	528.16	0.0031	565.72	0.0035	606.47	0.0046	651.40	0.0040

References

1. Turkish Standards Institute (1990) Soil retaining structures; properties and guidlines for design. Ankara, TS7994
2. Kaveh A, Jafari L, Farhoudi N (2015) Truss optimization with natural frequency constraints using a dolphin echolocation algorithm
3. Bekdaş G, Arama ZA, Kayabekir AE, Geem ZW (2020) Optimal design of cantilever soldier pile retaining walls embedded in frictional soils with harmony search algorithm. Appl Sci 10(9):3232

4. Kayabekir AE, Bekdaş G, Nigdeli SM, Temür R (2016) Optimum design of retaining walls under static and dynamic loads. In: International science symposium (ISS2016), pp 1–4
5. Aral S, Yılmaz N, Bekdaş G, Nigdeli SM (2020) Jaya optimization for the design of cantilever retaining walls with toe projection restriction. In: International conference on harmony search algorithm. Springer, Singapore, pp 197–206
6. Rao R (2016) Jaya: a simple and new optimization algorithm for solving constrained and unconstrained optimization problems. Int J Ind Eng Comput 7(1):19–34
7. Rao RV, Savsani VJ, Vakharia DP (2011) Teaching-learning-based optimization: a novel method for constrained mechanical design optimization problems. Comput Aided Des 43:303–315
8. Yang XS (2012) Flower pollination algorithm for global optimization. In: Durand-Lose J, Jonoska N (eds) Lecture notes in computer science, vol 27. Springer, London, 7445, pp 240–249
9. Geem ZW, Kim JH, Loganathan GV (2001) A new heuristic optimization algorithm: harmony search. SIMULATION 76(2):60–68
10. TBDY (2018) Türkiye Bina Deprem Yönetmeliği; Turkish Building Seismic Code, Ankara, Turkey
11. Celep Z (2014) Introduction to earthquake engineering and earthquake resistant building design. Istanbul, pp 299–305

Optimal Solving of Uninhabited Combat Air Vehicle Path Planning Using Neural Network Algorithm

Mojtaba Ershad and Ali Sadollah

Abstract Path planning of uninhabited combat air vehicle (UCAV) is a rather complicated global optimum problem which is about seeking a superior flight route considering different kinds of constrains under complex combat field environment. Neural network algorithm (NNA) is a recently developed optimizer inspired by the unique structure of artificial neural networks and biological nervous system. In this paper, we propose a discrete version of NNA based on discrete weight matrix for solving the UCAV path planning in various combat field environments, and the implementation procedure of our proposed discrete NNA is also described in detail. Series of experimental comparison results is represented to show the feasibility, effectiveness, and robustness of the proposed method.

Keywords Robotics · Uninhabited combat air vehicle · Path planning · Metaheuristics · Constrained optimization

1 Introduction

Unmanned combat air vehicle (UCAV) is a kind of unmanned air vehicle (UAV), which can operate some repetitive missions in a remote and hazardous environment that is impossible or dangerous for people to do missions [1]. Although UCAVs have been used in past, just for military missions nowadays, it could be used in civilian applications. The UCAV is also essential for the safety and security of a society. The output would achieve the maximum if the system operates autonomously.

Path planning is the process of finding a suitable flight path to arrive from the initial point to the destination, and this action is normally performed under specific constraints conditions, for instance, limited fuel, avoiding the threat area, limitation in turning angle, and so forth. The performance should be optimal or near-optimal. An ideal path planar is better to be optimal, complete, and also complex in computation [2].

M. Ershad · A. Sadollah (✉)
Department of Mechanical Engineering, University of Science and Culture, Tehran, Iran
e-mail: sadollah@usc.ac.ir

J. H. Kim et al. (eds.), *Proceedings of 7th International Conference on Harmony Search, Soft Computing and Applications*, Lecture Notes on Data Engineering and Communications Technologies 140, https://doi.org/10.1007/978-981-19-2948-9_8

Some actions will cause more cost, and the mission operator should consider and hesitate using them, for instance, using humans on board because of the weight and the capacity which are occupying by the humans and also the danger that threatens the humans. Other disadvantages of traditional aircraft by controlling via human abilities including fatigue, inattention, limited work hours can be achieved by using the UAVs for various operational tasks [3]. Although this cost for the UAVs can be fewer besides, having a smaller size and weight in comparison with traditional aircraft, the UAVs bring more effective energy consumption which permits them to be utilized for long-term tasks [4].

There are many optimization algorithms proposed to solve the problem of complicated multi-constrained path planning, and these algorithms utilize domain-specific knowledge for making decisions and accommodating different optimization criteria [5]. There are many path planning problems that have been investigated in literature. Low-altitude penetration path of the plane is arranged with the ant colony optimization (ACO) [6], particle swarm algorithm (PSO) [7], and genetic algorithms (GAs) [8].

In Ref. [9], bat algorithm (BA) with mutation for a single UCAV is proposed, and the obtained results are compared with other population-based optimization algorithms. In Ref. [10], using ergodicity and irregularity of chaotic variables helped the traditional artificial bee colony (ABC) to jump out of local optimum, and this algorithm could be used in a dynamic environment. In Ref. [11], the GAs with utilizing the periodic mutation applications based on vibrational mutation operation rather than using classical mutational application appeared that these applications provide classified but productive and efficient diversity in the population for optimal path planning of autonomous UAV.

Qu et al. [12] developed a hybrid of simplified gray wolf optimizer (SGWO) and modified symbiotic organisms search (MSOS) called HSGWO-MSOS for solving the UAV path planning problem. The optimization results verified the superiority of HSGWO-MSOS with respect to the GWO and SA algorithms.

Qu et al. [13] proposed reinforcement learning-based GWO algorithm called RLGWO for solving UAV path planning problem. In their proposed algorithm, the reinforcement learning was inserted that the individual was controlled to switch operations adaptively according to the accumulated performance. The results proved that the RLGWO could acquire a feasible route successfully in complicated environment. Furthermore, Aggarwal and Kumar [14] reviewed various path planning techniques for optimal solving of UAVs.

Zhang et al. [15] utilized the GWO to solve the UCAV two-dimension path planning problem. Simulation results demonstrated that the proposed method was more competent for the UCAV path planning scheme than other optimizers considering the quality, speed, and stability of final solutions.

In this paper, neural network algorithm (NNA) inspired by unique structure of artificial neural networks (ANNs) and biological nervous systems have been converted for solving discrete path planning of UCAV. Four scenarios have been considered. Afterward, the obtained optimization results have been compared with the state-of-the-art optimizers such as the SA and GAs.

The rest of this paper is organized as follows. Next section provides mathematical formulation of UCAV and its imposed constraints. Section 3 briefly describes the NNA and its searching operators. Afterward, Sect. 4 gives optimal results and discussion regarding the quality of obtained solution. Finally, conclusions are drawn at the end of this paper.

2 Mathematical Modeling of UCAV Path Planning

Optimization of an air robot path depends mostly on modeling the threat sources which can convert the problem into an easier problem. Generally, threats could be divided in two groups, the first is the natural threats, and the second is artificial threats. Artificial threats include the radars, missiles, artillery, and so forth. In this paper, threats are shown by circles in which their radii are the sample of the range of radars and missiles, and out of each circle, the UCAV cannot be attacked, and therefore, it is safe. However, to find and calculate the cost of a path, threat level is considered as an important factor [16].

Point S indicates the starting point, and point D indicates the destination point, and also k threats set $\{T_1, T_2, T_3, \ldots, T_k\}$. Firstly, the points S and D should be connected and divided into $(D + 1)$ segments. Draw a vertical line of SD at each segment point and also name these lines as $L_1, L_2, \ldots, L_k, \ldots, L_D$ take a discrete point at each line, create a discrete points collection $A = \{S, (x(1), y(1)), (x(2), y(2)), \ldots, (x(k), y(k)), \ldots, (x(D), y(D)), D\}$ and connect them to create a flight path, and now, the problem of path planning is turning into optimizing the coordinate series to achieve a superior fitness value of the objective function.

We can consider SD line as X axis and take the coordinate transformation on each discrete point $(x(k), y(k))$ and according to Eq. (1), where θ is the counterclockwise rotation to the original x axis and x_s and y_s represents the coordinates in the original coordinate system [17] as given follows:

$$\begin{bmatrix} x'(k) \\ y'(k) \end{bmatrix} = \begin{bmatrix} \cos\theta & \sin\theta \\ -\sin\theta & \cos\theta \end{bmatrix} \begin{bmatrix} x(k) - x_s \\ y(k) - y_s \end{bmatrix} \tag{1}$$

Using Eq. (2), the x coordinate of each point can easily be achieved by:

$$x' = \frac{\mathrm{SD}}{D+1}k \tag{2}$$

Therefore, A can be simplified to $A' = \{(0,0), (x'(1), (y'(1))), (x'(2), y'(2)), \ldots, (x'(k), y'(k)), \ldots, (x'(D), y'(D)), (|SD|,0)\}$ and then by normalizing Eq. (2), we have the following equation:

$$x'(k) = \frac{k}{D+1} \tag{3}$$

which can reduce the computational cost. Threat cost (J_t) and fuel cost (J_f) are the indicators of UCAV route performance, and threat and fuel costs can be calculated as follows [18, 19]:

$$J_t = \int_0^L w_t dl \tag{4}$$

$$J_f = \int_0^L w_f dl \tag{5}$$

where w_t and w_f are the variables related with the current path and changing along with 'l' and presenting the threat cost of the segments of the path, while L is the total length of the path. If we consider constant speed for the UCAV, then J_f is approximately equal to L. When two discrete points are connected, the line must be divided into 5 points, and at each point, the threat cost must be calculated. If the generated route passes the threat zone, then w_{t,l_i} must be applied as given follows [20, 21]:

$$w_{t,l_i} = \frac{L_i}{5}\sum_{k=1}^{N_t} t_k\left(\frac{1}{d^4_{0.1,i,k}} + \frac{1}{d^4_{0.3,i,k}} + \frac{1}{d^4_{0.5,i,k}} + \frac{1}{d^4_{0.7,i,k}} + \frac{1}{d^4_{0.9,i,k}}\right) \tag{6}$$

where N_t is the number of threatening areas, $d^4_{0.1,i,k}$ is the distance of 0.1 point on the edge ith to the threat kth, t_k indicates the threat level, and L_i is the length of sub-path ith. Finally, the total cost can be calculated from the summation of threat and fuel costs given as follows:

$$J_{\text{TOT}} = wJ_t + (1-w)J_f \quad 0 < w < 1 \tag{7}$$

where w is the weight coefficient, which has a relation to the fuel consumption and threat exposition degree, and the designer determines the coefficient and shows the importance of each cost. When the issue is preventing from radar's exposed threat, then the value of w should be close to 1. On the other hand, if the important issue is finding the shortest path, then the value of w should be chosen close to zero. In this paper, for giving equal chance for both costs, the amount of w is selected as 0.5. It is worth mentioning that a straight line directory from starting point to destination point, although has shortest pass, however, it is not feasible and practical.

3 Neural Network Algorithm

Neural network algorithm (NNA) benefits from complicated structure of the artificial neural networks (ANNs) and its operators in order to generate new candidate solutions [22]. Similar to other metaheuristic optimization algorithms, the NNA begins with an initial population so called population of pattern solutions. Inspired by the ANNs, in the NNA, the best obtained solution at each iteration is assumed as target data, and the aim is to reduce the error among the target data and other predicated pattern solutions.

The NNA resembles ANNs having N_{pop} input data having D dimension(s) and only one target data (response). After setting the target solution (X_{Target}) among the other pattern solutions, the target weight (W_{Target}), the weight corresponding to the target solution, has to be selected from the population of weight (weight matrix). Initial weights in ANNs are random numbers, and when the iteration number is increasing, they will be updated considering the calculated error of the network. Back to the NNA, initial weights are square matrix ($N_{\text{pop}} \times N_{\text{pop}}$) which generates random numbers uniformly between zero and one during iterations. However, there is a constraint for the weight values. The imposed constraint is the summation of weights for a pattern solution should not exceed one [22]. After forming the weight matrix (W), new pattern solutions (X_{New}) are calculated using the following equation inspired by the weight summation technique used in the ANNs:

$$\vec{X}_j^{\text{New}}(t+1) = \sum_{i=1}^{N\text{pop}} w_{ij}(t) \times \vec{X}_i(t), \quad j = 1, 2, 3, \ldots, N_{\text{pop}} \tag{8}$$

$$\vec{X}_i(t+1) = \vec{X}_i(t) + \vec{X}_i^{\text{New}}(t+1), \quad i = 1, 2, 3, \ldots, N_{\text{pop}} \tag{9}$$

where t is an iteration index. After creating the new pattern solutions from the previous population of patterns, based on the best weight value so called target weight, the weight matrix should be updated as well. The following equations suggest an updating equation for the weight matrix:

$$\vec{W}_i^{\text{Updated}}(t+1) = \vec{W}_i(t) + 2 \times \text{rand} \times \left(\vec{W}^{\text{Target}}(t) - \vec{W}_i(t)\right), \quad i = 1, 2, 3, \ldots, N_{\text{pop}} \tag{10}$$

As global searching operator, the bias operator in the NNA is another way to explore the search space (exploration process), and it acts similar to the mutation operator in the GAs. In general, the bias operator prevents the algorithm from premature convergence (especially at early iterations) and modifies a number of individuals in the population. The bias operator, β, is a modification factor, which determines the percentage of the pattern solutions that should be altered. The initial value of β is set to 1 (means 100% chance to modify all individuals in population), and its value adaptively has been reduced at each iteration using any reduction formulation as suggested follows:

$$\beta(t+1) = \beta(t) \times 0.99 \quad t = 1, 2, 3, \ldots, \text{Max\_ Iteration} \tag{11}$$

Accordingly, as local searching operator, in the NNA, unlike ANNs, transfer function operator transfers the new pattern solutions in the population from their current positions in the search space to new positions in order to update and generate better quality solutions toward the target solution. Therefore, the following equation is defined as a transfer function operator (TF) for the proposed method given as follows [22]:

$$\vec{X}_i^*(t+1) = \text{TF}\left(\vec{X}_i(t+1)\right) = \vec{X}_i(t+1) + 2 \times \text{rand} \times \left(\vec{X}^{\text{Target}}(t) - \vec{X}_i(t+1)\right) \tag{12}$$

In fact, using the NNA by its own was not enough for solving such these problems, and the optimization method should be changed for performing better performance and obtaining excellent efficiency. In the modified method, the algorithm generates a random path from the beginning till the destination, the cost of the random path must be calculated by cost function and then the path as an input proceeds to create neighbor function, then the algorithm chooses a point of the path by chance.

There are neighbors for each point of the path in which the algorithm can select from the neighbor's collection. After modification is applied, the new updated path will be proceeding for the cost function evaluation. In this step, the cost must be calculated, and the result has two separated aspects: (1) if the new path has a bigger cost in comparison with the previous path, the previous path would be chosen as better path, and its cost is preserved, (2) if the new path has less cost in comparison with the previous path, the new one would be chosen as the optimal path, and its cost is considered as the optimal cost. It is worth mentioning that once a neighbor point is selected, it cannot be chosen again in the path. Figure 1 shows the way that a new generated path is updated considering neighbors for a single iteration.

As can be seen in Fig. 1, the NNA has found a sample path from the starting point (i.e., 1) to destination point (i.e., 16) in red colors. By choosing point 7 and determining its neighbors, the NNA may find other optimal neighbors such as point 12. In this case, the updated path (in purple colors) may offer shortest pass. It is worth mentioning that the size of design variables may increase or reduce for every candidate solution.

4 Results and Discussions

To examine the NNA for optimal solving of UACV path planning, several scenarios are conducted to evaluate the results. In this paper, (0,0) location is set as starting point and (9,9) location as destination point, and the algorithm has freedom to reach destination without any limit in choosing the number of points. Indeed, the size of design variable for every candidate solution may vary. In this paper, we defined 8

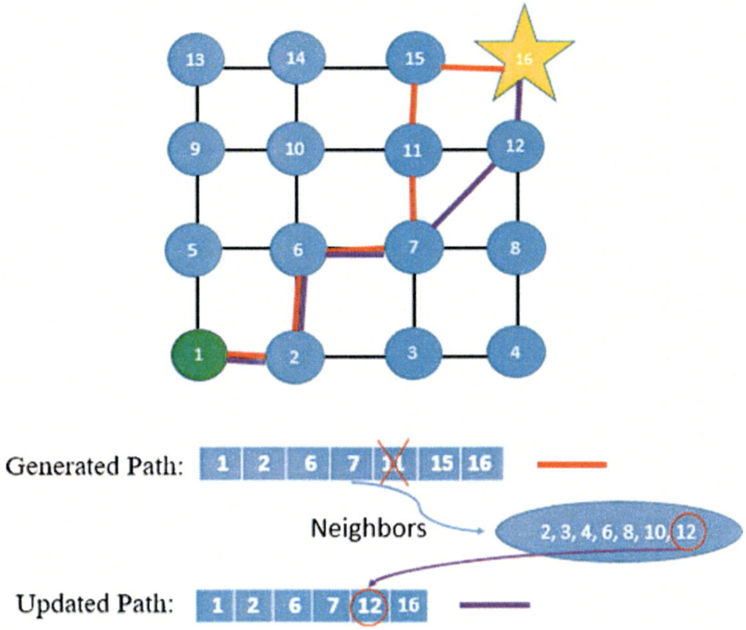

Fig. 1 Strategy utilized for updating a generated path using the NNA

different threats for every four scenarios. Each threat has its center and radius of danger along with its threat level shown in Table 1.

For validation and comparison purposes, the SA and GAs along with the NNA have been applied and examined over the path planning of UCAV. All the codes were written in MATLAB. For optimization task, 30 independent runs were carried out. Table 2 tabulates the attained statistical optimization results using three reported optimizers.

Looking at Table 2, the proposed method shows better stability in finding the shortest path among the compared optimizers. Furthermore, Fig. 2 demonstrates the optimal passes for each considered scenarios.

Looking at Fig. 2, the modified discrete NNA tried to reach the destination point with having minimum distance length and not surpassing danger area. Particularly, this trend can be shown for Scenario 3 which is very sensitive in finding the optimal path. Indeed, for Scenario 3, there are only 2 optimal paths.

5 Conclusions

This paper represented an application of modified discrete NNA for UCAV path planning problem in complicated combat field environment. Inspiring by unique structure of ANNs and biological nervous system, proposed discrete NNA has been

Table 1 Threat information for four defined scenarios

Threat information for Scenario 1								
Center	(3,7)	(7,3)	(7,6)	(1,5)	(2,3)	(3,6)	(5,8)	(6,5)
Level	2	13	5	1	12	3	7	9
Radius	10	12	7	12	9	9	9	10
Threat information for Scenario 2								
Center	(4,1)	(4,6)	(7,5)	(4,7)	(6,1)	(8,5)	(2,8)	(3,6)
Level	2	13	5	1	12	3	7	9
Radius	1	1	0.8	0.3	1.5	1.1	0.7	0.5
Threat information for Scenario 3								
Center	(8,8)	(7,8)	(6,7)	(5,6)	(4,4)	(3,3)	(2,2)	(1,1)
Level	2	13	5	1	12	3	7	9
Radius	1	0.9	0.8	1.4	0.5	0.5	0.7	0.7
Threat information for Scenario 4								
Center	(8,8)	(7,8)	(6,7)	(5,6)	(4,4)	(3,3)	(2,2)	(2,1)
Level	2	13	5	1	12	3	7	9
Radius	0.5	0.5	0.5	1.4	0.5	0.5	0.5	0.5

Table 2 Statistical optimization results obtained using three optimizers. SD means standard deviation

Method	Min. length	Ave. length	Max. length	SD	Friedman test (Ranking)	Alpha (0.05)
SA	9	11.23	21	4.78	2.46	1.12E–03
GAs	9	13.53	24	8.45	4.55	
NNA	9	9.40	19	1.84	1.10	

applied over four different scenarios having various threats with different levels. The simulation experiments shown that the proposed method is a feasible and effective way in the UCAV path planning, and it is also flexible, in that dynamic environments and pop-up threats are easily incorporated. The experimental comparison also shown the stability and superiority of the NNA method over the standard SA and GAs, which provides a more effective way for UCAV path planning.

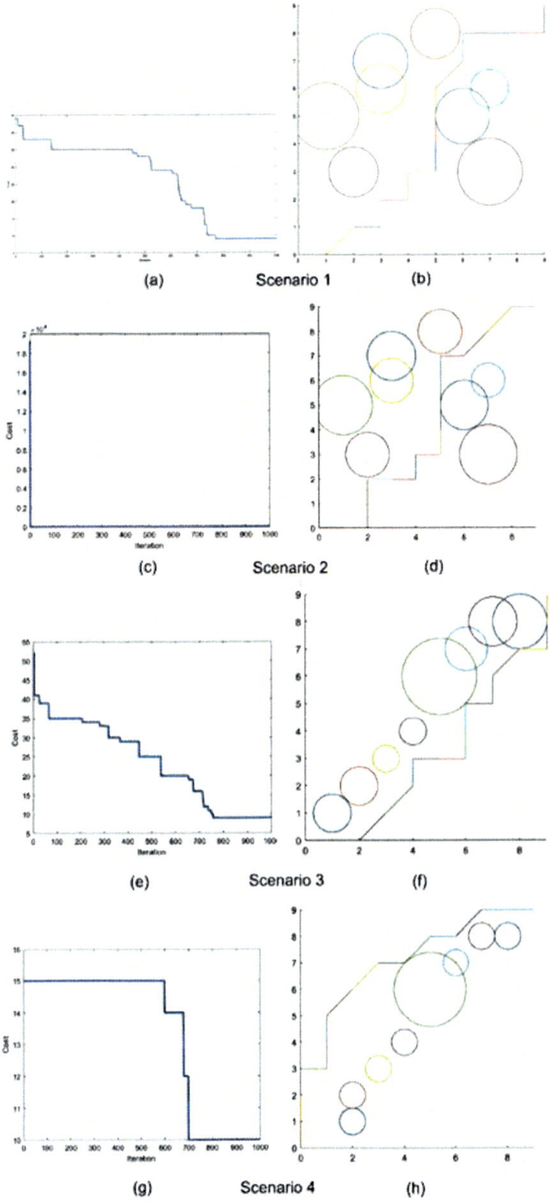

Fig. 2 Optimal path found by the NNA for 4 different scenarios along with convergence history

References

1. Foo JL, Knutzon J, Kalivarapu V, Oliver J, Winer E (2008) Path planning of un-manned aerial vehicles using B-splines and particle swarm optimization. J Aerosp Comput Inf Commun 6:271–290
2. Duan H, Yu Y, Zhang X, Shao S (2010) Three-dimension path planning for UCAV using hybrid meta-heuristic ACO-DE algorithm. Simul Model Pract Theory 18(8):1104–1115
3. Turker T, Sahingoz OK, Yilmaz G (2015) 2D path planning for UAVs in radar threatening environment using simulated annealing algorithm. In: 2015 International conference on unmanned aircraft systems (ICUAS). IEEE, Denver, CO, USA
4. Ozalp N, Sahingoz OK, Ayan U (2014) Autonomous multi unmanned aerial vehicles path planning on 3 dimensional terrain. In: IEEE 22nd signal processing and communications applications conference (SIU 2014), Trabzon, Turkey
5. Fu Y, Ding M, Zhou C (2012) Phase angle-encoded and quantum-behaved particle swarm optimization applied to three-dimensional route planning for UAV. IEEE Trans Syst Man Cybern Part A: Syst Hum 42(2):511–526
6. Ye W, Ma DW, Fan HD (2005) Algorithm for low altitude penetration aircraft path planning with improved ant colony algorithm. Chin J Aeronaut 18(4):304–309
7. Duan HB, Ma GJ, Luo DL (2008) Optimal formation reconfiguration control of multiple UCAVs using improved particle swarm optimization. J Bionic Eng 5(4):215–223
8. Wang YX, Chen ZJ (1999) Genetic algorithms (GA) based flight path planning with constraints. J Beijing Univ Aeronaut Astronaut 25(3):355–358
9. Wang G, Guo L, Duan H, Liu L, Wang H (2012) A Bat algorithm with mutation for UCAV path planning. Sci World J 2012:418946
10. Xu C, Duan H, Liu F (2010) Chaotic artificial bee colony approach to uninhabited combat air vehicle (UCAV) path planning. Aerosp Sci Technol 14(8):535–541
11. Volkan Pehlivanoglu Y (2012) A new vibrational genetic algorithm enhanced with a Voronoi diagram for path planning of autonomous UAV. Aerosp Sci Technol 16(1):47–55
12. Qu C, Gai W, Zhang J, Zhong M (2020) A novel hybrid grey wolf optimizer algorithm for unmanned aerial vehicle (UAV) path planning. Knowl-Based Syst 194:105530
13. Qu C, Gai W, Zhong M, Zhang J (2020) A novel reinforcement learning based grey wolf optimizer algorithm for unmanned aerial vehicles (UAVs) path planning. Appl Soft Comput 89:106099
14. Aggarwal S, Kumar N (2020) Path planning techniques for unmanned aerial vehicles: a review, solutions, and challenges. Comput Commun 149:270–299
15. Zhang S, Zhou Y, Li Z, Pan W (2016) Grey wolf optimizer for unmanned combat aerial vehicle path planning. Adv Eng Softw 99:121–136
16. Duan H, Qiao P (2014) Pigeon-inspired optimization: a new swarm intelligence optimizer for air robot path planning. Int J Intell Comput Cybern
17. Zhu W, Duan H (2014) Chaotic predator–prey biogeography-based optimization approach for UCAV path planning. Aerosp Sci Technol 32(1):153–161
18. Alison E, Arthur R (2007) Path-planning with avoidance using nonlinear branch-and bound optimization. AIAA Guidance, Navigation and Control Conference and Exhibit. AIAA, Hilton Head, South Carolina
19. Karimi J, Pourtakdoust SH (2013) Optimal maneuver-based motion planning over terrain and threats using a dynamic hybrid PSO algorithm. Aerosp Sci Technol 26(1):60–71
20. Liu CA, Li WJ, Wang HP (2004) Path planning for UAVs based on ant colony. J Air Force Eng Univ 2(5):9–12
21. Daniel R, Andries PE (2008) Social networks in simulated multi-robot environment. Int J Intell Comput Cybern 1(1):110–127
22. Sadollah A, Sayyaadi H, Yadav A (2018) A dynamic metaheuristic optimization model inspired by biological nervous systems: neural network algorithm. Appl Soft Comput 71:747–782

Cost Optimization and Comparison of Rectangular Cross-section Reinforced Concrete Beams Using TS500, Eurocode 2, and ACI 318 Code

Muhammed Çoşut, Gebrail Bekdaş, and Sinan Melih Niğdeli

Abstract In this study, Eurocode 2 (design of concrete structures), TS500 (reinforced concrete structures design and construction rules), and ACI 318 (building regulations for reinforced concrete structures) regulations for reinforced concrete rectangular beams were considered with the formulation of the design of metaheuristics using Jaya algorithm. The total cost of material and formwork labor is defined as the objective function, and the beam dimensions are defined as variables. It is aimed to reduce the total cost (concrete, reinforcement, formwork) to a minimum for a reinforced concrete beam with a rectangular section with an optimal design for three separate regulations. Therefore, with the optimal design, the cross-sectional dimensions of the beam should reach the optimum value. In the results of the analysis, it was observed that the beam breadth and height came very close to the Eurocode and TS500 regulations for C25/30 and C30/37 strength concrete, but the ACI 318 regulation came in different sizes. That is why cost differences have been formed. The breadth and height of the beam for C35/45 strength concrete have become the same in three regulations. It has been calculated that Eurocode and TS500 are approximately the same cost in terms of cost, while ACI 318 will be designed at a low cost of approximately 7.5%.

Keywords TS500 · Eurocode 2 · ACI 318 · Optimization · Optimum design · Jaya algorithm

M. Çoşut (✉) · G. Bekdaş · S. M. Niğdeli
Department of Civil Engineering, Istanbul University-Cerrahpaşa, 34320 Avcılar, İstanbul, Turkey
e-mail: muhammed.cosut@ogr.iuc.edu.tr

G. Bekdaş
e-mail: bekdas@iuc.edu.tr

S. M. Niğdeli
e-mail: melihnig@iuc.edu.tr

J. H. Kim et al. (eds.), *Proceedings of 7th International Conference on Harmony Search, Soft Computing and Applications*, Lecture Notes on Data Engineering and Communications Technologies 140, https://doi.org/10.1007/978-981-19-2948-9_9

1 Introduction

Housing is one of the most important requirements for the continuation of human life. This ensures that engineering studies and designs are constantly progressing and developing. Some conditions are not taken into account at the design stage [1]. The general purpose of engineering designs is to be carried out in such a way that building safety, ergonomics, economic, and esthetics are at the forefront. Due to problems such as increasing environmental pollution and depletion of natural resources, no matter how much we stay within the building safety limits, design with minimal costs and CO_2 emissions is required when designing structures [2]. In recent years, the emphasis has been on lightweight materials and a more efficient system of multi-story structures [3]. Although the economy is at the forefront, there is an application in engineering designs in which pre-sizing and structural elements are assigned and subsequent analyses are performed in general. The sizing of these structural elements is carried out in a way that depends on the hunch, engineering knowledge, and experience. In this way, performance analyses of the structure under loads are performed and re-analyses are performed by changing the section dimensions if necessary according to the analysis results, and this sizing process is extended until it remains on the safe side. In this way, the design works lead to a waste of time and prevent the most optimized cross-section from being obtained.

With the advancing technology, in engineering studies, by using mathematical programs that meet all the regulatory requirements, appropriate algorithms are written for the purpose, and the result is reached in a short time. The checks of the randomly assigned sections for compliance with the limit values and the checks of whether the regulatory requirements have been met are easily carried out with multi-step control. Thanks to these algorithms, safe dimensions of the cross-section of the structure are realized at minimum cost for Eurocode 2 (design of concrete structures) [4], TS500 (design and construction rules of reinforced concrete structures) [5], and ACI 318 (building rules necessary for reinforced concrete structures) [6] regulations.

2 The Design Methodology

2.1 *The Design Optimization*

There are many metaheuristic algorithms used in the optimization process. Some of them are: genetic algorithm (GA), harmony search method (HS), flower pollination algorithm (FPA), bat algorithm (BA), and Jaya algorithm. With the Jaya algorithm, the operations required for the process of approaching the objective function are less, and the ease of implementation shows its superiority compared to other algorithms. There are three stages of numerical optimization. These are the preliminary optimization stage, the analysis stage, and the optimization stage. At the preliminary optimization stage, design constants, variables, boundary values, population-specific parameters are defined, and the first solution matrix is created. At the analysis stage, the objective function is calculated, the design constraints are checked, and those who do not comply with the design constraints are penalized. At the optimization stage, variables are assigned using the algorithm formulation; as a result of calculations, the existing candidate solutions are changed, and the solution matrix is updated. The minimum-cost objective function can be defined by the formula Eq. (1).

$$\text{Min } f(x) = C_c \times V_c + C_s \times W_s + A_k \times (C_k + C_{ki}) \quad (1)$$

In Eq. 1, C_c refers to unit concrete cost (TL/m^3), C_s refers to unit reinforcement cost (TL/ton), C_k refers to the formwork material cost (TL/m^2), V_c refers to unit length concrete volume, W_s refers to unit length reinforcement steel weight, A_k refers to the calculation of the beam area for the cost of the formwork (TL/m^2).

In this optimization, the main goal is to calculate and compare the minimum cost of reinforced concrete beam design for each regulation using Eurocode, TS500, and ACI 318 regulations. The minimum-cost calculation varies according to the amount of concrete, the area of reinforcement, the formwork material, and the formwork workmanship. In optimization, it is performed using the Jaya [7] algorithm, which always tries to be closer to the approach of being optimal.

The rectangular-section reinforced concrete beam design is made as shown in Fig. 1. q is the distributed load, b is the beam breadth, h is the beam height, L is the beam length.

In the rectangular-section reinforced concrete beam optimization process, the assigned beam breadth (b) and beam height (h) are randomly assigned. By using random values of b and h in formulations for the optimization process, the purpose function is achieved by finding the pressure block depth (a), the required reinforcement area (A_s), checking the reinforcement area ratios, calculating the required amount of concrete, reinforcement area, and formwork costs.

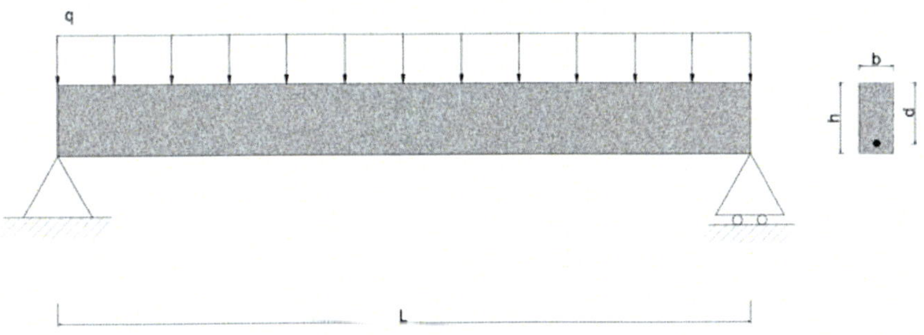

Fig. 1 Simply supported reinforced concrete beam

For Eurocode:

Under the influence of loads on the structure of the element, k is found relative to the dimensions of the beam Eq. 2. It is checked whether compressive reinforcement is required.

$$k = \frac{M}{f\text{ck} \times b \times d^2}, \text{ if } k > k_{\text{bal}} = 0.167, \text{ compressive reinforcement required.} \tag{2}$$

In order to calculate the reinforcement area, it is necessary to calculate the z value in Eq. 3, which is the internal force lever (moment lever), and a restriction has been imposed so that it is not greater than 0.95 * d.

$$z = d \times \left(0.5 + \sqrt{0.25 - \frac{k}{1.134}}\right) \tag{3}$$

The required reinforcement area (A_s) is made for the beam in Eq. 4.

$$A_s = \frac{M}{0.87 \times f\text{yk} \times z} \tag{4}$$

It is checked whether it provides maximum and minimum reinforcement ratios by finding the reinforcement area ratio (Eq. 5). If the value exceeds the maximum reinforcement ratio (Eq. 6), the objective function of the beam dimensions is penalized. If it is less than the minimum reinforcement ratio (Eq. 7), the reinforcement ratio is equalized to the minimum reinforcement ratio and the reinforcement area is calculated again.

$$\rho = \frac{A_s}{b \times d} \tag{5}$$

$$\frac{A_s}{b \times h} \leq \rho_{\text{max}} = 0.04 \tag{6}$$

$$100 \times \frac{A_s}{b \times d} \geq \rho_{\text{min}} = 26 \times \frac{f_{\text{ctm}}}{f_{\text{yk}}} (\%) \tag{7}$$

$$\rho_{\text{min}} = 26 \times \frac{f_{\text{ctm}}}{f_{\text{yk}}} (\%) \quad \text{not less than } \% \, 0.13 \tag{8}$$

For TS500:

By taking the moments of the forces generated in the compression region and the tensile region, the value a (stress block depth) is reached in Eq. 11. Concrete design compressive strength and reinforcement design yield strength calculation is in Eq. 9

$$fcd = \frac{f\text{ck}}{1.5}, f\text{yd} = \frac{f\text{yk}}{1.15} \tag{9}$$

$$Fc = Fs \rightarrow 0.85 \times f\text{cd} \times b \times a = \text{As} \times f\text{yd} \tag{10}$$

$$a = d - \sqrt{d^2 - \frac{2 \times M}{0.85 \times f\text{cd} \times b}} \tag{11}$$

The calculation of the reinforcement area is made, and the calculation of the required reinforcement area for the beam under loading is made in Eq. 12.

$$As = \frac{M}{f\text{yd} \times \left(d - \frac{a}{2}\right)} \tag{12}$$

It is checked whether it provides maximum and minimum reinforcement ratios by finding the reinforcement area ratio (Eq. 13). If the value exceeds the maximum reinforcement ratio (Eq. 15), the objective function of the beam dimensions is penalized. If it is less than the minimum reinforcement ratio (Eq. 14), the reinforcement ratio is equalized to the minimum reinforcement ratio and the reinforcement area is calculated again.

$$\rho = \frac{As}{b \times d} \tag{13}$$

$$\rho_{\min} = 0.8 \times \frac{f\text{ctd}}{f\text{yd}} \tag{14}$$

$$\rho_{\max} = 0.85 \times \rho_{\text{b}}, \rho_{\max} = 0.02, \rho_{\max} = 0.235 \times \frac{f_{\text{cd}}}{f_{\text{yd}}} \tag{15}$$

For ACI 318:

By taking the moments of the forces generated in the compression region and the tensile region, the value a (stress block depth) is reached in Eq. 17.

$$Fc = Fs \rightarrow 0.85 \times f\text{ck} \times b \times a = \text{As} \times f\text{yk} \tag{16}$$

$$a = d - \sqrt{d^2 - \frac{2 \times M}{0.85 \times f\text{ck} \times b}} \tag{17}$$

The calculation of the reinforcement area is made, and the calculation of the required reinforcement area for the beam under loading is made in Eq. 18.

$$As = \frac{M}{fyk \times \left(d - \frac{a}{2}\right)} \tag{18}$$

It is checked whether it provides maximum and minimum reinforcement ratios by finding the reinforcement area ratio (Eq. 19). If the value exceeds the maximum reinforcement ratio (Eq. 20), the objective function of the beam dimensions is penalized. If the reinforcement area is checked and the minimum reinforcement value (Eq. 21) is small, the minimum reinforcement value is selected (Table 1).

$$\rho = \frac{As}{b \times d} \tag{19}$$

$$\rho_{\max} = 0.75 \times 0.85 \times \beta 1 \times \frac{fck}{fyk} \times \frac{600}{600 + fyk} \tag{20}$$

$$A_{\mathrm{s,min1}} = \frac{\sqrt{fck}}{4 \times fyk} \times b \times d,\ A_{\mathrm{s,min2}} = \frac{1.4}{f_{\mathrm{yk}}} \times b \times d \tag{21}$$

Table 1 Design constraints and variables

Explanation	Symbol	Unit	Value or formula
Minimum section breadth	b_{min}	mm	250
Maximum section breadth	b_{maks}	mm	400
Minimum section height	h_{min}	mm	400
Maximum section height	h_{maks}	mm	600
Distributed load	q	kN/m	15
Beam length	L	m	6
Compressive strength of concrete	f_{ck}	MPa	25, 30, 35
Yield strength of concrete	f_{yk}	MPa	420
Specific gravity of steel	γ_{s}	t/m^3	7.86
Clear cover	Cc	mm	30
Cost of concrete per unit volume	C_{c}	TL/m^3	C25/30—253.63 C30/37—262.38 C35/45 – 278.63
Cost of steel per unit weight	C_{s}	TL/ton	8510
Cost of formwork material	C_{k}	TL/m^2	104
Cost of formwork labor	C_{ki}	TL/m^2	60

2.2 *The Optimization Process*

In this optimization process, the Jaya algorithm equation will be used. The variable values are assigned randomly so that they are between the maximum and minimum values. The objective function is achieved by using it in the formulas, and the optimization process continues by comparing the objective function at each step and choosing the best one. The optimization process takes place in four stages.

Step 1. At this stage, constants, constraint values (if they do not provide, the objective function is punished), variables necessary for the design are assigned, and thus, the objective function is reached, and the initial solution matrix is created.

Step 2. The best objective function value and the worst objective function value are recorded in the initial solution matrix. It is substituted into Eq. 22 using the Jaya equation.

$$X'_{i,\mathrm{new}} = X_{i,j} + r()\left(X_{i,g_{\mathrm{best}}} - \left|X_{i,j}\right|\right) - r()\left(X_{i,g_{\mathrm{worst}}} - \left|X_{i,j}\right|\right) \tag{22}$$

In Eq. 22, r() is the randomly assigned number, $X_{i,g_{\mathrm{best}}}$ is the best value of the initial solution matrix, $X_{i,g_{\mathrm{worst}}}$ is the worst value of the initial solution matrix, and $X_{i,j}$ is the value contained in the *i* row, *j* column.

The limit values of the generated variables are checked, and if they do not provide it, the equality is written, and the minimum value is equalized to the minimum value if it is small, the maximum value if it is large, and a new solution matrix is created by obtaining the objective function.

Step 3. By comparing the new solution matrix and the initial solution matrix objective function, the value that is good and the bad value are replaced, and the bad value is deleted.

Step 4. The maximum iteration value is checked and ended. The optimization process will be completed.

3 Numerical Examples

Using Jaya algorithm for reinforced concrete rectangular beam width, beam height, reinforcement area, and cost for Eurocode, TS500, ACI 318 regulations for C25/30 strength concrete in Table 2, for C30/37 concrete strength in Table 3, for C35/45 concrete strength in Table 4 are shown.

Table 2 For C25/30

	Eurocode	TS500	ACI 318
b (mm)	250	250	250
h (mm)	413.6419	413.7412	400
A_s (mm^2)	517.5039	517.6316	456.64
F (*x*) (TL)	454.61	454.73	421.73

Table 3 For C30/37

	Eurocode	TS500	ACI 318
b (mm)	250	250	250
h (mm)	403.8759	403.9722	400
A_s (mm^2)	525.8697	525.9997	452.6
F (*x*) (TL)	456.72	456.83	424.95

Table 4 For C35/45

	Eurocode	TS500	ACI 318
b (mm)	250	250	250
h (mm)	400	400	400
A_s (mm^2)	526.8990	527.18	449.8
F (*x*) (TL)	463.62	463.73	432.80

4 Conclusion

In this study, a rectangular-section reinforced concrete beam design was made for Eurocode 2, TS500, and ACI 318 regulations using Jaya algorithm. There may be some formulation differences due to the fact that the regulations have different studies and different laboratory result outputs. According to the regulation formulas and restrictions, the minimum-cost design was carried out according to different concrete classes. For all regulations, the beam breadth is calculated as 250 mm and the beam height is 400 mm for C35/45 strength concrete. For C25/30 and C30/37 strength concrete grades, the cross-sectional dimensions for Eurocode and TS500 are very close, and for ACI 318, the cross-sectional dimensions are 250–400 mm. However, due to some formulation differences, the reinforcement area and cost values have become different. While the Eurocode 2 design and the TS500 design differ very little in terms of equipment area and cost, the ACI 318 design was calculated to be approximately 7.5% less costly to implement.

References

1. O'Flaherty F, Browne E (2019) A proposal to modify the moment coefficient in Eurocode 2 for predicting the residual strength of corroded reinforced concrete beams. ScienceDirect, 324–339
2. Aral S, Yılmaz N, Bekdaş G, Nigdeli SM (2020) Jaya optimization for cantilever retaining walls with toe projection restriction. ICHSA 1275:197–206
3. Pierott R, Hammad AWA, Haddad A, Garcia S, Falcon G (2021) A mathematical optimisation model for the design and detailing of reinforced concrete beams. ScienceDirect 245:112861
4. British Standard (1992) Design of concrete structure. Eurocode 2
5. Turkish Standardization Institute (2000) Design and construction of concrete structures. Ankara, Turkey, TS500
6. ACI 318 (2005) Building code reqquirements for structural concrete and commentary
7. Rao R (2016) Jaya: a simple and new optimization algorithm for solving constrained and unconstrained optimization problems. Int J Ind Eng Comput 7(1):19–34

Training Neural Networks with Lévy Flight Distribution Algorithm

Mahdi Pedram, Seyed Jalaleddin Mousavirad, and Gerald Schaefer

Abstract In machine learning, multilayer perceptrons (MLPs) have long been used as one of the most popular and effective classifiers. With training the crucial process, the susceptibility of conventional algorithms such as back-propagation to get stuck in local optima has encouraged many researchers to opt for metaheuristic algorithms instead. In this paper, we propose a novel population-based metaheuristic algorithm for MLP training using Lévy flight distribution (LFD). In our approach, the optimum weights of the network are found via a population of agents moving through the search space either by Lévy flight motions or by random walks. Comparing the results of this algorithm on several datasets from the UCI repository with other population-based metaheuristic algorithms shows excellent results and superiority of the LFD algorithm.

Keywords Neural network training · Multilayer perceptron · Levy flight · Classification · Metaheuristic

1 Introduction

Among artificial neural networks (ANNs), which were initially inspired by human brain cells, multilayer perceptrons (MLPs) are simple feed-forward networks that

---

S. J. Mousavirad: On the date of publication of final version, Seyed Jalaleddin Mousavirad is affiliated with Universidade da Beira Interior, Covilha, Portugal, and is working on GreenStamp project.

M. Pedram
Lorestan University of Medical Sciences, Khorramabad, Iran

S. J. Mousavirad (✉)
Department of Computer Engineering, Hakim Sabzevari University, Sabzevar, Iran
e-mail: jalalmoosavirad@gmail.com

G. Schaefer
Department of Computer Science, Loughborough University, Loughborough, UK

J. H. Kim et al. (eds.), *Proceedings of 7th International Conference on Harmony Search, Soft Computing and Applications*, Lecture Notes on Data Engineering and Communications Technologies 140, https://doi.org/10.1007/978-981-19-2948-9_10

have proven their efficacy in tasks such as classification and regression [4]. The aim of training these networks is to achieve the desired output at the last layer of neurons [22]. This aim is performed by modifying the weights that connect the neurons of a layer to the neurons of the previous layer and is conducted in a supervised fashion based on the differences between network outputs and the desired results.

For a long time, the back-propagation (BP) algorithm has been the main approach to train MLPs [1, 3, 5]. BP adapts weights according to the mean squared error at the end nodes and which is then back-propagated to the previous layers. BP is essentially a gradient-based algorithm and thus shares the drawbacks of this family including a slow rate of convergence [40], getting stuck in local optima, and sensitivity to the initial weights.

Since finding the best weights for a network is an optimisation problem, metaheuristic approaches can be adapted to address the shortcomings of BP. The first use of a metaheuristic algorithm to train an ANN dates back 1989 [19] and employed a genetic algorithm. Since then, many metaheuristic algorithms have been used to train artificial neural networks [2, 20, 24–29] including, for example, particle swarm optimisation (PSO) [31], differential evolution (DE) [32], artificial bee colony (ABC) [12], ant lion optimiser (ALO) [15], sine cosine algorithm (SCA) [17] and human mental search (HMS) [23]. These population-based algorithms have been shown to work well and to be able to elude local optima [18].

Lévy flight distribution (LFD) [7] is a recently introduced metaheuristic algorithm based on Lévy flight that can be used for solving complex optimisation tasks. The algorithm tries to cover a large search space by adjusting the positions of neighbouring members of the population and diffusing them towards unexplored areas which increases the chance of finding global optima in search space.

The performance of the LFD algorithm motivates us to apply it for training MLP networks, as the main contribution of this paper, and compare its performance with nine other metaheuristic algorithms. Our results show excellent results and superiority of LFD over other approaches.

The remainder of the paper is organised as follows. Section 2 covers some background on MLP networks, while Sect. 3 details the LFD algorithm. In Sect. 4, we describe our proposed method to train MLP networks. Section 5 provides our experimental results and a comparison with other algorithms. Finally, Sect. 6 concludes the paper.

2 Multilayer Perceptrons

Multilayer perceptrons (MLPs), one of the most used feed-forward networks, can be used for different purposes, including classification and function approximation tasks. They are usually comprised of one input, some hidden, and an output layer.

It has been shown that only a single hidden layer is sufficient for any classification task, while the number of neurons in the input layer is determined by the feature space dimensionality, in classification tasks, the number of classes usually determines the number neurons in the output layer. Inputs of each neuron, aside from those in the input layer, are connected to the outputs of all neurons from the previous layer. The strengths of these connections, called weights, are multiplied by their respective inputs. Each neuron adds up its weighted inputs and then applies a nonlinear function to that sum, which is called the activation function. The output of a neuron with n inputs is calculated as

$$Y = f\left(\sum_{i=0}^{n} w_i X_i + \theta\right), \quad (1)$$

where X_i is the ith input and w_i is its respective weight, θ is a bias term which is also called the threshold for the activation function. The sigmoid function $f(x) = \frac{1}{(1+e^{-x})}$ is often used as activation function.

3 Lévy Flight Distribution

Lévy flight distribution (LFD) [7] is a recent population-based metaheuristic algorithm originally based on wireless sensor environments that makes use of Lévy flight (LF) motions for optimisation. In nature, many predators and other animals, when unable to find enough food, abandon their Brownian motion for Lévy flights [8, 34]. Inspired by this behaviour, the LFD algorithm is based on the superiority of LF over Brownian random walks [14] to explore large search spaces. Figure 1 shows examples of these two motions.

The algorithm begins with computing the distance between neighbouring nodes and then based on these distances decides whether to keep the nodes in their positions or move them to a more sparse location in the space. This movement, using LFs, prevents the overlapping among nodes and spreads them towards areas devoid of nodes.

In order to carry out an LF, two factors need to be determined: the first is the length of the flight and the second is the direction of the movement. The former is chosen by a Lévy distribution and the latter according to a uniform distribution. We follow Mantegna's algorithm for symmetric and stable Lévy distribution proposed

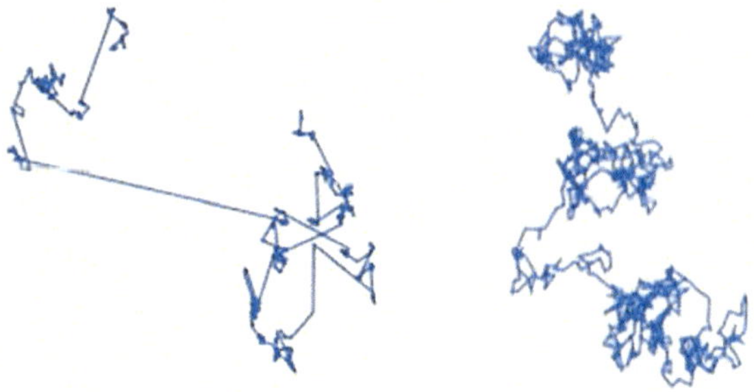

Fig. 1 Lévy flight (left) versus Brownian motion (right)

in [7]. The length of the flight is determined by

$$S = \frac{U}{|V|^{\frac{1}{\beta}}} \quad 0 < \beta \leq 2, \tag{2}$$

where β is the Lévy distribution index and U and V are derived from normal distributions

$$U \sim N(0, \sigma_u^2), \quad V \sim N(0, \sigma_v^2), \tag{3}$$

where the standard deviations are defined by

$$\sigma = \left\{ \frac{\Gamma(1+\beta)\sin(\frac{\pi\beta}{2})}{\Gamma\left[\frac{(1+\beta)}{2}\right]\beta 2^{\frac{\beta-1}{2}}} \right\}^{\frac{1}{\beta}} \tag{4}$$

with Γ the gamma function

$$\Gamma(z) = \int_0^\infty t^{z-1} \mathrm{e}^{-t} \mathrm{d}t. \tag{5}$$

Algorithm 1 shows the LF function in pseudo-code form. It takes four inputs: the current position of the node, the direction of the flight, and the lower and upper bounds of the search space, and returns the new position.

Algorithm 1 Lévy flight function

1: **Input**: CurrentPosition, DirectionPosition, LB, and UB.
2: **Output**: NewPosition.
3:
4: Calculate the step length S based on Mantegna's algorithm using Eq. (2).
5: Calculate the difference factor $DF = 0.01 \times S \times [CurrentPosition - DirectionPosition]$.
6: Execute the actual random walk or flight as $NewPosition = CurrentPosition + S \times rand() \times [size(CurrentPosition)]$.
7: Apply the bounds on NewPosition to ensure that it does not go outside the search space.
8: **return** NewPosition.

LFD is a population-based algorithm, where during each iteration the population gets updated. In each iteration, the Euclidean distance $\mathrm{ED}(X_i, X_J)$ between neighbouring nodes X_i and X_J is calculated. This distance is then compared with a predefined threshold (3 in this paper) and if the distance is less than the threshold, the positions need to be updated as the nodes are not spread enough in the search space. In case of X_J, the new position is then calculated as

$$X_J(t+1) = \text{Levy\_Flight}(X_J(t), X_{\text{Leader}}, \text{LB}, \text{UB}) \tag{6}$$

or by

$$X_J(t+1) = \text{LB} + (\text{UB} - \text{LB})\text{rand}(). \tag{7}$$

Here, t indicates the current iteration, Levy_flight is the Lévy flight function from Algorithm 1, X_{Leader} defines the position of the node with the least number of neighbours and determines the direction of the LF, and LB and LU lower and upper bounds. rand() is a random number generator steered by a uniform distribution bounded by 0 and 1.

The decision for choosing between Eqs. (6) and (7) is made by

$$R = \text{rand}(), \quad \text{CSV} = 0.5, \tag{8}$$

where R is a randomly generated number between 0 and 1, and CSV is the comparative scalar value. If the value of R is less than CSV, then (6) will update the position, otherwise (7) is used. This allows to explore the search space more and to minimise unvisited regions.

The position of X_i is also updated, by

$$X_i(t+1) = \text{TP} + \alpha_1 + \text{rand}()\alpha_2((\text{TP} + \alpha_3 X_{\text{Leader}})/2 - X_i(t)), \tag{9}$$

and

$$X_i^{\text{new}} = \text{Levy\_Flight}(X_i(t+1), \text{TP}, \text{LB}, \text{UB}), \tag{10}$$

where TP is the (target) position of the node that has the best objective function value. Parameters α_1, α_2 and α_3 are chosen randomly so that $0 < \alpha_1, \alpha_2, \alpha_3 \leq 10$. $\text{TF}_{\text{Neighbors}}$ is the sum of the target fitness of nodes adjacent to $X_i(t)$ and is given by

$$\text{TF}_{\text{Neighbors}} = \sum_{k=1}^{\text{NN}} \frac{D(k)X_k}{\text{NN}}, \tag{11}$$

where NN is the number of $X_i(t)$ neighbours and X_k denotes their positions.

$D(k)$ defines the fitness degree of these neighbours and is obtained as

$$D(k) = \frac{\partial_1(V - \text{Min}(V))}{\text{Max}(V) - \text{Min}(v)} + \partial_2, \tag{12}$$

where

$$V = \frac{\text{Fitness}(X_J(t))}{\text{Fitness}(X_i(t))}, \quad \text{and} \quad 0 < \partial_1, \quad \partial_2 \leq 1. \tag{13}$$

Algorithm 2 shows the pseudo-code of the LFD algorithm. Its merits are a good balance between exploration and exploitation, and its ability to avoid local optima.

Algorithm 2 Pseudo-code of the LFD algorithm

Initialise the population $X_i, i = 1, 2, ..., n$.
Initialise $LB, UB, Threshold$, and Max_{NFE}.
Calculate the fitness for each node.
T = best solution.
Output target position and target fitness.
while $NFE \leq Max_{NFE}$ **do**
  for each two adjacent nodes (main , neighbour) **do**
    Calculate Euclidean distance between the nodes
    if $Distance < Threshold$ **then**
      Update the neighbour node using Eqs. (6) and (7).
      Update the main node using Eq. (10).
      Constrain the nodes to the search space if necessary.
    end if
  end for
end while
return T.

4 MLP Training with LFD

To train an MLP network using the LFD algorithm, we first need to encode the neural network weights into a suitable search space for LFD. Figure 2 illustrates how this is performed to construct a vector of weights and bias terms.

As objective function we employ the classification error, defined as

$$E = 100\left(\sum_{p=1}^{P} \xi(\vec{p})/p\right), \tag{14}$$

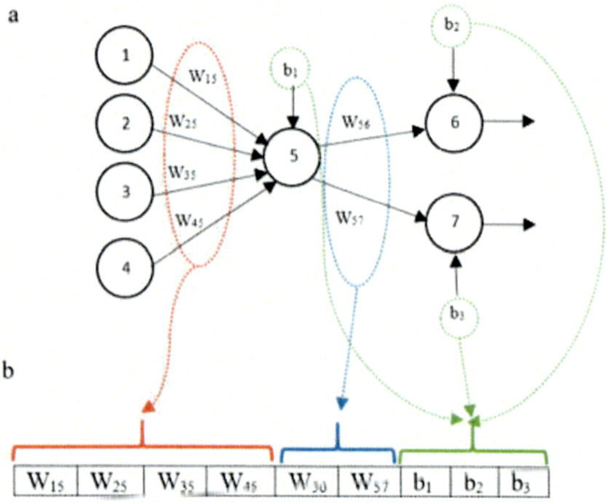

Fig. 2 **a** MLP with 4-1-2 structure, **b** the corresponding representation

with

$$\xi(p) = \begin{cases} 1, & \text{if } \overrightarrow{o_p} \neq \overrightarrow{d_p} \\ 0, & \text{otherwise} \end{cases}, \tag{15}$$

where $\overrightarrow{d_p} = (d_{p,1}, d_{p,2}, ..., d_{p,k})$ are target values and $\overrightarrow{o_p} = (o_{p,1}, o_{p,2}, ..., o_{p,k})$ are the output vectors of the, and P is the number of samples.

Based on the encoding and the objective function, the algorithm starts with a random population that then gets successively improved as outlined in Sect. 3 and terminates when a stopping criterion (in this paper, the maximum number of objective function evaluations) has been met upon which the best node gives the MLP weights and bias terms so that the network can be used for classification. LFD benefits from a high capability in exploration and exploitation due to Levy distribution.

5 Experimental Results

For evaluation, we train three-layer MLP networks where the number of neurons in the hidden layer is, following [21], set to $2N + 1$ where N is the number of input neurons.

We use five challenging datasets from the UCI repository, namely:

- Wine: contains chemical properties of 178 wines that are supposed to determine the origin of the wine. The dataset comprises 13 features and 3 classes.
- Seed: comprises 7 geometrical features of 3 classes of wheat. There are 210 samples in this dataset.
- Pima: contains 8 features of 768 patients determining whether subjects (Pima Indians) are symptomatic of diabetes or not.
- Breast: contains 699 samples with 9 features such as uniformity of cell size, clump thickness determining breast cancer malignancy.
- Iris: well-known dataset that comprises 4 features of 150 flowers from 3 categories of the iris family.

To evaluate the performance of the trained networks, we use tenfold cross-validation (10CV) by dividing the datasets into 10 randomly selected folds and using each fold once as the test set and the remaining 9 folds as training data. We repeated this process for all 10 folds and then report the average and standard deviation over the 10 these test folds.

To compare the performance of our LFD approach, we run equivalent experiments using 9 other population-based metaheuristic algorithms: PSO [6], DE [9], ABC [11], CS [39], BA [33], FA [36], ALO [35], DA [13], and SCA [30]. Some selected algorithms such as PSO and DE are well-established, while others such as DA and SCA are recently introduced algorithms. We set the population size for all algorithms to 50, and the maximum number of objective function evaluations to 25,000 times. Other algorithm-specific parameters are listed in Table 1.

Table 1 Parameter settings for the experiments

Algorithm	Parameter	Value
PSO [31]	Cognitive constant C_1	2
	Social constant C_2	2
	Inertia constant w	0 to 1
DE [32]	Scaling factor	0.5
	Crossover probability	0.2
ABC [10]	Limit	$n_e \times$ Variables
CS [38]	P_a	0.25
BA [37]	Loudness	0.5
	Pulse rate	0.5
FA [36]	Light absorption coefficient (γ)	1
	Attractiveness at $r = 0$ (β_0)	1
	Scaling factor	0.25
ALO [15]	No parameters	
DA [16]	No parameters	
SCA [17]	a	2
LFD	Comparative scalar value (CSV)	0.5
	Threshold	3

Table 2 shows these results for all algorithms and all datasets, while the resulting Friedman ranks are given in Table 3. The Friedman ranking assigns each algorithm an ordinal number with respect to the accuracy of the algorithm. If several algorithms have equal accuracy, the assigned Friedman rank is the average of the ranks.

Clearly, our LFD algorithm, outperforms all other algorithms yielding an average Friedman rank of 1.6. The second and third ranked algorithms are PSO with 2.8 and FA with 3.8, respectively. While the LFD algorithm achieves the highest accuracy for Seed, Wine and Breast (tied with PSO and ABC for the latter), it ranks second for Pima (FA ranked first) and Iris (PSO ranked first). Overall, LFD gives excellent results for MLP training, outperforming all other algorithms.

6 Conclusions

Multilayer perceptron networks are popularly employed for many classification tasks. While these networks are usually trained by gradient-based methods such as back-propagation, these suffer from drawbacks such as getting stuck in local optima. Population-based metaheuristic algorithms provide a useful alternative.

Table 2 10CV classification accuracy results for all datasets and algorithms

Algorithm		Wine	Seed	Pima	Breast	Iris
PSO	Mean	67.94	78.10	77.60	97.95	96.00
	Stddev	8.06	11.92	3.24	1.72	5.62
DE	Mean	62.35	70.00	76.94	97.36	92.00
	Stddev	3.74	11.01	4.97	2.06	5.26
ABC	Mean	61.21	72.38	78.26	97.95	84.67
	Stddev	3.38	8.03	4.45	1.03	9.45
CS	Mean	39.93	37.14	75.78	97.65	55.33
	Stddev	5.85	4.38	3.08	1.58	7.73
BA	Mean	23.07	32.86	76.69	97.22	44.67
	Stddev	6.80	6.90	5.13	1.76	10.45
FA	Mean	62.91	72.38	78.90	97.66	92.00
	Stddev	4.91	14.69	4.35	1.97	5.26
ALO	Mean	63.92	80.48	78.12	96.92	94.67
	Stddev	8.69	8.53	5.89	1.76	2.81
DA	Mean	62.32	70.48	77.85	97.51	92.67
	Stddev	6.37	7.03	5.40	1.8	3 5.84
SCA	Mean	61.18	71.43	74.47	97.08	90.67
	Stddev	3.85	8.98	4.20	1.82	7.83
LFD	Mean	79.35	82.85	78.50	97.95	95.33
	Stddev	9.33	11.49	4.89	0.36	3.22

Table 3 Friedman rank results

Dataset	PSO	DE	ABC	CS	BA	FA	ALO	DA	SCA	LFD
Wine	2	5	7	9	10	4	3	6	8	1
Seed	3	8	4.5	9	10	4.5	2	7	6	1
Pima	6	7	3	9	8	1	4	5	10	2
Breast	2	7	2	5	8	4	10	6	9	2
Iris	1	5.5	8	9	10	5.5	3	4	7	2
Average	2.8	6.5	4.9	8.2	9.2	3.8	4.4	5.6	8	1.6

In this paper, we have proposed using the Lévy flight distribution algorithm for MLP networks to overcome these shortcomings. Experimental results on a number of challenging classification problems and comparison with nine other metaheuristic algorithms illustrated the superiority of the proposed algorithm and yield excellent classification performance.

References

1. Amirsadri S, Mousavirad SJ, Ebrahimpour-Komleh H (2018) A Levy flight-based grey wolf optimizer combined with back-propagation algorithm for neural network training. Neural Comput Appl 30(12):3707–3720
2. Awad NH, Ali MZ, Suganthan PN, Reynolds RG (2016) Differential evolution-based neural network training incorporating a centroid-based strategy and dynamic opposition-based learning. In: IEEE congress on evolutionary computation. IEEE, pp 2958–2965
3. Bidgoli AA, Komleh HE, Mousavirad SJ (2015) Seminal quality prediction using optimized artificial neural network with genetic algorithm. In: 9th international conference on electrical and electronics engineering, pp 695–699
4. Boughrara H, Chtourou M, Amar CB, Chen L (2016) Facial expression recognition based on a MLP neural network using constructive training algorithm. Multimedia Tools Appl 75(2):709–731
5. Ebrahimpour-Komleh H, Mousavirad S (2013) Cuckoo optimization algorithm for feedforward neural network training. In: 21st Iranian conference on electrical engineering
6. Gudise VG, Venayagamoorthy GK (2003) Comparison of particle swarm optimization and backpropagation as training algorithms for neural networks. In: IEEE swarm intelligence symposium, pp 110–117
7. Houssein EH, Saad MR, Hashim FA, Shaban H, Hassaballah M (2020) Lévy flight distribution: a new metaheuristic algorithm for solving engineering optimization problems. Eng Appl Artif Intell 94:103, 731
8. Humphries NE, Queiroz N, Dyer JR, Pade NG, Musyl MK, Schaefer KM, Fuller DW, Brunnschweiler JM, Doyle TK, Houghton JD, Hays GC, Jones CS, Noble LR, Wearmouth VJ, Southall EJ, Sims DW (2010) Environmental context explains Lévy and Brownian movement patterns of marine predators. Nature 465(7301):1066–1069
9. Ilonen J, Kamarainen JK, Lampinen J (2003) Differential evolution training algorithm for feed-forward neural networks. Neural Process Lett 17(1):93–105
10. Karaboga D, Akay B (2009) A comparative study of artificial bee colony algorithm. Appl Math Comput 214(1):108–132
11. Karaboga D, Akay B, Ozturk C (2007) Artificial bee colony (ABC) optimization algorithm for training feed-forward neural networks. In: International conference on modeling decisions for artificial intelligence, pp 318–329
12. Karaboga D, Basturk B (2007) A powerful and efficient algorithm for numerical function optimization: artificial bee colony (ABC) algorithm. J Glob Optim 39(3):459–471
13. Khishe M, Safari A (2019) Classification of sonar targets using an MLP neural network trained by dragonfly algorithm. Wirel Pers Commun 108(4):2241–2260
14. Magdziarz M, Szczotka W (2016) Quenched trap model for Lévy flights. Commun Nonlinear Sci Numer Simul 30(1–3):5–14
15. Mirjalili S (2015) The ant lion optimizer. Adv Eng Softw 83:80–98
16. Mirjalili S (2016) Dragonfly algorithm: a new meta-heuristic optimization technique for solving single-objective, discrete, and multi-objective problems. Neural Comput Appl 27(4):1053–1073
17. Mirjalili S (2016) SCA: a sine cosine algorithm for solving optimization problems. Knowl Based Syst 96:120–133
18. Mirjalili S, Mirjalili S, Lewis A (2014) Let a biogeography-based optimizer train your multi-layer perceptron. Inf Sci 269:188–209
19. Montana DJ, Davis L (1989) Training feedforward neural networks using genetic algorithms. In: IJCAI, vol 89, pp 762–767
20. Moravvej SV, Mousavirad SJ, Moghadam MH, Saadatmand M (2021) An lstm-based plagiarism detection via attention mechanism and a population-based approach for pre-training parameters with imbalanced classes. In: International conference on neural information processing. Springer, Berlin, pp 690–701

21. Mousavirad SJ, Bidgoli AA, Ebrahimpour-Komleh H, Schaefer G (2019) A memetic imperialist competitive algorithm with chaotic maps for multi-layer neural network training. Int J Bio-Inspired Comput 14(4):227–236
22. Mousavirad SJ, Bidgoli AA, Ebrahimpour-Komleh H, Schaefer G, Korovin I (2019) An effective hybrid approach for optimising the learning process of multi-layer neural networks. In: International symposium on neural networks, pp 309–317
23. Mousavirad SJ, Ebrahimpour-Komleh H (2017) Human mental search: a new population-based metaheuristic optimization algorithm. Appl Intell 47(3):850–887
24. Mousavirad SJ, Jalali SMJ, Sajad A, Abbas K, Schaefer G, Nahavandi S (2020) Neural network training using a biogeography-based learning strategy. In: International conference on neural information processing
25. Mousavirad SJ, Rahnamayan S (2020) Evolving feedforward neural networks using a quasi-opposition-based differential evolution for data classification. In: IEEE symposium series on computational intelligence
26. Mousavirad SJ, Schaefer G, Ebrahimpour-Komleh H (2021) Optimising connection weights in neural networks using a memetic algorithm incorporating chaos theory. In: Metaheuristics in machine learning: theory and applications. Springer, Berlin, pp 169–192
27. Mousavirad SJ, Schaefer G, Jalali SMJ, Korovin I (2020) A benchmark of recent population-based metaheuristic algorithms for multi-layer neural network training. In: Genetic and evolutionary computation conference companion, pp 1402–1408
28. Mousavirad SJ, Schaefer G, Korovin I (2020) An effective approach for neural network training based on comprehensive learning. In: International conference on pattern recognition
29. Mousavirad SJ, Schaefer G, Korovin I, Oliva D (2021) RDE-OP: a region-based differential evolution algorithm incorporation opposition-based learning for optimising the learning process of multi-layer neural networks. In: 24th international conference on the applications of evolutionary computation
30. Sahlol AT, Ewees AA, Hemdan AM, Hassanien AE (2016) Training feedforward neural networks using sine-cosine algorithm to improve the prediction of liver enzymes on fish farmed on nano-selenite. In: 2016 12th international computer engineering conference. IEEE, pp 35–40
31. Shi Y, Eberhart R (1998) A modified particle swarm optimizer. In: IEEE international conference on evolutionary computation, pp 69–73
32. Storn R, Price K (1997) Differential evolution—a simple and efficient heuristic for global optimization over continuous spaces. J Glob Optim 11(4):341–359
33. Tuba M, Alihodzic A, Bacanin N (2015) Cuckoo search and bat algorithm applied to training feed-forward neural networks. In: Recent advances in swarm intelligence and evolutionary computation. Springer, Berlin, pp 139–162
34. Viswanathan GM, Buldyrev SV, Havlin S, DaLuz M, Raposo E, Stanley HE (1999) Optimizing the success of random searches. Nature 401(6756), 911–914
35. Yamany W, Tharwat A, Hassanin MF, Gaber T, Hassanien AE, Kim TH (2015) A new multi-layer perceptrons trainer based on ant lion optimization algorithm. In: Fourth international conference on information science and industrial applications. IEEE, pp 40–45
36. Yang XS (2010) Firefly algorithm, stochastic test functions and design optimisation. arXiv preprint arXiv:1003.1409
37. Yang XS (2010) A new metaheuristic bat-inspired algorithm. In: Nature inspired cooperative strategies for optimization. Springer, Berlin, pp 65–74
38. Yang XS, Deb S (2009) Cuckoo search via Lévy flights. In: World Congress on nature and biologically inspired computing, pp 210–214
39. Yi J, Xu W, Chen Y (2014) Novel back propagation optimization by cuckoo search algorithm. Sci World J 2014
40. Zhang JR, Zhang J, Lok TM, Lyu MR (2007) A hybrid particle swarm optimization-back-propagation algorithm for feedforward neural network training. Appl Math Comput 185(2):1026–1037

Pressure Management in Water Distribution Networks Using Optimal Locating and Operating of Pressure Reducing Valves

Peiman Mahdavi and Jafar Yazdi

Abstract Given the growing need for water and the crises caused by its scarcity, managing water resources is of great importance. High pressure in water networks can cause several problems, and the most important of which is the increase of water leakage from the network. In order to prevent water loss and also to manage the network pressure, pressure reducing valves (PRVs) can be used. The optimal location and adjustment of PRVs were determined by single- and two-objective differential evolution (DE) algorithms linked to the EPANET hydraulic solver. For this purpose, the network was investigated in two pressure dependent and demand dependent analysis, and finally, by applying this strategy in a real water distribution network in Tehran city, the maximum pressure of the network was decreased by 60%, while the total shortage of nodal demands was reduced by 58%. This significantly improves network performance.

Keywords Pressure management · Pressure reducing valve · Optimization · Differential evolution

1 Introduction

Water can be considered the most vital commodity of the present period, the value of which is becoming more and more visible and important in different human societies. Due to rapid industrial growth and infrastructure development, it is necessary to optimize and complete the management of infrastructures. Among these, infrastructures are water and sewage networks. With growing of time and increasing the crisis of water shortage, the need for water resources management and smartening and optimization of equipment and their continuous monitoring in the water and sewage

P. Mahdavi (✉) · J. Yazdi
Department of Water Resources Engineering, Faculty of Civil, Water and Environmental Engineering, Shahid Beheshti University, Tehran, Iran
e-mail: Peyman.mahdavy.pm@gmail.com

J. Yazdi
e-mail: j_yazdi@sbu.ac.ir

J. H. Kim et al. (eds.), *Proceedings of 7th International Conference on Harmony Search, Soft Computing and Applications*, Lecture Notes on Data Engineering and Communications Technologies 140, https://doi.org/10.1007/978-981-19-2948-9_11

networks is felt more than ever. Pressure is one of the most important factors in the amount of non-revenue water in urban water distribution networks and has the highest and fastest hydraulic effect on the amount of leakage. For this reason, the pressure in the water supply networks is one of the parameters that must be continuously monitored in the network and its value must be in the desired range, otherwise its increase will increase water loss and pipe failure and excessive decline will not meet the needs of consumers. The amount of leakage in the networks has been reported from 5 to 50% of the total produced water. According to studies conducted in Iran in the last ten years, the average amount of leakage from the networks is between 15 and 30% [1].

The solutions that can be mentioned to reduce the leak are: Finding the leak site directly and repairing it, replacing the damaged pipelines and pressure management (PM) and the latter of which is the most efficient and cost-effective option to achieve this goal [1]. Intelligent pressure control seems to be a suitable way to control leakage and reduce damage caused by high pressures in the network. PM may perform many activities with sections with different settings: pump control, reservoir operation, construction of pressure regulating ponds, implementation of pressure zoning plan (pressure zone) and PRV[1] automatic valves. In particular, use of PRV is a widely used method applied by companies to reduce excessive pressure at certain times of the day [2]. Pressure Reducing Valve has a low production cost, reduces network leakage and is easy to operate. Traditional pressure control systems rely mainly on PRVs with constant pressure outlets, and since the pressure in water distribution networks is a function of the consumption of the subscribers, this consumption is also a function of the hours of the day. The outlet in PRVs is generally adjusted to supply pressure during peak hours, therefore, unnecessary pressure is imposed on the network mainly in the middle of the night, which will increase ground leakage, breakdown and failure in the pipes. For this reason, intelligent PM by variable pressure PRVs can be a good way to adjust the network pressure. In design of engineering systems, there are conditions that the answer to the problem cannot be easily calculated and concluded with the usual methods, and the reason for this is the complexity of the problem and the existence of many variables affecting it. Therefore, various evolutionary algorithms based on random search have been developed to solve such problems, which have made it much easier to achieve optimal solutions and better states. One of these evolutionary algorithms that has been presented in the recent decades is the DE algorithm [3]. To date, this algorithm has been used in various issues and fields, and one of these fields is the water distribution networks. This algorithm has been used in water engineering problems, especially in pipeline network optimization problems, calibration of rainfall-runoff models, groundwater remediation, multi-reservoir systems and pressurized irrigation systems [3].

Latifi et al. [4] applied an optimization model to find optimal locations and pressure outlets for PRVs in existing networks. The use of this model showed that some PRVs can balance and limit the pressure in a certain range. In their work, an optimization code was prepared to estimate instantaneous water demand based on reported network

[1] Pressure Reducing Valve.

pressures. According to the estimation of instantaneous water demand, another optimization code based on DE algorithm is proposed to control the installed PRVs and VSPs. The results showed that using this method, network background leakage and energy consumption are decreased by 41.72% and 28.4%, respectively, compared to the non-managerial case [5]. Mahdavi et al. [6] have proposed an optimization model for pressure management to optimize the locations and settings of PRVs. The results show that this system is able to significantly reduce the high-pressure values and the total amount of drinking water that is lost annually. Tabesh and Vaseti [1] presented a method to reduce leakage by minimizing the sum of squares of additional pressure in the network nodes. Based on this method, by calculating the optimal value of the outlet pressure of the PRVs, the minimum possible pressure in the nodes was obtained which causes the greatest reduction in leakage and maintains the optimal service in the network. In the present study, the differential evolution algorithm has been used to find the strategies of pressure management in water distribution network with the help of optimal location and operation of PRVs, and the obtained results have been examined with the current situation of the studied network.

2 Materials and Methods

In order to find optimal situation and setting of PRVs in the network, DE was coded in MATLAB programming software (version 2018) and connected to EPANET 2.2 using EPANET-MATLAB Toolkit.

2.1 *Simulation Model*

The simulation model used in this research is the EPANET model. Water distribution networks can be analyzed in two ways: Demand dependent analysis (DDA) and pressure dependent analysis (PDA). Under DDA, nodal demands in various times are fixed values that must be delivered no matter how much is the nodal heads. This can result in situations where required demands are satisfied at nodes that have negative pressures—a physical impossibility. In the case of PDA, which is closer to reality, with decreasing nodal pressure from a certain value (P_{req}), the amount of withdrawal from the node will also decrease, and with further reduction of pressure to a certain extent, the amount of water that can be withdrawn will reach zero. The resulting Eq. (1) shows the relationship between demand flow and node pressure in water distribution networks [7].

$$d = D.\left(\frac{P - P_{min}}{P_{req} - P_{min}}\right)^{P_{exp}} \tag{1}$$

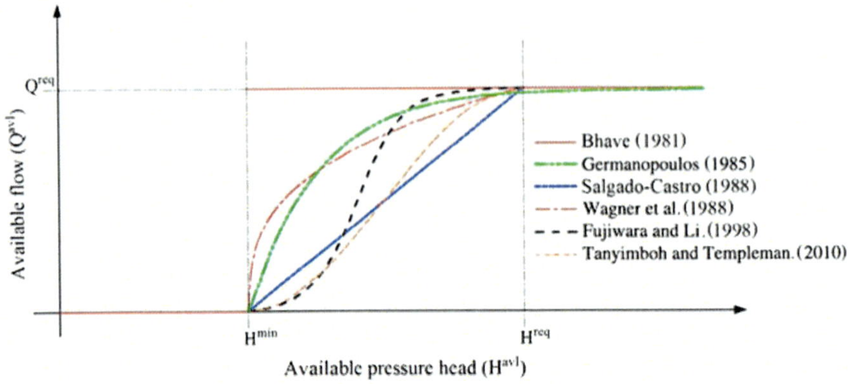

Fig. 1 Comparison of different head-flow relationships used in PDA

where

d	demand in PDA analysis
D	full node demand
P	node pressure
$P_{\min}$	minimum pressure
P_{req}	required pressure
$P_{\exp}$	pressure exponent.

In EPANET software, calculations related to PDA are performed by Eq. (1). Equation (1) is only one of the relationships between node pressure and demand flow in the pressure-based state, but there are other relationships, too. Figure 1 presents some common relationships of PDA [8].

2.2 *Optimization Model*

The algorithm used to find the optimal locations for PRVs and their outlet pressure in this research is DE algorithm. The reason for using meta heuristic algorithms such as DE algorithm in this problem is the non-convexity of its objective function. In this research, the network has been analyzed in two cases of PDA and DDA. In DDA case, single-objective DE algorithm and in PDA, multi-objective DE algorithm is used. In the case of a single-objective, only one goal is defined. Considering the amount of maximum pressure as an objective function and satisfying the minimum standard pressure, a wide range of solutions that have pressures slightly less than the standard pressure are considered as bad solutions. However, these solutions may have much less objective function than the optimal solutions obtained. This justifies the need to use a multi-objective algorithm. The optimization formulation in single-objective case is as follows:

$$\text{Min}\ \ F = \min\left(\max\left(P_{i,j}\right)\right) \tag{2}$$

Subject to:

$$\sum Q_{\text{in}} - \sum Q_{\text{out}} = Q_e \tag{3}$$

$$\sum_{k \in \text{loop1}} \Delta h_k = 0 \qquad \forall l \in Nl \tag{4}$$

$$P_{i,j} \geq P_{\min} \tag{5}$$

where

$P_{i,j}$ node i pressure in j time.
Q_{in} input flow to node.
Q_{out} output flow from node.
Δh_k head loss of pipe k.
Nl the total number of loops in the system.
$P_{\min}$ standard pressure (15 m is considered here).

The multi-objective optimization formulation is as follows:

$$\text{Min}\ F_1 = \max\left(P_{i,j}\right) \tag{6}$$

$$\text{Min}\ F_2 = \sum_{i=1}^{n}\sum_{j=1}^{m}\left(D_{f_{i,j}}\right) \tag{7}$$

$$D_{f_{i,j}} = D_{i,j} - d_{i,j} \tag{8}$$

where $D_{f_{i,j}}$ is the difference between full node demand and demand in PDA. The constraints here are the same as relations 3 and 4.

The reasonable range for the population size is between 3 and 10 times of the number of decision variables [9]; therefore, the amount of population size in this study was considered 10 times of the number of decision variables. In the study of Karaboğa and Demkdem, scaling factor (β) was considered as a variable which gave a good result [10]; therefore, here, the value of β was considered as a variable and a range from 0.2 to 0.8, and in four different cases of crossover probability (C_r) and a population size of 24 and a generation of 100. Each case was executed 3 times, and the minimum value was considered as the solution to that case.

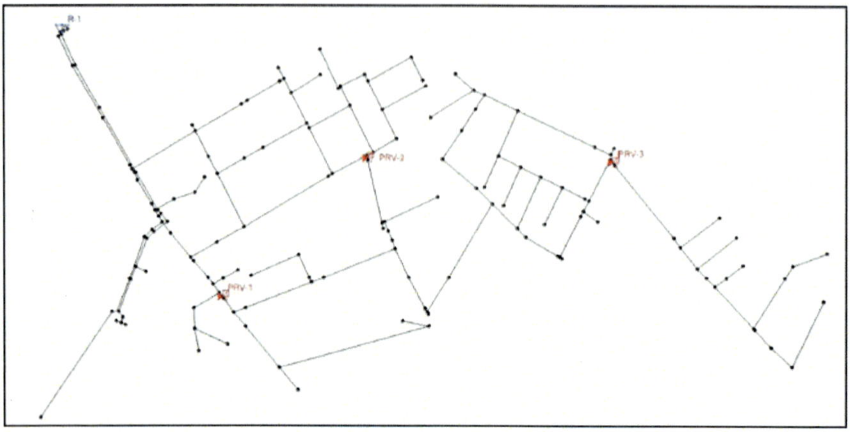

Fig. 2 Overview of the studied water distribution network in Tehran

2.3 *Case Study*

The studied network is related to an area of Tehran city water distribution network, presented in Fig. 2. This network has 176 pipes with a diameter of 25–300 mm. The network also has a storage tank at the highest point of the network at an elevation of 1753 m and the lowest part of the network has a height of 1625 m, which causes a difference of 128 m with the storage tank.

3 Results and Discussion

3.1 *Single-Objective DE Algorithm and DDA*

As mentioned, after the sensitivity of the C_r parameter, the value of $C_r = 0.2$ had the lowest value of the objective function that was used to perform the optimization.

Initially, three PRVs with a constant pressure pattern for the network were considered in the extended analysis, in which there will be a total of six decision variables. Three variables are for the valve location and the other three variables are related to the output pressure adjustment of each valve. The results of the minimization of the maximum pressure of the studied network in Tehran by the single-objective DE algorithm and DDA are given in Table 1.

For example, the convergence process during the simulation period in Scenario 3 with the constant adjustment of the PRVs is shown in Fig. 3.

The location of the PRVs in this scenario is shown in Fig. 4. Also, Fig. 5 shows the network pressure distribution at 3:00 AM compared to the current situation.

Table 1 Summary of the results of the studied network in Tehran

Scenarios	Number of PRVs	Number of decision variables	PRVs outlet pressure	Maximum pressure
Current status—without 3 default PRV	0	–	–	124.2
Current status—with 3 default PRV	3	–	Constant	71.9
1	3	6	Constant	68.2
	3	75	Variable	62.2
2	6	12	Constant	51.9
	6	150	Variable	48.5
3	10	20	Constant	45.8
	10	250	Variable	43.3

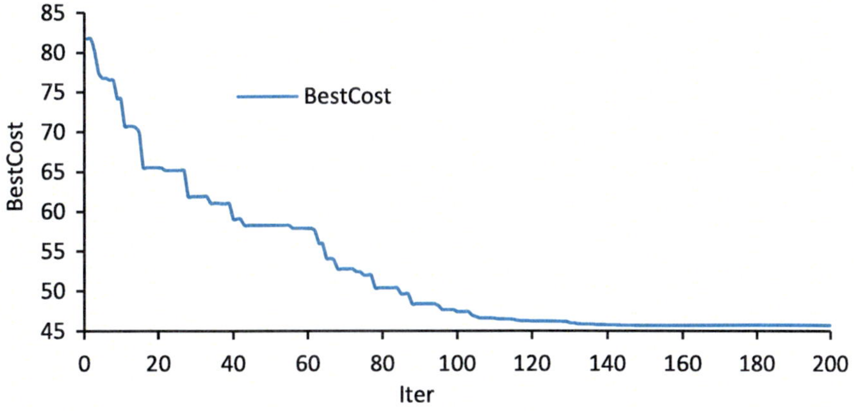

Fig. 3 Convergence process of the objective function by considering 20 decision variables

3.2 Multi-objective DE Algorithm and PDA

To further evaluate the optimal designs of network PM, it was also simulated as a PDA and the relevant Pareto front was extracted with the help of multi-objective optimization. Scenario 3 is considered as an example. In this case, the number of generations continues until the Pareto fronts in two consecutive repetitions fully overlap. To better compare these solutions (Fig. 6), the obtained Pareto front was clustered into four groups by K means clustering method a one representative solution was selected from each group. This was done for better analysis of the solutions as well as an overall estimate of the content of the set of obtained solutions (optimal location and settings of the PRVs). This approach is mentioned in the book [11].

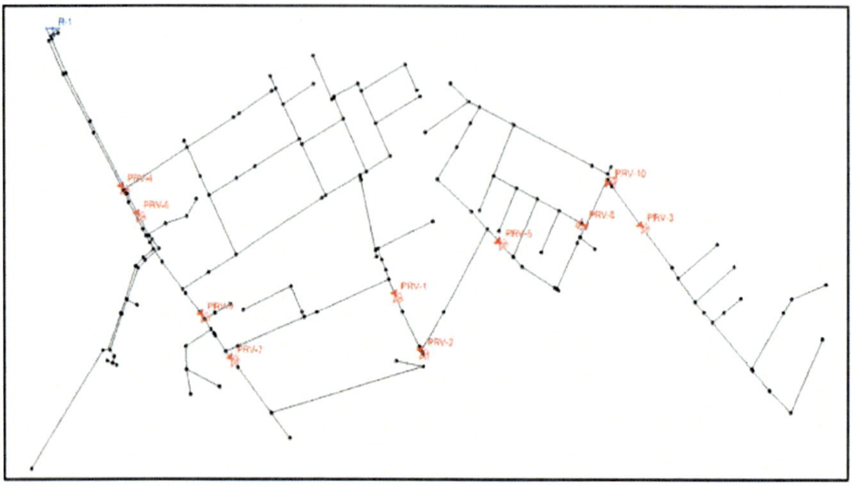

Fig. 4 Location of 10 PRV obtained by DE algorithm in the study network in Tehran

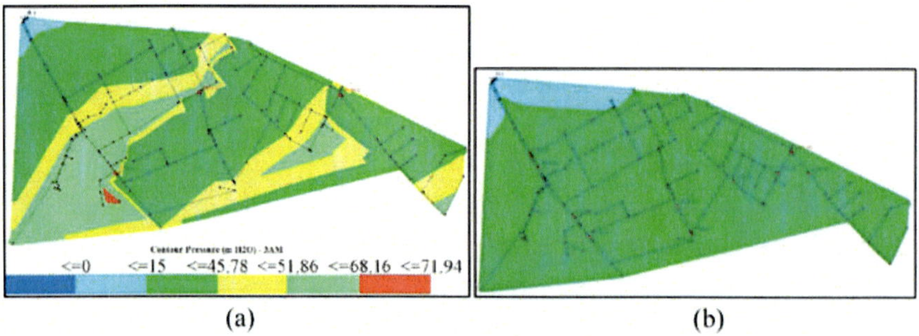

Fig. 5 **a** Current status pressure contour; **b** pressure contour with optimal location of 10 PRV; at 3:00 AM (minimum consumption)

Figures 7 and 8 show the Pareto front clustering diagram obtained along with the selected solutions.

Table 2 provides complete information of each representative solutions of the clusters.

According to the Pareto front obtained by two-objective DE algorithm in PDA case, compared to single-objective, we see a variety of optimal answers, each of which can be selected according to the preference of network officials and users over the two goals. Also, the optimal solutions outperform the current status.

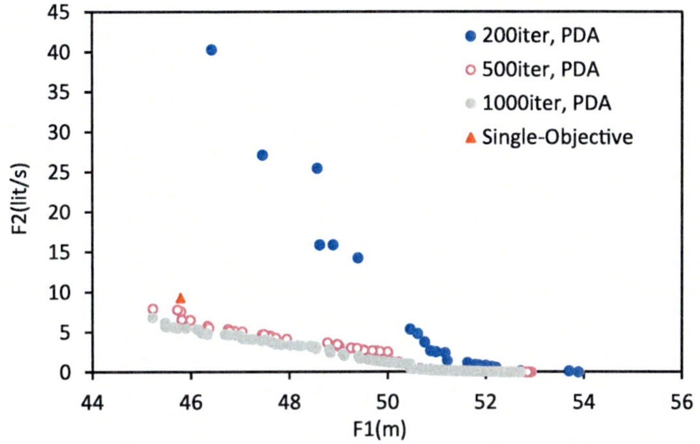

Fig. 6 Pareto front of different repetitions in the studied network in Tehran with 10 PRVs

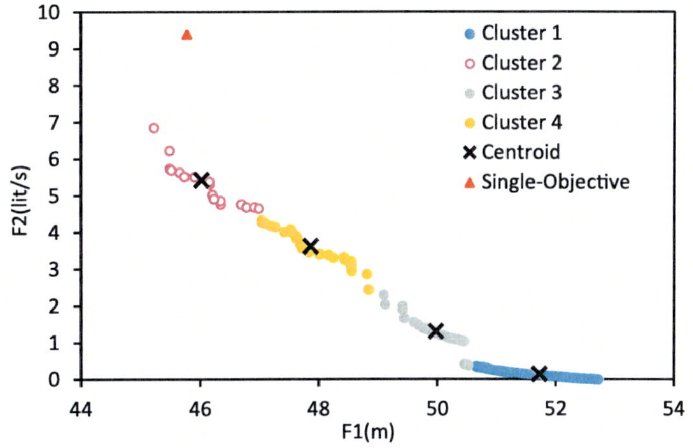

Fig. 7 Pareto front clustering diagram and selecting representative solutions

4 Conclusion

According to the result of adopting optimal location and adjustment strategies of PRVs, network performance improved. Considering the solution 3 as an example (10 PRVs with fixed adjustment) compared to the current state of the studied network without the three default PRVs, the maximum pressure was decreased 60%, the nodes with demand shortage reduced from 10 to 7 nodes and the total shortage of nodal demands reduced by 58%. In comparison of solution 3 with the current situation of studied network with three default PRVs, maximum pressure was decreased 31%, the nodes with demand shortage have been reduced from 13 to 7 nodes and the

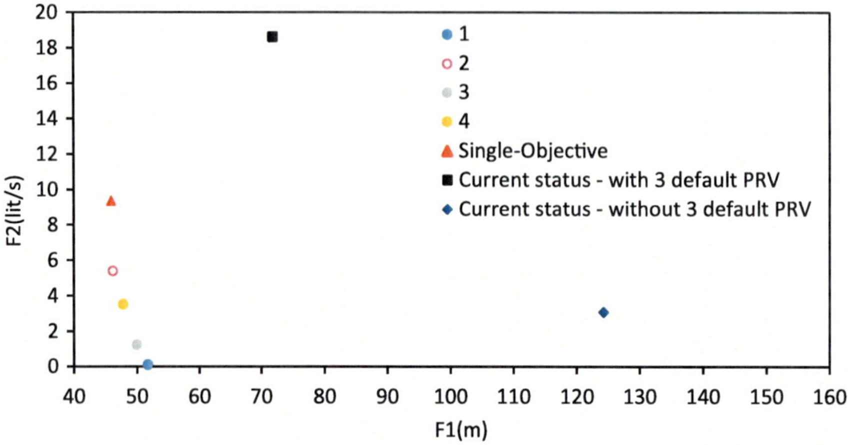

Fig. 8 A close view of the optimal solutions selected from the Pareto front

Table 2 The results of the PM strategies in studied network in Tehran obtained by two-objective DE and PDA

Solutions	F1	F2	Number of nodes with a lack of pressure	Volume of water shortage (lit)
1	51.7	0.1	2	360
2	46.1	5.4	9	19,440
3	50	1.3	7	4680
4	47.8	3.6	7	12,960
Single-objective	45.8	9.4	10	33,840
Current status—without 3 default PRV	71.9	18.6	13	66,960
Current status—with 3 default PRV	124.2	3.1	10	11,160

total shortage has been reduced by 93%. Other solutions also outperformed than the current state.

Therefore, it can be concluded that by optimally locating and adjusting the PRVs using the results of the optimization model, it is possible to reduce the maximum pressure of network to a very better status. Due to this, the amount of water loss is also reduced, damage due to high network pressure such as pipe failure and fittings is minimized, water consumption will decrease and also water will reach consumers with the desired pressure.

References

1. Tabesh M, Vaseti M (2006) Leakage reduction in water distribution networks by minimizing the excess pressure. Iran-Water Resour Res 2(2):53–66 [cited 2022 January 17]. Available from https://www.sid.ir/en/journal/ViewPaper.aspx?id=82346
2. Vicente D, Garrote L, Sánchez R, Santillán D (2016) Pressure management in water distribution systems: current status, proposals, and future trends. J Water Resour Plan Manag 142(2):04015061
3. Mansouri R, Torabi H (2015) Application of differential evolution (DE) algorithm for optimizing water distribution networks (case study: Ismail Abad pressurized irrigation network). Water Soil Sci (Agriculture Science) 25(4/2):81–95 [cited 2022 January 17]. Available from https://www.sid.ir/en/journal/ViewPaper.aspx?id=504979
4. Latifi M, Naeeni ST, Gheibi MA (2018) Upgrading the reliability of water distribution networks through optimal use of pressure-reducing valves. J Water Resour Plan Manag 144(2):04017086
5. Monsef H, Naghashzadegan M, Farmani R, Jamali A (2018) Pressure management in water distribution systems in order to reduce energy consumption and background leakage. J Water Supply: Res Technol AQUA 67(4):397–403
6. Mahdavi MM, Hosseini K, Behzadian K, Ardehsir A, Jalilsani F (2010) Leakage control in water distribution networks by using optimal pressure management: a case study. Water distribution systems analysis, pp 1110–1123
7. Wagner JM, Shamir U, Marks DH (1988) Water distribution reliability: simulation methods. J Water Resour Plan Manag 114(3):276–294
8. Mahmoud HA, Savić D, Kapelan Z (2017) New pressure-driven approach for modeling water distribution networks. J Water Resour Plan Manag 143(8):04017031
9. Storn R, Price K (1997) Differential evolution—a simple and efficient heuristic for global optimization over continuous spaces. J Global Optim 11(4):341–359
10. Karaboğa D, Ökdem S (2004) A simple and global optimization algorithm for engineering problems: differential evolution algorithm. Turk J Electr Eng Comput Sci 12(1):53–60
11. Deb K (2001) Multi-objective optimization using evolutionary algorithms. Wiley

The Optimal Location and Dimensions of Flood Control Detention Dams at Kan River Basin, Tehran, Iran

Mehrsa Pouladi, Jafar Yazdi, and Mohammad Shahsavandi

Abstract Detention dams are one of the structures used to deal with the destructive effects of floods, especially in mountainous rivers. These structures control the discharge of surface water flowing into or down a "dry watercourse" in a storm so the discharge can be carried by the waterway. These structures not only affect the peak discharge, but also increase the travel time. Based on "Integrated Flood Management", the impact of a set of measures should be considered together. Therefore in this study, the location and dimensions of several detention dams in the extent of Kan basin are determined simultaneously in an integrated approach. For this purpose, multi-objective harmony search algorithm (NSHS) is used to determine the location and dimensions of these structures by minimizing the initial cost and outflow peak discharge as objective functions. The results of this study show that using only two detention dams, the peak discharge is reduced by 64.44%, while the use of these structures, in almost all sub-basins (seven sub-basins), reduces the peak discharge by only 9.7%.

Keywords Flood · Integrated Flood Management · Detention dam · Multi-objective harmony search algorithm (NSHS)

1 Introduction

Mountainous rivers have a high potential for flash floods due to their steep slopes. These types of floods have a very high destructive power and can cause a lot of financial and human losses. To reduce the risk of floods, two types of structural and non-structural measures are generally used. One of the proper methods for reducing severe floods in mountainous rivers is to control flood by watershed management

M. Pouladi · M. Shahsavandi
Faculty of Civil, Water and Environmental Engineering, Shahid Beheshti University, Tehran, Iran

J. Yazdi (✉)
Department of Water Resources Engineering, Faculty of Civil, Water and Environmental Engineering, Shahid Beheshti University, Tehran, Iran
e-mail: j_yazdi@sbu.ac.ir

J. H. Kim et al. (eds.), *Proceedings of 7th International Conference on Harmony Search, Soft Computing and Applications*, Lecture Notes on Data Engineering and Communications Technologies 140, https://doi.org/10.1007/978-981-19-2948-9_12

measures and the use of detention dams in tributary rivers. Detention dams are short structures with a small reservoir volume that are constructed solely for flood control and to delay and mitigate flood peak discharge in the downstream. In this study, the problem of location and dimensions of these structures in an approach of "Integrated Flood Management" has been studied.

Determination of the proper location and dimensions of detention dams have a significant impact on the performance and effectiveness of the flood management project. This is especially important when the system consists of several branches with several consecutive reservoirs. This is due to the possibility of the routed discharge to be coincide with the flood wave of other branches downstream and has a reinforcing effect. Despite the importance of the topic, studies in this field are limited. In the following, these researches are briefly introduced.

Yazdi et al. [6] optimally designed the detention reservoirs using a simulation–optimization approach. This method includes the integration of ant colony optimization (ACO), artificial neural networks with objective functions of risk indicators including: VaR and CVaR. Optimal solutions have specific capabilities at the level of performance and cost, and this allows stakeholders and decision makers to choose the proper design when there are different approaches and interests to flood risk management. Heiner and MacKenzie [1] examined the outflow of detention dams as a function of the upstream water level based on field data. They also developed a spreadsheet that can accurately determine the outflow of the detention dam. Thuy Ngo et al. [2] proposed a method for determining the optimal design of detention dams (in terms of their size and location) using analytical and experimental relationships for a region and linear reservoir theory. The results show that the optimal solution reduces the downstream peak discharge by 33%, which is related to two scenarios with a return period of $T = 100$ years. Thuy Ngo et al. [3] couple the EPA-SWMM hydraulic model with an evolutionary algorithm (EPSO) to minimize downstream flood damage while considering the risk of flooding in the reservoir. Optimal design is applied in an urban case study in Seoul, Korea for severe historical flood events and designed rainfall scenarios. Optimal design performance in terms of reducing downstream flood intensity is much better than the existing design (79% in events in 2010 and 20% in events in 2011). Yazdi and Khazae [5] identified detention reservoirs as an effective strategy for underdevelopment (LID) effects to reduce the negative effects of urban floods. The model is developed using the evolutionary harmonic search (HS) algorithm under a significant number of artificial rainfalls and is implemented to find the desired location and size of online/offline detention dams in the network. The results of this approach show that offline construction of dams generally performs better than online construction. An optimal design and arrangement of dams effectively reduces flooding in the entire network by up to 12.5%.

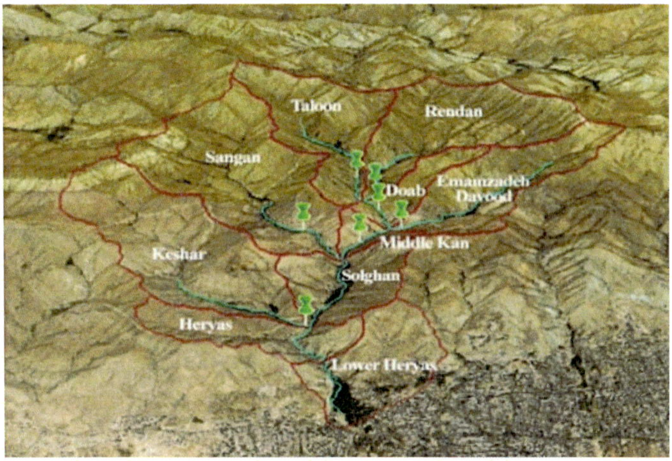

Fig. 1 Topography of the Kan basin and the location of the detention dams

2 Experimental Procedure

2.1 Case Study

The Kan river basin in Tehran is one of the mountainous basins with steep rivers and short concentration time (about one hour), whose longitude and latitude coordinates are between 51°10′ to 51°23′ and 35°45′ to 35°58′, respectively. This basin with an area of 216 square kilometers consists of ten small sub-basins (Fig. 1). The average elevation of this basin is 2377 m above sea level. The highest and lowest elevation of this basin is 3822 and 1327, respectively. As can be seen, the elevation changes in the area are very large, and the area has a rugged topography.

2.2 Location of Detention Dams

To locate the detention dams, two parameters have been considered. Lower the slope of the dam and the width of the mountain gorge, the greater the volume of the reservoir and the lower the cost of construction. Figure 1 shows the suitable position of the detention dams in terms of the above two parameters in the catchment area. The hydrological/hydraulic suitability of these sites should be controlled by the hydraulic model.

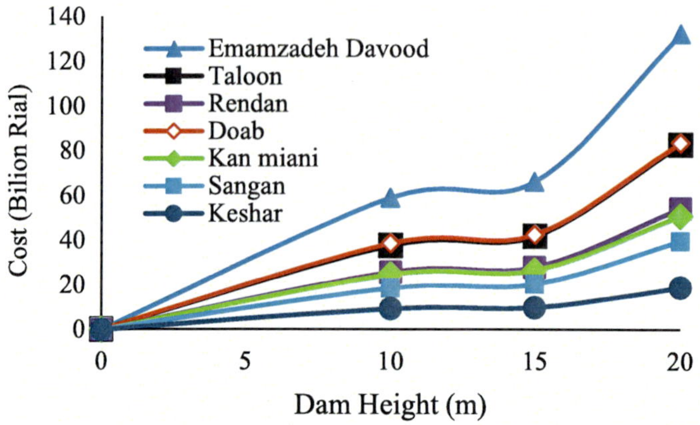

Fig. 2 Height cost curve of dams in different sub-basins

2.3 Design Criteria of Dams

In order to design a dam, various criteria must be met, such as slip, overturning, and allowable bearing pressure. In this regard, the relevant coefficients must be kept within the allowable range. In addition to hydraulic criteria, cost analysis should be considered in the process of designing. In this study, dimensions of the detention dams in different sub-basins have been calculated, and the price analysis has been performed under three different heights of 10, 15, and 20 m. Figure 2 shows the cost curve for the three mentioned sizes.

2.4 Materials and Methods

In order to evaluate the performance of detention dams, the SWMM hydraulic model has been used, and the effect of considered dam(s) on the peak discharge has been investigated. A 50 year return period flood is used for designing. This hydrograph is selected according to the proposed criteria in the "comprehensive plan of surface waters of Tehran". The data of this hydrograph is based on the studies of "comprehensive flood management studies of Kan River catchment area" [4].

SWMM model has been used for hydraulic modeling of the river networks. In order to evaluate the performance of the dams in different sub-basins, four different scenarios have been defined. The first scenario is the base form, without using detention dam, the second scenario is based on modeling the rivers network with all seven structures (as illustrated in Fig. 1), and the third scenario evaluates the performance of the dams in the case that the location and heights of the dams are calculated in the optimization model. Finally, the fourth scenario is the same as the third one, except that the overflow width is also considered as a decision variable.

2.5 Formulation of the Optimization Problem

The optimization model of this study is a dual-objective optimization model. With a fixed level of investment, the peak discharge can be reduced to a certain value. Further, reduction requires an increase in the level of investment. Therefore, two independent objective functions were used to minimize the investment cost and flood peak discharge. The general formulation of the objective functions is as below:

$$\text{Min } F_1 = \sum_{d=1}^{m} \text{Cost}_d = \sum_{d=1}^{m} f(H_d) \tag{1}$$

$$\text{Min } F_2 = Q_{p,\text{out}} \tag{2}$$

In which m represents the number of dams, Cost_d refer to the initial cost of the dth dam, H_d is the detention height, and $Q_{P,\text{out}}$ is the outflow discharge of the basin. The constraints of the problem are as below, where W_d is the dth overflow width.

$$H_{d,\min} < H_d < H_{d,\max} \tag{3}$$

$$W_{d,\min} < W_d < W_{d,\max} \tag{4}$$

In this study, based on engineering judgment, the search space of the height and width of the detention dams are considered between 5 to 20 m and 3 to 30 m, respectively. To solve the above objective functions, the NSHS algorithm has been used in which HS operators along with NS criteria for generating new Pareto fronts. The main steps of the NSHS algorithm are summarized as follows [6]:

1. Generate as many random vectors ($X^1, \ldots, X^{\mathbf{hms}}$) as the harmony memory size (HMS), and then, store them in harmony memory (HM).
2. Generate as many new harmonies as the repository size (RS), and then, store them in the repository.
3. Perform additional work if the value in step 2 came from HM.
4. With probability pitch adjusting rate (PAR; $0 \leq \text{PAR} \leq 1$), change X_i' by a small amount.
5. With probability 1-PAR, do nothing.
6. Create a temporary HM of size (HMS + RS) by merging the HM and repository.
7. Sort the temporary HM based on non-domination.
8. Sort the harmonies in each front ($F_1, F_2, \ldots, F_l$) based on the crowding distance,
9. Update the HM of size HMS by selecting the non-dominated solutions starting from the first ranked non-dominated front (F_1) and proceed with the subsequently ranked non-dominated fronts ($F_2, F_3, \ldots, F_l$), until the size exceeds HMS.

10. Repeat steps 2–7 until the termination criterion (e.g., maximum number of iterations) is satisfied,

3 Results

Figure 3 shows the outflow hydrograph in the first two scenarios (as explained in Sect. 2.4). The peak discharge of the basin in the first scenario (without any detention dam) is 195.15, and in the second scenario (with all seven dams) is 176.25 m^3/s, which shows a 9.7% drop in the peak discharge. Figure 4 also shows the set of optimal Pareto front solutions for the third and fourth scenarios.

As it turns out, in the fourth scenario, the initial cost is lower than three others, and hence, the Pareto front was selected to find the optimal design. From this, three

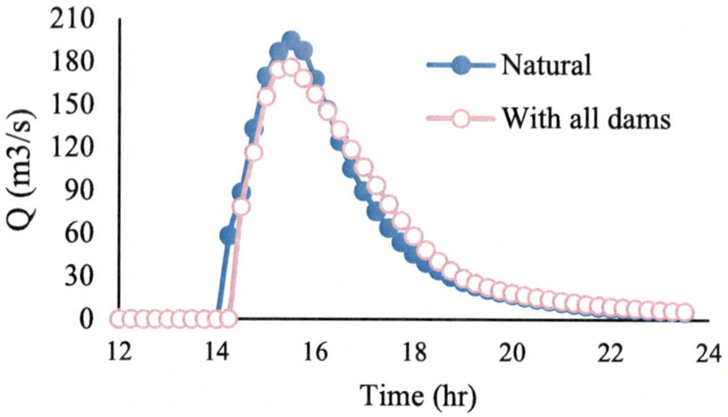

Fig. 3 Outflow hydrograph in the first two scenarios

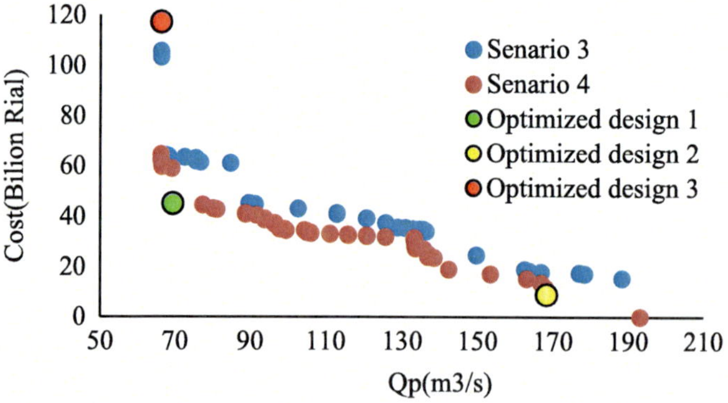

Fig. 4 Set of solutions obtained from the implementation of the model in the third and fourth scenarios with three selected optimal designs

Table 1 Specifications of the optimal designs selected from the Pareto front

Case 1							
Dam name	Emamzadeh	Rendan	Doab	Taloon	Kan	Sangan	Keshar
Location	–	–	–	–	✓	–	✓
Dam height	–	–	–	–	13	–	20
Overflow width	–	–	–	–	5	–	20
peak discharge	69.4 (m^3/s)						
Cost	45.04 (Billion Rial)						
Case 2							
Dam name	Emamzadeh	Rendan	Doab	Taloon	Kan	Sangan	Keshar
Location	–	–	–	–	–	–	✓
Dam height	–	–	–	–	–	–	5
Overflow width	–	–	–	–	–	–	7
peak discharge	168.2 (m^3/s)						
Cost	8.96 (Billion Rial)						
Case 3							
Dam name	Emamzadeh	Rendan	Doab	Taloon	Kan	Sangan	Keshar
Location	✓	–	–	–	✓	✓	✓
Dam height	7	–	–	–	11	13	19
Overflow width	6	–	–	–	6	3	29
peak discharge	66.2 (m^3/s)						
Cost	117.21 (Billion Rial)						

efficient solutions were selected as representatives and for further study, which are numbered from 1 to 3. Detailed information on each of these three solutions is provided in Table 1. Solutions 2 and 3 are the marginal points of the Pareto front, and solution 1 is in the middle. Comparison of these three options shows that solution 1 is suitable for the optimal design. This is because more expensive designs make little change in peak discharge. For example, the peak discharge in solutions 1 and 3 is not much different, but it can be seen that the construction cost in solutions 3 has increased by about 1.6%, which is not desirable. Therefore, solutions 1 will be the optimal design from the engineering and technical point of view and is proposed as the best design.

Comparison of outflow hydrographs in first three scenarios, show that, in scenario 1, using only two dams in the sub-basins of middle Kan and Keshar reduces the peak discharge by 64.44%, while the presence of dams at all seven proposed sub-basins, in addition to the high cost of construction, will results in 9.7% reduction in peak flow discharge. Figure 5 shows the hydrographs of the various scenarios mentioned above. As can be seen, scenario 1 has been able to effectively reduce the peak discharge of the hydrograph compared to the presence of dam in all locations.

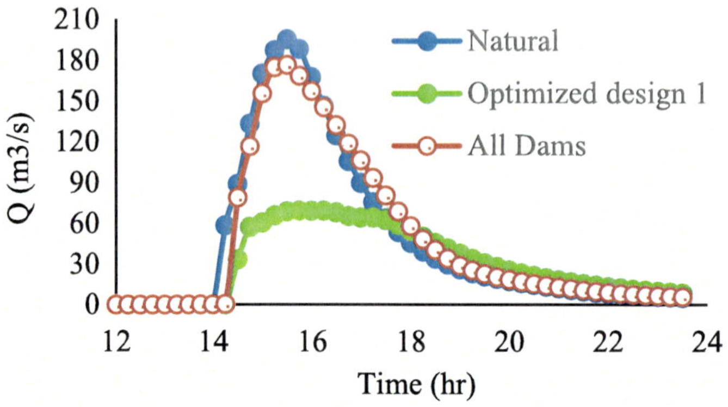

Fig. 5 Comparison of hydrographs in first three scenarios ($T = 50$ years)

In the selected solution, changing the dimensions of the dam overflows has not reduced the peak discharge, and it shows that overflow size has no impact on the peak discharge. After selecting the optimal design for flood discharge with a return period of 50 years, the selected solution was executed for three more return period of 10, 25, and 100 years. The results of these three return period show that the optimal solution could reduce the peak discharge by 67.03, 65.8, and 53.62%, respectively. Figures 6 and 7 illustrate this as well.

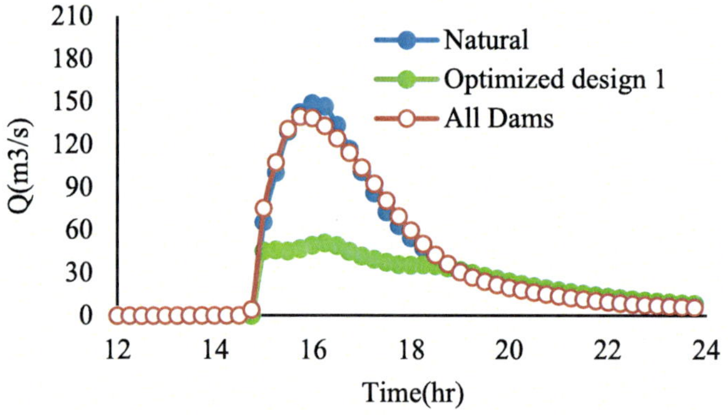

Fig. 6 Comparison of hydrographs in first three scenarios ($T = 25$ years)

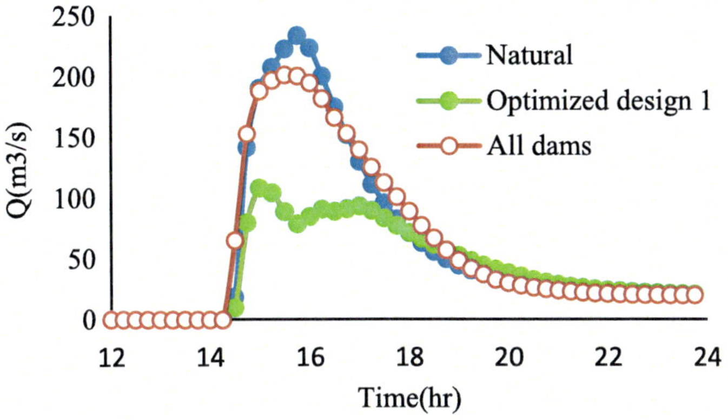

Fig. 7 Comparison of hydrographs in first three scenarios ($T = 100$ years)

4 Conclusions

In this research, a simulation–optimization model was developed for the specifying the location and optimal dimensions of detention dams. After implementing the model and analyzing the results, the following important points were obtained:

- The use of detention dams in all sub-basins did not lead to the lowest peak discharge among the scenarios. According to studies, the peak discharge in this case has decreased by only about 9.7% compared to the base scenario.
- Using an optimal design, the flood discharge can be reduced by about 64.44% compared to base scenario.
- Optimal design can reduce flood peak discharge for 25, 50, and 100 return periods by about 67.03, 65.8, and 53.62%, respectively.
- In this research, the optimal solution shows the location of detention dams in the two sub-basins, Keshar and middle Kan. The optimal height and width of the dams in these two sub-basins are 13 and 20 m and also 5 and 20, respectively.
- The dimensions of the overflows had no effect on the peak flow discharge.

To continue this issue in the future research, the design method can be developed as a risk-based design method and the results can be compared. It is also possible to study the operations of dam outflows and the reservoir in flood conditions.

References

1. Heiner BJ, MacKenzie K (2015) Extended detention stormwater basins outlet structure flows Physical Model Study
2. Thuy Ngo T, Yazdi J, Mousavi SJ, Kim JH (2016) Linear system theory-based optimization ofdetention basin's location and size at watershed scale. https://doi.org/10.1061/(ASCE)HE.1943-5584.0001451

3. Thuy Ngo T, Yoo DG, Lee YS, Kim JH (2016) Optimization of upstream detention reservoir facilities for downstream flood mitigation in urban areas. https://doi.org/10.3390/w8070290
4. Water Research Institute (2012) Comprehensive flood management studies of kan river catchment area. IFM project, Ministry of Energy, Tehran, Iran
5. Yazdi J, Khazae P (2019) Copula-based performance assessment of online and offline detention ponds for urban stormwater management. https://doi.org/10.1061/(ASCE)HE.1943-5584.0001810
6. Yazdi J, Sadollah A, Lee EH, Yoo DG, Kim JH (2015) Application of multi-objective evolutionary algorithms for the rehabilitation of storm sewer pipe networks.https://doi.org/10.1111/jfr3.12143

Minimization of the CO_2 Emission for Optimum Design of T-Shape Reinforced Concrete (RC) Beam

Melda Yücel, Sinan Melih Nigdeli, and Gebrail Bekdaş

Abstract In structural engineering, either structure safety or sustainable design of structure is an important parameter, which must be considered. On the other hand, the issue based on the creation of an optimum model toward any design in the structural engineering area becomes possible by fulfilling the mentioned design conditions. In this regard, in the present study, a reinforced concrete (RC) beam model with a T-shape cross-section was tried to optimize in the way of providing the minimum carbon dioxide (CO_2) emission from the structural materials comprised of both concrete and steel. In this direction, optimization applications were provided according to different concrete grades in terms of compressive strength, besides that optimal cost levels were also obtained for all of the design combinations. While the optimization analyses were conducted to minimize CO_2 emission, a well-known metaheuristic algorithm called as flower pollination algorithm (FPA) was benefited. Therefore, both optimization outcomes and statistical evaluations of them were presented concerning the eco-friendliest, sustainable, and also conveniently cost designs through decreasing one of the detrimental factors as CO_2 emission.

Keywords Reinforced concrete (RC) structures · Optimization · CO_2 emission · Eco-friendly design · Flower pollination algorithm

M. Yücel (✉) · S. M. Nigdeli · G. Bekdaş
Department of Civil Engineering, Istanbul University-Cerrahpaşa, 34320 Avcılar, Istanbul, Turkey
e-mail: melda.yucel@ogr.iu.edu.tr

S. M. Nigdeli
e-mail: melihnig@iuc.edu.tr

G. Bekdaş
e-mail: bekdas@iuc.edu.tr

J. H. Kim et al. (eds.), *Proceedings of 7th International Conference on Harmony Search, Soft Computing and Applications*, Lecture Notes on Data Engineering and Communications Technologies 140, https://doi.org/10.1007/978-981-19-2948-9_13

1 Introduction

Optimization is known as the updating, namely the recovery process for any problem or model according to the desired and defined conditions. For example, generating a new and more effective medicine for breast cancer disease requires some limitations for health or different conditions in terms of ingredients of drug substances. From a different viewpoint, this process also accompanies different aims. For example, when an automobile is designed, both creating the most comfortable and safe design and decreasing the consumption for the production process are significant targets. On the other side, a bridge-road must be constructed with the aim of usage by pedestrians safely and being practical in terms of movement of vehicles. According to these statements, there are various optimization techniques, which can be also utilized to realize the processes. However, nowadays, metaheuristic algorithms are more preferred thanks to the efficiency and usability of their performance. In this respect, from the past to the present, some properties of the well-known metaheuristics, which contain naturally inspired methodologies that were utilized, in this study, are summarized in Table 1.

However, it can be detected that FPA has been utilized for the solution of most of the structural optimization problems in civil engineering that it exists within the mentioned methodologies and is beneficial for the present study. In this meaning, many researches and applications conducted via FPA can be given as examples about the generation of cost-effective or safe structural models [21, 22], designing the most safety retaining wall models [23–25], adjusting the mechanical parameters of vibration control devices to minimize the hazards [26–28], etc.

In the current study, the principal target includes finding the best design for reinforced concrete (RC) beam with T-shape section in the way of providing the minimum CO_2 emission coming from structural materials. In this scope, the optimal cost level for structural designs was also detected with the optimization applications. Additionally, to determine the minimized emission and optimal cost level, several grades for concrete compressive strengths were utilized besides that total of 25 independent cycles were carried out for each optimization process with 3000 iterations and 25 candidate solutions. Thus, it can be possible to observe the success of optimization processes for sustainable and eco-friendly structural designs by determining some statistical measurements together with the mentioned results.

2 Minimization of CO_2 Emission for T-Shape Beam

2.1 Design Methodology

Design methodology of optimization application is comprised of minimizing the emission amount of CO_2 from the structural materials. Thus, firstly, optimization applications were arranged according to the principal parameters containing the

Table 1 The chronological representation of the mostly-used metaheuristics

Name of methods	Abbreviation	Year	Phenomena	Inventor	References
Simulated Annealing	SA	1983	The physical annealing process for any solid material	S. Kirkpatrick, C.D. Gelatt, and M.P. Vecchi	[1]
Particle Swarm Optimization	PSO	1995	Vital activities of creatures of community in the population such as escaping from predators, foraging, and arranging some environmental conditions like temperature	J. Kennedy and R. Eberhart	[2]
Ant Colony Optimization	ACO	1999	The feature of ants to find the shortest way for food source	M. Dorigo and G. Di Caro	[3]
Glowworm Swarm Optimization	GSO	2005	The capability of glowworms to glow at different intensities	K.N. Krishnanand and D. Ghose	[4]
Artificial Bee Colony Algorithm	ABC	2005	The foraging behavior of honey bees	D. Karaboga	[5]
Big Bang-Big Crunch	BB-BC	2006	Some events realized during the formation and development process of universe	O.K. Erol and I. Eksin	[6]
Invasive Weed Optimization	IWO	2006	The colonization strategy of invasive weeds	A.R. Mehrabian and C. Lucas	[7]
Firefly Algorithm	FA	2007	The flashing ability of fireflies affecting in the communication with the other ones together generation a protection mechanism across predators	X.S. Yang	[8]

(continued)

Table 1 (continued)

Name of methods	Abbreviation	Year	Phenomena	Inventor	References
Cuckoo Search	CS	2009	The reproduction strategy of cuckoo birds realizing via parasitism	X.S. Yang and S. Deb	[9]
Bat Algorithm	BA	2010	The echolocation behavior of bats, which benefits to perform about some activities such as detection of preys' location, avoidance from obstacles, etc.	X.S. Yang	[10]
Flower Pollination Algorithm	FPA	2012	The pollination process of flowery plants	X.S. Yang	[11]
Black Hole Algorithm	BH	2013	The formation of a celestial body known as black hole	A. Hatamlou	[12]
Gray Wolf Optimization	GWO	2014	The leadership hierarchy between gray wolfs living a swarm and the hunting strategy realized with this way	S. Mirjalili, S.M. Mirjalili, and A. Lewis	[13]
Colliding Bodies Optimization	CBO	2014'	The physics rules managing the unidimensional collision realizing between two objects, namely momentum and energy laws	A. Kaveh and V.R. Mahdavi	[14]

(continued)

Table 1 (continued)

Name of methods	Abbreviation	Year	Phenomena	Inventor	References
Shark Smell Optimization	SSO	2014	The attempt of sharks based on reaching to preys, namely the attempt to find the best solution	O. Abedinia, N. Amjady, and A. Ghasemi	[15]
Tree-Seed Algorithm	TSA	2015	An approach depending on propagation of trees in nature through seeds, which is a natural relationship between trees and seeds	M.S. Kiran	[16]
Hydrological Cycle Algorithm	HCA	2017	The motion of water continuously in nature	A. Wedyan, J. Whalley, and A. Narayanan	[17]
Golden Eagle Optimizer	GEO	2020	The hunting activities of golden eagles, which is a predatory bird, and some special strategies used in this process	A. Mohammadi-Balani, M.D. Nayeri, A. Azar, and M. Taghizadeh-Yazdi	[18]
Archimedes Optimization Algorithm	AOA	2021	A physics law as Archimedes' principle, which mimics the buoyant force	F.A. Hashim, K. Hussain, E.H. Houssein, M.S. Mabrouk and W. Al-Atabany	[19]
Jellyfish Search	JS	2021	The behavior of jellyfish in the ocean including some activities like following the ocean current, moving within a swarm, and controlling the changing mechanism between among these movements	J.S. Chou and D.N. Truong	[20]

design variables, constants, constraints of the structural model, and also algorithm requirements. In this respect, if the objective function is needed to express, it can be explained as minimizing the total amount of CO_2 spreading from concrete and steel reinforcements existing the inside of RC beam structure with T cross-section. Here, optimization processes were also conducted with the consideration of three different concrete grades for compressive strength as 25, 30, and 35 MPa in terms of recognizing the emission changing. While in these processes, structural beam designs were analyzed according to the rules and constraint requirements described in the regulation of TS500-Turkish Standard Requirements for Design and Construction of Reinforced Concrete Structures [29].

2.2 Structural Model

Structural model of beam contains a T-shaped cross-section besides that it also has two parts as span and cantilever, which are expressed via L_s and L_c, respectively. Moreover, values of the applied dead loads are 10 kN/m for span part (g_s) and 15 kN/m for cantilever part (g_c) together with live loads (q) have the same value for each beam part as 20 kN/m. The structural design of the beam was subjected to one force as a bending moment. The longitudinal section details of the beam can be seen in Fig. 1.

Also, Fig. 2 shows the details of T cross-section. Additionally, the design parameters as variables are comprised of the breadth of beam (b_w) together with the height of beam cross-section (h). Here, b expresses the effective breadth of the slab, and slab thickness is described as h_f, which reflect the constant parameters of design. The value of the concrete cover thickness (d') was handled as 60 mm, too.

As to the constraints expressed with the regulation requirements for design (TS500 Standard), g_1 limitation is considered and applied to beam structure that it is related with the breadth of beam slab (Eq. (1)). All of the values and some expressions arranged for design can be seen in Table 2.

$$g_1 = \frac{b - b_w}{2} \leq 6h_f \tag{1}$$

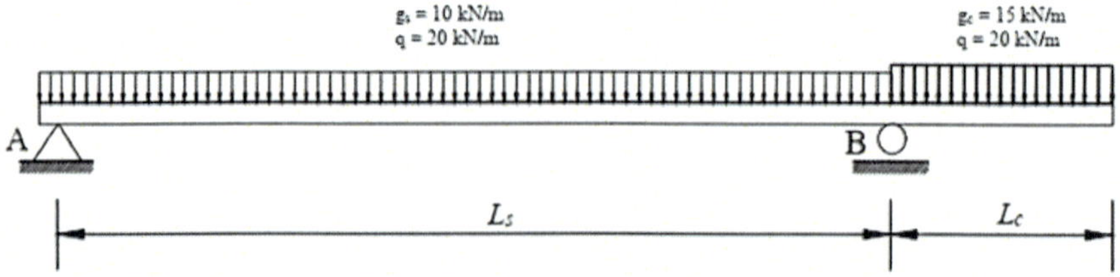

Fig. 1 Structural model and longitudinal section of the beam

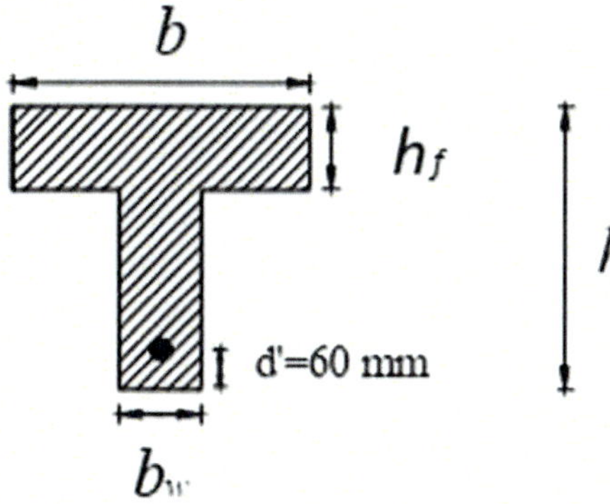

Fig. 2 Details of T-shape cross-sections for RC beam

In Table 2, unit amounts of CO_2 emission for concrete and steel materials are benefited from several scientific researches [30, 31], besides that unit costs for each concrete grades and steel reinforcement are determined from the guideline of Republic of Turkey Ministry of Environment and Urbanization [32] generated for the construction and installation processes. All of the costs were also converted to US dollar ($) with the usage of exchange rate occurred in 02.11.2021.

2.3 *Details of Optimization Process*

Mainly, the optimization process was fictionalized by considering the objective function as minimization of total CO_2 spreading from the structural materials. For this reason, the optimal designs of beam structures were provided by using the Eq. (2).

$$\text{Minimum } CO_2 = V_c \, CO_{2_{\text{concrete}}} + W_s \, CO_{2_{\text{steel}}} \tag{2}$$

Here V_c and W_s show the concrete volume and total weight of steel reinforcement; also, $CO_{2_{\text{concrete}}}$ and $CO_{2_{\text{steel}}}$ reflect the unit CO_2 emission for concrete and steel, respectively.

3 Metaheuristic Method: Flower Pollination Algorithm

Flower pollination algorithm (FPA) is one of the natural-inspired and population-based algorithms that was developed by X.S. Yang. FPA was proposed in the way of a natural behavior belonging to flowery plants known as pollination, which is realized with some pollinator creatures (such as animals like bee, fish, bat, or some out sources as water, wind, etc.). Besides, this process can be carried out in two different ways known as cross-pollination or self-pollination [10]. Self-pollination realizes within the self-structure of a specific flower. Cross-pollination is also applied

Table 2 Design properties and requirements for the minimization of CO_2 emission

	Property	Symbol	Ranges and values	Increment	Unit
Design parameters	Breadth of beam	b_w	200–500	–	mm
	Height of beam	h	3 h_f–600		
	Slab breadth	b	1250	–	mm
	Slab thickness	h_f	150		
	Yield strength of steel	f_{yk}	420	-	MPa
	Concrete compressive strength	f_{ck}	25–35	5	
	Design value for yield strength of steel	f_{yd}	$\frac{f_y}{1.15}$	–	
	Design value for compressive strength of concrete	f_{cd}	$\frac{f_{ck}}{1.5}$		
	Amount of live load	q	20	–	kN/m
	Amount of dead load for beam span	g_s	10		
	Amount of dead load for cantilever part of beam	g_c	15		
	Applied total load	P	1.4g + 1.6q		
	Span length	L_s	4		M
	Cantilever length	L_c	1.5		
	Concrete cover	d'	60		Mm
Design constants	Bending bar diameter	$\varnothing_{bendingbar}$	12–24	2	Mm
	Minimum value for reinforcement ratio	ρ_{min}	$0.8\frac{f_{ctd}}{f_{yd}}$	–	–
	Maximum value for reinforcement ratio	ρ_{max}	0.02		
	Weight per unit of volume of steel	γ_s	7850	–	kg/m^3
	Concrete unit cost	$U_{cost\,c}$	21.36 (25 MPa)	–	\$/m^3
			22.19 (30 MPa)		
			23.75 (35 MPa)		
	Steel unit cost	$U_{cost\,s}$	499.06 (8–24 ∅)		\$/ton
	Concrete unit CO_2 emission	CO_c^2	224.34 (25 MPa)		kg/m^3
			224.94 (30 MPa)		
			265.28 (35 MPa)		

(continued)

Table 2 (continued)

	Property	Symbol	Ranges and values	Increment	Unit
	Steel unit CO_2 emission	CO_s^2	3.01		kg/kg

between different flower members by pollinators, which fly with a distribution as *LÉvy* function (Eq. (3)). The function expression can be seen in Eq. (4).

$$X_{i,\text{new}} = \begin{cases} \text{if sp} > \text{rand}(),\ X_{i,j} + \text{Lévy}\left(X_{i,g_{\text{best}}} - X_{i,j}\right) & \text{Cross-pollination} \\ \text{else,} \qquad X_{i,j} + \text{rand}()\left(X_{i,m} - X_{i,k}\right) & \text{Self-pollination} \end{cases} \tag{3}$$

$$\text{Lévy} = \left(\frac{1}{\sqrt{2\pi}}\right)(\text{rand}())^{-1.5} e^{\left(-\frac{1}{2\,\text{rand}()}\right)} \tag{4}$$

where $X_{i,\text{new}}$ is the updated value of *j*th corresponds to *i*th design variable. $X_{i,j}$ is the value for this variable within the initial solution matrix. $X_{i,g_{\text{best}}}$ expresses the best candidate within all solutions in terms of the objective function as minimum CO_2 emission. Moreover, $X_{i,k}$ and $X_{i,m}$ reflect the two different *m*th and *k*th solution values, which possess to randomly selected candidates. As to *sp*, which is a parameter specific to FPA known as switch probability, and rand() is a random number generation function, too.

4 Numerical Applications

In this section, some numerical applications together with statistical calculations are explained in the way of optimization processes for beam models. For this respect, the minimized CO_2 emissions are detected with respect to each concrete compressive strength containing 25, 30, and 35 MPa. Also, optimum cost levels are reflected for them together with the most convenient beam sections as design parameters (h and b_w). Moreover, the mean value and standard deviation of emissions are detected for the usage of 25 independent optimization cycles (Table 3).

Also, the changing of emission and convergence to the minimum level for all grades are reflected via Fig. 3. Here, each of the graphs is reflected for the best iterations among whole cycles.

5 Conclusion

In the present study, the minimization of CO_2 emission for T-shaped RC beam was tried to realize. As the determined results, they were ensured that optimum beam

Table 3 Optimization evaluations for minimization of CO_2 emission in terms of 25 cycles

	Design properties	Expression	Grades of concrete		
			C25	C30	C35
Optimization results	Beam height	h	450		
	Beam breadth	b_w	200		
	Optimal level of total cost	$Cost_{opt}$	38.9299	40.0071	42.0944
	Minimum CO_2 emission	$MinCO_2$	364.8131	365.3122	420.0066
For 25 cycles	Mean of CO_2 emissions	$MeanCO_2$	364.8131	365.3122	420.0066
	Standard deviation of CO_2 emissions	$Std.Dev.ofCO_2$	5.57e–14	5.57e–14	0.00e+00

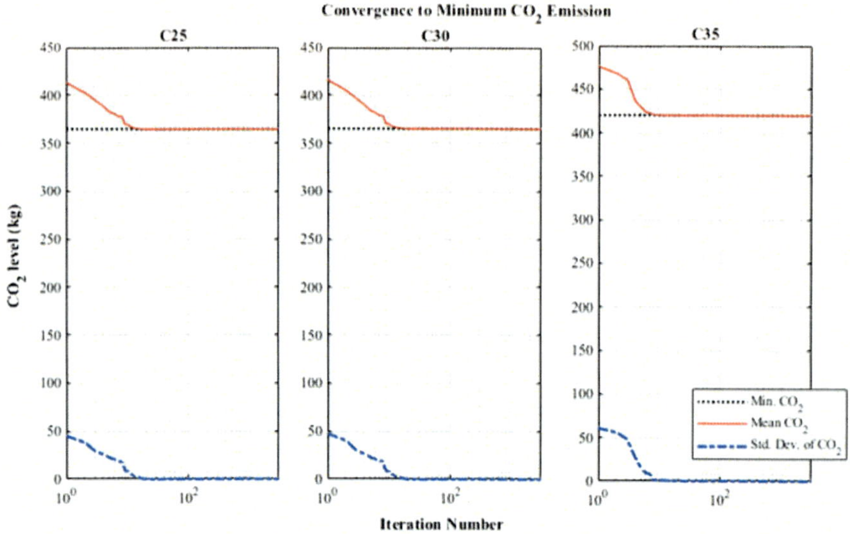

Fig. 3 Convergence to minimum CO_2 emission for all iterations provided in the best cycles

designs were generated for each grade of concrete compressive strengths. Also, the minimum amounts of CO_2 emission can be observed successfully due to that error values for them are extremely small in the way of usage of multiple cycles.

As a summarization, it can be seen that minimum CO_2 emission was found in terms of 25 MPa compressive strength. Additionally, the optimal level of the total material cost was also detected for this concrete grade. On the other hand, while the amplitude of concrete grade was increasing, the difference values between minimum emission and optimal cost level were also rising. For this reason, C25 concrete grade can be accepted as more eco-friendlier material strength in terms of minimum CO_2 emission.

Also, for the present processes, FPA can be accepted as an extremely efficient methodology to realize the minimization of emission amounts by deviating with pretty small error rates from the optimized point.

References

1. Kirkpatrick S, Gelatt CD, Vecchi MP (1983) Optimization by simulated annealing. Science 220(4598):671–680
2. Kennedy J, Eberhart R (1995) Particle swarm optimization. In: Proceedings of ICNN'95-International conference on neural networks. Perth-Australia, pp 1942–1948
3. Dorigo M, Di Caro G (1999) Ant colony optimization: a new meta-heuristic. In: Proceedings of the 1999 congress on evolutionary computation-CEC99 (Cat. No. 99TH8406). IEEE, Wahington DC-USA, pp 1470–1477
4. Krishnanand KN, Ghose D (2009) Detection of multiple source locations using a glowworm metaphor with applications to collective robotics. In: Proceedings 2005 IEEE Swarm intelligence symposiumm (SIS 2005). IEEE, California-USA, pp 84–91
5. Karaboga D (2005) An idea based on honey bee swarm for numerical optimization. Technical Report-tr06, Computer Engineering Department, Engineering Faculty, Erciyes University, Talas, Turkey, pp 1–10
6. Erol OK, Eksin I (2006) A new optimization method: big bang–big crunch. Adv Eng Softw 37(2):106–111
7. Mehrabian AR, Lucas C (2006) A novel numerical optimization algorithm inspired from weed colonization. Eco Inf 1(4):355–366
8. Yang XS (2010) Nature-inspired metaheuristic algorithms, 2nd edn. Luniver Press, United Kingdom
9. Yang XS, Deb S (2009) Cuckoo search via lévy flights. In: World congress on nature & biologically inspired computing (NABIC 2009). IEEE Publications, India, pp 210–214
10. Yang XS (2010) A new metaheuristic bat-inspired algorithm. In: Nature inspired cooperative strategies for optimization (NICSO 2010). Springer, Berlin, Heidelberg, pp 65–74
11. Yang XS (2012) Flower pollination algorithm for global optimization. In: International conference on unconventional computing and natural computation. Springer, Berlin, Heidelberg, pp 240–249
12. Hatamlou A (2013) Black hole: a new heuristic optimization approach for data clustering. Inf Sci 222:175–184
13. Mirjalili S, Mirjalili SM, Lewis A (2014) Grey wolf optimizer. Adv Eng Softw 69:46–61
14. Kaveh A, Mahdavi VR (2014) Colliding bodies optimization: a novel meta-heuristic method. Comput Struct 139:18–27
15. Abedinia O, Amjady N, Ghasemi A (2014) A new metaheuristic algorithm based on shark smell optimization. Complexity 21(5):97–116
16. Kiran MS (2015) TSA: tree-seed algorithm for continuous optimization. Expert Syst Appl 42(19):6686–6698
17. Wedyan A, Whalley J, Narayanan A (2017) Hydrological cycle algorithm for continuous optimization problems. J Optim, 1–25
18. Mohammadi-Balani A, Nayeri MD, Azar A, Taghizadeh-Yazdi M (2020) Golden eagle optimizer: a nature-inspired metaheuristic algorithm. Comput Ind Eng 152:107050
19. Hashim FA, Hussain K, Houssein EH, Mabrouk MS, Al-Atabany W (2021) Archimedes optimization algorithm: a new metaheuristic algorithm for solving optimization problems. Appl Intell 51(3):1531–1551

20. Chou JS, Truong DN (2021) A novel metaheuristic optimizer inspired by behavior of jellyfish in ocean. Appl Math Comput 389:125535
21. Yucel M, Bekdas G, Nigdeli SM, Sevgen S (2018) Artificial neural network model for optimum design of tubular columns. Int J Theor Appl Mech 3:82–86
22. Kayabekir AE, Nigdeli SM (eds) (2020) Metaheuristic approaches for optimum design of reinforced concrete structures: emerging research and opportunities: emerging research and opportunities. IGI Global
23. Mergos PE, Mantoglou F (2020) Optimum design of reinforced concrete retaining walls with the flower pollination algorithm. Struct Multidiscip Optim 61(2):575–585
24. Mevada A, Patel V, Arekar V (2021) Cost optimization of cantilever retaining wall using flower pollination algorithm. Int J Adv Res Sci Commun Technol (IJARSCT) 6(1):915–930
25. Yücel M, Bekdaş G, Nigdeli SM, Kayabekir AE (2021) An artificial intelligence-based prediction model for optimum design variables of reinforced concrete retaining walls. Int J Geomech 21(12):04021244
26. Nigdeli SM, Bekdaş G (2019) Optimum design of multiple positioned tuned mass dampers for structures constrained with axial force capacity. Struct Design Tall Spec Build 28(5):e1593
27. Yucel M, Bekdaş G, Nigdeli SM, Sevgen S (2019) Estimation of optimum tuned mass damper parameters via machine learning. J Build Eng 26:100847
28. da Silva CAX, Taketa E, Koroishi EH, Lara-Molina FA, Faria AW, Lobato FS (2020) Determining the parameters of active modal control in a composite beam using multi-objective optimization flower pollination. J Vibr Eng Technol 8(2):307–317
29. TS500: Turkish Standard Requirements for Design and Construction of Reinforced Concrete Structures. Ankara-Turkey (2020)
30. Paya-Zaforteza I, Yepes V, Hospitaler A, Gonzalez-Vidosa F (2009) CO_2-optimization of reinforced concrete frames by simulated annealing. Eng Struct 31(7):1501–1508
31. Yepes V, Gonzalez-Vidosa F, Alcala J, Villalba P (2012) CO_2-optimization design of reinforced concrete retaining walls based on a VNS-threshold acceptance strategy. J Comput Civ Eng 26(3):378–386
32. Republic of Turkey Ministry of Environment and Urbanization, Directorate of Higher Technical Board, 2021 Construction and Installation Unit Prices. https://webdosya.csb.gov.tr/db/yfk/icerikler//bf2021-turkce-20210129113217.pdf. Last accessed 31 Oct 2021

Prediction of Minimum CO_2 Emission for Rectangular Shape Reinforced Concrete (RC) Beam

Melda Yücel, Gebrail Bekdaş, and Sinan Melih Nigdeli

Abstract Machine learning methodologies help to directly determine the desired design parameters of any structural design problem. In this regard, with the present study, a simply supported reinforced concrete (RC) beam with rectangular cross-section was handled for the prediction of the design outcomes as minimum carbon dioxide (CO_2) emission and optimal cost level of structural materials. In this respect, firstly, an optimization process was performed with a well-known metaheuristic algorithm, and then a prediction application was realized via artificial neural networks (ANNs) to detect the mentioned parameters. Also, different concrete compression strengths, beam lengths, and concrete cover thicknesses were benefited to generate a dataset for training the ANNs. Finally, to validate the model's success, test data was created and evaluated in comparison with optimal results. Thus, with these processes, both generating of eco-friendly, cost-effective, and optimal designs were made possible to provide, and rapid, effective, and reliable decision-making applications were presented to determine the parameters in terms of the mentioned structural designs.

Keywords Reinforced concrete (RC) structures · CO_2 emission · Eco-friendly design · Machine learning · Artificial neural networks

M. Yücel (✉) · G. Bekdaş · S. M. Nigdeli
Department of Civil Engineering, Istanbul University-Cerrahpaşa, 34320 Avcılar, Istanbul, Turkey
e-mail: melda.yucel@ogr.iu.edu.tr

G. Bekdaş
e-mail: bekdas@iuc.edu.tr

S. M. Nigdeli
e-mail: melihnig@iuc.edu.tr

J. H. Kim et al. (eds.), *Proceedings of 7th International Conference on Harmony Search, Soft Computing and Applications*, Lecture Notes on Data Engineering and Communications Technologies 140, https://doi.org/10.1007/978-981-19-2948-9_14

1 Introduction

Structural engineering is an engineering discipline that contains many design problems within itself. To be able to resolve these problems, analysis processes required that they need a long time. This case causes both cost loss and also endeavoring the redundant effort and labor force. However, about the generation of any structural design conveniently, some significant objectives should exist, and each one should be provided in simultaneous time. These objectives should be both that structure carries the conditions of strength, safety, serviceability, etc., and following a cost and energy effective way. While these kinds of conditions are also provided via the methods called optimization, the projected most proper design can be created. But, this should be known that when the analysis of large-scale structural designs is tried to realize with the mentioned methodologies, they can take days even weeks. In this issue, especially nowadays, machine learning methods, which are frequently preferred in all kinds of fields, help to remove such problems. In this respect, various studies are realized about structural engineering and structural design, and following ones can be given as examples that are the creation of the most economic, the most resisted or the lightest structure design by determining the most convenient/optimum section sizes for any structural member [1–4], the forecasting minimum usage of energy in the context of generating eco-friendly structural designs [5–7], creating of the control mechanism, which can work in the most correct way to control vibrations occurred during a dynamic motion (such earthquake), which can cause to damage of structure [8–10], the directly prediction of some parameters such characteristic strength, ingredient properties of concrete which is the most principal material of structural design [11–13].

In this study, to directly determine the minimum CO_2 emissions together with optimal cost levels for beam designs, two separate phases were performed as optimization and machine learning process, respectively. In this scope, a metaheuristic algorithm, which is known as flower pollination algorithm (FPA) was utilized to carry out the optimization applications, and then, artificial neural networks (ANNs) were benefited to predict the desired parameters as outputs. Also, to validate the success and performance of the prediction model, a test dataset containing different cases for beam structure was created and evaluated, besides that the predictions were compared with optimum results considering some error metrics. Thereby, an intelligence predictive model can be created to determine the structural parameters of an RC beam, and this model can be used for designing the structures rapidly and reliable way.

2 Prediction Methodology for Minimum CO_2 Emission

2.1 Optimization Process

An optimization process was conducted to realize the minimizing CO_2 emission for RC beams with rectangular sections. During the applications, a well-known meta-heuristic as flower pollination algorithm (FPA), which was developed by Yang [14], was considered. Also, four concrete compressive strengths (25, 30, 35, and 40 MPa) were utilized for generating a dataset in terms of the usage in prediction processes. On the other side, while all of the optimization cases were carried out, beam structures were designed with the consideration of the rules in TS500-Turkish Standard Requirements for Design and Construction of Reinforced Concrete Structures regulation [15]. Additionally, the main structure is illustrated with Fig. 1 that it is a simply supported beam, which has one span (L_s). The applied loads are also symbolized via g and q for dead and live loads, respectively. The beam model is subjected to both bending moment and shear force, too.

In Fig. 2, the cross-section can be also seen for the model with the breadth (b_w) and height of beam (h). Here, d and d' express the effective beam height and concrete cover thickness, respectively.

In Table 1, the design parameters of the optimization process are shown. The unit CO_2 emissions [16, 17] and costs [18] are also reflected that cost amounts were changed according to the ratio of US dollar within the date of 02.11.2021.

The main objective function of the applications can be defined with Eq. (1). This expression meant to minimization of CO_2 emission occurred with the usage of structural materials.

$$\text{Min}\,CO_2 = V_{\text{concrete}}\,CO_2^{\text{concrete}} + W_{\text{steel}}\,CO_2^{\text{steel}} \tag{1}$$

where V_{concrete} is total volume of used concrete besides that W_{steel} defines the steel weight. CO_2^{concrete} and CO_2^{steel} also symbolize the unit values of CO_2 emission in terms of concrete and steel materials, respectively.

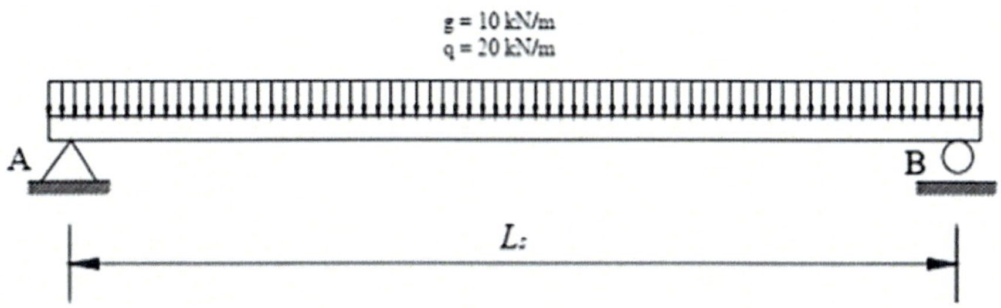

Fig. 1 Structural model and longitudinal section of beam

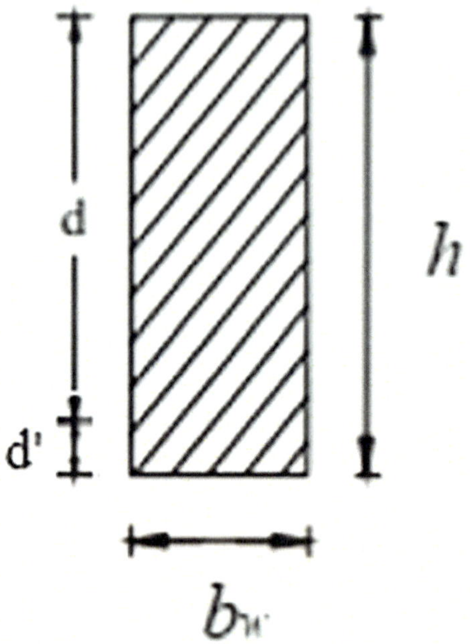

Fig. 2 Cross-section details of RC beam

2.2 *Details of Prediction Model*

For the prediction process, which was operated via ANNs, the main dataset was generated for training the related prediction model. In this scope, some design parameters expressed as constants were handled as variable values to observe multiple optimum results for beam design by using an application within MATLAB R2018a [19]. According to these statements, all of the parameters and their properties can be seen in Table 2.

When Table 2 was considered, a total of 484 different optimization cases were benefited to create a prediction model based on machine learning methodology.

3 Numerical Applications

In Table 3, calculated values of errors in terms of several metrics were represented for the training process of the main model. Also, in Fig. 3, correlation and fitting of predicted results can be seen according to actual data as optimum values, which were provided via FPA.

As the consideration of the greatness of correlations and smallness of error values, it can be expressed that the main prediction model can be applied for any mentioned beam design to determine the structural properties directly. For this respect, a test dataset was generated with different input values, and the determined optimum results

Table 1 Design parameters and rules utilized in optimization

	Property	Notation	Ranges/value	Increment	Unit
Design variables	Cross-section breadth	b_w	300–750	–	mm
	Cross-section height	h	250–600		
Design constants	Yield strength of steel bending bar	f_{yk}	420	–	MPa
	Yield strength of steel stirrup	f_{ywk}			
	Compressive strength of concrete	f_{ck}	25–40	5	
	Tensile strength of concrete	f_{ctk}	$0.35\sqrt{f_{ck}}$	–	
	Design yield strength of steel bending bar	f_{yd}	$\frac{f_{yk}}{1.15}$		
	Design yield strength of stirrup	f_{ywd}	$\frac{f_{ywk}}{1.15}$		
	Design compressive strength of concrete	f_{cd}	$\frac{f_{ck}}{1.5}$		
	Design tensile strength of concrete	f_{ctd}	$\frac{f_{ctk}}{1.5}$		
	Live load	q	20	–	kN/m
	Dead load	g	10		
	Applied load value	P	$1.4g + 1.6q$		
	Diameter of reinforcement bars	$\varnothing_{\text{bending bar}}$	12–24	2	mm
		$\varnothing_{\text{stirrup}}$	8–12		
	Minimum value of reinforcement ratio	ρ_{min}	$0.8\frac{f_{ctd}}{f_{yd}}$	–	–

(continued)

Table 1 (continued)

	Property	Notation	Ranges/value	Increment	Unit
	Maximum value of reinforcement ratio	$\rho_{\max}$	0.02		
	Minimum value of reinforcement ratio for stirrup	$\rho_{w,\min}$	$0.3\frac{f_{ctd}}{f_{ywd}}$		
	Maximum spacing between stirrups (confinement zone)	$s_{c\max}$	$\min\left(\frac{h}{4}, 150, 8\varnothing_{\text{longitudinal bar}}\right)$	–	mm
	Minimum spacing between stirrups (confinement zone)	$s_{c\min}$	50		
	Maximum stirrup spacing (middle zone)	$s_{m\max}$	$\min\left(\frac{h}{2}, 200\right)$		
	Minimum stirrup spacing (middle zone)	$s_{m\min}$	100		
	Distance of first stirrup to column-face (confinement zone)	s_{initial}	50		
	Weight per unit of volume of steel	γ_s	7850	–	kg/m^3
	Unit concrete cost	U_{concrete}	21.36 (25 MPa)	–	$\$/m^3$
			22.19 (30 MPa)		
			23.75 (35 MPa)		
			25.11 (40 MPa)		
	Unit steel cost	U_{steel}	680.87 (8–24 ∅)	–	\$/ton
	Unit value of CO_2 emission for concrete	CO_2^{concrete}	224.34 (25 MPa)	–	kg/m^3
			224.94 (30 MPa)		
			265.28 (35 MPa)		
			265.28 (40 MPa)		

(continued)

Table 1 (continued)

	Property	Notation	Ranges/value	Increment	Unit
	Unit value of CO_2 emission for steel	CO_2^{steel}	3.01	–	kg/kg

Table 2 Parameters and properties of the main prediction model

Expression	Property	Notation	Ranges	Increment	Unit
Concrete compressive strength	Inputs	f_{ck}	25–40	5	MPa
Span length of beam		L_s	4–6	0.2	m
Thickness of concrete cover		d'	20–30	1	mm
Optimum cost level	Outputs	Opt. cost	–	–	$
Amount of the minimum CO_2		Min CO_2			kg

Table 3 Error measurements for main prediction model to actual data

Error metric	Opt. cost	Min CO_2
Mean absolute error (MAE)	0.3188	0.7387
Mean absolute percentage error (MAPE)	0.6062	0.2007
Mean square error (MSE)	0.1653	0.8783
Root mean square error (RMSE)	0.4066	0.9372

for the model can be seen in Table 4. Additionally, the same error metrics were calculated for the test model in terms of each design couple and presented in Tables 5 and 6 for optimal cost level and minimum CO_2 emission, respectively.

4 Conclusion

In the current study, the prediction of minimum CO_2 emission together with optimal cost level can be tried to find directly and rapidly via ANNs. In this respect, firstly, an optimization process was realized with different structural design couples. According to results determined for the mentioned designs, which generate the main training and prediction model, it can be said that the process was accepted extremely effective application to utilize for the determination of the desired results. The correlation of the main training model according to actual optimum data (over % 99) besides error measurements between actual and predicted values (smaller than 0.94) was detected quite low. ANNs are successful to determine the desired values for the optimization process.

On the other side, all of the same performance measurements were observed for the generated test model. According to obtained results, each error rate was

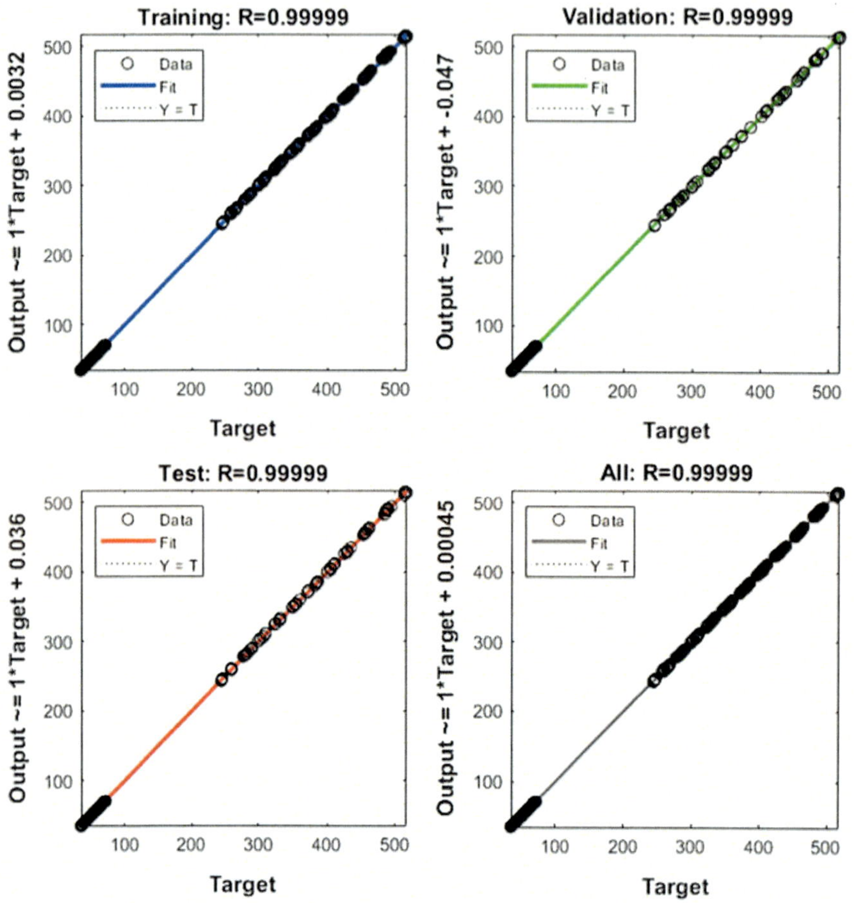

Fig. 3 Correlation results for the training process of the main model

Table 4 Optimization results for test model

Inputs			Outputs	
f_{ck}	L_s	d'	Opt. cost	Min CO_2
25	4.50	22	42.0160	297.3596
30	3.85	26	32.6020	231.7089
35	5.00	18	50.7935	371.3371
25	6.10	30	70.6629	508.4170
45	4.40	32	41.9741	301.5314
40	5.20	20	55.0025	395.5196

Table 5 Comparisons of actual and predicted values for optimal cost in terms of test model

Prediction of ANNs	Error rates for FPA			
Opt. cost	Error	Absolute error	Absolute percentage (%) error	Squared error
41.7568	0.2592	0.2592	0.6169	0.0672
32.7329	– 0.1309	0.1309	0.4015	0.0171
51.0986	– 0.3051	0.3051	0.6007	0.0931
70.9241	– 0.2612	0.2612	0.3696	0.0682
42.1924	– 0.2183	0.2183	0.5202	0.0477
54.9431	0.0595	0.0595	0.1081	0.1081
		MAE	**MAPE**	**MSE**
Mean		0.2057	0.4362	0.0495
RMSE			0.0224	

Table 6 Comparisons of actual and predicted values for minimum CO_2 emission in terms of the test model

Prediction of ANNs	Error rates for FPA			
Min CO_2	Error	Absolute error	Absolute percentage (%) error	Squared error
298.4650	– 1.1054	1.1054	0.3717	1.2220
230.4475	1.2614	1.2614	0.5444	1.5912
370.8551	0.4820	0.4820	0.1298	0.2323
507.7225	0.6944	0.6944	0.1366	0.4822
300.0183	1.5131	1.5131	0.5018	2.2895
396.1636	– 0.6440	0.6440	0.1628	0.4148
		MAE	**MAPE**	**MSE**
Mean		0.9501	0.3079	1.0387
RMSE			1.0191	

determined small for each output parameter. While the test data was investigated, it can be recognized that several design couples exceed the main data limitations for the input parameter.

In this respect, according to all of the error rates and correlation measurements, both output parameters can be predicted via the created prediction model by ANNs. Also, this model can be accepted as really effective to convergence to actual data for different test couples, which did not exist within the main training model and were not used for the prediction process. As a summarization, to detect the objective values and generate the eco-friendly, clean besides cost-effective designs rapidly, the prediction structure generated with ANNs can be benefited for any RC beam with a rectangular section as in the current study.

References

1. Khalilzade Vahidi E, Rahimi F (2016) Investigation of ultimate shear capacity of RC deep beams with opening using artificial neural networks. Adv Comput Sci Int J 5(4):57–65
2. Pham AD, Ngo NT, Nguyen TK (2020) Machine learning for predicting long-term deflections in reinforce concrete flexural structures. J Comput Design Eng 7(1):95–106
3. Bekdaş G, Yucel M, Nigdeli SM (2021) Evaluation of metaheuristic-based methods for optimization of truss structures via various algorithms and Lèvy flight modification. Buildings 11(2):49
4. Yücel M, Bekdaş G, Nigdeli SM, Kayabekir AE (2021) An artificial intelligence-based prediction model for optimum design variables of reinforced concrete retaining walls. Int J Geomech 21(12):04021244
5. Yücel M, Namli E (2018) Yapay zekâ modelleri ile betonarme yapılara ait enerji performans sınıflarının tahmini. Uludağ Univ J Faculty Eng 22(3):325–346
6. Liu T, Tan Z, Xu C, Chen H, Li Z (2020) Study on deep reinforcement learning techniques for building energy consumption forecasting. Energy Build 208:109675
7. Nabavi SA, Aslani A, Zaidan MA, Zandi M, Mohammadi S, Hossein Motlagh N (2020) Machine learning modeling for energy consumption of residential and commercial sectors. Energies 13(19):5171
8. Yucel M, Öncü-Davas S, Nigdeli SM, Bekdas G, Sevgen S (2018) Estimating of analysis results for structures with linear base isolation systems using artificial neural network model. Int J Control Syst Robot 3:50–56
9. Moeindarbari H, Taghikhany T (2018) Novel procedure for reliability-based cost optimization of seismically isolated structures for the protection of critical equipment: a case study using single curved surface sliders. Struct Control Health Monit 25(1):e2054
10. Yucel M, Bekdaş G, Nigdeli SM, Sevgen S (2019) Estimation of optimum tuned mass damper parameters via machine learning. J Build Eng 26:100847
11. Yaman MA, Abd Elaty M, Taman M (2017) Predicting the ingredients of self compacting concrete using artificial neural network. Alex Eng J 56(4):523–532
12. Feng DC, Liu ZT, Wang XD, Chen Y, Chang JQ, Wei DF, Jiang ZM (2020) Machine learning-based compressive strength prediction for concrete: An adaptive boosting approach. Constr Build Mater 230:117000
13. Ahmad A, Ostrowski KA, Maślak M, Farooq F, Mehmood I, Nafees A (2021) Comparative study of supervised machine learning algorithms for predicting the compressive strength of concrete at high temperature. Materials 14(15):4222
14. Yang XS (2012) Flower pollination algorithm for global optimization. In: International conference on unconventional computing and natural computation. Springer, Berlin, Heidelberg, pp 240–249
15. TS500: Turkish Standard Requirements for Design and Construction of Reinforced Concrete Structures. Ankara-Turkey (2020)
16. Paya-Zaforteza I, Yepes V, Hospitaler A, Gonzalez-Vidosa F (2009) CO_2-optimization of reinforced concrete frames by simulated annealing. Eng Struct 31(7):1501–1508
17. Yepes V, Gonzalez-Vidosa F, Alcala J, Villalba P (2012) CO_2-optimization design of reinforced concrete retaining walls based on a VNS-threshold acceptance strategy. J Comput Civ Eng 26(3):378–386
18. Republic of Turkey Ministry of Environment and Urbanization, Directorate of Higher Technical Board, 2021 Construction and Installation Unit Prices. https://webdosya.csb.gov.tr/db/yfk/icerikler//bf2021-turkce-20210129113217.pdf. Last accessed 31 Oct 2021
19. MATLAB Mathworks, Matlab 2018a. Neural Net Fitting. https://www.mathworks.com/help/deeplearning/ref/neuralnetfitting-app.html. Last accessed 01 Nov 2021

Metaheuristics Applied to Pattern-Based Portuguese Relation Extraction

Luiz Felipe Manke and Leandro dos Santos Coelho

Abstract Relation extraction is an important part of the natural language processing field that has been receiving increasing attention due to the massive growth of the information available on the web, which makes its tasks impossible through manual means. Although for domain-specific relation extraction tasks, pattern-based methods have a long and established history as a successful approach, and they can suffer from precision and recall problems and may require a lot of manual effort. To work around these issues, this paper proposes the application of well-known metaheuristics to select patterns that maximize performance metric. This approach was applied to a binary sentence-level relation extraction problem in Portuguese language, and the results were compared using statistical tests and $F1$ score, reaching a significant value of 0.67 with the harmony search algorithm. The other algorithms evaluated are genetic algorithm and simulated annealing.

Keywords Metaheuristics · Relation extraction · Natural language processing · Harmony search

1 Introduction

Relation extraction (RE) is an important part of the natural language processing (NLP) field that aims to automatically identify semantic associations (relations) between named entities in text. This field has been receiving increasing attention

L. F. Manke (✉)
Neoway Business Solutions, Florianopolis, Santa Catarina, Brazil
e-mail: luizmanke@gmail.com

L. F. Manke · L. dos Santos Coelho
Department of Electrical Engineering, Federal University of Parana (UFPR), Curitiba, Parana, Brazil

L. dos Santos Coelho
Industrial and Systems Engineering Graduate Program (PPGEPS), Pontifical Catholic University of Parana (PUCPR), Curitiba, Parana, Brazil

J. H. Kim et al. (eds.), *Proceedings of 7th International Conference on Harmony Search, Soft Computing and Applications*, Lecture Notes on Data Engineering and Communications Technologies 140, https://doi.org/10.1007/978-981-19-2948-9_15

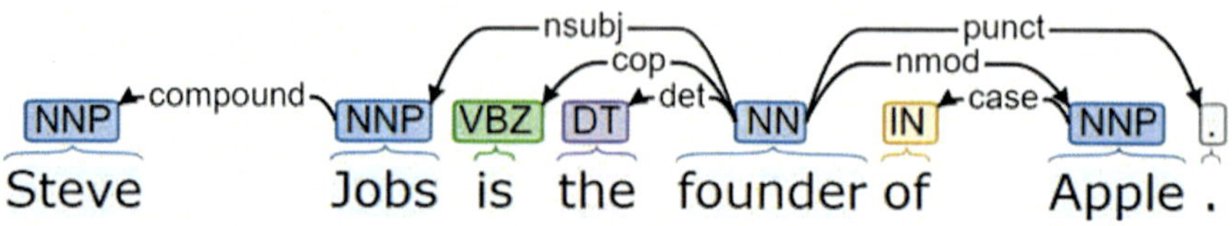

Fig. 1 Dependency parse and part of speech example

due to the massive growth of the information available on the web, which makes RE tasks impossible through manual means.

For domain-specific RE, pattern-based methods have a well established successful history [1, 2]. Patterns are rules that define a specific relation, where the relation is assigned to the entities in a sentence if this sentence matches at least one of the rules. For example, the pattern "*is the founder of*" would be useful to extract the CEO-COMPANY relation in the sentence "Steve Jobs *is the founder of* Apple."

Although patterns are useful for extracting relations, it can suffer from two distinct problems: (i) Recall problem, where only a few selected patterns are able to find newer sentences, failing extract a large number of sentences with relations; and (ii) precision problem, where a large number of extracted relations are incorrect, due to the patterns not being reliable enough [3].

This paper proposes a framework to automatically select the best set of patterns that balance out precision and recall using metaheuristics, which are higher-level algorithms designed to find a heuristic that may provide a sufficiently good solution to an optimization problem. To evaluate this framework, the metaheuristics are applied to a benchmark containing relation extraction samples in Portuguese language. They are then optimized in terms of $F1$ score and compared using the statistical Wilcoxon Signed-rank test.

The remainder of this article is organized as follows. Section 2 presents pattern relation extraction and its related works. Section 3 details the metaheuristic algorithms used in this paper. Section 4 describes the Portuguese benchmark and the performance metrics used. Section 5 explains the application of metaheuristics to pattern-based RE. The results are discussed in Sect. 6. And Sect. 7 is the conclusion of this work.

2 Pattern-Based Relation Extraction

Patterns were the initial form of structure used to extract relations from text. They are sets of rules used as conditional statements to judge whether or not entities within a sentence belong to a specific relation. Depending on the application, this approach can require a lot of manual work and domain knowledge to create a useful set of patterns.

The simplest form of pattern is the word matching rule, which only looks for exact keyword matching in sentences. In Ravichandran and Hovy [2], a question answering system was created using an optimal set of matching word patterns that were automatically extracted from the Internet, based only on a few handcrafted examples.

To increase generalization, NLP tools can be incorporated into the patterns, such as dependency parsers and part of speech (POS) taggers, illustrated in Fig. 1. The former is the process to analyze the grammatical structure in a sentence and find out related words, while the latter tags all the words in a sentence as to their part of speech, e.g., noun, verb, and adjective. Jijkoun et al. [4] used dependency parse methods in a question answering system to correctly answer more questions and improve the system's recall. Wu and Weld [5] also improved recall in an open extraction system using dependency parse and POS features. In Portuguese language, Ferreira et al. [6] created a relation extraction system based solely on POS patterns from human-annotated examples.

In Boujelben et al. [7], a combination of pattern relation extraction with a genetic algorithm to discover more interesting rules for a given benchmark was proposed. The patterns were formed by the named entities (NE) tags, the POS of words surrounding the NEs, and the number of words before, after, and between the NEs. The crossover was performed by combining the elements of different rules, and the mutation consisted in excluding one constraint of a given rule. Compared to previous works, the proposed method increased the score by 8% in terms of precision and recall.

3 Metaheuristics

Metaheuristics are higher-level algorithms designed to find a heuristic that may provide a sufficiently good solution to an optimization problem. It samples a subset of solutions which is otherwise too large to be completely explored [8].

3.1 Simulated Annealing

Simulated annealing (SA) is a single-solution method inspired by the physical process of heating a material and then slowly lowering its temperature reduce the system energy and decrease defects. Similar techniques have been independently introduced on several occasions [9–11], where the idea of slow cooling is implemented by slowly decreasing the probability of accepting worse solutions as the solution space is explored, reducing the extent of its search.

3.2 Genetic Algorithm

Genetic algorithm (GA) is a well-known algorithm, which is inspired from biological evolution process. It was proposed by Holland [12] and mimics the Darwinian theory, reflecting the process of natural selection, where the fittest individuals of a population are selected for reproduction in order to produce offspring of the next generation.

3.3 Harmony Search

Harmony search (HS) was presented by Geem et al. [13]. It is an algorithm inspired by the process of music harmony improvisation, while musicians search for the most beautiful music harmony as specified by the standard tunes; this algorithm searches for the global solution as specified by an objective function. This optimization process is based on a population, where each new population is formed by new solutions coming from the possible boundaries, or improved solutions coming from the past population.

4 Resources and Metrics

4.1 Benchmark

This paper makes use of the Portuguese benchmark generated by Batista et al. [14]. It uses a distant supervision methodology to combine the diversified texts from Wikipedia with the structured content from DBpedia to form a massive relation extraction dataset.

Wikipedia is a free online encyclopedia, created and edited by volunteers around the world, that has a vast amount of diversified content. DBpedia is a project aiming to extract structured content from the information created in Wikipedia and store it as *relation(entity_1, entity_2)*, e.g., *founderOf(Steve Jobs, Apple)*.

By extracting sentences from the Wikipedia texts that contain two entities, one can assume that it expresses the same relation found in DBpedia for these entities. Although this kind of process certainly contains noises, since sentences containing both entities may not express the relation stored in DBpedia, the dataset is still valuable given its large amount of relations extracted.

The total number of sentences extracted with this methodology was 97,988 and the relations were aggregated in 10 different groups, including *locatedInArea*, *origin* and *notRelated*. After the automatic creation of the benchmark, a few samples were manually checked, concluding that the precision of the data was around 80%.

4.2 *Performance Metrics*

Shapiro-Wilk Test The Shapiro-Wilk test examines if a variable is normally distributed in some population. It was introduced by Shapiro and Wilk [15]. This test can also be used to assure a safer use of parametric tests.
Wilcoxon Signed-Rank Test The Wilcoxon Signed-rank test is a statistical hypothesis test, proposed by Wilcoxon [16], to compare the locations of two populations using a set of matched samples. It is a nonparametric test, which does not assume that the differences between paired samples are normally distributed.

5 Application

In this section, the application of metaheuristics to the supervised binary sentence-level relation extraction benchmark described previously is proposed. The goal is to select the best relation extraction patterns to improve the system performance in detecting whether or not entities are related to each other in a sentence.

Since Batista's benchmark has 10 different relations, it must be adapted to a binary output. To do that, samples tagged as *notRelated* will be assigned as one group, while all the other tags will be assigned as another group.

Another benchmark modification is the random under-sampling of the data. Since the total number of samples reaches more than 90 thousand, it must be done to speed up the evaluation process of multiple runs of the algorithms. The total number of samples used is 2000, where 70% were used for training and 30% for evaluation.

To create the initial patterns for the algorithms, we take advantage of the existing positive training examples. The patterns are composed of the part of speech tags of the words between entities. If any sentence matches a pattern, the entities are considered related in that sentence.

Each solution to our problem consists of a group of patterns. To select the best group, we propose the use of metaheuristics. Although each algorithm has different strategies to find the best combination of patterns, all of them have the intention of gradually relax its constraints, increasing generalization. This is done by randomly toggling one of the pattern's tags, i.e., accepting any POS tag, or restoring to the original POS tag. This way, the examples with relations that were already detected by the pattern will continue to, while other similar positive relations can be found as well.

To optimize our problem, the objective function of the algorithm used is defined as the $F1$ score of the whole training set for a group of patterns. $F1$ score is a popular metric for RE tasks evaluations. It is the harmonic mean of the precision and recall, which means that having either a low score in precision or recall will result in a low $F1$ score.

The complete list of hyperparameters used in this application is presented in Table 1, which were chosen empirically. In the SA algorithm, *TempInit* (initial temperature)

Table 1 Metaheuristics hyperparameters settings

SA	GA	HS
TempInit = 100	NumElite = 10	HMCR = 0.9
TempDecay = 0.99	MutationRate = 0.05	PAR = 0.5
StepChance = 0.3		

and *TempDecay* (temperature decay) are responsible for randomly accepting worse solutions, and *StepChance* is the probability of pattern generalization. In the GA, *NumElite* is the number of the most fittest individuals that won't be mutated between generations, and *MutationRate* is the probability of pattern generalization. Finally, in HS, *HMCR* is the probability of selecting a solution from the current population, while *PAR* is the chance of pattern generalization.

6 Results Analysis

Three approaches were applied to our problem: simulated annealing, genetic algorithm, and harmony search. For each one, 100 executions were performed using 500 generations with 50 solutions, being a solution composed of 15 relation extraction patterns. It means that, in every generation, 50 possible solutions are available, and during evaluation, entities in a sentence are assigned as *related* if their pattern matches at least one of the 15 patterns in the solution. In this section, the results of this application are discussed.

The algorithms were optimized using the training set, composed of 1400 samples. Figure 2 shows the convergence plot for each algorithm. It was calculated by taking the mean of the best $F1$ score of each run for every generation. Since they have different optimization strategies, they are expected to vary as well. For instance, the single-solution SA algorithm seems to get stuck in a local minimum, while the population-based algorithms, GA and HS, seem able to explore a greater search space in early generations before fine-tuning the solutions later on.

In Fig. 3, an analysis of the diversity of the solutions was performed during the optimization. The box plots were computed using the $F1$ scores of the solutions of every run. Here, different patterns can be seen. Although HS has a small diversity, its strategy is always to replace the worse solution in every iteration, gradually improving the solutions. In the other hand, GA has a larger diversity, which is important to create better solutions in the offspring. Lastly, the single-solution SA has a constant wide range of solutions, but the solutions do not improve as fast as the others.

Figure 4 shows the distribution of the best $F1$ score reached during the training process for the 100 runs. The normality of the sample distribution is checked using the Shapiro-Wilk test. Table 2 shows both the W statistics and the p-values computed. The null hypothesis for this test is that the data is normally distributed. The SA has

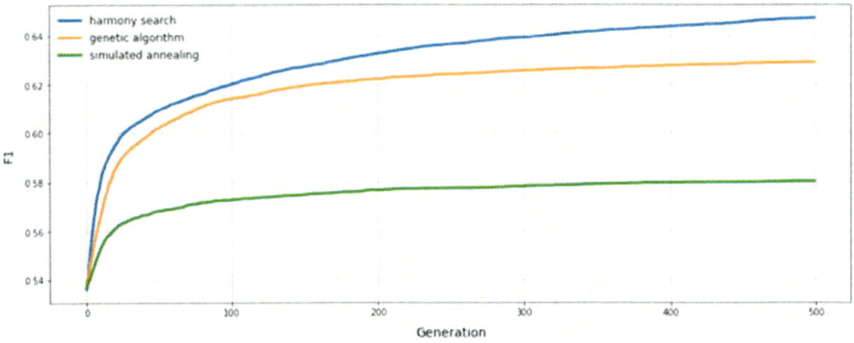

Fig. 2 Metaheuristics convergence

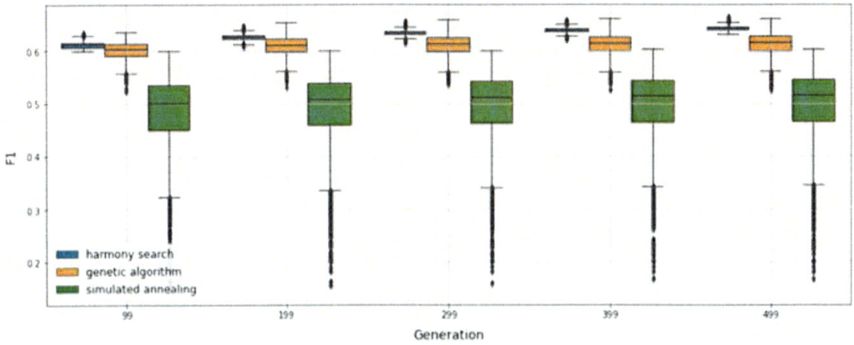

Fig. 3 Diversity of solutions by metaheuristic

Table 2 Shapiro-Wilk test of the metaheuristics

	W	*p*-value
SA	0.9836	0.2527
GA	0.9868	0.4215
HS	0.9762	0.0668

a level of significance $\alpha = 0.2527$ (W critical = 0.9836). The GA has a level of significance $\alpha = 0.4215$ (W critical = 0.9868). And the HS has a level of significance $\alpha = 0.0668$ (W critical = 0.9762). One would accept the null hypothesis only from GA and HS, concluding that there is no information to discard normality in the data.

Since not every distribution of the best solutions of the metaheuristics were proven to be Gaussian with the Shapiro-Wilk test, nonparametric statistical tests should be more reliable. In this case, we used Wilcoxon Signed-rank test to check if the distributions are equivalent. The null hypothesis for this test is that the median of the population of differences between the paired data is zero. Table 3 presents the W

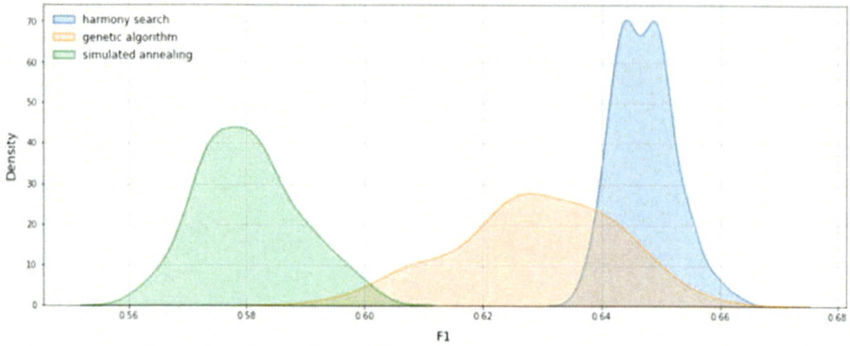

Fig. 4 Metaheuristics distribution

Table 3 Wilcoxon Signed-rank test of the metaheuristics

	W	p-value
HS-GA	196	0.0
HS-SA	0	0.0
GA-SA	1	0.0

Table 4 $F1$ score comparison of harmony search algorithms

	$F1$ mean	$F1$ std[a]	Elapsed time[b]
SA	0.5858	**0.01614**	1 h 29 min
GA	0.6501	0.02456	1 h 28 min
HS	**0.67**	0.01886	1 h 52 min

[a]std = standard deviation
[b]Configuration = Intel Core i5-10300H, 8 GB RAM

statistics and the p-values of the pairwise tests. Since all the p-values are below 0.05, one can conclude that there is enough information to reject the null hypothesis, i.e., none of the distributions are equivalent.

Finally, to test the final results, 600 unseen test samples were reserved. Using only the best final solution for each run, the mean and standard deviation of the $F1$ score were computed and are presented in Table 4. Although all the algorithms have a low standard deviation, in terms of $F1$ mean HS reached a significantly better value, almost 2% and 9% higher than GA and SA, respectively. Combined with the previous statistical tests, the HS showed better performances for our application, from early generation to the final results.

7 Conclusion and Future Research

In this paper, we explored the application of metaheuristics to a relation extraction problem in Portuguese language. The goal was to select patterns, formed by part of speech tags, that maximize the $F1$ score when detecting relations in sentences.

Among the algorithms tested, the harmony search had the best performance overall, reaching an $F1$ mean of 0.67. It also achieved a statistically Gaussian distribution in its final scores and led the convergence plot since early generations, even maintaining a less diverse population than the other approaches. This is another application where HS produce exceptional results due to its simplicity and the ability to reach a balance between exploration (diversification) and exploitation (intensification) ranges.

The idea of using metaheuristics to automatically find the best set of patterns for RE tasks was successfully achieved. For future research, we propose to fine-tune the hyperparameters of the algorithms, as well as the extension of other metaheuristics. It will also be important to implement other RE techniques to this benchmark for further comparison.

References

1. Hearst MA (1992) Automatic acquisition of hyponyms from large text corpora. In: Coling 1992 volume 2: the 15th international conference on computational linguistics, Nantes, France
2. Ravichandran D, Hovy E (2002) Learning surface text patterns for a question answering system. In: Proceedings of the 40th annual meeting of the Association for Computational Linguistics, Philadelphia, United States, pp 41–47
3. Mandya A, Bollegala D, Coenen F, Atkinson K (2017) Frame-based semantic patterns for relation extraction. In: International conference of the Pacific Association for Computational Linguistics. Springer, Singapore, pp 51–62
4. Jijkoun V, Mur J, de Rijke M (2004) Information extraction for question answering: improving recall through syntactic patterns. In: Proceedings of the 20th international conference on computational linguistics, Stroudsburg, United States, pp 1284–1290
5. Wu F, Weld DS (2010) Open information extraction using Wikipedia. In: Proceedings of the 48th annual meeting of the Association for Computational Linguistics, Uppsala, Sweden, pp 118–127
6. Ferreira J, Oliveira HG, Rodrigues R (2019) NLPyPort: named entity recognition with CRF and rule-based relation extraction. IberLEF@ SEPLN, pp 468–477
7. Boujelben I, Jamoussi S, Hamadou AB (2014) Genetic algorithm for extracting relations between named entities. In: 6th language and technology conference, Poznanń, Poland, pp 484–488
8. Bianchi L, Dorigo M, Gambardella LM, Gutjahr WJ (2009) A survey on metaheuristics for stochastic combinatorial optimization. Nat Comput 8(2):239–287
9. Pincus M (1970) A Monte Carlo method for the approximate solution of certain types of constrained optimization problems. Oper Res 18(6):1225–1228
10. Khachaturyan A, Semenovsovskaya S, Vainshtein B (1981) The thermodynamic approach to the structure analysis of crystals. Acta Crystallogr Sect A Crystal Phys Diffr Theor Gen Crystallogr 37(5):742–754
11. Kirkpatrick S, Gelatt CD, Vecchi MP (1983) Optimization by simulated annealing. Science 220(4598):671–680

12. Holland JH (1975) Adaptation in natural and artificial systems: an introductory analysis with applications to biology, control, and artificial intelligence. MIT Press, Cambridge
13. Geem ZW, Kim JH, Loganathan GV (2001) A new heuristic optimization algorithm: harmony search. Simulation 76(2):60–68
14. Batista DS, Forte D, Silva R, Martins B, Silva M (2013) Extracçao de relaçoes semânticas de textos em português explorando a dbpédia e a wikipédia. Linguamatica 5(1):41–57
15. Shapiro SS, Wilk MB (1965) An analysis of variance test for normality (complete samples). Biometrika 52(3/4):591–611
16. Wilcoxon F (1992) Individual comparisons by ranking methods. In: Breakthroughs in statistics. Springer, Berlin, pp 196–202

A New Hybrid Method for Text Feature Selection Through Combination of Relative Discrimination Criterion and Ant Colony Optimization

Majid Hemmati, Seyed Jalaleddin Mousavirad, Ehsan Bojnordi, and Mostafa Shaeri

Abstract Text categorization plays a significant role in many information management tasks. Due to the increasing volume of documents on the Internet, automated text categorization has been more considered for classifying documents in pre-defined categories. A major problem of text categorization is the high dimensionality of feature space. Most of the features are irrelevant and redundant impacting the classifier performance. Hence, feature selection is used to reduce the high dimensionality of feature space and increase classification efficiency. In this paper, we proposed a hybrid two-stage method for text feature selection based on Relative Discrimination Criterion (RDC) and Ant Colony Optimization (ACO). To this end, we applied RDC method, at first, in order to rank features based on their values. Features, then, which their values are lower than a threshold are removed from the feature set. In the second stage, as a wrapper method, an ACO-based feature selection method is applied, to select redundant or irrelevant features that have not been removed in the first stage. Finally, to assess the proposed methods, we have conducted several experiments on different datasets to indicate the superiority of our proposed algorithm. We aim to propose a hybrid approach which is computationally more efficient in much the same way as it is more accurate compared to the other embedded or wrapper methods. The obtained results endorse that the proposed method is of remarkable performance in text feature selection.

Keywords Text categorization · Feature selection · Ant Colony Optimization · Relative Discrimination Criterion · Document frequency · Term count

---

M. Hemmati
Department of Computer Engineering, University of Kurdistan, Sanandaj, Iran

S. J. Mousavirad
Computer Engineering Department, Hakim Sabzevari University, Sabzevar, Iran

E. Bojnordi (✉)
Information Technology Department, Iranian National Tax Administration, Bojnord, Iran
e-mail: bojnordi.ehsan@gmail.com

M. Shaeri
Department of Mathematical Sciences, University of Mazandaran, Babolsar, Iran

J. H. Kim et al. (eds.), *Proceedings of 7th International Conference on Harmony Search, Soft Computing and Applications*, Lecture Notes on Data Engineering and Communications Technologies 140, https://doi.org/10.1007/978-981-19-2948-9_16

1 Introduction

Text categorization has become a key technology to organize text data, a major problem of text categorization is the high dimensionality of feature space. The performance of machine learning algorithms for the categorization of documents reduces, due to the high dimensionality of feature space. Hence, selecting the proper subset of features for increasing the efficiency and precision of a classifier is necessary. These years, several approaches have been proposed for text feature selection. Yang and Pedersen [1] have compared five feature selection criteria used for text categorization, such as information gain (IG), χ^2 statistic, document frequency (DF), term strength (TS), and mutual information (MI). Besides, two feature selection evaluation metrics applied on text datasets including multi-class odds ratio (MOR) and class discriminating measure (CDM) were offered by Chen et al. [2]. Bahassine et al. [3] proposed an improved Chi-square for Arabic text classification. Moreover, Cekik and Uysal [4] developed a filter feature selection method called proportional rough feature selector (PRFS). They used the rough set for a regional distinction according to the value set of terms so as to identify documents that exactly or possibly belong to a class.

In addition to the conventional feature selection methods, population-based metaheuristic algorithms such as Differential Evolution (DE) [5, 6], Particle Swarm Optimization (PSO) [7], and Human Mental Search (HMS) algorithms [8, 9] can be used. Furthermore, some population-based optimization algorithms such as Firefly Algorithm (FA) [10], Gray Wolf Optimization (GWO) [11], Imperialist Competition Algorithm (ICA) [12], Ant Colony Optimization (ACO) [13], and Genetic Algorithm (GA) [14] have been used in this area. Population-based optimization algorithms, like ACO, focus on optimizing general combinatorial optimization problems. ACO was firstly proposed for text feature selection by Aghdam et al. [13]. Chen et al. proposed a method in which some features were selected through rough sets and then ACO was employed to select the remaining features [15]. Also, Paniri et al. introduced a heuristic learning approach for the ACO heuristic function, which learned directly from experiences [16]. It utilized the temporal difference reinforcement learning algorithm in high-dimensional feature selection.

In this paper, a two-stage hybrid method for text feature selection is proposed. In the first stage, a filtering method based on the relative discrimination criterion is applied for removing redundant and irrelevant features. It leads to reducing the computational cost of the next stage. In the second stage, Ant Colony Optimization is employed to reduce the features passed from the first stage. To be exact, the main contribution of this paper is to employ the Relative Discrimination Criterion (rather than other filter methods) in conjunction with ACO algorithm, and we indicate that this modification can improve the results.

The rest of the paper is organized as follows. Section 2 describes a brief overview of feature selection methods. Section 3 presents proposed feature selection algorithms. Experimental results are shown in Sect. 4, and the paper is concluded in Sect. 5.

2 Feature Selection

The high dimensionalities of features pose challenges to learning algorithms. In the presence of many irrelevant features, learning models tend to overfit and become less comprehensible. Thus, feature selection can be considered as an effective means to identify relevant features for dimensionality reduction [17–19], and it is proved that some features can be removed without performance deterioration [20]. The feature selection algorithms designed with different strategies broadly fall into three categories: filter, wrapper, and embedded models [21]. If the feature selection technique performs independently from the learning algorithm, it follows the filter approach. Otherwise, it follows a wrapper approach. In embedded models, the feature selection method can be merged with the learning algorithm. In these methods, the qualities of filter and wrapper methods are integrated. Generally, the filter approach is computationally more efficient. However, it might fail to select the right subset of features, if the criterion we use deviates from the one used for training the learning machine. Wrapper models, on the other hand, utilize the learning algorithm as a fitness function and search for the best subset of features in the space of all feature subsets. In this paper, a hybrid optimization method which is a combination of both filter and wrapper methods is used for text feature selection.

3 Proposed Method

In this section, we aim to introduce our two-stage text feature selection method. At the first stage, we filtered some features through RDC and then passed them to the next stage. More precisely, we used a backward elimination strategy for feature selection. Afterward, ACO, as a wrapper method, was applied to select some features from those not removed in the former stage. Indeed, in the latter stage, we employed a forward selection strategy for feature selection. Furthermore, we applied a k-nearest neighbor classifier to evaluate all feature subsets. To better understand all aforementioned parts of our proposed method, we elaborate on them in the following.

3.1 Relative Discrimination Criterion

RDC method is an effective univariate filter criterion, which has been recently proposed to reduce the dimensionality of text data [22]. The RDC metric uses information of term counts for filtering terms. In this method, high scores are assigned to features that frequently occur in a specific class compared to others. RDC considers the difference between document frequencies (DF) for respective term counts of a term in the positive and negative classes. The RDC of term t is defined as follows,

$$\mathrm{RDC}(t) = \frac{\left(\left|df_{\mathrm{pos}} - df_{\mathrm{neg}}\right|\right)}{\min\left(df_{\mathrm{pos}}.df_{\mathrm{neg}}\right) \times \mathrm{tc}} \quad (1)$$

where df_{pos} is the document frequency in the positive class and df_{neg} is the document frequency in the negative class for the particular term count called tc.

It is worth mentioning that although RDC is an effective filter method for feature selection, the correlation between features is ignored in this way. Thus, it cannot identify redundant features properly. In order to find a high relevance feature related to class labels with low redundancy, we use ACO finding correlated features and add them to the features subset to enhance the quality of text classifiers. It should be noted that the feature subset used in ACO step is comprised of the half of the one fed to RDC. To put it simply, before starting the ACO step, we firstly sort the subset in descending order based on the features' RDC scores, then we eliminate the bottom-half of it so as to remove ineffective features, and consequently decrease the run time of ACO step.

3.2 *ACO for Feature Selection*

In order for a feature selection problem to be solved by ACO, it should be mapped into a graph, so that features are considered as nodes, and edges between them are added once the next features are selected. Subsets of features are evolved as long as ants are traversing the graph and the termination condition is not satisfied. To put it mildly, the process of ACO feature selection begins with generating ants that each ant with transition rule chooses nodes. When each ant selects a node, local pheromone updating is taken on the selected node. The feature selection process for each ant ends when the stopping condition is satisfied. In this paper, the stopping condition is the number of features each ant must choose, that have been set according to the number of features received from the first stage of the proposed algorithm. Afterward, ants complete their solutions by evaluating the subset of features for each ant in respect to their performance and then the global update is performed. It is noteworthy that in this paper, ACO continues until the half number of features chosen in the first stage is selected.

Moreover, the use of heuristic information, which is readily available for many problems can guide the ants toward the most promising solutions. Heuristic information is modified by the ants during the algorithm's lifetime. It measures the heuristic preference of moving from one node to another node. In our proposed method, we use the classifier performance as the heuristic information for feature selection.

In the ACO approach, for constructing solutions, a number of ants with a feature as the starting point are considered. Then, an ant selects the next feature from unselected features with a given probability. Let K denotes an ant located at node i, $q_0 \in [01]$ be a parameter, and q a random value in [0 1]. The next node j is randomly chosen according to the following [23, 24].

$$\begin{aligned}
&\text{If } q \le q_0 \quad P_j^k = \begin{cases} 1 \text{ if } j = \operatorname{argmax}_{\mu \in N_k(j)} \left\{ \tau_j^\alpha . \eta_j^\beta \right\} \\ 0 \qquad\qquad \text{otherwise} \end{cases} \\
&\text{Else} \quad P_j^k(t) = \begin{cases} \frac{[\tau_j(t)]^\alpha \cdot [\eta_j]^\beta}{\sum_{\mu \in N_k(j)} [\tau_\mu(t)]^\alpha \cdot [\eta_\mu]^\beta} \text{ if } j \in N_k(j) \\ 0 \qquad\qquad \text{otherwise} \end{cases}
\end{aligned} \tag{2}$$

where $N_k(j)$ is the set of feasible nodes that ant k from a node can move to them, $\alpha$and$\beta \in R$ are two parameters indicating the relative importance of the pheromone trail and the heuristic information, τ_j and η_j are, respectively, the pheromone value and heuristic information associated with the feature j. The mentioned transition rule in Eq. (2) establishes a trade-off between the exploration of new connections and the exploitation of the information available at that moment [25, 26].

Two kinds of pheromone evaporation are performed in the algorithm. Once each ant chooses ith feature, the pheromone local updating is performed, and when all ants in the colony complete their solutions the global pheromone updating is applied to the best-found solution.

Once an ant travels from node i to j, the online step-by-step pheromone trail update is done that encourages generating solutions different from those found earlier. The purpose of the local pheromone update rule is to make the visited nodes less and less attractive as they are visited by ants, indirectly favoring the exploration of not yet visited nodes. Each ant applies the local pheromone update rule to the last node selected according to Eq. (3),

$$\tau_j(t) = \left(1 - \rho'\right)\tau_j(t) + \rho'\tau_0 \tag{3}$$

where ρ' denotes the local pheromones decay coefficient, $\tau_j(t)$ is the pheromone value of the selected node, and τ_0 is the initial value of the pheromones.

Once all ants in the colony construct their solutions, these solutions are evaluated, and the global pheromone update is applied. In the proposed algorithm, elitist strategy is used where only the best ant leaves pheromone in its own features. At first, all pheromones are evaporated and then some extra pheromone is added to the pheromones corresponding to the best solution in each iteration as it has shown in Eq. (4),

$$\tau_j(t+1) = (1-\rho) \cdot \tau_j(t) + \rho S_t^{k\text{best}} \tag{4}$$

where $\tau_j(t)$ and $\tau_j(t+1)$ is the pheromone value of jth node in iteration t and $t + 1$, respectively, $\rho \in [01]$ is a parameter called evaporation rate, and $S_t^{k\text{best}}$ is the performance of the best ant in each iteration.

4 Experimental Results and Analysis

4.1 Dataset

There are various standard datasets that can be used as a test collection for text categorization. The classic datasets such as *Reuters-21578 collection* and *20newsgroup* are the main benchmark for text classification evaluation. In this paper, we evaluated our proposed method by using these datasets.

4.2 Parameters and Evaluation Metrics

Parameters that have been set for ACO are given in Table 1. It is obvious that the number of features we use in each subset is not pre-defined. The bigger subset we use, the longer calculations take. Moreover, during the implementation, we noticed that the function evaluations, regardless of the number of features in the subset, are converged to the same results after iterating about 200. Therefore, we decided to use an iteration parameter as a termination condition, and set it to 200 in order for a reduction in the run time.

Also, we employed the k-nearest neighbor classifier for documents classification. To evaluate the documents by the classifier, a standard representation for documents content needs to be uniformly applied in which a document d_j will generally represent as a vector of term weights:

$$d_j = \{w_{1j}.w_{2j} \dots w_{|Tj|}\} \tag{5}$$

where w_{ij} represents the weight of the term i in the document j, T denotes the term set, and $|T|$ is the cardinality of T. Each entry in the term vector represents the occurrences of a word in a document. Weights are determined by using the normalized *tfidf* function [24].

In a text classification task, precision, recall, and $F1$-measure are three widely used metrics in evaluating an unordered set of documents. Precision is the proportion of the correctly retrieved documents to all retrieved documents, while recall is the proportion of the correctly retrieved documents to the test data that have to be retrieved. $F1$-measure is a harmonic combination of the precision and recall values used in information retrieval [25].

Table 1 ACO parameters

Population	Iteration	Initial pheromone	α	β	ρ	ρ'
10	200	0.01	1	0.1	0.3	0.05

4.3 Experimental Results

To demonstrate the superiority of our proposed method, we aim to compare its results with the following methods including standard RDC, ACO, GINI [14], and Chi-square [26] which is called CHI. Moreover, in order to pinpoint the crucial role of RDC in our hybrid approach, we deployed our implementation with some other filter methods such as GINI or CHI instead of RDC. The combination of RDC with ACO is an effective way in the vast majority of cases, as can be seen in the following tables. Tables 2, 3, and 4 illustrate the performance of the classifier on Reuters-21578 in terms of average precision, recall, F1-measure, respectively. In the same way, Tables 5, 6 and 7 showcase that of on 20 new group dataset.

In the following, to assess the quality of RDC-ACO in comparison with a well-known metaheuristic hybrid method, we carried out another experiment comparing the proposed method with the IG-GA method introduced by Uğuz [27]. IG-GA as a two-stage method is composed of a filter-based method, information gain (IG), and a wrapper-based method, Genetic Algorithm (GA). Since GA is genuinely stochastic,

Table 2 Performance (average value of precision) of KNN on Reuters-21578

Method	Number of features							
	10	20	50	100	250	500	1000	1500
RDC-ACO	**0.5592**	0.4969	0.5199	**0.5925**	**0.6708**	**0.7384**	**0.7795**	**0.7902**
CHI-ACO	0.3236	**0.5392**	**0.6088**	0.554	0.5039	0.6302	0.7541	0.7803
GINI-ACO	0.1444	0.1445	0.3961	0.4444	0.6309	0.7378	0.7202	0.7314
ACO	0.3458	0.3403	0.454	0.5226	0.6060	0.6328	0.6916	0.7351
RDC	0.2131	0.2590	0.3087	0.3467	0.5014	0.5651	0.6187	0.6872
CHI	0.0828	0.2633	0.4008	0.3926	0.4055	0.4605	0.5888	0.6375
GINI	0.0667	0.0443	0.0443	0.2751	0.2800	0.4339	0.5858	0.6268

Table 3 Performance (average value of Recall) of KNN on Reuters-21578

Method	Number of features							
	10	20	50	100	250	500	1000	1500
RDC-ACO	**0.1960**	**0.3317**	**0.3566**	**0.5448**	**0.6043**	**0.6611**	**0.8137**	**0.8388**
CHI-ACO	0.1236	0.1952	0.2669	0.3892	0.5447	0.6597	0.8055	0.8132
GINI-ACO	0.1056	0.1024	0.1479	0.1855	0.4185	0.6201	0.7359	0.7426
ACO	0.1392	0.1767	0.2130	0.2872	0.4038	0.5443	0.7663	0.7598
RDC	0.1341	0.2291	0.3082	0.3724	0.4922	0.5663	0.6186	0.7972
CHI	0.1067	0.1268	0.2096	0.2636	0.4051	0.5019	0.6423	0.7251
GINI	0.1043	0.1	0.1	0.1232	0.2055	0.4443	0.5805	0.6746

Table 4 Performance (average value of F1-measure) of KNN on Reuters-21578

Method	Number of features							
	10	20	50	100	250	500	1000	1500
RDC-ACO	**0.2903**	**0.3978**	**0.4230**	**0.5677**	**0.6358**	**0.6976**	**0.7962**	**0.8138**
CHI-ACO	0.1789	0.2866	0.3711	0.4572	0.5235	0.6446	0.7789	0.7964
GINI-ACO	0.1220	0.1198	0.2154	0.2618	0.5033	0.6739	0.7280	0.7370
ACO	0.1983	0.2326	0.29	0.3706	0.4847	0.5852	0.7270	0.7472
RDC	0.1646	0.2431	0.3084	0.3591	0.4967	0.5657	0.6186	0.7381
CHI	0.0972	0.1712	0.2752	0.3154	0.4052	0.4803	0.6144	0.6785
GINI	0.0813	0.0614	0.0614	0.1702	0.2371	0.4390	0.5831	0.6498

Table 5 Performance (average value of precision) of KNN on 20newsgroup

Method	Number of features							
	10	20	50	100	250	500	1000	1500
RDC-ACO	**0.4058**	0.6177	**0.6392**	**0.6650**	**0.6744**	**0.6714**	**0.6889**	**0.6928**
CHI-ACO	0.3546	0.4353	0.3307	0.4330	0.4565	0.5132	0.6103	0.6212
GINI-ACO	0.3672	**0.6417**	0.6209	0.6366	0.6274	0.6362	0.6538	0.6564
ACO	0.2528	0.3453	0.4518	0.4437	0.3531	0.4140	0.4329	0.4644
RDC	0.2819	0.4212	0.5375	0.5767	0.6180	0.6640	0.6808	0.6917
CHI	0.1188	0.1268	0.1726	0.2574	0.3559	0.4801	0.5502	0.5885
GINI	0.2996	0.3696	0.5128	0.5667	0.5850	0.6162	0.6264	0.6373

Table 6 Performance (average value of Recall) of KNN on 20newsgroup

Method	Number of features							
	10	20	50	100	250	500	1000	1500
RDC-ACO	0.1441	**0.3524**	**0.4858**	**0.5914**	**0.6369**	**0.6635**	**0.6803**	**0.6849**
CHI-ACO	0.0691	0.1248	0.2205	0.3366	0.4338	0.5046	0.6011	0.6148
GINI-ACO	**0.1557**	0.2754	0.3782	0.5210	0.5966	0.6274	0.6477	0.6513
ACO	0.0722	0.0827	0.1072	0.1335	0.2235	0.3143	0.3686	0.4202
RDC	0.1248	0.2070	0.4131	0.5291	0.6054	0.6503	0.6737	0.6810
CHI	0.0850	0.1289	0.1789	0.2595	0.3593	0.4747	0.5442	0.5838
GINI	0.1321	0.2164	0.3238	0.4911	0.5569	0.6073	0.6226	0.6353

the number of output selected features is variable in size. So, this is why we represented our experimental results separately in Tables 8 and 9. As we can see in these tables, RDC-ACO is superior to IG-GA.

Table 7 Performance (average value of F1-measure) of KNN on 20newsgroup

Method	Number of features							
	10	20	50	100	250	500	1000	1500
RDC-ACO	0.2127	**0.4488**	**0.5520**	**0.6260**	**0.6551**	**0.6674**	**0.6846**	**0.6888**
CHI-ACO	0.1157	0.1940	0.2646	0.3787	0.4449	0.5089	0.6057	0.6180
GINI-ACO	**0.2187**	0.3854	0.4701	0.5730	0.6116	0.6318	0.6507	0.6538
ACO	0.1123	0.1335	0.1733	0.2052	0.2737	0.3573	0.3981	0.4412
RDC	0.1730	0.2776	0.4672	0.5519	0.6116	0.6571	0.6772	0.6863
CHI	0.0991	0.1279	0.1757	0.2584	0.3576	0.4774	0.5472	0.5861
GINI	0.1834	0.2730	0.3970	0.5262	0.5706	0.6117	0.6245	0.6363

Table 8 Performance of KNN on Reuters-21578 using RDC-ACO and IG-GA

Method	Number of features						
		9	16	26	60	128	268
RDC-ACO	Precision	**0.5445**	**0.4436**	**0.4640**	**0.5612**	**0.5486**	**0.6175**
IG-GA	Precision	0.2429	0.3792	0.4971	0.4682	0.6331	0.6028
RDC-ACO	Recall	**0.1913**	**0.3087**	**0.3826**	**0.5067**	**0.5813**	**0.6020**
IG-GA	Recall	0.1084	0.1414	0.1656	0.2455	0.4524	0.5762
RDC-ACO	F1-measure	**0.2831**	**0.3641**	**0.4194**	**0.5326**	**0.5645**	**0.6096**
IG-GA	F1-measure	0.1499	0.2059	0.2485	0.3221	0.5277	0.5892

Table 9 Performance of KNN on 20 news group using RDC-ACO and IG-GA

Method	Number of features						
		8	17	36	58	150	283
RDC-ACO	Precision	**0.3280**	**0.5456**	**0.631**	**0.6918**	**0.6461**	**0.6285**
IG-GA	Precision	0.1478	0.3420	0.2377	0.2459	0.2814	0.3323
RDC-ACO	Recall	**0.1371**	**0.2883**	**0.4490**	**0.4791**	**0.5646**	**0.602**
IG-GA	Recall	0.0530	0.062	0.0872	0.1303	0.2293	0.3105
RDC-ACO	F1-measure	**0.1934**	**0.3772**	**0.5247**	**0.5661**	**0.6026**	**0.615**
IG-GA	F1-measure	0.0781	0.1053	0.1276	0.1704	0.2527	0.3210

5 Conclusion and Future Work

In this paper, we have introduced a two-stage text feature selection to reduce the high dimensionality of feature space through combining a filter model, RDC, with a wrapper model, ACO. The obtained results demonstrate that the performance of the proposed method is higher than its opponents. In addition, the importance of using RDC as an appropriate filter in our hybrid method has been shown. More precisely, it is an integral part of our method, so that its role cannot be ignored in this study. Besides, we used the k-nearest neighbor as a classic classifier. Hence, for future work, we intend to use more complex and effective classifiers like support vector machines (SVM) and artificial neural networks (ANN).

References

1. Yang Y, Pedersen JO (1997) A comparative study on feature selection in text categorization. In: ICML, vol 412–420, p 35. Nashville, TN, USA
2. Chen J, Huang H, Tian S, Qu Y (2009) Feature selection for text classification with Naïve Bayes. Expert Syst Appl 36(3):5432–5435
3. Bahassine S, Madani A, Al-Sarem M, Kissi M (2020) Feature selection using an improved Chi-square for Arabic text classification. J King Saud Univ-Comput Inf Sci 32(2):225–231
4. Cekik R, Uysal AK (2020) A novel filter feature selection method using rough set for short text data. Expert Syst Appl 160:113691
5. Mousavirad SJ, Schaefer G, Korovin I, Moghadam MH, Saadatmand M, Pedram M (2021) An enhanced differential evolution algorithm using a novel clustering-based mutation operator. In: 2021 IEEE international conference on systems, man, and cybernetics (SMC), pp 176–181. https://doi.org/10.1109/SMC52423.2021.9658743
6. Mousavirad SJ, Rahnamayan S (2020) One-array differential evolution algorithm with a novel replacement strategy for numerical optimization. In: 2020 IEEE international conference on systems, man, and cybernetics (SMC), pp 2514–2519. https://doi.org/10.1109/SMC42975.2020.9283154
7. Mousavirad SJ, Rahnamayan S (2020) CenPSO: a novel center-based particle swarm optimization algorithm for large-scale optimization. In: 2020 IEEE international conference on systems, man, and cybernetics (SMC), pp 2066–2071. https://doi.org/10.1109/SMC42975.2020.9283143
8. Bojnordi E, Mousavirad SJ, Schaefer G, Korovin I (2021) MCS-HMS: a multi-cluster selection strategy for the human mental search algorithm. arXiv preprint arXiv:2111.10676
9. Mousavirad SJ, Schaefer G, Korovin I, Saadatmand M (2021) HMS-OS: improving the human mental search optimisation algorithm by grouping in both search and objective space. arXiv preprint arXiv:2111.10188
10. Marie-Sainte SL, Alalyani N (2020) Firefly algorithm based feature selection for Arabic text classification. J King Saud Univ-Comput Inf Sci 32(3):320–328
11. Purushothaman R, Rajagopalan S, Dhandapani G (2020) Hybridizing Gray Wolf Optimization (GWO) with Grasshopper Optimization Algorithm (GOA) for text feature selection and clustering. Appl Soft Comput 96:106651
12. Mousavirad SJ, Ebrahimpour-Komleh H (2013) Feature selection using modified imperialist competitive algorithm. ICCKE 2013:400–405. https://doi.org/10.1109/ICCKE.2013.6682833
13. Aghdam MH, Ghasem-Aghaee N, Basiri ME (2009) Text feature selection using ant colony optimization. Expert Syst Appl 36(3):6843–6853
14. Shang W, Huang H, Zhu H, Lin Y, Qu Y, Wang Z (2007) A novel feature selection algorithm for text categorization. Expert Syst Appl 33(1):1–5
15. Chen Y, Miao D, Wang R (2010) A rough set approach to feature selection based on ant colony optimization. Pattern Recogn Lett 31(3):226–233
16. Paniri M, Dowlatshahi MB, Nezamabadi-pour H (2021) Ant-TD: Ant colony optimization plus temporal difference reinforcement learning for multi-label feature selection. Swarm Evol Comput 64:100892
17. Jayaprakash A, KeziSelvaVijila C (2019) Feature selection using ant colony optimization (ACO) and road sign detection and recognition (RSDR) system. Cogn Syst Res 58:123–133

18. Guyon I, Elisseeff A (2003) An introduction to variable and feature selection. J Mach Learn Res 3(Mar), 1157–1182
19. Liu H, Yu L (2005) Toward integrating feature selection algorithms for classification and clustering. IEEE Trans Knowl Data Eng 17(4):491–502
20. Ng AY (2004) Feature selection, L 1 vs. L 2 regularization, and rotational invariance. In: Proceedings of the twenty-first international conference on Machine learning, p 78
21. Mladenić D (2005) Feature selection for dimensionality reduction. In: International statistical and optimization perspectives workshop "Subspace, Latent Structure and Feature Selection". Springer, pp 84–102
22. Rehman A, Javed K, Babri HA, Saeed M (2015) Relative discrimination criterion—a novel feature ranking method for text data. Expert Syst Appl 42(7):3670–3681
23. Cordón García O, Herrera Triguero F, Stützle T (2002) A review on the ant colony optimization metaheuristic: basis, models and new trends. Mathware Soft Comput 9(2) [–3]
24. Salton G, Buckley C (1988) Term-weighting approaches in automatic text retrieval. Inf Process Manage 24(5):513–523
25. Van Rijsbergen C (1979) Information retrieval: theory and practice. In: Proceedings of the Joint IBM/University of Newcastle upon Tyne Seminar on Data Base Systems, pp 1–14
26. Imani MB, Keyvanpour MR, Azmi R (2013) A novel embedded feature selection method: a comparative study in the application of text categorization. Appl Artif Intell 27(5):408–427
27. Uğuz H (2011) A two-stage feature selection method for text categorization by using information gain, principal component analysis and genetic algorithm. Knowl-Based Syst 24(7):1024–1032

Page Level Input for Handwritten Text Recognition in Document Images

Lalita Kumari, Sukhdeep Singh, and Anuj Sharma

Abstract Important information present all over the world in the form of libraries need to be digitized. An end-to-end handwritten text recognition system covers every aspect, from capturing an image as input to predict text written inside it. In the present study, we have presented a document image-based handwritten text recognition system with three parts as: page level image as input with multiple lines, recognition of text of input image and attention level of text. This model is trained and tested on various split ratios of labelled words of the IAM handwriting dataset. The recognition model evaluation is performed on two publicly available datasets that are IAM handwriting dataset and handwritten short answer scoring dataset. We have used popular CTC word beam search technique by Scheidl et al. [1] in recognition phase, and attention of text has been incorporated using established sentic data by Ref. [2]. Our findings at recognition level result in 8.99% validation character error rate on IAM handwriting dataset and 35.92 BLEU score on the handwritten short answering scoring dataset. The post recognition findings suggest text attention role in acceptance of recognized words useful for real life business documents.

Keywords Handwritten text recognition · Segmentation · Word attention · Convolutional neural network · Word beam search

L. Kumari · A. Sharma (✉)
Department of Computer Science and Applications, Panjab University, Chandigarh, India
e-mail: anujs@pu.ac.in
URL: https://anuj-sharma.in

L. Kumari
e-mail: lalita@pu.ac.in

S. Singh
D. M. College (Affiliated to Panjab University, Chandigarh), Moga, Punjab, India
e-mail: sukha13@ymail.com

J. H. Kim et al. (eds.), *Proceedings of 7th International Conference on Harmony Search, Soft Computing and Applications*, Lecture Notes on Data Engineering and Communications Technologies 140, https://doi.org/10.1007/978-981-19-2948-9_17

1 Introduction

Converting handwritten text into digital form is a prominent research problem in the domain of pattern recognition. Handwritten text recognition (HTR) converts handwritten text from sources such as physical documents, pictures or directly using handwriting as input into digital. In general, the quality of paper and clarity of handwritten text are bound to degrade over time. The HTR converts the handwritten text into digital form. Its application varies such as digitization of library scripts, documents presenting key historical events and medical prescription translation to understand drug usage better. The HTR poses challenges such as different writing styles, multilingual scripts and wear and tear of the manuscripts, especially in recognition of historical documents. The HTR system requires building components such as pre-processing, line segmentation and word segmentation. In this paper, a tightly coupled system of pre-processing, line segmentation followed by word segmentation and convolutional recurrent neural network (CRNN)-based model to recognize those words presented in a document. Sentic analysis of word acceptance score has been done for real life use documents.

Segmentation and recognition are critical steps of any end-to-end HTR system. Most of the earlier works solve these problems independently. In contrast, this paper discussed for these techniques jointly in an end-to-end manner. Blob like features used to segment pages into lines using projection analysis techniques [3]. Various models like hidden Markov models and convolutional neural networks (CNNs) are used to predict the text inside an segmented image [4]. In one such study, effect of different CNN-based models such as SimpleHTR, Bluche and Puigcerver in prediction and classification of the name of cities are analysed [5]. Typically, machine learning-based models able to improve performance of miscellaneous applications in pattern recognition [6, 7]. A related work has been done on publicly available handwritten short answer scoring dataset [8]. The motivation to perform experiments in this paper based on notion "human write useful text in business documents". Therefore, attention level of recognized text is an important parameter to know real life usage of documents. The key contributions of the present study are:

1. Input image as a page input with multiple lines (closely related to real life usage) and segmenting at word level.
2. Improved BLEU score in study for new publicly available handwritten short answer scoring dataset [8].
3. Attention score of recognized text as a post-processing step to know acceptance of text in business documents.

The rest of this paper is organized as follows. In Sect. 2, we have described the overview of the system, followed by the algorithm explanation in Sect. 3. In Sect. 4, experimental set-up has been illustrated. Finally, Sect. 5 concludes this study with discussion on results and future scope.

2 System Overview

The HTR system consists of several steps [9]. The initial step is identification of region of interest (ROI). Later, extracted image is segmented into lines followed by segmentation of lines into words. Text recognition of these words is done by 5 CNN and 2 Long Short-Term Memory (LSTM) RNN layers. At last, Connectionist Temporal Classification (CTC) layers either calculate the loss values while training or decode the final characters sequence of text in an image during testing by constrained it with dictionary and language model (Refer Fig. 1) [10].

2.1 Pre-processing

Pre-processing involves all the steps followed in the process of getting word images from a given input page of the dataset. In this section, pre-processing steps are discussed sequentially.

Identifying the ROI—It is required for the layout analysis of the page. Each page of the IAM handwriting dataset contains the transcription of the handwritten text at the top. That transcription part needs to be removed prior to further processing of the image. In this study, cropping and resizing are being used as identifying ROI since its computation cost is negligible, and overall architecture still produces acceptable accuracy. The image containing the handwritten text is extracted from the input image by removing 1/5th (obtained by trial and error) from the top and bottom of an IAM handwriting dataset page. In Fig. 2, M is the total height of a page in pixels. A light balance image has been obtained by applying the light illumination technique discussed in [11]. Further, denoising is achieved by binarization of image [12]. The obtained image is sent for line segmentation.

Line Segmentation—It is the most critical step in a handwritten text recognition system. It segments the documents into lines, and all these lines are subjected to word segmentation. The system uses a statistical approach for line segmentation [13]. This method is able to handle skewed documents, and lines that are running into each other. First, initial candidates of lines have been found out by using piece-wise projection. Any connected component presents between the two lines either associated with upper line or lower line, decision regarding such connected components is done either by probability obtained from distance metric or by evaluating the lines as bivariate Gaussian densities (Refer Fig. 2).

Post-Processing Line Segmentation—While developing the system, initially, word segmentation was directly applied to line images. Hence, the obtained result gave a large error rate due to the slant and slope present in the image. Hence, the present system is used [14] to remove slanting and slopping of the image which gets better results than without deslanting the image. The algorithm first removes the slope by

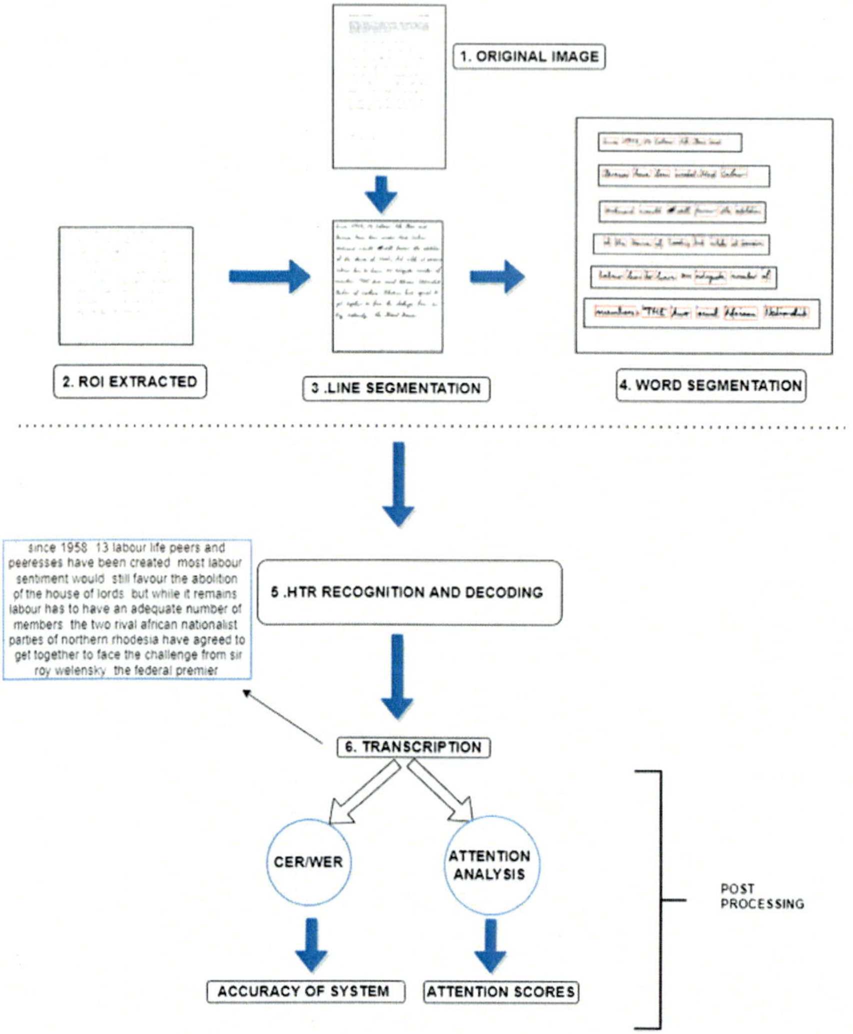

Fig. 1 Overview of text recognition and word acceptance

applying the rotation of the image, and then the core region is obtained by studying the probability density distribution of the pixel values.

Word Segmentation—After line segmentation, to extract the words, present work examines each line individually and for each line, word segmentation is achieved by blob like representation of words [3]. A blob is described as connected region in space. For blob analysis, differential expression is used that is similar to Laplacian of Gaussian (LOG) for generating a multi-scale representation. Hence, anisotropic

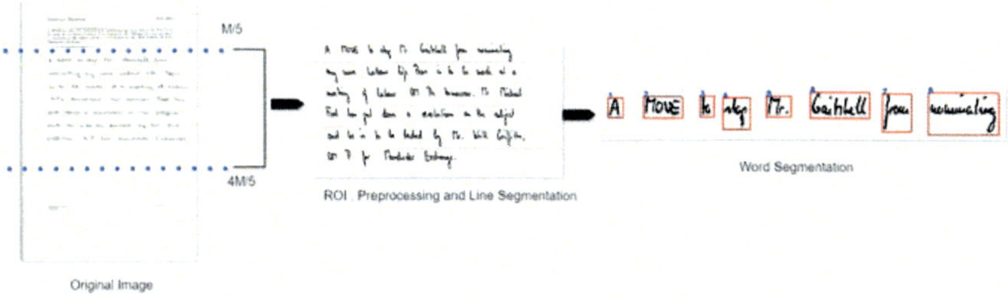

Fig. 2 Identifying ROI, pre-processing, line and word segmentation

filters are used instead of isotropic. After finding of right scale and creating blobs, the blobs are mapped to the real images to locate the words (refer Fig. 2).

2.2 Handwritten Text Recognition Model

The HTR model used in this study consists of 5 CNN layers followed by 2 LSTM RNN layers [15]. The CTC loss is used while training the model and decoding layers for prediction of final output (refer Fig. 3). The different CNN layers extract the input image features, then the output of these layers or extracted features feeds into RNN. The RNN propagates important information through this feature sequence. This sequence contains 256 features per time step. The RNN output is mapped into matrix size 32×80. The CTC layer computes the loss value of the RNN output matrix and ground truth text while training, and while recognition it decodes final text from the RNN output matrix using best path decoding or word beam search decoding [15]. CRNN model is trained with the help of CTC loss values. The output of RNN layer contains the character probabilities. In this decoding algorithm, a prefix tree and optionally language model are used to query a character. The prefix tree is queried for the next possible characters and decoded word restraint to a dictionary word that allows any number of non-word characters between them.

3 Algorithm Discussion

This section presents algorithm of the HTR system (shown in Fig. 1). We have discussed Algorithm 1 as follows, a handwritten text page is given as input which contains multiple lines of texts. ROI has been extracted from the image if needed. As a pre-processing step for the recognition, denoising of the image is done by applying popular images pre-processing techniques such as light illumination [11] followed by binarization [12]. After this step, input image in the form of page is segmented into lines [13]. To further improve the recognition accuracy, deslanting operation is

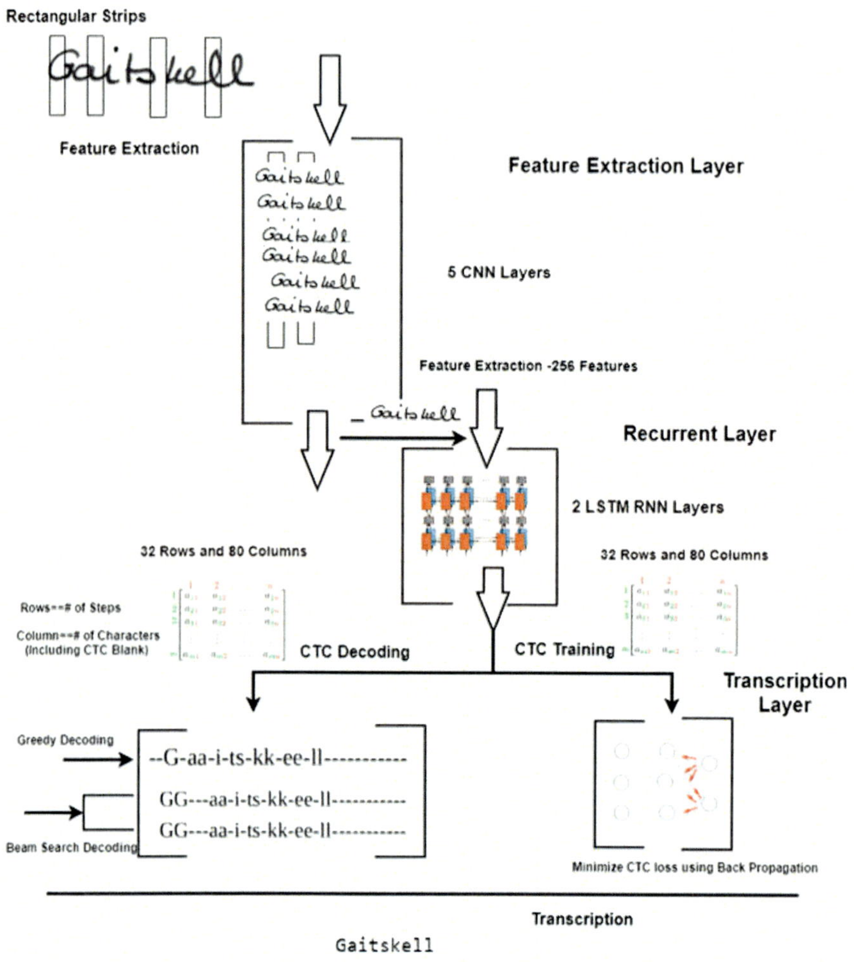

Fig. 3 HTR model overview

performed on each line image of input page. Since we have word level annotation of images available with IAM dataset, we further go in our experiments by doing word segmentation of line images obtained earlier. These set of handwritten text of images feed back into a HTR model, which predicts the text inside a given word image of a page. We also calculated the attention score of predicted text that shows the acceptance possibility of our HTR model in real world business applications. We have used Scheidl [1] toolbox of CTC word beam search (CTC-WBS) at step 10 [16].

Algorithm 1 HTR Algorithm

Input-Handwritten text Image.
Output-Recognized handwritten text with accuracy and attention score of recognition.

```
Reading of the Input Image
Getting ROI from input Image (I_ROI)(if required)
I_Binary=binarization(I_ROI)
Lines=lineSegmentation(I_Binary)
newLines=deslanting(Lines)
for each line in newLines
  words=wordSegmentation(line)
  allWords+= words
end for
Predict= HTR(WordImages) [9]
accuracy=calAccuracy(transcription, pagePrediction)
attention=calAttention(pagePrediction)
```

4 Experiments

Dataset—In this study, two different datasets are used. First, *IAM offline* [17] dataset is used to evaluate architecture. It is a state-of-the-art dataset for testing handwritten recognition models. It consists of 1539 pages of scanned text contributed by 657 different writers. It also has 13,353 isolated and labelled text lines and 115,320 isolated and labelled words. In this study, the CNN model is trained on different sizes of train and validation set of word images.

Publicly available *handwritten short answer scoring* dataset is used to evaluate the HTR system [8]. It has 186 number of handwritten images, for which the corresponding transcription is given.

4.1 Accuracy Parameters Used

In the present study, three different types of accuracy parameters are used, namely Character Error Rate (CER), Word Error Rate (WER) and BLEU. In this section, a brief discussion about them is presented.

Error Rates—In this study, CER and WER are used to study the accuracy of the HTR system. It is based upon concept of the Levenshtein Distance (LD). The CER and WER differ only by the fact of whether LD is applied at word level or character level. It is formulated as below:

$$\text{WER} = \frac{S + D + I}{N} = \frac{S + D + I}{S + D + C} \tag{1}$$

where, N is total number of words present in transcription, S is total number of substituted words, I is total number of insertions, D is total number of deletions and C is total number of correct words.

BLEU—The BLEU is a method of comparing how close HTR recognition results in comparison with actual translation. Its score varies from 0 (No matching) to 100 (original text) [18]. It counts matching n-gram in the recognized text to the n-gram of the referenced text. By default, it calculates for a score up to 4 n-grams applying uniform weights. Its also called BLEU-4.

4.2 *Implementation Details*

In this study, the model present in [15] is trained over two different optimizers that is RMSProp Optimizer and Adam Optimizer. The learning rate controls how rapidly the given NN model is adopted to the problem. A high value of learning rate may cause the model to converge very quickly to a suboptimal solution. While having a too-small value may take very large numbers of epoch to converge the model. Thus the learning rate value 0.001 is used while training the model with both optimizers. The model is trained over the label words presented in IAM handwriting dataset. The total number of words is divided into training and validation set using different split ratios as presented in Table 1 thus, validation character error rate (VCER) is obtained. A split ratio of 0.95 means that a total of 95% of words presented in IAM handwriting dataset is being considered in the training set, and rest 5% words are in the validation set. After training validation, CER is calculated from the element of the validation set. Table 1 gives the various validation CER (VCER) and WER values obtained from training.

Further, a visual comparison among the optimisers can be seen in Fig. 4a, In this, the blue line represents the VCER values of RMSProp optimiser. Similarly, Fig. 4b shows the word accuracy plot. Word accuracy represents how many % of words of validation set is correctly recognized by the model at the time of training. Prior to training IAM handwriting dataset, word images are not directly loaded, rather first converted into Lightning Memory-Mapped Database (LMDB) format. In this, word images are stored in the form of key-value pair, for faster memory access. The

Table 1 Shows training result of IAM offline dataset words at different split ratios

Split ratio	VCER (in %)		Word accuracy (in %)	
	RMSProp	Adams	RMSProp	Adams
0.95	10.32	10.18	75.16	75.23
0.85	11.74	11.42	73.69	74.39
0.75	12.11	11.91	72.57	72.97
0.65	12.43	11.85	70.87	72.13

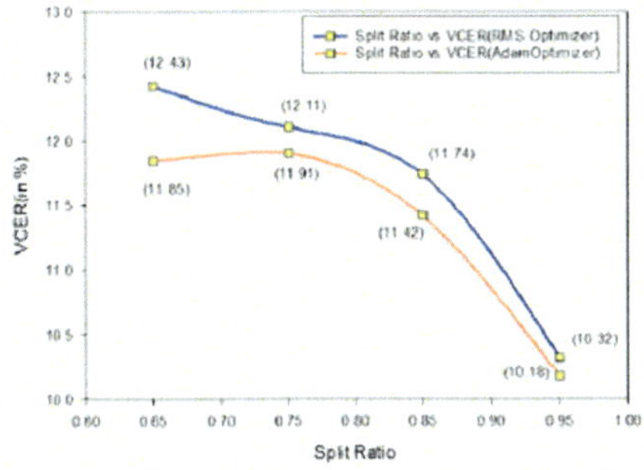

(a) VCER and Split ratio

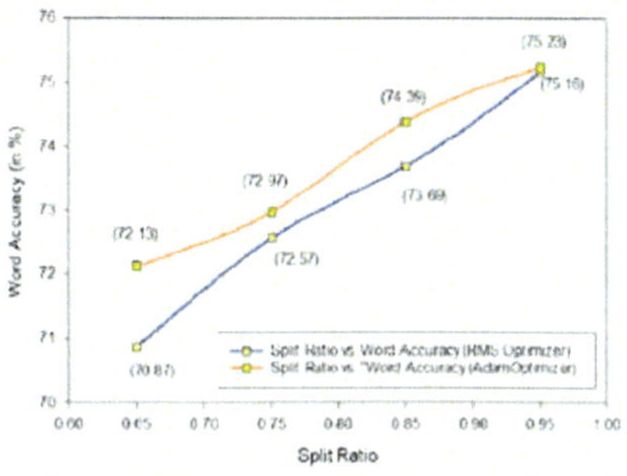

(b) Word Accuracy and Split ratio

Fig. 4 Different plot of split ratios

Table 2 CER and WER values in % of the HTR system on *IAM offline* [17] dataset

Split ratios	Original				Superior				Supreme			
	RMSProp		Adams		RMSProp		Adams		RMSProp		Adams	
	CER	WER	CER	WER	CER	WER	CER	WER	CER	WER	CER	WER
0.95	11.68	32.20	12.16	32.49	9.71	21.04	10.31	22.12	9.52	20.39	10.05	21.33
0.85	11.54	32.08	11.16	31.10	9.48	21.15	9.14	19.74	9.26	20.48	8.99	19.24
0.75	12.01	32.46	11.68	31.03	10.03	21.69	9.89	21.23	9.79	20.97	9.71	20.76
0.65	12.75	33.33	11.76	31.47	10.91	23.50	9.77	20.56	10.64	22.71	9.54	20.04

The table is divided into three types of post analysis. **Original** is calculating accuracy of the HTR system as it is. **Superior** is calculating accuracy by removing punctuation marks from recognized as well as actual text. **Supreme** is superior mode with all the text converted into lower case characters

stopping criteria of the training was consecutive 25 epochs in which CER is not improved.

A randomly chosen set of 50 images of pages is being provided as input to the HTR system, as per algorithm 1 described in previous section. Pages are first divided into lines, lines are segmented into words and the words obtained by this process are send to earlier trained NN model, and CER and WER was calculated under different post-processing scenario. In this study, three post-processing scenarios are discussed as follows,

1. **Original**—In this, obtained recognition result is directly processed for accuracy against actual transcription. This is applicable in the scenarios where exact matching is required. Tables 2 and 4 give the average CER 11.99%, WER 32.51% and 48.53 BLEU on RMSProp optimizer and CER 11.69%, WER 31.52% and 49.89 BLEU on Adam optimizer, respectively, on IAM handwriting dataset. Meanwhile, in Tables 3 and 4, the average CER 20.19%, WER 44.29% and BLEU 35.75 on RMSProp optimizer and CER 19.67%, WER 43.55% and BLEU 36.10 on Adam optimizer, respectively, on handwritten short answer scoring dataset.

Table 3 CER and WER values in % of the HTR system on *handwritten short answer scoring* [8] dataset

Split ratios	Original				Superior				Supreme			
	RMSProp		Adams		RMSProp		Adams		RMSProp		Adams	
	CER	WER	CER	WER	CER	WER	CER	WER	CER	WER	CER	WER
0.95	20.40	44.51	20.06	43.61	18.69	36.01	18.53	34.91	18.04	33.37	18.11	33.66
0.85	19.26	43.06	19.11	42.94	17.56	34.25	17.30	33.31	17.11	32.86	16.74	30.94
0.75	20.87	45.47	19.32	42.52	19.29	37.11	17.64	34.33	18.74	35.16	17.02	31.95
0.65	20.24	44.12	20.19	45.13	18.79	35.84	18.50	36.42	18.18	33.74	17.75	33.87

The table is divided into three types of post analysis same as in Table 2

Table 4 BLEU scores of the HTR system on *D-1 IAM handwriting dataset* [17] dataset and *D-2 Handwritten short answer scoring dataset* [8]

Dataset name	Original		Superior		Supreme	
	RMSProp	Adams	RMSProp	Adams	RMSProp	Adams
D-1	48.53	49.89	51.95	54.04	52.72	54.63
D-2	35.75	36.10	42.54	43.40	45.81	46.96

The table is divided into three types of post analysis as in Table 2

2. **Superior**—In this post-processing, mode accuracy is calculated by removing punctuation marks from recognized as well as actual text. Tables 2 and 4 show the average CER 10.03%, WER 21.80% and 51.95 BLEU on RMSProp optimizer and CER 9.77%, WER 20.91% and 54.04 BLEU on Adam optimizer, respectively, on IAM handwriting dataset. Meanwhile, in Tables 3 and 4, the average CER 18.58%, WER 35.80% and 42.54 BLEU on RMSProp optimizer and CER 17.99%, WER 34.74% and 43.40 BLEU on Adam optimizer, respectively, on handwritten short answer scoring dataset.
3. **Supreme**—Accuracy further improved by converting all the recognized and actual text in lower case characters of the superior mode in post-processing. Tables 2 and 4 show the average CER 9.80%, WER 21.13% and 52.72 BLEU on RMSProp optimizer and CER 9.57%, WER 20.34 and 54.63 BLEU on Adam optimizer, respectively, on IAM handwriting dataset. Meanwhile, in Tables 3 and 4, the average CER 18.01%, WER 33.78% and 45.81 BLEU on RMSProp optimizer and CER 17.40%, WER 32.60% and 46.96 BLEU on Adam optimizer, respectively, on handwritten short answer scoring dataset.

There is very less literature available on the handwritten short answer scoring dataset. The latest available literature on the short answer scoring dataset is [8]. In the present study, the same experiments have been conducted and its outperformed the existing study by 12.26% that is BLEU score is increased from 32 to 35.92 and for IAM handwriting dataset BLEU reported value is 49.21 in original mode of post-processing. Apart from this, in this study, we have also explored the domain of semantic analysis by providing the IAM word attention score and word acceptance

Table 5 Comparison table

Paper	Input	Segmentation	CER (%)	WER (in %)
Liwicki et al. [19]	Line	–	18.2	–
Louradour et al. [20]	Word /Line	–	16.85	–
Chen et al. [21]	Line	–	11.15	34.55
Krishnan et al. [22]	Word/Line	–	9.78	32.89
Our paper	Page	Yes (Line seg. and Word seg.)	10.18	24.77

based upon available sentic data. In particular, attention score values of IAM dataset words. Attention score is eloquence of a given word. For example, in present study as per available sentic data 'excise' is having a attentive value of 0.358 and 'excite' is having a attention score of 0.936. 'kill_brain_cell' having a attention score of −0.99. High value of 'excite' is a indication that human tends to be more attentive towards that word and for a real world business application, we may developed our algorithm to predict the word having the high attention value as accurately as possible. Thus, reduce computational cost by only focusing of words with high attention value, which is typically most suitable scenario for a real world business application. Similarly, for 'kill_brain_cell', low attention value indicates less word acceptance rate, that is people are less attentive to the given word.

4.3 Comparison

In this section, present study is compared with earlier works. In Table 5, input column refers at what level input image is given in HTR architecture. Segmentation column provides details of segmentation techniques used.

5 Conclusion and Future Scope

The present study provided an efficient and improved end-to-end system for handwritten text recognition, and the experimentation proved that results for BLEU for handwritten short answer scoring dataset is 35.92. In this study, the CRNN model [15] is also trained over various split ratios of test and validation set, and analysis of the same is presented. Evaluation of performance of given HTR model is done by calculating CER, WER and BLEU values over IAM handwritten dataset and handwritten short answer scoring dataset. As a post-processing step, we also find attention score of various words present in IAM dataset based upon available sentic data. Future studies incorporate robust pre-processing approaches, along with NN algorithms with lesser training time.

References

1. Scheidl H, Fiel S, Sablatnig R (2018) Word beam search: a connectionist temporal classification decoding algorithm. In: 16th international conference on frontiers in handwriting recognition. IEEE, pp 253–258
2. SenticNet: Senticnet: helping machines to learn, leverage, love (May 2021). https://sentic.net/
3. Manmatha R, Srimal N (1999) Scale space technique for word segmentation in handwritten documents. In: Proceedings of the second international conference on scale-space theories in computer vision. In: SCALE-SPACE '99. Springer, Berlin, pp 22–33
4. Singh S, Sharma A (2019) Online handwritten Gurmukhi words recognition: an inclusive study. ACM Trans Asian Low-Resour Lang Inf Process 18(3)
5. Nurseitov D, Bostanbekov K, Kanatov M, Alimova A, Abdelrahman Abdallah GA (2020) Classification of handwritten names of cities and handwritten text recognition using various deep learning models. Adv Sci Technol Eng Syst J 5(5):934–943
6. Kumari L, Sharma A (2022) A review of deep learning techniques in document image word spotting. Arch Comput Methods Eng 29(2):1085–1106
7. Singh S, Chauhan VK, Smith EHB (2020) A self controlled rdp approach for feature extraction in online handwriting recognition using deep learning. Appl Intell 50(7):2093–2104
8. Gold C, Zesch T (2020) Exploring the impact of handwriting recognition on the automated scoring of handwritten student answers. In: 2020 17th international conference on frontiers in handwriting recognition (ICFHR), pp 252–257
9. Scheidl H (2021) Build a handwritten text recognition system using tensorflow. https://towardsdatascience.com/build-a-handwritten-text-recognition-system-using-tensorflow-2326a3487cd5
10. Graves A, Fernández S, Gomez F, Schmidhuber J (2006) Connectionist temporal classification: labelling unsegmented sequence data with recurrent neural networks. In: ICML '06. Association for Computing Machinery, pp 369–376
11. Chen KN, Chen CH, Chang CH (2012) Efficient illumination compensation techniques for text images. Digit Signal Process 22:726–733
12. Su B, Lu S, Tan CL (2010) Binarization of historical document images using the local maximum and minimum, pp 159–166
13. Arivazhagan M, Srinivasan H (2007) A statistical approach to line segmentation in handwritten documents. In: Proceedings of SPIE. The International Society for Optical Engineering
14. Vinciarelli A, Luettin J (2001) A new normalization technique for cursive handwritten words. Pattern Recogn Lett 22(9):1043–1050
15. Scheidl H (2018) Handwritten text recognition in historical document. Diploma-Ingenieur in Visual Computing. Master's thesis, Technische Universität Wien, Vienna
16. Scheidl H (2021) githubharald/ctcwordbeamsearch. https://github.com/githubharald/CTCWordBeamSearch
17. Marti UV, Bunke H (2002) The IAM-database: an English sentence database for offline handwriting recognition. Int J Doc Anal Recogn 5(1):39–46
18. Papineni K, Roukos S, Ward T, Zhu WJ (2002) Bleu: a method for automatic evaluation of machine translation
19. Liwicki M, Graves A, Bunke H (2012) Neural networks for handwriting recognition. In: Computational intelligence paradigms in advanced pattern classification. Springer, Berlin, pp 5–24
20. Louradour J, Kermorvant C (2013) Curriculum learning for handwritten text line recognition. In: Proceedings—11th IAPR international workshop on document analysis systems, DAS 2014

21. Chen Z, Wu Y, Yin F, Liu CL (2017) Simultaneous script identification and handwriting recognition via multi-task learning of recurrent neural networks. In: 2017 14th IAPR international conference on document analysis and recognition (ICDAR), vol 01, pp 525–530
22. Krishnan P, Dutta K, Jawahar C (2018) Word spotting and recognition using deep embedding. In: 2018 13th IAPR international workshop on document analysis systems (DAS), pp 1–6

Evolutionary Population Dynamic Mechanisms for the Harmony Search Algorithm

Seyedeh Zahra Mirjalili, Shelda Sajeev, Ratna Saha, Nima Khodadadi, Seyed Mohammad Mirjalili, and Seyedali Mirjalili

Abstract Evolutionary algorithms have been widely adopted in science and industry for optimizing challenging problems mainly due to their black box nature and high local optima avoidance. As popular soft computing techniques, they benefit from several stochastic operators, including but not limited to, selection, recombination, mutation, elitism, population diversity, and population dynamics. Among such operators, some have been extensively used and analyzed in different algorithms, while others are yet to be explored in different algorithms. This motivated our attempts to integrate Evolutionary Population Dynamics (EPD) in the Harmony Search (HS) algorithm. EPD is an evolutionary mechanism that excludes and/or replaces a set of the poor solutions in each generation and prevents them from reducing the quality of other solutions. EPD has been used in three different ways in HS to impact 10%, 30%, or 50% of the population to see its impact on the performance of this algorithm. It was observed that 10% is a reasonable portion of the population in HS to

S. Z. Mirjalili · S. Sajeev · R. Saha · S. Mirjalili (✉)
Centre for Artificial Intelligence Research and Optimisation, Torrens University Australia, Brisbane, QLD, Australia
e-mail: ali.mirjalili@gmail.com

S. Z. Mirjalili
e-mail: zahra.mirjalili@torrens.edu.au

S. Sajeev
e-mail: shelda.sajeev@torrens.edu.au

R. Saha
e-mail: ratna.saha@torrens.edu.au

N. Khodadadi
Department of Civil and Environmental Engineering, Florida International University, Miami, FL, USA

S. M. Mirjalili
Department of Engineering Physics, Polytechnique Montreal, Montréal, Canada

S. Mirjalili
Yonsei Frontier Lab, Yonsei University, Seoul, South Korea

J. H. Kim et al. (eds.), *Proceedings of 7th International Conference on Harmony Search, Soft Computing and Applications*, Lecture Notes on Data Engineering and Communications Technologies 140, https://doi.org/10.1007/978-981-19-2948-9_18

improve its performance on IEEE Congress of Evolutionary computation (CEC) test functions, which effectively mimic challenging real-world optimization problems.

Keywords Evolutionary algorithm · Evolutionary operator · Evolutionary Population Dynamics · Harmony Search · Optimization · Algorithm

1 Introduction

When it comes to the field of computational intelligence and soft computing, there are three active branches: evolutionary computation, neural networks, and fuzzy logic. In evolutionary computation, algorithms are developed by mimicking problem-solving in nature. Algorithms in each of these classes are called soft due to the accommodation of imperfection, uncertainties, and approximation. By contrast, exact algorithms are usually considered complicated computing methods that find the exact optimal solutions for optimization problems. Despite the reliability of complex computing algorithms, they tend to suffer from degraded performance on challenging problems and when computational resources are limited. Soft computing has been around since the 1980s and recently skyrocketed due to the ever-increasing complexity of problems and the inefficiency of hard computing algorithms in such problems.

A classification of optimization algorithms in the literature that effectively categorizes hard and soft optimization algorithms include two categories: deterministic versus non-deterministic. In the first category, algorithms deliver the same output when we provide them with the same input. Such algorithms are very efficient on problems with the complexity class of P (P-complete). In fact, a deterministic algorithm is usually designed to solve P-complete problems in polynomial time. The main drawback of deterministic algorithms is that they cannot solve NP-complete and NP-hard problems in polynomial time.

To solve such challenging problems in polynomial time, non-deterministic algorithms are needed. An algorithm in this class lands on different outputs given the same input. The non-deterministic outcome may occur due to random components, parallel execution of an algorithm's operator, or the use of probabilistic models. Due to the nature of such algorithms, they find approximated solutions for problems. Obviously, we sacrifice accuracy over feasibility here when there is a deterministic way of solving a problem. These concepts directly translate into all three branches of evolutionary computation, neural networks, and fuzzy logic. Due to the scope of this work, we only focus on evolutionary algorithms.

The overall framework of algorithms in evolutionary computation tends to be similar. A set of candidate solutions is first randomly created. This set is then evolved during several steps, which is often called generation, iteration, or epoch. Several operators are then used to combine these solutions and generate new ones to form a new set as a new generation. What makes algorithms different is the operators used to select, combine, change, protect, and/or remove the solutions in each generation [1]:

- SELECTION: The selection process is made based on the quality of solutions; the better, the higher chance of being selected to contribute to the creation of the next generation.
- CROSSOVER: The combination is performed on the solution so that we increase the diversity of solutions when making a new generation.
- MUTATION: Slight changes are also done randomly to each solution as well to create entirely new solutions different from the existing ones.
- ELITISM: Some of the best solutions are protected from changes to move a new generation intact. This will help an algorithm to avoid losing the best solution found in each generation.
- EPD: The removal is done to exclude some of the poor solutions in each generation and prevent them from reducing the quality of other solutions.

Some of these evolutionary operators have been widely used in different evolutionary algorithms. However, the EPD operator has not been explored as a simple, computationally cheap, and substantial potential in improving the performance of evolutionary algorithms. It can be easily used in other categories of optimization algorithms, including swarm intelligence, physics-based, memetic, hybrid, and so on. This motivated our attempts to integrate the EPD operator in the Harmony Search algorithm for the first time in the literature to improve its performance and investigate the impact of this operator on the search behavior of optimization algorithms. The rest of the paper is organized as follows:

Section 2 presents the Harmony Search algorithm. The Evolutionary Population Dynamics are introduced and integrated into the HS algorithm in Sect. 3. The results of the proposed HS_EPD are presented and compared on several CEC composited test functions in Sect. 4. Finally, Sect. 5 concludes the work and suggests future directions.

2 Harmony Search

Harmony Search (HS) was proposed in 2001 by Geem et al. as an optimization algorithm [2]. It belongs to the family of meta-heuristics, which can be considered as black box optimization techniques. There are several classifications in the literature, but one of the most popular ones is based on their inspiration. Evolutionary algorithms mimic evolutionary phenomena in nature. Some of the most popular algorithms are Genetic Algorithm (GA) [3], Evolution Strategy (ES) [4], and Differential Evolution (DE) [5]. The second class includes swarm intelligence algorithms. Some of the popular or recent ones are Particle Swarm Optimization (PSO) [6, 7], African Vulture Optimization Algorithm (AVOA) [8], Artificial Gorilla Troops Optimizer (AGTO) [9], and Artificial Hummingbird Algorithm (AHA) [10]. The last class includes human-related algorithm and some examples are: Imperialist Competitive Algorithm (ICA) [11], Teaching learning-Based Optimization (TLBO) [12], and Social Group Optimization (SGO) [13].

The HS algorithm belongs the third class discussed above and mimics the process of composing music by a musician, which was referred to as the perfect state of harmony in the original paper. In fact, HS simulates the process of searching to find an "optimal" harmony, which is a combination of sounds. In HS, the search/optimization process starts with randomly creating a set of solutions called harmony memory. New solutions are created using this set to update the news harmony set. In HS, each *harmonic* is equivalent to *chromosomes* in the GA algorithm. The *pitches* in harmonies are similar to *genes* in chromosomes in GA. The h*armony memory* is identical to the *population* in GA.

In the HS algorithm, the process of creating new solutions is done using one of the following principles:

- Only utilizing the harmony memory
- Pitch adjusting
- Randomization.

The first method mimics how a musician replicates music from an existing piece of music. This method is replicated in HS by the probability of accepting new harmonics in the harmony memory. It is similar to elitism in evolutionary algorithms, in which the best solutions are transferred to the latest population to avoid missing them when using other evolutionary operators (e.g., crossover or mutation). A parameter called acceptance rate ($r_{\text{acceptance}}$) is used in the HS algorithm in this regard, in the range [0,1]. Small values for these parameters will lead toward less transition of best solutions from the current harmony memory to the next. High values will cause more identically best solutions in the next harmony memory.

The second mechanism inspires from the process of slight variation to play something a little different from original music. Finally, the last technique is similar when a musician creates a completely new piece of music. This is done using the following equation [14]:

$$\overrightarrow{X_{\text{new}}} = \overrightarrow{X_{\text{old}}} + \overrightarrow{b_{\text{range}}} \cdot \vec{\epsilon}, \tag{1}$$

where $\overrightarrow{X_{\text{new}}}$ is the new harmonic that we want to create, $\overrightarrow{X_{\text{old}}}$ is the harmonic undergoing changes including pitches, $\overrightarrow{b_{\text{range}}}$ represents pitch bandwidth, and $\vec{\epsilon}$ includes random numbers in the range of [−1,1].

In summary, the main controlling parameters in HS are as follows:

- The acceptance rate in the harmony memory ($r_{\text{acceptance}}$)
- Update pitch adjustment rate (r_{pa})
- Pitch limits (min and max of each pitch)
- Pitch bandwidth (b_{range}).

The flowchart of the HS algorithm is presented in Fig. 1. It can be seen that the optimization/search process starts with a random set of harmonics to form the first harmony memory. The main controlling parameters (pitch adjustment rate, pitch bandwidth, and acceptance rate) are updated afterward. The HS algorithm then

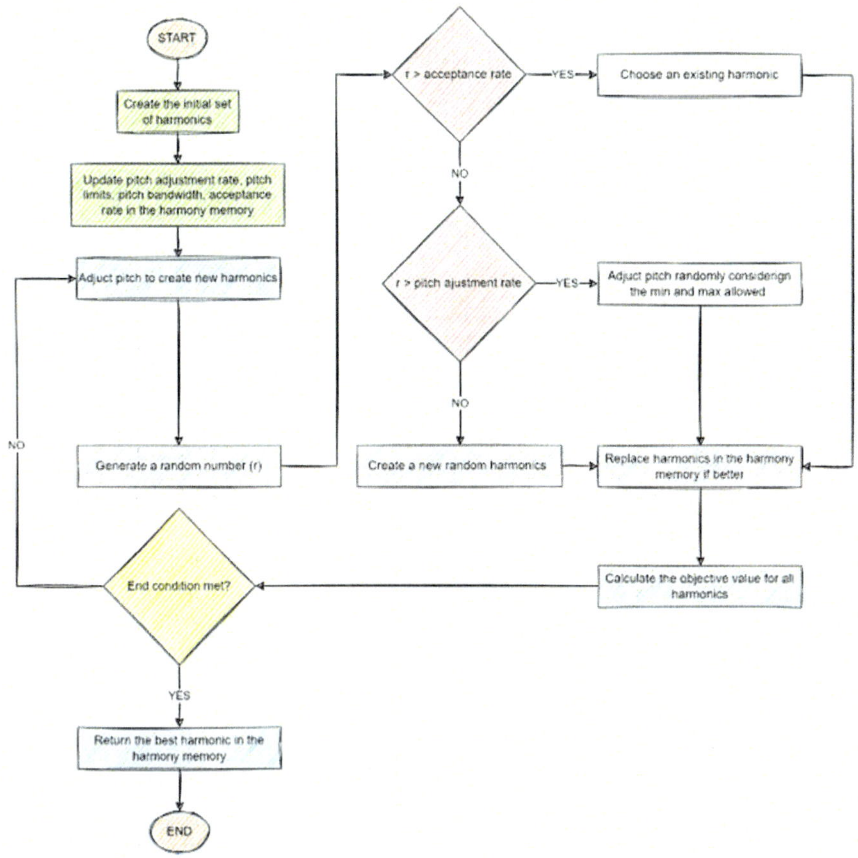

Fig. 1 Flowchart of the HS algorithm

creates/accepts new harmonics until the end criterion is satisfied. This process is done by adjusting the pitches and deciding whether to move them to the recent harmony memory or not, using a random number and the probabilities of acceptance or pitch adjustment rate.

In the next section, the EPD is introduced and integrated into the HS algorithm.

3 Integrating EPD to HS

The EPD operator is derived from the self-organized criticality theorem [15], which highlights the fact that small changes in a population (e.g., a handful of sand, snow, etc.) are able to provide balances in a balanced population without external forces [16]. This is observed in the evolutionary processes in nature too. Nature tends to impact a whole population by removing poorly fit organizations. This is often referred to as Evolutionary Population Dynamics (EPD).

Interestingly, an algorithm that uses EPD might affect the behavior in an opposite manner to GA. In GA, the fittest individual is given a higher chance of survival to contribute to the production of the following population. In the EPD-based algorithm, poor individuals have a higher chance of being completely excluded from the current population and contributing to the next population. Some of the most popular algorithms that use this principle in the literature are Extremal optimization [17], evolutionary programming using self-organizing criticality [18], and Gray Wolf Optimizer with EPD [19].

The use of EPD in meta-heuristics was extensively investigated by Lewis et al. in [16]. According to [5], in any population-based algorithm, the worst individuals could be eliminated and replaced by other individuals. We follow the same principle in this work by finding a portion of the worst harmonics in each harmony memory and replacing them with entirely random harmonics. This is shown in Fig. 2. It can be seen that n harmonics are chosen to face removal and randomized in initialization to mimic the EPD in HS.

To investigate the impact of EPD on the performance of HS, we have considered three scenarios in this work as:

- HS_EPD1: n is chosen in a way that EPD will impact 10% of harmonics
- HS_EPD2: n is selected in a way that 30% of harmonics will be affected by EPD
- HS_EPD3: n is chosen in a way that EPD will impact 50% of harmonics.

In all these cases, the worst harmonics will be chosen. We have selected these three cases to see how the performance of HS is when EPD is applied slightly (10%), moderately (30%), or extensively (50%).

We argue that removing the worst harmonics in each harmony memory will increase the overall quality of harmony memory over time and assist to improve the performance of HS. Note that, the EPD does increase the computational complexity of HS, so any improvement will be achieved without compromising other performance metrics in this algorithm. The following section will provide the results and discussions to investigate these.

4 Experimental Results and Discussions

To test the performance of the proposed HS_EPD Algorithms, six challenging test functions from CEC composite benchmark suite have been used. The details of these test functions can be found in [20]. They provide some of the most challenging testbeds in the literature to effectively test the performance of algorithms. To verify the results, the three HS_EPD variables are compared with the original HS. We have used 100 harmonics over 1000 iterations. The statistical results, including mean, standard deviation, and median, are calculated after running each algorithm 30 times. The results are provided in Table 1.

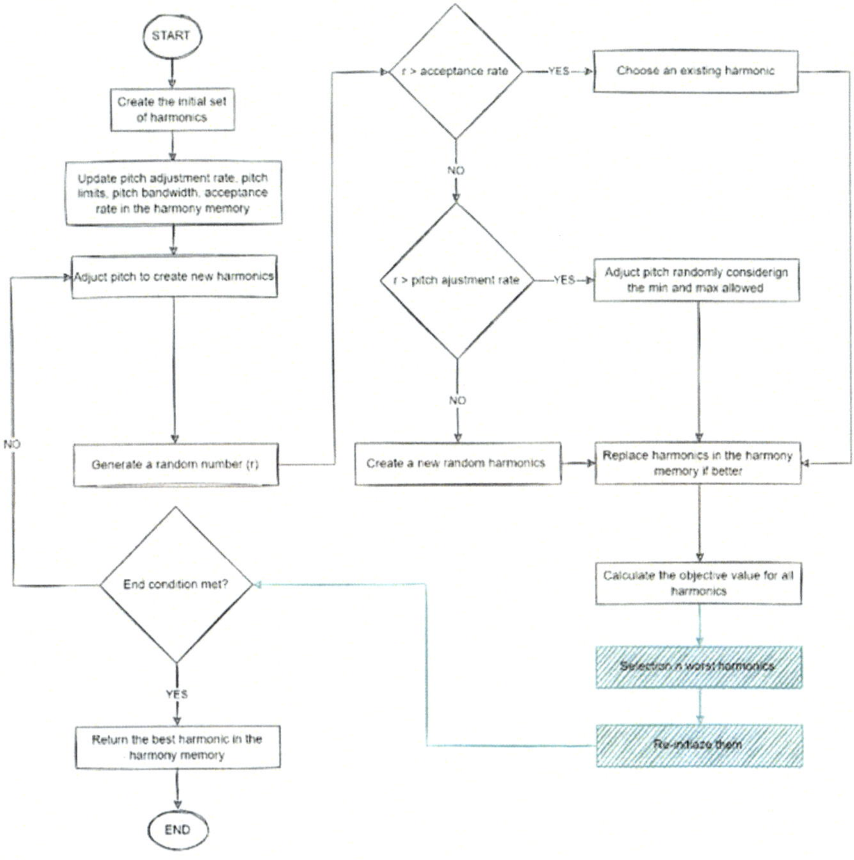

Fig. 2 Flowchart of HS with EPD

This table gives that the vanilla HS provides the best results in the first composite test (CF). HS_EPD1, HS_EPD2, and HS_EPD3 follow this. This is the easiest test function in this benchmark suit, so this is expected. However, looking at other test functions, it is evident that HS_EPD variants clearly outperform the vanilla HS. The discrepancy of results increased from CF2 to CF6. This is correlated with the increased complexity from CF2 to CF6 as well, which demonstrates that EPD has the potential to help HS with solving challenging problems.

Out of the three HS_EPD variants, it seems that the HS_EPD1 algorithm performs best on CF2 and CF3; whereas, HS_EPD2 achieves the best results on CF5 and CF6. Another interesting observation shows that the higher portion of the population impacted by EPD, the better is the performance on challenging test problems. This is due to the fact that poor harmonics are replaced by random ones more frequently, which leads to improved exploratory behavior of HS. This is essential in solving the challenging problem with a large number of locally optimal solutions.

Table 1 Results on composite test functions (100 solutions, 1000 iterations, and 30 runs)

CF	Metric	HS	HS_EPD1	HS_EPD2	HS_EPD3
CF1	Mean	**0.266513389**	0.400399996	0.451715499	1.125426732
	Std	**0.387661357**	0.614628944	0.334102054	1.659997413
	Median	**0.128577392**	0.129691337	0.398418957	0.585904992
CF2	Mean	77.20892544	**44.22172301**	78.64062473	68.9507034
	Std	46.09133159	**28.40700684**	45.8725902	33.12098604
	Median	70.70250086	**46.31843659**	83.01876222	63.65508893
CF3	Mean	32.06941904	**19.37234211**	27.07047651	37.32525094
	Std	50.04861834	**34.50993415**	37.42768877	58.67583226
	Median	6.269464695	**5.97360965**	6.255385208	10.96346256
CF4	Mean	239.2180953	254.4525668	**210.8037491**	219.5229331
	Std	75.58900946	64.37841561	**54.31510458**	44.27119203
	Median	206.9532918	224.9458951	**206.7245143**	218.3074891
CF5	Mean	127.0344637	111.1679693	102.1229179	**98.42914364**
	Std	99.78565134	28.40389198	42.34221428	**44.57338186**
	Median	103.0754954	101.5351978	103.9831019	**108.2716011**
CF6	Mean	390.5489762	399.3180773	400.9524994	**365.197451**
	Std	74.38885218	33.41422394	45.99829993	**98.44863091**
	Median	408.7654725	410.0972846	414.3900272	**416.0381519**

The above results demonstrate that EPD could be exceptionally efficient operators for the HS algorithm. However, the question is how the convergence rate will be changed. To investigate this further, Fig. 3 is provided. This figure shows that all algorithms provide similar convergence patterns on CF1, which aligns with the observation made in Table 1. The convergence patterns of HS_EPD algorithms are different from other test functions.

In CF2, for instance, the HS_EPD1 provides the fastest convergence among others. This function tends to be unimodal in most areas and slight variations of HS's population seem to be more efficient. In CF6, however, the HS_EPD3 shows substantially better convergence behavior. This is once again due to the more remarkable change of removing poor harmonics in the HS_EPD3 algorithm.

Considering all these, the results demonstrate the merits of incorporating the EPD operator in the HS algorithm, since the original algorithm inherently does not have any mechanism to deal with poor harmonics. In fact, the acceptance rate mimics the elitism as an effective operator in avoiding losing the best harmonics. However, the results of this study also demonstrated that the poor individuals are worthy of being taken care of using other evolutionary operators, including EPD.

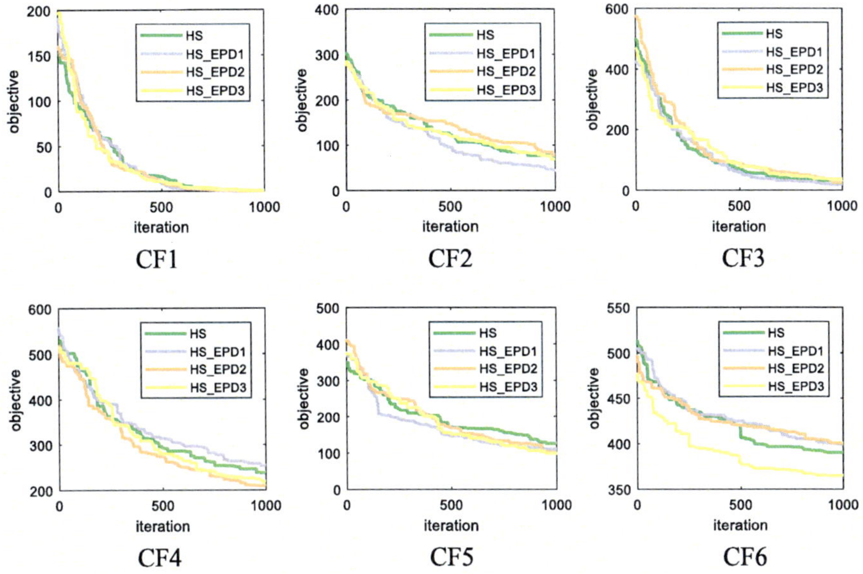

Fig. 3 Convergence curves of HS, HS_EPD1, HS_EPD2, and HS_EPD3 on the test functions

5 Conclusion and Future Works

This work focused on the computationally inexpensive method to improve the performance of the HS algorithm, which is one of the most well-regarded techniques in the optimization community. It was discussed that the HS algorithm does have a mechanism to maintain the best harmonies and transfer them to the next harmony memory. However, the worst harmonics are not affected by any means. This motivated us to select a portion of such harmonics and randomly initialize them at the end of each iteration using the EPD operator. Three cases of using EPD (slightly, moderately, and extensively) were considered to investigate the impact of this novel mechanism. It was observed that the performance of the HS algorithm could be improved using such mechanisms by targeting poor harmonics, without increasing the overall computation cost of the algorithm.

For future work, it is recommended to test EPD on other variants of the HS algorithm. Also, the merits of such a mechanism on multi-objective optimization will be worthy of investigation. In addition, EPD in HS will be analyzed in medical application. We are planning to investigate the application of our integrated optimization technique to the segmentation of dense breast mammography images in order to enhance segmentation and, as a result, cancer diagnosis. Cancer detection in dense breast is challenging even for experienced radiologist, when tumor is present in dense tissue and both tumor and overlaying tissue appear white in the mammogram.

References

1. Yu X, Gen M (2010) Introduction to evolutionary algorithms. Springer Science & Business Media
2. Geem ZW, Kim JH, Loganathan GV (2001) A new heuristic optimization algorithm: harmony search. Simulation 76(2):60–68
3. Holland JH (1992) Genetic algorithms. Sci Am 267(1):66–73
4. Rechenberg I (1989) Evolution strategy: nature's way of optimization. In: Optimization: methods and applications, possibilities and limitations. Springer, pp 106–126
5. Eltaeib T, Mahmood A (2018) Differential evolution: a survey and analysis. Appl Sci 8(10):1945
6. Kennedy J, Eberhart R (1995) Particle swarm optimization. In: Proceedings of ICNN'95-international conference on neural networks, vol 4. IEEE, pp 1942–1948
7. Althobiani F, Khatir S, Brahim B, Ghandourah E, Mirjalili S, Wahab MA (2021) A hybrid PSO and Grey Wolf optimization algorithm for static and dynamic Crack identification. Theoretical and applied fracture mechanics, p 103213
8. Abdollahzadeh B, Gharehchopogh FS, Mirjalili S (2021) African vultures optimization algorithm: a new nature-inspired metaheuristic algorithm for global optimization problems. Comput Ind Eng 158:107408
9. Abdollahzadeh B, Soleimanian Gharehchopogh F, Mirjalili S (2021) Artificial gorilla troops optimizer: a new nature-inspired metaheuristic algorithm for global optimization problems. Int J Intell Syst 36(10):5887–5958
10. Zhao W, Wang L, Mirjalili S (2022) Artificial hummingbird algorithm: a new bio-inspired optimizer with its engineering applications. Comput Methods Appl Mech Eng 388:114194
11. Atashpaz-Gargari E, Lucas C (2007) Imperialist competitive algorithm: an algorithm for optimization inspired by imperialistic competition. In: 2007 IEEE congress on evolutionary computation. IEEE, pp 4661–4667
12. Rao RV, Savsani VJ, Vakharia D (2011) Teaching–learning-based optimization: a novel method for constrained mechanical design optimization problems. Comput Aided Des 43(3):303–315
13. Satapathy S, Naik A (2016) Social group optimization (SGO): a new population evolutionary optimization technique. Complex Intell Syst 2(3):173–203
14. Yang X-S (2009) Harmony search as a metaheuristic algorithm. In: Music-inspired harmony search algorithm. Springer, pp 1–14
15. Bak P, Tang C, Wiesenfeld K (1987) Self-organized criticality: an explanation of the 1/f noise. Phys Rev Lett 59(4):381
16. Lewis A, Mostaghim S, Randall M (2008) Evolutionary population dynamics and multi-objective optimisation problems. In Multi-objective optimization in computational intelligence: theory and practice. IGI Global, pp 185–206
17. Boettcher S, Percus AG (1999) Extremal optimization: methods derived from co-evolution. arXiv preprint math/9904056
18. Lewis A, Abramson D, Peachey T (2003) An evolutionary programming algorithm for automatic engineering design. International conference on parallel processing and applied mathematics. Springer, pp 586–594
19. Saremi S, Mirjalili SZ, Mirjalili SM (2015) Evolutionary population dynamics and grey wolf optimizer. Neural Comput Appl 26(5):1257–1263
20. Liang J-J, Suganthan PN, Deb K (2005) Novel composition test functions for numerical global optimization. In: Proceedings 2005 IEEE Swarm intelligence symposium, SIS 2005. IEEE, pp 68–75

Chaotic Stochastic Paint Optimizer (CSPO)

Nima Khodadadi, Seyed Mohammad Mirjalili, Seyedeh Zahra Mirjalili, and Seyedali Mirjalili

Abstract Optimization of engineering problems requires addressing several common difficulties in the optimization problem, including but not limited to a large number of decision variables, multiple often conflicting objectives, constraints, locally optimal solutions, and expensive objective functions. It is pretty common that an algorithm performs very well on test functions but struggles when applying to real-world problems. This paper proposes a chaotic version of the recently proposed algorithm called chaotic stochastic paint optimizer (CSPO). A comparative study with other meta-heuristics demonstrates the merits of this algorithm and the change applied in this work.

Keywords Stochastic paint optimizer · Optimization · Engineering problems · Chaotic stochastic paint optimizer

N. Khodadadi
Department of Civil and Environmental Engineering, Florida International University, Miami, FL, USA
e-mail: nkhod002@fiu.edu

S. M. Mirjalili
Department of Engineering Physics, Polytechnique Montréal, Montreal, Canada

S. Z. Mirjalili · S. Mirjalili (✉)
Centre for Artificial Intelligence Research and Optimisation, Torrens University Australia, Brisbane, QLD, Australia
e-mail: ali.mirjalili@gmail.com

S. Z. Mirjalili
e-mail: zahra.mirjalili@torrens.edu.au

S. Mirjalili
Yonsei Frontier Lab, Yonsei University, Seoul, South Korea

J. H. Kim et al. (eds.), *Proceedings of 7th International Conference on Harmony Search, Soft Computing and Applications*, Lecture Notes on Data Engineering and Communications Technologies 140, https://doi.org/10.1007/978-981-19-2948-9_19

1 Introduction

Optimization problems can be found in a wide range of fields. Despite the diversity and commonality, they have similar components: decision variables, constraints, and objectives. The decision variables are the unknowns of optimization problems. Some examples are the length/width of a beam, the diameter of an aircraft fuselage, connection weights in neural networks, the order of classes in a timetable for a given teacher, etc. An optimization algorithm will determine optimal values for these parameters. Every unique combination of values for the inputs creates a potential solution for the optimization problem on hand. The set of all possible solutions is called solution set or search space, which can be finite or infinite.

To evaluate solutions in the solution set and compare them, we need a measure. This is where an objective function comes into play. An objective function, often called cost or merit function, maps the solution set to an objective set. This means that there is a unique member in the objective set for every member of the solution set. If the problem has more than one objective, there is a set for each objective independently. Some examples of objective functions are efficiency, fuel consumption, distance, etc. The combination of the solution set and objective set(s) search defines a search landscape in an optimization problem [1].

The last component of an optimization problem is the set of constraints. Constraints create a subset of all possible solutions. If a solution leads to a violation of any of the constraints, it is considered infeasible and should be excluded as it does not belong to the subset of feasible solutions. With the three elements: decision variables, objectives, and constraints, an optimization problem can be formulated as a minimization single-objective problem without the loss of generality as follows [2]:

$$P\text{: Minimize} \quad f(x) \tag{1}$$

$$\text{Subject to: } x \in X \tag{2}$$

$$\text{where: } f\text{: } R^n \rightarrow R \text{ and } X \text{ is a subset of } R^n \tag{3}$$

In the above formulation, R^n is the set of all possible solutions, X represents the set of all feasible solutions (search space), x is a solution in X, and f is the objective function.

A solution (e.g., x^*) is called optimal solution if it is a member of $X(x^* \in X)$ and attains the minimum (or maximum in case of maximization) value of the objective function $(\forall x \in X\text{: } f(x^*) \leq f(x))$.

The first step in solving any optimization problem is to formulate and perhaps program it, which is essential to the next step: to find or develop an optimization algorithm to solve it. There are different classes of optimization algorithms in the literature. One of the most practical classifications is to categorize them based on the differentiability of the problem.

If the objective function(s) are differentiable, we can use or develop a derivative-based (gradient-based) optimization algorithm [3]. One of the most popular algorithms in this category is gradient descent. This algorithm is a first-order technique that takes repetitive steps in the opposite direction of the gradient of the objective function. This leads to an initial solution toward a local minimum [4].

On the other hand, for the non-differentiable objective function, we need other types of algorithms. There are different algorithms in this category, but there is no doubt that the most popular ones are heuristics or meta-heuristics. An example of such an algorithm is greedy search. An optimization algorithm with the greedy mindset makes a sequence of changes in a solution and chooses the best result. This iterative process is continued until no further improvement can be made to the existing solution. Heuristic algorithms are problem-specific, meaning that they are designed for a particular set of problems.

However, meta-heuristics are generic optimization algorithms. This means that they consider a problem as a black box and evaluate solutions mainly based on the value of the objective function. The literature shows that they have been widely used in both science and industry to solve the optimization problem. They mostly benefit from stochastic mechanisms to help with avoiding locally optimal solutions. As a drawback, they tend to be more computationally expensive than other optimization algorithms. Due to the stochastic nature, they also estimate the global optimum for optimization problems, so there should be quality assurance measures and techniques in place to ensure the quality of the final solution obtained by them [5].

There are so many meta-heuristics proposed lately in the literature with unprecedented acceleration pace [6]. Some of the most popular and recent of them are particle swarm optimization (PSO) [7], genetic algorithm (GA) [8], harmony search (HS) [9], dynamic water strider algorithm (DWSA) [10], gray wolf optimizer (GWO) [10], advanced charged system search (ACSS) [11], neural network algorithm (NNA) [12], hybrid invasive weed optimization-shuffled frog-leaping algorithm (IWOSFLA) [13], moth flame optimization (MFO) [14], dynamic arithmetic optimization algorithm (DAOA) [15], colliding bodies optimization (CBO) [16], African vulture optimization algorithm (AVOA) [17], artificial gorilla troops optimizer (AGTO) [18], and artificial hummingbird algorithm (AHA) [19]. According to the No Free Lunch (NFL) [20] theorem, none of these algorithms will be able to solve all optimization problems. This means that one or one set of algorithms might show good performance of a set of the problem but poor on the other. This reinforces the need to have comparative studies on different problems areas. Such a comparative analysis also provides new insights on the performance of meta-heuristics and potential ways to improve them. In the present paper, a chaotic version of the newly established art-based algorithm, so-called stochastic paint optimizer (SPO) [21], is introduced and applied to several engineering examples.

The literature shows that the chaos theory can be used to generate random values for meta-heuristics and achieve improved performance [22]. This motivates our attempts to provide a comparison of several meta-heuristics on a very challenging problem that are in engineering. The rest of the paper is organized as follows:

Section 2 provides a summary of the SPO algorithm. Section 3 discussed the details of the chaotic version of SPO. The definition of problems with experimental results is discussed in Sect. 4. Finally, Sect. 5 concludes the work and suggests future directions.

2 Stochastic Paint Optimizer (SPO)

Stochastic paint optimizer (SPO) [21] is a new meta-heuristic method developed in 2020 based on color theory. In fact, the solution in the SPO algorithm is combined to create a new population in a say manner that colors are mixed to create new colors. Fig. 1 shows four basic techniques for combining colors based on the color wheel. All these techniques are utilized to create new colors (X^i_{new}).

In the following sub-sections, the mathematical models used in SPO to mimic color mixing patterns in Fig. 1 are presented.

2.1 Analogous Combination Technique

In this type of mixing colors, two to four colors next to each other in the color wheel are used (see Fig. 1a). In SPO, this is done using the following equation:

$$X^1_{\text{new}} = X_i + \text{rand}.(X_{i+1} - X_{i-1}) \quad (4)$$

where X_{i-1}, X_i, and X_{i+1} are three solutions from the population and *rand* is a random vector in the range [0, 1].

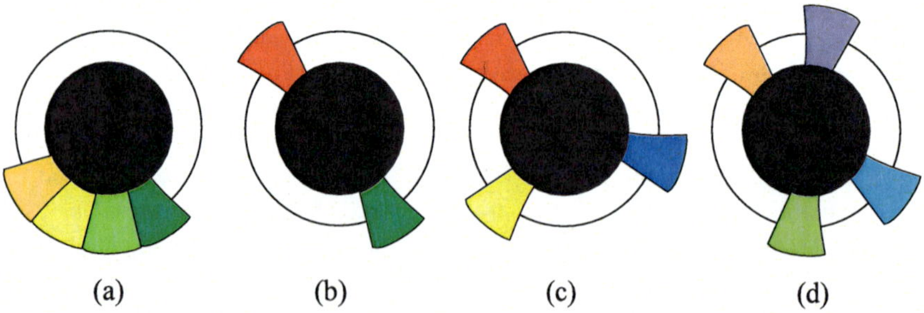

Fig. 1 Four basic techniques for combining colors: **a** analogous combination **b** complementary combination **c** triadic combination **d** tetradic combination

2.2 Complementary Combination Technique

When mixing colors in this fashion, complementary colors are used, as can be seen in Fig. 1b. This is simulated in SPO by randomly choosing a primary color (X_p), a tertiary color (X_T), and an existing color (X_i). The following equation is used for this purpose:

$$X_{\text{new}}^2 = X_i + \text{rand}.(X_P - X_T) \tag{5}$$

where X_p represents a primary color, X_T indicates a tertiary color, X_i is an existing color, and *rand* is a random vector in the range [0, 1].

2.3 Triadic Combination Technique

In this mechanism, new colors are made from the evenly spaced colors around the color wheel. This is done in SPO as follows:

$$X_{\text{new}}^3 = X_i + \text{rand}.\left(\frac{X_P + X_S + X_T}{3}\right) \tag{6}$$

where X_p represents a primary color, X_S is a secondary color, X_T indicates a tertiary color, X_i is an existing color, and *rand* is a random vector in the range [0, 1].

2.4 Tetradic Combination Technique

In this type of color mixing, four complementary colors are chosen from rectangular patterns on the color wheel. In SPO, this is done using the following equation:

$$\begin{aligned} X_{\text{new}}^4 &= X_i + \left(\frac{\text{rand}_1.X_P + \text{rand}_2.X_S + \text{rand}_3.X_T + \text{rand}_4.X_{\text{rand}}}{4}\right) \\ X_{\text{rand}} &= \text{LB} + \text{rand}.(\text{UB} - \text{LB}) \end{aligned} \tag{7}$$

where LB and UB are the lower and upper limits of the design variable and rand_1, rand_2, rand_3, and rand_4 are four random vectors in the range [0, 1].

A question that may arise now is how the colors are represented and mixed in SPO. In this algorithm, each solution is considered as a color with *d* dimension where *d* is the number of variables in the problem. The objective value of those solutions defined their 'colors'. In other words, the objective function plays the role of the color wheel. Therefore, a primary color is the best solution, secondary color is solutions with reasonably good objective values, and tertiary colors are the solutions with poor objective values.

3 Chaotic Stochastic Paint Optimizer

In this section, the chaotic maps are employed in SPO for obtaining better performance. Ten chaotic maps, namely Chebyshev [23], circle [24], Gauss/mouse [25], iterative [23], logistic [26], piecewise [27], sine [27], singer [27], sinusoidal [27], and tent [26], are used that can be observed in Fig. 2.

This challenging set of maps with a different behavior was selected, with an initial point of 0.7. One can choose any number from 0 to 1 for the initial point. It should be mentioned that the initial value could have essential impacts on some of the chaotic maps' fluctuation patterns. In CSPO, chaotic maps are used instead of random vectors to create new colors. This manipulation enhanced the exploration and exploitation

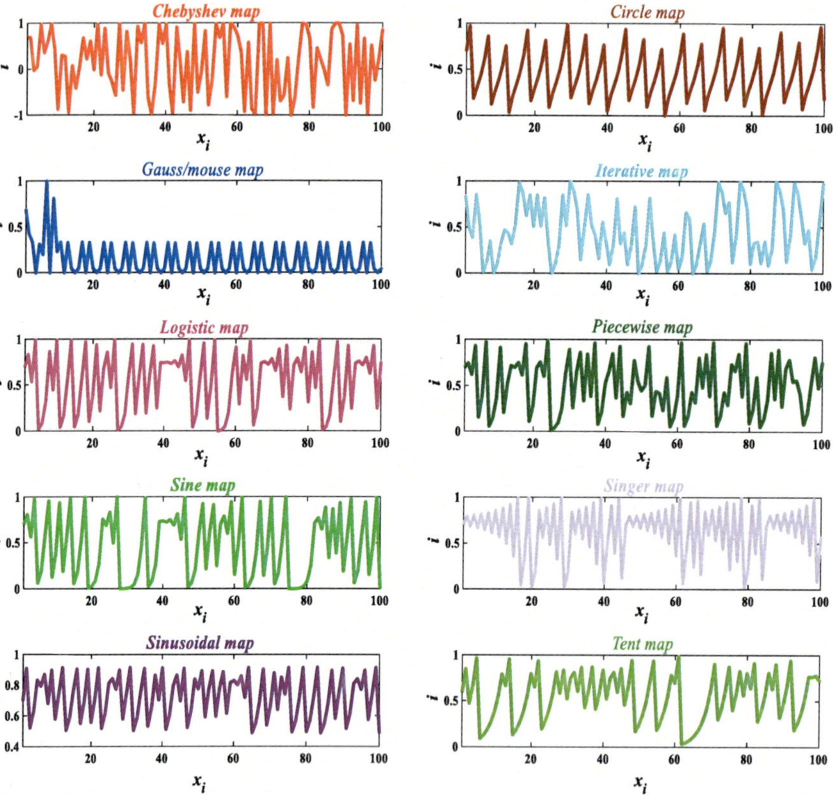

Fig. 2 Chaotic map value distribution

in SPO. All new combination techniques are as follow:

$$\text{Chaotic Analogous } X_{\text{new}}^1 = X_i + \text{chaos}.(X_{i+1} - X_{i-1}) \tag{8}$$

$$\text{Chaotic Complementary } X_{\text{new}}^2 = X_i + \text{chaos}.(X_P - X_T) \tag{9}$$

$$\text{Chaotic Triadic } X_{\text{new}}^3 = X_i + \text{chaos}.\left(\frac{X_P + X_S + X_T}{3}\right) \tag{10}$$

$$\text{Chaotic Tetradic } X_{\text{new}}^4 = X_i + \left(\frac{\text{chaos}_1.X_P + \text{chaos}_2.X_S + \text{chaos}_3.X_T + \text{chaos}_4.X_{\text{rand}}}{4}\right) \tag{11}$$

where: $X_{\text{rand}} = \text{LB} + \text{chaos}.(\text{UB} - \text{LB})$.

4 Results and Discussion

In this section, numerous widespread engineering design problems are examined to prove the efficiency of CSPO. For the sake of comparison, the outcomes of WSA [28], GWO [29], NNA [12], CBO [16], and MFO [14] algorithms are additionally reported. The optimization technique is performed using the CSPO and SPO algorithms to satisfy this aim. We have run each algorithm 30 times independently and use the same number of solutions and iteration in all algorithms. Other controlling parameters are identical to the references cited in each of the following sub-sections.

4.1 Cantilever Beam

This problem was first presented and discussed in [30]. The objective is to reduce the weight of the beam that consists of five hollow square-shared blocks. The decision variables of this problem are the length of a side for these squares (x_1 to x_5) after running. The results of CSPO and other algorithms are presented in Table 1. This table shows the optimal design variable values obtained by CSPO and various algorithms. As seen, the solution superior to any other approach has been obtained by CSPO.

4.2 Compound Gear

This problem's details can be found in [31], in which there are four variables that defined the number of teeth in the gears. The objective is to minimize the gear ratio. The results of CSPO, SPO, WSA, GWO, NNA, CBO, and MFO are presented in Table 2.

Table 1 Results of a cantilever beam with other methods

Algorithm	x_1	x_2	x_3	x_4	x_5	Optimal cost
CSPO	5.97875	4.875564	4.466371	3.479394	2.139049	1.3032512
SPO	5.977432	4.874808	4.469891	3.478552	2.138456	1.303252
WSA	5.975921	4.880778	4.465427	3.477822	2.139194	1.303252
GWO	5.971406	4.881808	4.473385	3.476943	2.135661	1.303256
NNA	5.969628	4.88762	4.46466	3.477317	2.139987	1.303257
CBO	5.96968	4.878843	4.460389	3.486275	2.144072	1.303259
MFO	5.997765	4.874964	4.454762	3.483629	2.128316	1.30327

Table 2 Results of compound gear with other methods

Algorithm	x_1	x_2	x_3	x_4	Optimal cost
CSPO	49	19	16	43	2.70E−12
SPO	43	16	19	49	2.70E−12
WSA	43	16	19	49	2.70E−12
GWO	49	19	16	43	2.70E−12
NNA	49	16	19	43	2.70E−12
CBO	53	13	20	34	2.31E−11
MFO	51	30	13	53	2.31E−11

Table 2 describes the statistical details. In terms of optimal cost, CSPO, SPO, WSA, GWO, and NNA methods were superior to CBO and MFO algorithms. Table 2 lists the optimal designs in which four variables (number of teeth) are acquired as the optima by various permutations.

4.3 Welded Beam

In a welded beam system, four design parameters should be optimized to provide the least construction cost. This is a classical problem with seven constraints, and more details can be found in [32]. Table 3 presents the variables and optimal outcome of CSPO in comparison with other algorithms. As seen here, CSPO, WSA, NNA, and MFO found the best results.

4.4 Three-bar Truss

This problem is another popular one in the literature [33]. The design of a three-bar truss with two parameters should be optimized to minimize its weight. Despite

Table 3 Results of the welded beam with other methods

Algorithm	$x_1 = h$	$x_2 = l$	$x_3 = t$	$x_4 = b$	Optimal cost
CSPO	0.20572963	3.47048866	9.03662391	0.20572964	1.724852
SPO	0.205729	3.470493	9.036623	0.205729	1.724853
WSA	0.20573	3.470489	9.036624	0.20573	1.724852
GWO	0.205676	3.478377	9.03681	0.205778	1.72624
NNA	0.20573	3.470489	9.036624	0.20573	1.724852
CBO	0.205728	3.47052	9.036627	0.20573	1.724862
MFO	0.20573	3.470489	9.036624	0.20573	1.724852

Table 4 Results of three-bar truss with other methods

Algorithm	x_1	x_2	Optimal cost
CSPO	0.7886811	0.408231412	263.895843
SPO	0.788625	0.408389	263.895845
WSA	0.788683	0.408227	263.895843
GWO	0.788648	0.408325	263.896006
NNA	0.788639	0.40835	263.895844
CBO	0.788504	0.408733	263.895882
MFO	0.788601	0.408458	263.895847

having a small number of variables, there are so many constraints that make it quite challenging for optimization problems. The results of the seven comparative algorithms in this work are presented in Table 4. It can be seen in this table that both CSPO and WSA find a design with the best cost of 263.895843.

Taken together, the results demonstrated that the use of chaotic maps improved the performance of CSPO and assisted it to provide superior results on the majority of case studies in this work.

5 Conclusion

Chaotic maps have been widely used to improve the performance of meta-heuristics. It is a way to increase the stochastic behaviors of such algorithms to better avoid locally optimal solutions and/or balance exploration and exploitation. We integrated ten chaotic maps into the recently proposed SPO algorithm and compared with other algorithms such as vanilla SPO, GWO, WSA, NNA, CBO, and MFO. All algorithms were compared on four classical optimization problems in mechanical engineering. It was observed that the CSPO algorithm outperforms other algorithms on the majority of case studies.

References

1. Khodadadi N, Azizi M, Talatahari S, Sareh P (2021) Multi-objective crystal structure algorithm (MOCryStAl): introduction and performance evaluation. IEEE Access
2. Kaveh A, Talatahari S, Khodadadi N (2019) The hybrid invasive weed optimization-shuffled frog-leaping algorithm applied to optimal design of frame structures. Period Polytech Civ Eng 63(3):882–897
3. LeCun Y, Bottou L, Bengio Y, Haffner P (1998) Gradient-based learning applied to document recognition. Proc IEEE 86(11):2278–2324
4. Ruder S (2016) An overview of gradient descent optimization algorithms. arXiv Prepr. arXiv1609.04747
5. Khodadadi N, Mirjalili S (2022) Truss optimization with natural frequency constraints using generalized normal distribution optimization. Appl. Intell, 1–14
6. Kaveh A, Khodadadi N, Talatahari S (2021) A comparative study for the optimal design of steel structures using CSS and ACSS algorithms. Iran Univ Sci Technol 11(1):31–54
7. Kennedy J, Eberhart R (1995) Particle swarm optimization. In: Proceedings of ICNN'95-international conference on neural networks, vol 4, pp 1942–1948
8. Holland JH (1992) Genetic algorithms. Sci Am 267(1):66–73
9. Geem ZW, Kim JH, Loganathan GV (2001) A new heuristic optimization algorithm: harmony search. SIMULATION 76(2):60–68
10. Kaveh A, Eslamlou AD, Khodadadi N (2020) Dynamic water strider algorithm for optimal design of skeletal structures. Period Polytech Civ Eng 64(3):904–916
11. Kaveh A, Khodadadi N, Azar BF, Talatahari S (2020) Optimal design of large-scale frames with an advanced charged system search algorithm using box-shaped sections. Eng Comput, pp 1–21
12. Sadollah A, Sayyaadi H, Yadav A (2018) A dynamic metaheuristic optimization model inspired by biological nervous systems: neural network algorithm. Appl Soft Comput 71:747–782
13. Kaveh A, Talatahari S, Khodadadi N (2019) Hybrid invasive weed optimization-shuffled frog-leaping algorithm for optimal design of truss structures. Iran J Sci Technol Trans Civ Eng 44(2):405–420
14. Mirjalili S (2015) Moth-flame optimization algorithm: a novel nature-inspired heuristic paradigm. Knowl-Based Syst 89:228–249
15. Khodadadi N, Vaclav S, Mirjalili S (2022) Dynamic arithmetic optimization algorithm for truss optimization under natural frequency constraints. IEEE Access, 1. https://doi.org/10.1109/ACCESS.2022.3146374
16. Kaveh A, Mahdavi VR (2014) Colliding bodies optimization: a novel meta-heuristic method. Comput Struct 139:18–27
17. Abdollahzadeh B, Gharehchopogh FS, Mirjalili S (2021) African vultures optimization algorithm: a new nature-inspired metaheuristic algorithm for global optimization problems. Comput Ind Eng 158:107408
18. Abdollahzadeh B, Soleimanian Gharehchopogh F, Mirjalili S (2021) Artificial gorilla troops optimizer: a new nature-inspired metaheuristic algorithm for global optimization problems. Int J Intell Syst
19. Zhao W, Wang L, Mirjalili S (2022) Artificial hummingbird algorithm: a new bio-inspired optimizer with its engineering applications. Comput Methods Appl Mech Eng 388:114194
20. Wolpert DH, Macready WG (1997) No free lunch theorems for optimization. IEEE Trans Evol Comput 1(1):67–82
21. Kaveh A, Talatahari S, Khodadadi N (2020) Stochastic paint optimizer: theory and application in civil engineering. Eng Comput, 1–32
22. Sheikholeslami R, Kaveh A (2013) A survey of chaos embedded meta-heuristic algorithms. Int J Optim Civ. Eng 3(4):617–633
23. He D, He C, Jiang L-G, Zhu H, Hu G (2001) Chaotic characteristics of a one-dimensional iterative map with infinite collapses. IEEE Trans Circ Syst I Fundam Theor Appl 48(7):900–906

24. Devaney RL (1989) An introduction to chaotic dynamical systems. Chapman and Hall/CRC
25. Bucolo M, Caponetto R, Fortuna L, Frasca M, Rizzo A (2002) Does chaos work better than noise? IEEE Circ Syst Mag 2(3):4–19
26. Ott E (2002) Chaos in dynamical systems. Cambridge University Press
27. Peitgen H-O, Jürgens H, Saupe D, Feigenbaum MJ (2004) Chaos and fractals: new frontiers of science, vol 106. Springer
28. Kaveh A, Bakhshpoori T (2016) A new metaheuristic for continuous structural optimization: water evaporation optimization. Struct Multidiscip Optim 54(1):23–43
29. Mirjalili S, Mirjalili SM, Lewis A (2014) Grey wolf optimizer. Adv Eng Softw 69:46–61
30. Chickermane H, Gea HC (1996) Structural optimization using a new local approximation method. Int J Numer Methods Eng 39(5):829–846
31. Kannan BK, Kramer SN (1994) An augmented Lagrange multiplier based method for mixed integer discrete continuous optimization and its applications to mechanical design
32. Coello CAC (2000) Use of a self-adaptive penalty approach for engineering optimization problems. Comput Ind 41(2):113–127
33. Cheng M-Y, Prayogo D (2014) Symbiotic organisms search: a new metaheuristic optimization algorithm. Comput Struct 139:98–112

The Investigation of Optimization of Eccentricity in Reinforced Concrete Footings

Sinan Melih Nigdeli and Gebrail Bekdaş

Abstract In this study, it is aimed to investigate the consideration of the orientation of the mounted column to the footings in the optimum design. In that case, a good balance on critical sections moments and the design can be done optimally. The results including eccentricity of the mounted column are compared with the case of footings that have columns in the mid-point. The optimization is done via harmony search algorithm, and multiple cases of bending moment of the mounted column are investigated. According to the results, the optimum orientation of the column has up to 71.49% cost reduction, and the gap between the two designs closes if the axial force increases.

Keywords Reinforced concrete · Footing · Optimization · Harmony search · Eccentricity

1 Introduction

In an engineering design, if the optimum design dimensions are needed to be found, it is not possible to find via a mathematical method. The design variables as dimensions are needed to check the design constraints. In that case, the unknowns of the optimization problem are used in the design steps, and these values are assumed by a designer and checked for the constraints. This process can be iteratively done to find the optimum design via assumed values. The automatic process of this can be done via metaheuristic methods.

The optimum design of reinforced (RC) members is also in these types of problem. In the design, several constraints that are proposed in the design regulations must be checked.

S. M. Nigdeli · G. Bekdaş (✉)
Department of Civil Engineering, Istanbul University-Cerrahpaşa, 34320 Avcılar, Istanbul, Turkey
e-mail: bekdas@iuc.edu.tr

S. M. Nigdeli
e-mail: melihnig@iuc.edu.tr

J. H. Kim et al. (eds.), *Proceedings of 7th International Conference on Harmony Search, Soft Computing and Applications*, Lecture Notes on Data Engineering and Communications Technologies 140, https://doi.org/10.1007/978-981-19-2948-9_20

RC beams that are controlled via the ductile requirements of balanced reinforcements have been optimized via several metaheuristics such as genetic algorithm (GA) [1–3], simulated annealing (SA) [4], harmony search (HS) [5–8], bat algorithm (BA) [8, 9], teaching–learning-based-optimization (TLBO) [10], and Jaya algorithm (JA) [11]. Also, different algorithms were evaluated for T-shaped RC beams [12, 13]. Also, RC columns have been optimized via GA [14], HS [15, 16], BA [17], and TLBO [18]. As the combination of beam and columns members, RC frames were also optimized via metaheuristics [19–23]. All of the RC members are checked according to structural design limits, but soil structures are additionally designed via geotechnical design limits. Metaheuristics have been used in the optimum design of RC footings [24–29] and RC retaining walls [30–37]. In addition to optimization studies, metaheuristics have been used to generate optimization data for RC members that are used in machine learning of artificial intelligence estimation models [38–40].

In the present study, the music-inspired HS [41] was used in the optimization of RC footings according to ACI 318 [42]. The effect of consideration of the optimum column orientation is investigated and compared with the cases without optimization of orientation in which the column is positioned at the mid-point of the footing.

2 The Optimization Problem

The design variables of RC footings are shown in Fig. 1. The optimization method covers the optimum design of footings dimensions (X_1–X_3), column orientation (X_4 and X_5), reinforcement as sizes (X_6 and X_7), and gaps between them (X_8 and X_9). For practical use in the construction size, dimensions are discrete variables, and the reinforcement sizes are also existing ones in the market.

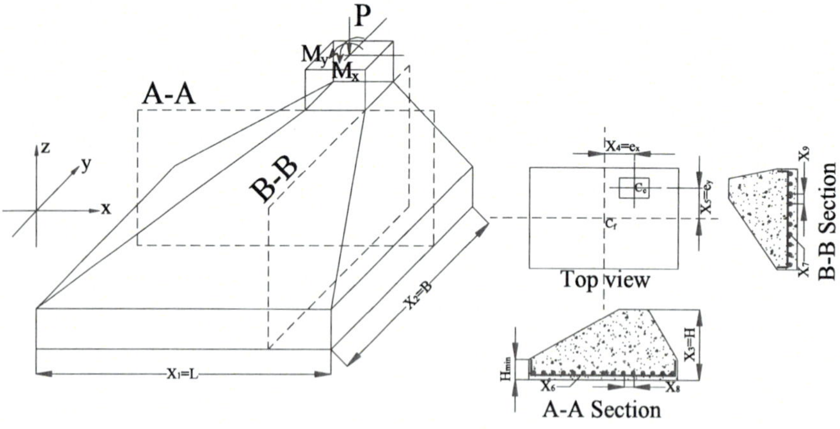

Fig. 1 Optimization problem with design variables

In the present study, different cases of loadings were investigated. Three cases of axial force (P) are taken, and moment values (M_x and M_y) are also multiple cases for all axial force cases. The design variable ranges and design constants are given in Table 1.

In the optimization iteration, firstly, X_1–X_5 are generated. Then, the bearing pressures ($q_{1,2,3,4}$) and the settlement (δ) are checked. As formulated in Eq. (1), W_f is the total weight of footing including the upper soil. e_x and e_y are the dimensions for the eccentric orientation of the column. If the center of footing (C_f) and center of column (C_c) are at the same point, these values are constant as zero.

$$q_{1,2,3,4} = \frac{P + W_f}{BL} \pm \frac{6(M_y - Pe_x)}{BL^2} \pm \frac{6(M_x - Pe_y)}{B^2L} \quad (1)$$

Table 1 Design constants and ranges of design variables

Definition	Symbol	Unit	Value
Yield strength of steel	f_y	MPa	420
Compressive strength of concrete	f'_c	MPa	25
Concrete cover	c_c	mm	100
Max. aggregate diameter	D_{max}	mm	16
Elasticity modulus of steel	E_s	GPa	200
Specific gravity of steel	γ_s	t/m^3	7.86
Specific gravity of concrete	γ_c	kN/m^3	23.5
Cost of concrete per m^3	C_c	\$/m^3	40
Cost of steel per ton	C_s	\$/t	400
Internal friction angle of soil	ϕ'	°	35
Unit weight of base soil	γ_B	kN/m^3	18.5
Poisson's ratio of soil	ν	–	0.3
Modulus of elasticity of soil	E	MPa	50
Maximum allowable settlement	δ	Mm	25
Safety for bearing capacity	FS	–	3.0
Minimum footing thickness	h_{min}	m	0.25
Column breadth in two direction	b/h	mm	500/500
Axial force	P	kN	250–500–750
Flexural moments	M_x, M_y	kNm	0–0.25P–0.5P
Range of width of footing	B	m	2.0–5.0
Range of length of footing	L	m	2.0–5.0
Range of height of footing	H	m	h_{min}–1.0
Range of diameter of reinforcement bars of two direction	ϕ	mm	16–24
Range of distance between reinforcement bars	s	mm	5ϕ–250

About the bearing capacity, Eqs. (1) and (2) are checked as constraints. q_{ult} is the ultimate bearing capacity that is calculated via Eq. (4) by using factors formulated as Eqs. (5)–(10).

$$q_{1,2,3,4} \geq 0 \tag{2}$$

$$\text{FS} < \frac{q_{\text{ult}}}{\max(q_{1,2,3,4})} \tag{3}$$

$$q_{\text{ult}} = \gamma D N_q F_{qs} F_{qd} + 0.5\gamma B N_\gamma F_{\gamma s} F_{\gamma qd} \tag{4}$$

$$N_q = e^{\pi \tan\varphi'} \tan^2\left(\frac{\pi}{4} + \frac{\varphi'}{2}\right) \tag{5}$$

$$N_\gamma = 2(N_q + 1)\tan\varphi' \tag{6}$$

$$F_{qs} = 1 + \frac{B}{L}\tan\varphi' \tag{7}$$

$$F_{\gamma s} = 1 - 0.4\frac{B}{L} \tag{8}$$

$$F_{qd} = \begin{cases} 1 + 2\tan\varphi'(1 - \sin\varphi')^2\left[\arctan\left(\frac{D}{B}\right)\right] & \text{if } D > B \\ 1 + 2\tan\varphi'(1 - \sin\varphi')^2\left(\frac{D}{B}\right) & \text{if } D < B \end{cases} \tag{9}$$

$$F_{\gamma d} = 1.0 \tag{10}$$

The settlement is calculated according to Eq. (11) [43], and the shape factor (β_z) is shown as Eq. (12).

$$\delta = \frac{P + W_f(1 - \nu^2)}{\beta_z E\sqrt{BL}} \tag{11}$$

$$\beta_z = -0.0017\left(\frac{L}{B}\right)^2 + 0.0597\left(\frac{L}{B}\right) + 0.9843 \tag{12}$$

If the constraints [Eqs. (2), (3), and (11)] are violated, the objective is assigned with 10^6 \$ which is a very big value. If not, the optimization continues by assigning the other variables with values and controlling the other constraints related to moment capacity in the critical sections, one-way and two-way shear capacity according to ACI 318 [42]. The objective function ($f(x)$) is the total cost of footing as given in Eq. (13), where V_{concrete} and W_{steel} represent the volume of concrete and weight of

Table 2 Optimum results for $P = 250$ kN

M_x, M_y	$e_x \neq 0, e_y \neq 0$			$e_x = 0, e_y = 0$		
	0	0.25P	0.5P	0	0.25P	0.5P
B_x	2	2	2	2	2.55	3.95
B_y	2	2	2	2	2.65	3.95
H	0.45	0.55	0.7	0.45	0.5	0.75
ϕ_x	16	16	18	16	18	16
S_x	250	200	220	250	240	250
A_{sx}	1608.495	1809.557	2290.221	1608.495	2544.69	3015.929
ϕ_y	16	16	16	16	16	16
S_y	250	180	170	250	220	250
A_{sy}	1608.495	2010.619	2211.681	1608.495	2412.743	3015.929
$f(x)$	73.97108	84.71566	99.44781	73.97108	135.2948	348.8304
e_x	– 0.05	0.3	0.7	0	0	0
e_y	– 0.15	0.5	0.65	0	0	0

steel, respectively.

$$f(x) = V_{\text{concrete}}\, C_c + W_{\text{steel}}\, C_s \tag{13}$$

The classical optimization process continues for the generation of all design variables and checking the design constraint by assigning 10^6 \$ for violating values. By the elimination of worst results, the optimum results are obtained.

3 Numerical Examples

The optimum results are investigated for three axial forces. The results are compared with the results that have constant zero values for e_x and e_y. The optimum results are given in Tables 2, 3, and 4.

4 Conclusion

According to the optimum results, the mentioned optimization has no additional effect on the cost if the flexural moment values are zero. For $P = 250$ kN axial force, the orientation provides 37.38% and 71.49% cost reduction compared to zero column eccentricity cases for flexural moments equal to 0.25P and 0.5P, respectively.

By the increase of axial force to 500 kN, the relative reduction percentages reduce to 32.45% and 58.26% for moments equal to 0.25P and 0.5P, respectively.

Table 3 Optimum results for $P = 500$ kN

M_x, M_y	$e_x \neq 0, e_y \neq 0$			$e_x = 0, e_y = 0$		
	0	0.25P	0.5P	0	0.25P	0.5P
B_x	2	2.55	2.65	2	2.25	3.95
B_y	2	2.7	2.65	2	2	4.75
H	0.85	1	1	0.85	1	0.95
ϕ_x	16	16	16	16	18	18
S_x	250	200	150	250	240	190
A_{sx}	1608.495	2211.681	3418.053	1608.495	2544.69	5089.38
ϕ_y	16	16	20	16	16	16
S_y	250	180	240	250	210	230
A_{sy}	1608.495	2010.619	3455.752	1608.495	2412.743	4021.239
$f(x)$	101.9711	130.8789	212.7412	101.9711	193.7435	509.6304
e_x	0	0.6	0.75	0	0	0
e_y	−0.3	0.3	0.75	0	0	0

Table 4 Optimum results for $P = 750$ kN

M_x, M_y	$e_x \neq 0, e_y \neq 0$			$e_x = 0, e_y = 0$		
	0	0.25P	0.5P	0	0.25P	0.5P
B_x	3	3.2	3.9	2.95	3.2	4.5
B_y	3.05	3.45	3.9	2.95	3.45	4.7
H	1	1	1	1	1	1
ϕ_x	16	20	20	16	18	20
S_x	180	230	190	190	190	200
A_{sx}	3216.991	4084.07	6283.185	3015.929	5089.38	6911.504
ϕ_y	16	18	16	16	18	20
S_y	250	200	120	190	190	200
A_{sy}	2412.743	4325.973	6232.92	3015.929	5089.38	6911.504
$f(x)$	253.5478	326.6178	477.715	246.5325	449.42	646.0704
e_x	−0.55	0.4	0.9	0	0	0
e_y	0	0.75	0.85	0	0	0

For $P = 750$ kN, a small cost increase for column orientation is seen for zero flexural moments. Due to this, it is more useful to take the eccentricity as constant zero if there are no flexural moments of the column. By the increase of the axial force, the percentages of reduction of cost are near 27%. In that case, the advantages of optimum column orientation are more important for low axial force values.

Since footings under columns are mostly preferred as a foundation system for low-rise structures with low axial force, the optimization of column orientation is very effective in the reduction of the cost of the design.

References

1. Coello CC, Hernandez FS, Farrera FA (1997) Optimal design of reinforced concrete beams using genetic algorithms. Expert Syst Appl 12:101–108
2. Govindaraj V, Ramasamy JV (2005) Optimum detailed design of reinforced concrete continuous beams using Genetic Algorithms. Comput Struct 84:34–48. https://doi.org/10.1016/j.compstruc.2005.09.001
3. Fedghouche F, Tiliouine B (2012) Minimum cost design of reinforced concrete T-beams at ultimate loads using Eurocode2. Eng Struct 42:43–50. https://doi.org/10.1016/j.engstruct.2012.04.008
4. Leps M, Sejnoha M (2003) New approach to optimization of reinforced concrete beams. Comput Struct 81:1957–1966. https://doi.org/10.1016/S0045-7949(03)00215-3
5. Akin A, Saka MP (2010) Optimum detailed design of reinforced concrete continuous beams using the harmony search algorithm. In: The Tenth international conference on computational structures technology, paper 131
6. Bekdaş G, Nigdeli SM (2012) Cost optimization of T-shaped reinforced concrete beams under flexural effect according to ACI 318. In: 3rd European conference of civil engineering
7. Bekdaş G, Nigdeli SM (2013) Optimization of T-shaped RC flexural members for different compressive strengths of concrete. Int J Mech 7:109–119
8. Ulusoy S, Kayabekir A, Bekdaş G, Nigdeli S (2020) Metaheuristic algorithms in optimum design of reinforced concrete beam by investigating strength of concrete. Challenge J Concr Res Lett 11(2):26–30. https://doi.org/10.20528/cjcrl.2020.02.001
9. Bekdaş G, Nigdeli SM, Yang X, Metaheuristic optimization for the design of reinforced concrete beams under flexure moments
10. Bekdaş G, Nigdeli SM (2015) Optimum design of reinforced concrete beams using teaching-learning-based optimization. In: 3rd international conference on optimization techniques in engineering (OTENG'15), pp 7–9
11. Kayabekir AE, Bekdaş G, Nigdeli SM (2019) Optimum design of T-beams using Jaya algorithm. In: 3rd international conference on engineering technology and innovation (ICETI), Belgrad, Serbia
12. Kayabekir AE, Bekdaş G, Nigdeli SM (2020) Evaluation of metaheuristic algorithm on optimum design of T-beams. In: International conference on harmony search algorithm. Springer, Singapore, pp 155–169
13. Kayabekir AE, Nigdeli M (2021) Statistical evaluation of metaheuristic algorithm: an optimum reinforced concrete T-beam problem. In: Advances in structural engineering—optimization. Springer, Cham, pp 299–310
14. Rafiq MY, Southcombe C (1998) Genetic algorithms in optimal design and detailing of reinforced concrete biaxial columns supported by a declarative approach for capacity checking. Comput Struct 69:443–457
15. Bekdaş G, Nigdeli SM (2012) Cost optimization of T-shaped reinforced concrete beams under flexural effect according to ACI 318. In: 3rd European conference of civil engineering, pp 2–4
16. Nigdeli SM, Bekdas G, Kim S, Geem ZW (2015) A novel harmony search based optimization of reinforced concrete biaxially loaded columns. Struct Eng Mech Int J 54(6):1097–1109
17. Bekdas G, Nigdeli SM (2016) Bat algorithm for optimization of reinforced concrete columns. PAMM 16(1):681–682
18. Bekdaş G, Nigdeli SM (2016) Optimum design of reinforced concrete columns employing teaching learning based optimization. Challenge J Struct Mech 2(4):216–219

19. Camp CV, Pezeshk S, Hansson H (2003) Flexural design of reinforced concrete frames using a genetic algorithm. J Struct Eng-ASCE 129:105–115
20. Govindaraj V, Ramasamy JV (2007) Optimum detailed design of reinforced concrete frames using genetic algorithms. Eng Optimiz 39(4):471–494
21. Bekdaş G, Nigdeli SM (2017) Modified harmony search for optimization of reinforced concrete frames. In: International conference on harmony search algorithm. Springer, Singapore, pp 213–221
22. Ulusoy S, Kayabekir AE, Bekdaş G, Nigdeli SM (2018) Optimum design of reinforced concrete multi-story multi-span frame structures under static loads. Int J Eng Technol 10(5):403–407
23. Rakıcı E, Bekdaş G, Nigdeli SM (2020) Optimal cost design of single-story reinforced concrete frames using Jaya Algorithm. In: International conference on harmony search algorithm. Springer, Singapore, pp 179–186
24. Nigdeli SM, Bekdaş G, Yang XS (2018) Metaheuristic optimization of reinforced concrete footings. KSCE J Civ Eng 22(11):4555–4563
25. Chaudhuri P, Maity D (2020) Cost optimization of rectangular RC footing using GA and UPSO. Soft Comput 24(2):709–721
26. Khajehzadeh M, Taha MR, El-Shafie A, Eslami M (2011) Modified particle swarm optimization for optimum design of spread footing and retaining wall. J Zhejiang Univ, Sci, A 12(6):415–427
27. Khajehzadeh M, Taha MR, El-shafie A, Eslami M (2012) Optimization of shallow foundation using gravitational search algorithm. J Appl Eng Technol 4(9):1124–1130
28. Camp CV, Assadollahi A (2013) CO_2 and cost optimization of reinforced concrete footings using a hybrid big bang-big crunch algorithm. Struct Multidiscip Optim 48(2):411–426
29. Khajehzadeh M, Taha MR, Eslami M (2013) A new hybrid firefly algorithm for foundation optimization. Natl Acad Sci Lett 36(3):279–288
30. Ceranic B, Fryer C, Baines RW (2001) An application of simulated annealing to the optimum design of reinforced concrete retaining structures. Comput Struct 79:1569–1581
31. Camp CV, Akin A (2012) Design of retaining walls using Big Bang–Big Crunch optimization. J Struct Eng-ASCE 138(3):438–448
32. Kaveh A, Abadi ASM (2011) Harmony search based algorithms for the optimum cost design of reinforced concrete cantilever retaining walls. Int J Civ Eng 9(1):1–8
33. Talatahari S, Sheikholeslami R, Shadfaran M, Pourbaba M, Optimum design of gravity retaining walls using charged system search algorithm. Math Prob Eng 2012, Article ID 301628.
34. Aral S, Yılmaz N, Bekdaş G, Nigdeli SM (2020) Jaya optimization for the design of cantilever retaining walls with toe projection restriction. In: International conference on harmony search algorithm. Springer, Singapore, pp 197–206
35. Yılmaz N, Aral S, Nigdeli SM, Bekdaş G (2020) Optimum design of reinforced concrete retaining walls under static and dynamic loads using Jaya Algorithm. In: International conference on harmony search algorithm. Springer, Singapore, pp 187–196
36. Kayabekir AE, Yücel M, Bekdaş G, Nigdeli SM (2020) Comparative study of optimum cost design of reinforced concrete retaining wall via metaheuristics. Challenge J Concr Res Lett 11:75–81
37. Kayabekir AE, Arama ZA, Bekdaş G, Nigdeli SM, Geem ZW (2020) Eco-friendly design of reinforced concrete retaining walls: multi-objective optimization with harmony search applications. Sustainability 12(15):6087
38. Yücel M, Nigdeli SM, Kayabekir AE, Bekdaş G (2021) Optimization and artificial neural network models for reinforced concrete members. In: Nature-inspired metaheuristic algorithms for engineering optimization applications. Springer, Singapore, pp 181–199
39. Bekdaş G, Yücel M, Nigdeli SM (2021) Estimation of optimum design of structural systems via machine learning. Front Struct Civ Eng, 1–12
40. Yücel M, Bekdaş G, Nigdeli SM, Kayabekir AE (2021) An artificial intelligence-based prediction model for optimum design variables of reinforced concrete retaining walls. Int J Geomech 21(12):04021244
41. Geem ZW, Kim JH, Loganathan GV (2001) A new heuristic optimization algorithm: harmony search. SIMULATION 76:60–68

42. American Concrete Institute (2011) Building code requirements for structural concrete and commentary. ACI 318–11
43. Poulos HG, Davis EH (1974) Elastic solutions for soil and rock mechanics. Wiley

WSAR with Levy Flight for Constrained Optimization

Adil Baykasoğlu and Mümin Emre Şenol

Abstract The Lévy distribution, which represents a form of random walk (Lévy flight) consisting of a series of consecutive random steps, has recently been demonstrated to improve the performance of metaheuristic algorithms. Through consecutive random steps, Lévy flight is particularly beneficial for undertaking massive "jump" operations that allow the search to escape from a local optimum and restart in a different region of the search space. We examine this concept in this work by applying Lévy flight to weighted superposition attraction–repulsion algorithm (WSAR), a basic but effective swarm intelligence optimization method that was recently introduced in the literature. The experiments are performed on several constrained design optimization problems. The performance of the proposed Levy flight WSAR algorithm is compared with several metaheuristics including harmony search algorithm (HS) and plain WSAR. The computational results revealed that Levy flight WSAR (LF-WSAR) is able to outperform the other algorithms. HS and WSAR show competitive performance with each other. In addition, performance of the LF-WSAR is statistically compared with other algorithms through nonparametric statistical tests. According to the statistical results, the difference between LF-WSAR and other algorithm is statistically significant.

Keywords WSAR algorithm · Harmony search · Lévy flight · Constrained design optimization problems

1 Introduction

In real world, the vast majority of engineering problems can be classified as optimization problems. In the related literature, developing metaheuristic algorithms

A. Baykasoğlu · M. E. Şenol (✉)
Department of Industrial Engineering, Faculty of Engineering, Dokuz Eylül University, Izmir, Turkey
e-mail: emre.senol@deu.edu.tr

A. Baykasoğlu
e-mail: adil.baykasoglu@deu.edu.tr

J. H. Kim et al. (eds.), *Proceedings of 7th International Conference on Harmony Search, Soft Computing and Applications*, Lecture Notes on Data Engineering and Communications Technologies 140, https://doi.org/10.1007/978-981-19-2948-9_21

for solving optimization problems has been the topic of considerable research. Baykasoğlu and Akpinar [1, 2] introduced the weighted superposition attraction (WSA) algorithm, which is an effective metaheuristic algorithm for solving variety of problems. The WSA algorithm was first suggested for continuous optimization problems and has since been successfully applied to a variety of real world and hypothetical complex optimization problems [3, 4]. Furthermore, Baykasoğlu recently proposed the WSAR algorithm as an extension of WSA [5].

A recent trend in metaheuristic community is to combine metaheuristic algorithms with random walk mechanisms like Lévy flights in order to enhance their searching ability further. Based on this motivation, Levy flight-based random walk mechanism is incorporated into the WSAR algorithm in this study. This version of WSAR algorithm that is termed as LF-WSAR is applied to well-known engineering design optimization problems in order to observe its performance.

The remaining of the paper is organized as follows. In Sect. 2, the WSAR algorithm is briefly introduced, and in Sect. 3, the test functions are given and experimental results are reported. Finally, concluding remarks are presented in Sect. 4.

2 The WSAR Algorithm

Baykasoğlu [5] has proposed the WSAR algorithm as the WSA algorithm's successor. In contrast to the WSA algorithm, the WSAR algorithm contains two superpositions. Attractive superposition (AS) is the first superposition, and it may attract solution vectors (particles, agents) to it. Repulsive superposition (RS) is the second superposition, and it repels solution vectors (particles, agents) from its position. Attraction and/or repulsion mechanisms explicitly or implicitly employed in various forms in many metaheuristic algorithms like differential evolution, particle swarm optimization, Jaya algorithm, firefly algorithms, etc. However, some of these algorithms like firefly algorithm require pairwise comparisons between solution vectors that considerably increases computational load; particle swarm optimization requires remembering of local and global optimal solutions, etc. Moreover, none of the previous metaheuristics utilized the superposition principle, which reduces computational load in comparison with firefly type algorithms. There is no need to keep memory records of previous good performing solutions which is the case in particle swarm type algorithms, etc. Superpositions can inherently determine search directions towards good solutions, which is one of the most desired search principles in metaheuristic algorithms. The WSAR algorithm has three steps, namely initialization, superposition determination, and moving solution vectors. The nomenclature of the WSAR algorithm is given in Table 1.

The WSAR algorithm's *initialization phase* consists of "parameter setting" and "creation of initial solutions". In the superposition determination phase, a biased randomized sampling approach is used to determine superpositions [6]. One can see the details of the superposition determination process in the following study [5]. In the moving solution vectors phase, solution vectors' next positions are delineated

Table 1 Nomenclature of the WSAR algorithm

Symbol	Description
$x_{ij}(t)$	The value of the position of solution vector ***i*** on dimension ***j*** at iteration ***t***
$x_{AS}(t)$	AS at iteration t
$x_{RS}(t)$	RS at iteration t
$\lvert.\rvert$	Absolute function
$x_{(r1)j}(t)$	Randomly selected solution vector from the population at iteration t
ss(t)	Random walk step sizing function
φ_{step}	User defined parameter for step sizing function
μ	User defined parameter for step sizing function

according to their fitness. If the AS has better fitness than the solution vector, the solution vector will move towards to AS and escape from the RS by using Eq. 1. On the other hand, it can also move towards to AS and escape from the RS with a probability, which calculated through the following equation: $e^{(f(i)-f(\mathrm{AS}))}$, where $f(i)$ represents the fitness of the solution vector i and $f(\mathrm{AS})$ represents the fitness of the AS. If the generated random number is less than the calculated probability, the corresponding solution vector will move towards to AS and escapes from the RS. Besides, the solution vector will move randomly by using Eq. 2.

$$\begin{aligned} x_{ij}(t+1) = x_{ij}(t) + (\mathrm{rand}(0,1) * (x_{AS}(t) - |x_{ij}(t)|) \\ - \mathrm{rand}(0,1) * (x_{RS}(t) - |x_{ij}(t)|)) \end{aligned} \tag{1}$$

$$x_{ij}(t+1) = x_{ij}(t) + \mathrm{ss}(t) * \mathrm{unifrnd}(-1,1) * \left|x_{(r1)j}(t) - x_{(r2)j}(t)\right| \tag{2}$$

ss(t) is defined as follows in Eq. 3.

$$\mathrm{ss}(t+1) = \begin{cases} \mathrm{ss}(t) - \mathrm{e}^{-\frac{t}{t+1}} * \varphi_{step} * \mathrm{ss}(t) \text{ if } \mathrm{rand}(0,1) \le \mu \\ \mathrm{ss}(t) + \mathrm{e}^{-\frac{t}{t+1}} * \varphi_{step} * \mathrm{ss}(t) \text{ if } \mathrm{rand}(0,1) > \mu \end{cases} \tag{3}$$

2.1 *LF-WSAR Modifications*

In this section, the differences between WSAR algorithm and LF-WSAR algorithm are given. In the LF-WSAR algorithm, instead of using random numbers (rand(0, 1)) in Eq. 1 and ss(t) in Eq. 2, a group of numbers that are generated from Lévy flight equation [7] are used. The Lévy flight equation that is used in LF-WSAR is given by Eqs. 4–7.

$$\mathrm{Distribution}_{\text{Lévy}} = U/|V|^{1/\beta} \tag{4}$$

$$\sigma = \left(\frac{\Gamma(1+\beta) * \sin\left(\frac{\Pi * \beta}{2}\right)}{\Gamma\left(\frac{1+\beta}{2}\right) * \beta * 2^{\frac{\beta-1}{2}}} \right) \tag{5}$$

$$x_{ij}(t+1) = x_{ij}(t) + \left(\text{Lévy1} * \left(x_{\text{AS}}(t) - \left|x_{ij}(t)\right|\right) - \text{Lévy2} * \left(x_{\text{RS}}(t) - \left|x_{ij}(t)\right|\right)\right) \tag{6}$$

$$x_{ij}(t+1) = x_{ij}(t) + \text{Lévy3} * \text{unifrnd}(-1, 1) * \left|x_{(r1)j}(t) - x_{(r2)j}(t)\right| \tag{7}$$

where β denotes the power-law index, V denotes a random number sampled from the Gaussian distribution $N(0, 1)$, and U is a random number sampled from the Gaussian distribution $N(0, \sigma 2)$, and σ is the standard deviation.

2.2 Constraint Handling Mechanism

Inverse-tangent-constraint-handling (ITCH) method is utilized in this research to handle constraints in the constrained design optimization problems. The ITCH technique was first proposed in the literature by Kim et al. [8], who demonstrated its effectiveness on a variety of constrained design optimization problems. ITCH is given in Eq. 8.

$$(\vec{x}) = \begin{cases} \hat{g}(\vec{x}) = g_{\max}(\vec{x}) & \text{if } g_{\max}(\vec{x}) > 0 \\ \hat{f}(\vec{x}) = a\tan\left[f(\vec{x})\right] - \pi/2 & \text{otherwise} \end{cases} \tag{8}$$

where $g_{\max}(\vec{x}) = \max_{g_i(\vec{x})}\left[g_1(\vec{x}), g_2(\vec{x}), g_3(\vec{x}), \ldots, g_m(\vec{x})\right]$ and $a\tan[.]$ represent the inverse tangent. Note that $\hat{g}(\vec{x}) < 0$ for any $\vec{x}$, and thus $\hat{f}(\vec{x}) < \hat{g}(\vec{x})$ is guaranteed.

3 Test Functions and Experimental Results

In this section, firstly engineering design problems are introduced afterwards computational results are presented.

3.1 Engineering Design Problems

In this study, three well-known engineering design problems are used to test the performance of the LF-WSAR. These problems are introduced in the following sub-sections.

3.2 Design of a Tension–Compression Coil Spring

The ideal design of the tension–compression coil spring is the first test problem. In this problem, there are three design variables. They are defined as follows:

$$x = (x_1, x_2, x_3)^{\mathrm{T}} := (N, D, d)^{\mathrm{T}} \tag{9}$$

where N is the number of spring coils, D is the diameter of the winding, and d is the diameter of the wire. The goal of the objective function is to reduce the spring's weight while maintaining minimum deflection, shear stress, surge frequency, and outside diameter restrictions. In the literature, the design optimization problem is formulated as a constrained nonlinear programming problem and one can find the details of the problem and the mathematical formulation in the following studies [9, 10].

3.3 Design of a Pressure Vessel

The second test problem is to find the best design for a cylindrical vessel with hemi-spherical heads on both ends. The goal of design is to reduce total cost, which includes material, forming, and welding expenses [8]. T_{s} (shell thickness), T_{h} (head thickness), R (inner radius), and L (length) are the four design variables (length of the cylindrical section of the vessel, not including the head). The available thicknesses of rolled steel plates are represented by the first two variables, which are integer multiples of 0.0625; the other variables are continuous valued. The following are the formal definitions of the design variables:

$$x = (x_1, x_2, x_3, x_4)^{\mathrm{T}} := (T_{\mathrm{s}}, T_{\mathrm{h}}, R, L)^{\mathrm{T}} \tag{10}$$

In the literature, the design optimization problem is described as a constrained nonlinear programming problem and one can find the details of the problem along with mathematical formulation in the following studies [8, 10].

3.4 Design of a Welded Beam

The welded beam is the third design optimization problem, and it deals with the lowest-cost design of a structural welded beam. Weld stress, buckling load, beam deflection, and beam bending stress are the problem's constraints. This design optimization problem includes four continuous design variables: h (weld thickness), l (weld length), t (bar thickness), and b (bar width) (bar breadth). The following are the formal definitions of design variables:

$$x = (x_1, x_2, x_3, x_4)^{\mathrm{T}} := (h, l, t, b)^{\mathrm{T}} \tag{11}$$

The main focus in welded beam design optimization problem is to eliminate the beam's structural cost as well as the key cost aspects of the welded beam, such as set-up work, welding labour, and material cost. In the literature, the presented problem is classified as a constrained nonlinear programming problem and readers can find the details of the problem with mathematical formulation in the following study [8].

3.5 Experimental Results

Results of the LF-WSAR and other metaheuristic algorithms, namely adaptive firefly algorithm (AFA) [10], artificial bee colony algorithm (ABC) [11], evolutionary algorithm with infeasible solutions (ISEA) [12], simple evolutionary algorithm (SEA) [13], upgraded ABC (UABC) [14], chaos embedded great deluge algorithm (CEGDA) [15], constrained particle swarm optimization (COPSO) [16], HS [17], WSA [2], and plain WSAR on the selected engineering design optimization problems are shown in Table 2. The best, average, and standard deviation of 20 runs are reported in this table. According to experimental results, LF-WSAR outperforms the other metaheuristic algorithms. Moreover, nonparametric statistical tests are employed in order to analyse the LF-WSAR performance further. Based on the results obtained from nonparametric statistical results, LF-WSAR performs better than the other algorithms. Plain WSAR and HS provide the second best performance. On the other hand, the difference between LF-WSAR and all other compared algorithms is found statistically significant as the $p < 0.1$ (Table 3).

4 Conclusions

In this research, a random walk mechanism, namely Lévy flight is incorporated into a recent metaheuristic algorithm, WSAR in order to enhance its performance further. It is observed Lévy flight-based random walk mechanism has a positive effect on the performance of the WSAR algorithm based on the computational results that are performed on selected constrained design optimization problems. Obtained results are also verified by employing two nonparametric statistical tests. Enhancing and analysing the performance of the WSAR algorithm with the help of some other step sizing mechanisms is planned as a future research. Moreover, further experimental test will also be performed with more complex recent CEC benchmark problems.

Table 2 Experimental results

Algorithm		Coil spring problem	Pressure vessel problem	Welded beam problem
AFA	Best	0.0126653049	6059.71427196	1.724852
	Worst	0.0000128058	6090.52614259	1.724852
	Average	0.01266770446	6064.33605261	1.724852
ABC	Best	0.12665	6059.714736	1.724852
	Worst	–	–	–
	Average	0.12709	6245.308144	1.741913
ISEA	Best	0.127210	6059.7143	1.7448523
	Worst	–	–	–
	Average	–	6379.938037	–
SEA	Best	0.12688	6059.714355	1.76558
	Worst	0.017037	6846.628418	2.844060
	Average	0.013014	6355.343115	1.968200
UABC	Best	0.12665	6059.714335	1.724852
	Worst	–	–	–
	Average	0.01266523	6192.116211	1.741913
CEGDA	Best	0.0126652296	6059.83905683	1.724852
	Worst	0.0140793687	6823.60245024	1.724852
	Average	0.0128750789	6149.72760669	1.724852
COPSO	Best	0.012665	6059.714335	1.724852
	Worst	–	–	–
	Average	0.012665	6071.013366	1.724881
HS	Best	0.012664	6059.72188231	1.72485254
	Worst	0.012665	6068.75680753	3.82321834
	Average	0.0126644	6066.41025196	2.12575148
WSA	Best	0.01266523	6059.832672	1.724852
	Worst	0.01271790499	6071.65725	1.724852
	Average	0.0126922992	6068.74682	1.724852
WSAR	Best	0.012665	6059.709	1.724852
	Worst	0.01275466	6067.82943	1.724852
	Average	0.01269197	6066.5481	1.724852
LF-WSAR	Best	0.012662	6059.705	1.724852
	Worst	0.012665	6059.82943	1.724852
	Average	0.012664	6059.758	1.724852

Table 3 Nonparametric test results

Friedman test average rankings		Wilcoxon signed-rank test between LF-WSAR and other metaheuristic algorithms	
Algorithms	Sum of ranks	LF-WSAR versus	p-value
AFA	5.8333 (9)	AFA	0.085
ABC	8.6667 (11)	ABC	0.085
ISEA	6.0000 (10)	ISEA	0.085
SEA	4.6667 (4)	SEA	0.085
UABC	5.5000 (8)	UABC	0.085
CEGDA	4.6667 (4)	CEGDA	0.085
COPSO	4.8333 (6)	COPSO	0.085
HS	3.8333 (2)	HS	0.085
WSA	5.3333 (7)	WSA	0.085
Plain WSAR	3.8333 (2)	WSAR	0.085
LF-WSAR	1.8333 (1)		

References

1. Baykasoğlu A, Akpinar Ş (2017) Weighted superposition attraction (WSA): a swarm intelligence algorithm for optimization problems—part 1: unconstrained optimization. Appl Soft Comput 56:520–540
2. Baykasoğlu A, Akpinar Ş (2015) Weighted superposition attraction (WSA): a swarm intelligence algorithm for optimization problems—part 2: constrained optimization. Appl Soft Comput 37:396–415
3. Baykasoğlu A, Ozsoydan FB, Senol ME (2020) Weighted superposition attraction algorithm for binary optimization problems. Oper Res Int J 20(4):2555–2581
4. Baykasoğlu A, Şenol ME (2019) Weighted superposition attraction algorithm for combinatorial optimization. Expert Syst Appl 138:112792
5. Baykasoğlu A (2021) Optimizing cutting conditions for minimizing cutting time in multi-pass milling via weighted superposition attraction-repulsion (WSAR) algorithm. Int J Prod Res. http://doi.org/10.1080/00207543.2020.1767313
6. Baykasoğlu A, Akpinar Ş (2020) Enhanced superposition determination for weighted superposition attraction algorithm. Soft Comput 1–26
7. Viswanathan GM, Buldyrev SV, Havlin S, Da Luz MGE, Raposo EP, Stanley HE (1999) Optimizing the success of random searches. Nature 401(6756):911–914. Mirjalili S, Gandomi AH, Mirjalili SZ, Saremi S, Faris H, Mirjalili SM (2017) Salp swarm algorithm: a bio-inspired optimizer for engineering design problems. Adv Eng Softw 114:163–191
8. Kim TH, Maruta I, Sugie T (2010) A simple and efficient constrained particle swarm optimization and its application to engineering design problems. Proc Inst Mech Eng C J Mech Eng Sci 224(2):389–400
9. Arora JS (1989) Introduction to optimum design. McGraw-Hill, New York
10. Baykasoğlu A, Ozsoydan FB (2015) Adaptive firefly algorithm with chaos for mechanical design optimization problems. Appl Soft Comput 36:152–164
11. Akay B, Karaboga D (2012) Artificial bee colony algorithm for large-scale problems and engineering design optimization. J Intell Manuf 23(4):1001–1014
12. Coello Coello CA, Becerra RL (2004) Efficient evolutionary optimization through the use of a cultural algorithm. Eng Optim 36(2):219–236

13. Mezura-Montes E, Coello CC, Landa-Becerra R (2003) Engineering optimization using simple evolutionary algorithm. In: Proceedings. 15th IEEE international conference on tools with artificial intelligence. IEEE, pp 149–156
14. Brajevic I, Tuba M (2013) An upgraded artificial bee colony (ABC) algorithm for constrained optimization problems. J Intell Manuf 24(4):729–740
15. Baykasoglu A (2012) Design optimization with chaos embedded great deluge algorithm. Appl Soft Comput 12(3):1055–1067
16. Aguirre AH, Zavala AM, Diharce EV, Rionda SB (2007) COPSO: constrained optimization via PSO algorithm. Center for Research in Mathematics (CIMAT). Technical report no. I-07-04/22-02-2007, 77
17. Geem ZW, Kim JH, Loganathan GV (2001) A new heuristic optimization algorithm: harmony search. Simulation 76(2):60–68

Spatiotemporal Clustering of Groundwater Depth in Ardabil Plain

Vahid Nourani, Mahya Gholizadeh Ansari, and Parnian Ghaneei

Abstract One of the effective applications of clustering is to obtain a suitable pattern and limits a large amount of data to facilitate the assessment process. To benefit from the features of clustering methods in groundwater evaluation, the groundwater depth data of the Ardabil plain were divided into an optimal number of clusters using different clustering methods [Fuzzy C-means (FCM), K-means, and self-organizing maps (SOM)]. In this research, quantitative parameters of groundwater of 30 wells in Ardabil plain have been sampled. Each well was sampled 12 times per year during the years 1365–1397 and has been studied and compared. The trend of spatiotemporal changes of groundwater depth was investigated using four clustering methods, including FCM, K-means, SOM, and WARD. The optimal number of clusters was obtained from the SOM clustering method due to its ability to recognize complex relationships. Finally, according to the results of the best clustering method, clusters and members of each cluster were identified.

Keywords Groundwater depth · Cluster analysis · Ardabil plain

1 Introduction

The uncontrolled increase of population in Iran, the limitation of water resources, and the over-exploitation of aquifers have caused irreparable damage to the country's natural resources in recent years. Communities in low-rainfall areas, where food production is largely controlled, and rainfall and groundwater distribution are controlled, are driven by people's efforts to manage limited water resources. With the increase in population and human activity, water consumption also increases. At

V. Nourani · M. G. Ansari (✉) · P. Ghaneei
Faculty of Civil Engineering, Center of Excellence in Hydroinformatics, University of Tabriz, 29 Bahman Ave., Tabriz, Iran
e-mail: mahyaansari933@yahoo.com

V. Nourani
Faculty of Civil and Environmental Engineering, Near East University, Near East Boulevard, 99138 Nicosia, North Cyprus

J. H. Kim et al. (eds.), *Proceedings of 7th International Conference on Harmony Search, Soft Computing and Applications*, Lecture Notes on Data Engineering and Communications Technologies 140, https://doi.org/10.1007/978-981-19-2948-9_22

the same time, the amount of water on the planet is constant, and as a result, human needs and dependence on water increase. The expansion of civilization leads to the over-extraction of groundwater resources. A large part of our country, Iran, is located in arid and semi-arid regions. Groundwater sources are the main and only reliable and permanent source of water supply in arid and semi-arid areas, especially in the case of droughts. However, the quantitative protection of groundwater is important. Solving groundwater problems is very costly and takes a lot of time. Therefore, identifying and analyzing these valuable resources are vital. In this regard, one of the problems in examining an aquifer through intelligent methods is the multiplicity of observation wells in the aquifer; analyzing all of the observation wells is very time-consuming [1].

One of the statistical methods for presenting the distribution map and quantitative multi-parameter distribution in groundwater aquifers is cluster analysis. Clustering techniques have been used in various researches of water engineering, including quantitative and qualitative studies of surface water, groundwater, river classification, and catchment. One of the essential steps to extract the most homogenous patterns is finding efficient methods to recognize the complex relationships among various parameters. For example, one of the robust multivariate analyses is clustering used to classify multidimensional inputs into homogeneous groups and create a pattern from datasets. This method can extract features from an unlabeled input to form clusters having maximum within-group-object similarity and between group-object dissimilarity [2, 3].

Researchers have taken advantage of various clustering algorithms to assess water resource quantity and quality for specific purposes, for example, industry, drinking, and irrigation. One of the most popular linear clustering methods is K-means cluster analysis, which has a simple linear algorithm that classifies inputs into separate K clusters and has been applied to several hydrology fields [4]. The fuzzy c-means (FCM) is one of the most widely used fuzzy-based clustering methods in which a member can belong to two or more clusters with various membership degrees. Also, self-organizing map (SOM) as a neural clustering technique is a useful unsupervised tool to assess groundwater quality and quantity (e.g., [3, 5]).

The present study represents an attempt to identify spatiotemporal groundwater depth changes using three practical clustering algorithms (K-means, FCM, and SOM) to extract the most homogenous structures of clusters.

2 Materials and Methods

2.1 Case Study

Ardabil plain is located in northwestern Iran in the latitude of 38° and 3′ to 38° and 27′ of north and longitude of 47° and 55′ to 48° and 20′ of the east. The area of Ardabil plain is about 990 km^2. This study used monthly groundwater depth data

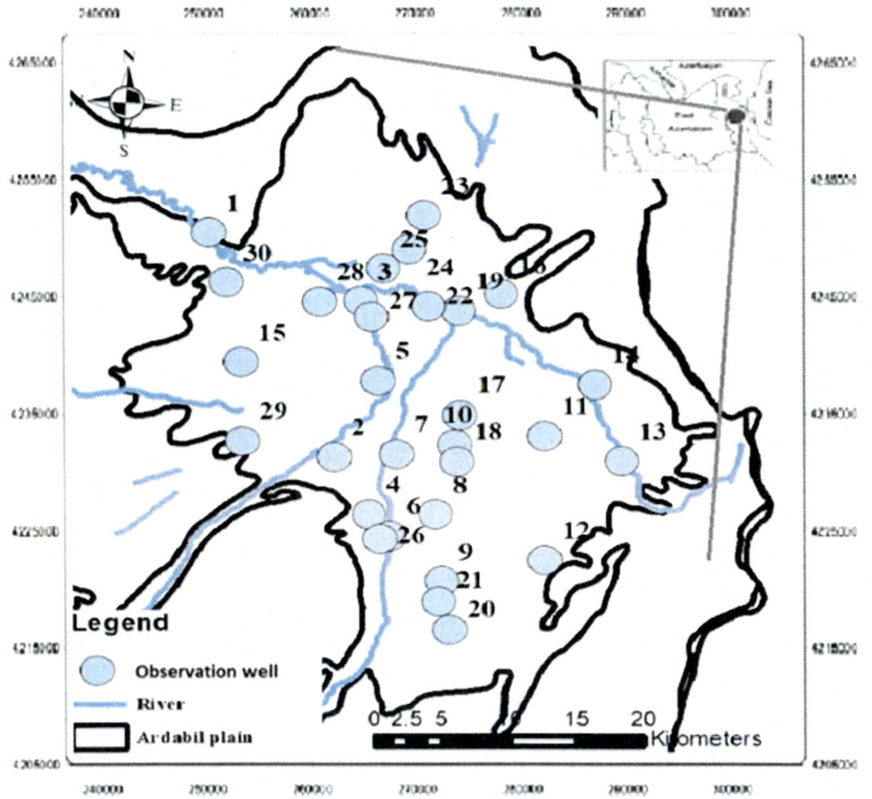

Fig. 1 Location of Ardabil plain and observation wells

(from 1998 to 2017) of the Ardabil aquifer. The location map and spatial distribution of the piezometers are shown in Fig. 1.

2.2 *K-means Clustering Method*

K-means is one of the clustering methods which has shown a decent performance in hydrological research. K-means classifies inputs into clusters in which the distance between the members and the centroid of clusters is at the minimum degree [6].

2.3 Fuzzy C-Means Clustering (FCM)

The FCM is one of the soft clustering algorithms in the fields of hydrology. This clustering method creates overlapping cluster boundaries and uses an iterative optimization process to minimize the following objective function (*J*), which is the weighted sum of errors within groups.

$$J_{\mathrm{FCM}} = (M.C) = \sum_{i=1}^{c} \sum_{k=1}^{n} U_{ik}^{m} \| X_k - V_i \|^2 \quad (1)$$

M is the membership matrix; *C* is the cluster center matrix; *c* is the number of clusters; *m* is a fuzzy exponent, and U_{ik}^{m} is the membership degree of data point *k* in cluster *i* [7].

2.4 Self-Organizing Map (SOM)

The SOM algorithm performs an iterative procedure called learning that organizes neurons according to their similarities with preserving the dataset's inherent topological characteristics.

The learning algorithm of SOM consists of the following steps:

1. entering *n*-dimensional input vector *x* into the network
2. measuring the Euclidean distances between each input and all neurons
3. finding the closest neuron to the input sample (BMU)
4. reducing the distance between the weights and the BMU, continue by changing the weight vector *w_v* at each time.

After completing learning, each input is assigned to its nearest output neuron or BMU. Samples with similar properties are assigned to the same BMU, which is regarded as a center for clustering in SOM [8].

2.5 Clustering Performance Criterion

The silhouette coefficient (SC) is measured to assess and compare the performance of clustering methods. To calculate SC for clustering structures, firstly, the *S*(*i*) (Eq. 2) for all members are calculated as silhouette index:

$$S(i) = \frac{b(i) - a(i)}{\max\{a(i).b(i)\}} \quad (2)$$

$a(i)$ is an average Euclidean distance between member i and all members of cluster A, and $b(i)$ is the least average dissimilarity of member i to the members within a cluster distinct from cluster A. The overall quality of a clustering method can then be validated using the average silhouette width for the entire dataset, which is defined as:

$$\text{SC} = \text{Max}\left(\frac{1}{n}\sum_{1}^{n} S(i)\right) \tag{3}$$

3 Results and Discussions

In this present study, with the aim of comparing different clustering algorithms, three individual clustering methods (FCM, K-means, and SOM) were used for groundwater quantity assessment. The monthly data of groundwater depth were divided into two-time spans (from 1998–2007 to 2008–2017). The datasets of time spans were separately clustered by proposed methods. The optimal number of clusters and results of clustering methods are shown in Table 1, and the spatial distributions of clusters are shown in Fig. 2.

The comparison results showed that the SOM clustering method had the highest performance in clustering the groundwater depth data compared to other methods (Table 1). The main reason for the superiority of SOM against FCM and K-means is the way that this flexible method can recognize the nonlinear relationships between members. In contrast, the rigid structure of the other methods limits their applicability in dealing with complex datasets. According to the results of the SOM method (Fig. 2), between the first and second-time spans, the groundwater depth of cluster 1 experienced an increase of 30%. The stability of members of cluster 1 between the first and second-time spans indicates the high degree of similarity and strong relationships among the members of cluster 1. In the first time span, cluster 2 had only two members (P17 and P15) which in the second period, P15 joined cluster 1. It is noteworthy that two members P4 and P26 of cluster 3, in the second-time span, showed a heterogenous behavior with their members of cluster 3 and moved

Table 1 Results of individual clustering methods for the groundwater depth based on SC

Clustering method	Time span			
	1		2	
	NC	SC	NC	SC
FCM	5	0.647	5	0.695
K-means	3	0.651	4	0.695
SOM	5	0.674	4	0.691

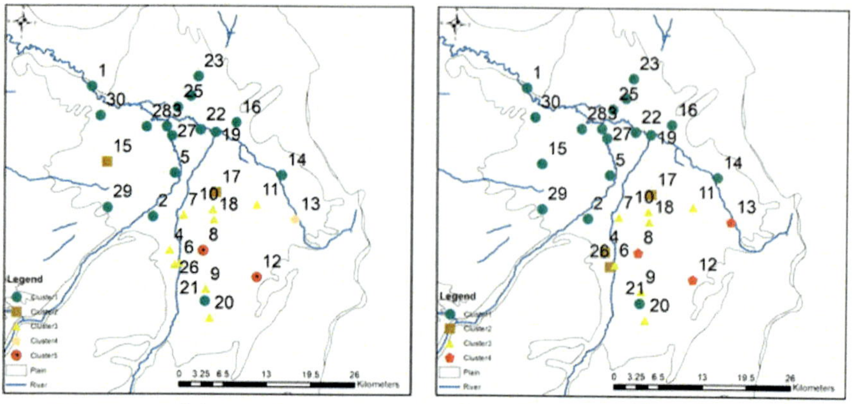

(a) (b)

Fig. 2 Spatial distribution of clusters in the **a** first and **b** second-time spans

to cluster 2 with a 26% increase in average depth. Cluster 3 in the first and second-time spans was one of the clusters with homogeneous behavior and retained all its previous members during the period with a 25% increase in average depth. Finally, cluster 4 and cluster 5 were combined in the second-time span and formed a new cluster with three members and an average water depth of 45.2 m.

Recognizing and forming new patterns from a cluster in the following years could result from heterogeneous changes between members. Understanding these changes among the members of a cluster is a successful spatial cluster ensemble performance because it makes the time and place of occurred uncommon events approximately recognizable. For instance, noticeable changes in pumping, cultivation, industrial, and drinking patterns can result in inconsistency among a cluster's members.

4 Conclusion

Assessment of hydrological phenomena has complicated procedures where analysis requires powerful applications of enhanced methods and to that end. In the present study, first, to explore and assess the spatiotemporal changes of groundwater quantity, the study was conducted over two-time spans. Three different cluster analyses methods (FCM, K-means, and SOM) were used and compared to recognize the patterns of groundwater depth over the Ardabil plain. The best results were obtained by the SOM clustering method due to its ability to recognize complex relationships among piezometers. Finally, the spatiotemporal changes of clusters were evaluated via the assessment of the SC. Overall, the results of this study indicate that cluster analysis is an effective method for the spatiotemporal analysis of groundwater depth changes.

References

1. Nourani V, Alami MT, Vousoughi FD (2016) Self-organizing map clustering technique for ANN-based spatiotemporal modeling of groundwater quality parameters. J Hydroinformatics 18:288–309. https://doi.org/10.2166/hydro.2015.143
2. Nourani V, Parhizkar M (2013) Conjunction of SOM-based feature extraction method and hybrid wavelet-ANN approach for rainfall-runoff modeling. J Hydroinformatics 15:829–848. https://doi.org/10.2166/hydro.2013.141
3. Sharghi E, Nourani V, Soleimani S, Sadikoglu F (2018) Application of different clustering approaches to hydroclimatological catchment regionalization in mountainous regions, a case study in Utah State. J Mt Sci 15:461–484. https://doi.org/10.1007/s11629-017-4454-4
4. Javadi S, Hashemy SM, Mohammadi K, Howard KWF, Neshat A (2017) Classification of aquifer vulnerability using K-means cluster analysis. J Hydrol 549:27–37. https://doi.org/10.1016/j.jhydrol.2017.03.060
5. Choi B, Yun S, Kim K, Kim J, Mi H, Koh Y (2014) Hydrogeochemical interpretation of South Korean groundwater monitoring data using self-organizing maps. J Geochem Explor 137:73–84. https://doi.org/10.1016/j.gexplo.2013.12.001
6. Fabbrocino S, Rainieri C, Paduano P, Ricciardi A (2019) Cluster analysis for groundwater classification in multi-aquifer systems based on a novel correlation index. J Geochem Explor 204:90–111. https://doi.org/10.1016/j.gexplo.2019.05.006
7. Lee KJ, Yun ST, Yu S, Kim KH, Lee JH, Lee SH (2019) The combined use of self-organizing map technique and fuzzy c-means clustering to evaluate urban groundwater quality in Seoul metropolitan city, South Korea. J Hydrol 569:685–697. https://doi.org/10.1016/j.jhydrol.2018.12.031
8. Chen I, Chang L, Chang F (2018) Exploring the spatio-temporal interrelation between groundwater and surface water by using the self-organizing maps. J Hydrol 556:131–142. https://doi.org/10.1016/j.jhydrol.2017.10.015

Assessing the Performance of a Machine Learning-Based Hybrid Model in Downscaling Precipitation Data

Nazak Rouzegari, Vahid Nourani, Ralf Ludwig, and Patrick Laux

Abstract In this paper, the performance of Statistical Downscaling (SD) for average monthly precipitation time series for Mahabad located in the north western part of Iran was quantified. The structure of the proposed approach is composed of two parts: input selection and average monthly precipitation simulation. Data from the Intercomparison Project Phase 5 (CMIP5) was chosen, and the input selection based on a Machine Learning-based (ML) method, i.e. the decision tree model (M5), was applied to select the most important predictors amongst many potential large-scale climate variables of the CMIP5 model. Next, the Random Forest (RF) model was trained, using the observed precipitation time series using the most important predictor variables, to downscale the precipitation time series of the CMIP5 model for the historical period (1992–2005) over the study area. Using the hybrid M5-RF model led to a reliable performance in the prediction of precipitation time series as the resulting Correlation Coefficient (R) and Root Mean Square Error (RMSE) values were 0.78 and 0.1 for calibration and 0.75 and 0.13 for validation, respectively.

Keywords Statistical downscaling · Machine learning · M5 model · Random forest · Iran

N. Rouzegari (✉) · V. Nourani
Faculty of Civil Engineering, Department of Water Resources Engineering, Centre of Excellence in Hydroinformatics, University of Tabriz, Tabriz, Iran
e-mail: nazak_rouzegari@yahoo.com

R. Ludwig
Department of Geography, Ludwig Maximilian University of Munich, Munich, Germany

P. Laux
Institute of Meteorology and Climate Research Atmospheric Environmental Research (IMK-IFU), Campus Alpin, Garmisch-Partenkirchen, Germany

J. H. Kim et al. (eds.), *Proceedings of 7th International Conference on Harmony Search, Soft Computing and Applications*, Lecture Notes on Data Engineering and Communications Technologies 140, https://doi.org/10.1007/978-981-19-2948-9_23

1 Introduction

CMIP5 models are the primary tool for climate change impact studies. However, they cannot be used directly due to their coarse resolution and need to be downscaled for fine-scale studies. Thus, downscaling techniques are vital to obtain data at a local-scale. SD techniques have the advantages of reliable accuracy, relatively simple implementation, and low computational costs. Therefore, they may be convenient if the study focuses on the downscaling of precipitation at basin scale [1]. In downscaling, CMIP5 model-simulated predictors are used to establish the relationship with the observed predictand. One of the main steps in the development of predictive models is identifying the predictors of a given outcome. In many empirical analyses, the presence of some data in a set of variables does not play a significant role in explaining the phenomenon under analysis but also is able to generate random noise preventing the detection of the most related predictors and main effects [2]. Therefore, it is essential to identify the most dominant predictors of a given result. For this purpose, various techniques have been purposed. For example, Principal Component Analysis, Linear Discriminant Analysis, Latent Semantic Analysis, Independent Component Analysis [3]. Regarding the downscaling process based on the predictor selected, it has been demonstrated that ML approaches like Artificial Neural Network and Support Vector Machine are superior compared to other techniques like conventional linear regression models, and they have proven to be efficient for modelling highly nonlinear relationships [1, 4]. Moreover, another ML method, RF, has been considered as a robust and competent algorithm for representing complex relationships in precipitation downscaling as it can implement different types of input variables and operates flexibly. However, little information is available about the downscaling using ML models at basin scale, and there are no studies regarding the capacity of the prediction of the CMIP5 models precipitation using a combination of M5-based predictor screening and RF downscaling strategy in Mahabad. Hence, the major objectives of this study were to develop a hybrid ML downscaling model based on M5 and RF to downscale the precipitation simulation of a CMIP5 model using historical data for the years 1992–2005 in Mahabad, Iran. The performance of the approach was studied and evaluated using statistical performance metrics like R and RMSE.

The remainder of the study is organized as follows: first, an overview regarding the main steps in developing the approach is provided. Next in the Experimental Procedure section, a concise summary about the study area and its climatic condition, the observational station data, and the selected CMIP5 model are described, followed by the applied methods, i.e. M5 for input selection, RF for downscaling of the model, and the evaluations metrics. Then, the results are given and discussed in more detail in the Results and Discussion sections, followed by the Conclusion section.

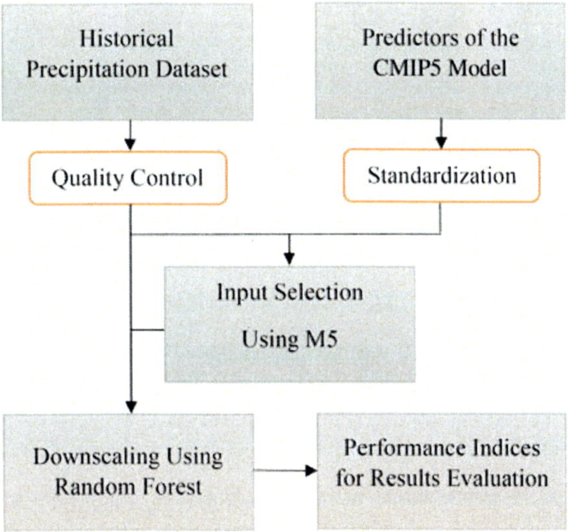

Fig. 1 Flowchart of the proposed methodology

1.1 Overview

The proposed method consists of two main steps illustrated in Fig. 1. Mahabad located in north-west of Iran was chosen as the study area. Considering Fig. 1, first, daily observed precipitation data provided by the Islamic Republic of Iran Meteorological Organization (IRIMO) was quality checked, and a CMIP5 model was selected and standardized. This was followed by identifying the most dominant predictors of the CMIP5 model using M5. Then, climatic data (average monthly precipitation) were statistically downscaled based on the historical precipitation time series, selected inputs, and the RF method. In the final step, the results were assessed using different performance indices.

2 Experimental Procedure

2.1 Study Area

Mahabad (36.77° N, 45.73° E) is a city and the capital of Mahabad County in West Azerbaijan Province, Iran (see Fig. 2). It has situated in a mountainous area and has cold winters and temperate summers. Maximum temperature in summers can reach 32.86 °C whilst in winters the minimum temperature may drop to −7.15 °C. The annual rainfall is around 350.05 mm, and the average relative humidity is 57.5%. In this research, the 14-year monthly average precipitation data for the study period

Fig. 2 The geographical position of Mahabad

(1992–2005) were obtained from IRIMO website (www.irimo.ir) and used in this study (see Table 1).

In this study, a procedure was introduced and assessed on the observational precipitation data of Mahabad and the historical precipitation data of a CMIP5 model to evaluate the accuracy of the purposed method in the prediction of precipitation climatic data. CMIP5 models simulate large-scale climate data considering the impact of changes in greenhouse gases. The outputs of these models cannot be directly applied in local-scale climate studies as they were developed in rough spatial resolutions. As a result, the models outputs should be downscaled to fine-scale climate information with appropriate techniques. Selecting a suitable CMIP5 models for the study area was done according to the R calculated between monthly observed precipitation and the precipitation of CMIP5 models for the base period (1992–2005). In this study, amongst several CMIP5 models, one model, i.e. MPI-ESM-LR was selected and used based on its relatively higher R value, and the predictors from this model

Table 1 Mean values of 14-year monthly data of precipitation

Month	Jan	Feb	Mar	Apr	May	Jun	Jul	Aug	Sep	Oct	Nov	Dec
Precipitation (mm)	16	11	35	48	53	10	3	1	1	8	46	37

Table 2 Information of the CMIP5 model

Model	Resolution (lat° × lon°)	Country	Research centre
MPI-ESM-LR	1.9° × 1.9°	Germany	Max Planck Institute for Meteorology (MPI-M)

were chosen in the current research. Table 2 indicates the information about the CMIP5 model used.

The potential predictors applied before the downscaling process are provided in Table 3, as well. As shown in this table, there are several potential predictors for the downscaling model, and selecting the most dominant inputs is essential which depends on time and location, yet the accuracy of downscaling procedures depends

Table 3 List of predictors applied in this research [3]

1. Cloud Area Fraction (clt)
2. Atmosphere Mass Content of Cloud Ice (clivi)
3. Cloud Condensed Water Content (clwvi)
4. Surface Evaporation (evspsbl)
5. Surface Latent Heat Flux (hfls)
6. Specific Humidity at Various Pressure Levels (hus10, hus20, hus30, hus50, hus70, hus100, hus200, hus300, hus400, hus500, hus600, hus700, hus850)
7. Relative Humidity at Various Pressure Levels (hur10, hur20, hur30, hur50, hur70, hur100, hur200, hur300, hur400, hur500, hur600, hur700, hur850)
8. Near-Surface Relative Humidity (hurs)
9. Near-Surface Specific Humidity (huss)
10. Precipitation (pr)
11. Convective Precipitation (prc)
12. Surface air Pressure (ps)
13. Surface Upwelling Longwave Radiation (rlus)
14. Surface Toa Outgoing Longwave Radiation (rlut)
15. Surface Downwelling Shortwave Radiation (rsds)
16. Surface Toa Outgoing Shortwave Radiation (rsut)
17. Toa Incoming Shortwave Radiation (rsdt)
18. Surface Net Downward Radiative Flux at the Top of Atmosphere (rtmt)
19. Near-Surface Wind Speed (sfcWind)
20. Temperature at Various Pressure Levels (ta10, ta20, ta30, ta50, ta70, ta100, ta200, ta300, ta400, ta500, ta600, ta700, ta850)
21. Minimum Near-Surface Air Temperature (tasmin)
22. Maximum Near-Surface Air Temperature (tasmax)
23. Sea surface temperature (ts)
24. Near-Surface Air Temperature (tas)
25. Eastward Near-Surface Wind Speed (uas)
26. Meridional (Northward) Wind Various Pressure Levels (va10, va20, va30, va50, va70, va100, va200, va300, va400, va500, va600, va700, va850)
27. Northward Near-Surface Wind Speed (vas)

10, 20, 30, 50, 70, 100, 200, 300, 400, 500, 600, 700 and 850 are various pressure levels measured in hectopascal (hpa)

on the relationship between the predictands (precipitation at the weather station) and predictors (from the CMIP5 model). So, in this research, the M5 model was used for this purpose as it is introduced as one of the reliable methods in selecting the dominant parameters [5].

2.2 Predictor Screening

In this section, the method for obtaining the selected predictors' data is mainly discussed. The goal of predictors selection is to decrease dimensionality without losing much information [1]. The M5 model is one of the most practical tools of data mining presented by Quinlan [6]. In this model, the multi-dimensional parameter space is split, and the model is created considering the overall quality criterion. This tree-structured model splits the domain of the training data into several classes (sub-area) and a regression equation is defined for each of them. The splitting criterion for this tree structure model is based on the standard deviation of each sub-area as a calculation of the error at that node and measuring the expected error as a result of testing each attribute at that node. The Standard Deviation Reduction (SDR) is calculated as Eq. (1).

$$\mathrm{SDR} = \mathrm{Sd}(T) - \sum_{i=1}^{N} \frac{T_i}{T} \mathrm{Sd}(T_i) \quad (1)$$

where N represents the number of the data, Sd is the standard deviation, T is a set of data entering each node, and T_i indicates a subset of data that has the ith potential test result. After examining all the possible splits, the M5 model tree selects the one that could maximize the error reduction. It is obvious that the upper node, which is called "root", is the best node for classification and other variables in the lower nodes have less significance.

As it can be seen in Table 3, there are many potential predictors provided at surface and pressure levels, so the selection of dominant inputs is an essential step to develop models like RF [2]. There are several input selection methods for selecting the most dominant predictors of CMIP5 models, but amongst all of them, the M5 model, as a feature extraction tool, in combination with RF has been rarely applied for downscaling the CMIP5 models data. Nourani et al. [5] introduced this model as an effective tool in the prediction of precipitation. So, it was used in this research to determine the dominant predictors by calculating multi-linear relationships between the predictors (see Table 3) and the predictand (precipitation). It should also be mentioned that the data were standardized to eliminate the influence of single-sample data before input selection. A large number of the CMIP5 model precipitation variables were considered as the inputs to M5, and the first few predictors located in the upper nodes of the tree were selected as the inputs of the ML downscaling model.

The Weka software [7] was applied to create the M5 model to select the dominant inputs of the RF model.

2.3 Statistical Downscaling

The ML method applied in this study was the RF model to relate the large-scale predictor parameters of the CMIP5 model and the observational precipitation of the case study. RF was proposed by Breiman [8] as a novel ML algorithm consisting of multiple independent classification and regression decision trees that can regulate different types of input variables, avoid overfitting, and make a final decision. Each decision tree consists of a root node, sub-node, and leaf node. The models' steps are roughly divided into the following:

(1) using a bootstrapping method for extracting the training dataset (two-third of the original dataset by default);
(2) building many regressions decision trees based on the minimum mean square error to build a forest;
(3) providing the result by averaging the predicted values of all trees.

The number of trees in the forest and the maximum depth of trees are the main RF hyperparameters. For more details about RF, refer to Breiman et al. [9]. It has been mentioned that the robustness of RF in classification can help develop a hybrid model for statistical downscaling of precipitation time series [10].

In this study, the data standardization method was performed before the SD via RF to decrease systematic error in the average and variance of the CMIP5 model outputs relative to the observational data. It should be noted that the dominant predictors of the selected CMIP5 model extracted via the M5 model in the closest grid to the study area, and the precipitation (predictand) values of the historical period (1992–2005) were used as the RF model inputs, and the capability of proposed hybrid model was verified by downscaling average monthly precipitation of Mahabad.

2.4 Evaluation Criteria

Both calibration and validation performances in downscaling process were evaluated by calculating the R and RMSE indices. The R value between two variables is defined as Eq. (2). It shows the ability of the model in replicating the observed data [5].

$$R = \frac{\sum(X - \overline{X})\sum(Y - \overline{Y})}{\sqrt{\sum(X - \overline{X})^2\sum(Y - \overline{Y})^2}} \tag{2}$$

where X, $\overline{X}$, Y, and Y are predictand, the mean values of predictand, predictor, and the mean values of predictor, respectively. RMSE is defined as Eq. (3) which shows the error of the model:

$$\text{RMSE} = \sqrt{\frac{1}{n}\sum_{i=1}^{n}(X - Y)^2} \tag{3}$$

here X, Y, and n are observed, simulated and the number of observations data, respectively. These indices are usually applied to evaluate the accuracy of different downscaling models [5].

3 Results and Discussion

3.1 Dominant Predictor Selection for CMIP5 Model Downscaling

In this paper, one ML model namely, M5 was used in the context of classification, with an application to the average monthly precipitation dataset of Mahabad in the historical period. The precipitation simulation was evaluated using MPI-ESM-LR from the CMIP5 models. Based on the grid data of MPI-ESM-LR, the nearest grid points to the case study were used for extracting the most dominant inputs via M5 and then, the selected inputs, along with the average monthly precipitation values of the historical period (1992–2005) were used as the downscaling model inputs. Table 4 provides the selected inputs using the M5 model.

The model was created using the Weka software, and the first few predictors located in the upper nodes of the tree were selected. The *pr* and variables from the humidity type (*hur (50)* and *huss (8500)*) are amongst the dominant parameters that were expected to be chosen because of the obvious and important role of humidity in precipitation formation. Regarding the *tas*, climate warming increases surface temperature, which raises the energy transferred to storms as an important factor in climate change and precipitation formation [11]. According to the high values of R via the pre-processing M5 model for precipitation values in MPI-ESM-LR that is shown in Table 4, it is obvious that the M5 model tree resulted in a good performance in selecting the most dominant predictors of the CMIP5 model as it benefits from the linear methods and shows multi-linear function [11].

Table 4 The selected dominant predictors based on the M5 model

Model	*R*	Dominant predictors
MPI-ESM-LR	0.81	clt, hur (50), huss (8500), pr, tas

3.2 CMIP5 Model-Based Precipitation Downscaling

After selecting the effective predictors, the RF model was applied using Python programming to downscale the 14-year average monthly precipitation parameters of the weather station before downscaling, the data standardization was done over the period 1992–2005. 75% of the data were used for calibration and 25% for validation purposes. The process of hyperparameter choice of RF was conducted based on trial-and-error, (e.g. the number of estimators = 130, random state = 10), and the results were compared to the 14-year average monthly observed precipitation of the ground weather station for the base period as illustrated in Figs. 3 and 4.

Figure 3 shows scatter plot between the downscaled results and the observed precipitation in the study area. It can be observed that data are relatively concentrated on the line in some months whilst there are relative differences between simulated

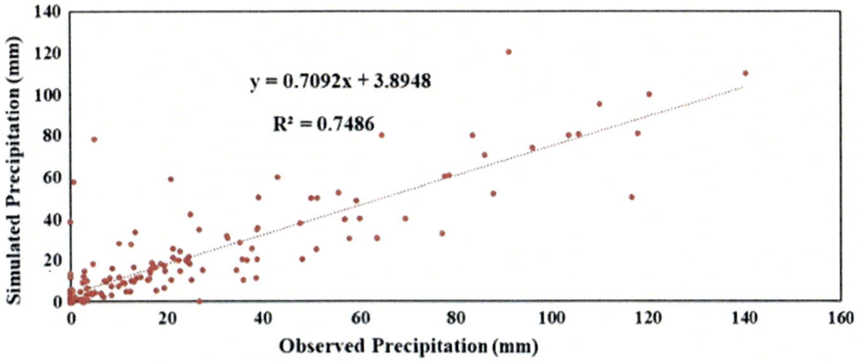

Fig. 3 Scatter plot between downscaled results and observed precipitation for historical period

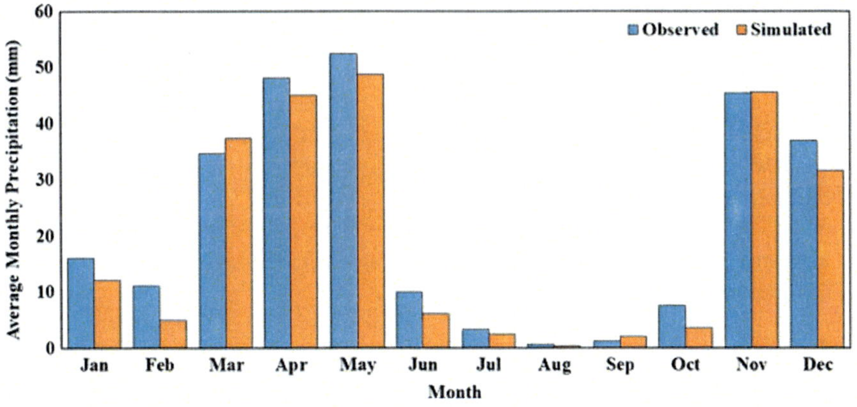

Fig. 4 Observed and simulated average monthly precipitation, downscaled by RF model for the base period (1992–2005)

and downscaled precipitation data according to the performance quantified by R and RMSE. Results in Fig. 4 indicates that the simulated average monthly precipitation time series experienced a drop in most months, except for March, September, and November compared to the observed monthly precipitation values. According to the figures, the downscaling results for the base period showed a relatively good performance in predicting the historical precipitation using RF, and the plots clearly show that the modelled data are close to the observed data.

4 Discussion

The performances of the model obtained from the downscaling via the hybrid model for the monthly precipitation simulation of MPI-ESM-LR for Mahabad denote that the RF model, with the dominant variables determined by the M5 model, produced acceptable climate predictions and showed relatively good performance according to the evaluating criteria for this study region in the historical period (R and RMSE (standardized) for calibration were 0.78 and 0.1 and R and RMSE for and validation were 0.75 and 0.13, respectively). However, some uncertainties remain. It is advisable to study the improvement and applicability of the hybrid model for other areas, subsequently considering the impact of different climatic conditions. Moreover, it is suggested to consider different AI techniques such as Long Short-Term Memory and Emotional Artificial Neural Network for downscaling CMIP5 models and use different optimization techniques to choose the model's hyperparameters to improve the results.

5 Conclusion

Improving downscaling through a hybrid ML-based statistical downscaling model was investigated in this study. The main objective was to determine the most important predictors of MPI-ESM-LR using the M5 model and to study the performance of the RF method in accurately simulating the average monthly precipitation of Mahabad located Iran for a historical period of 1992–2005. R and RMSE reached 0.78 and 0.1 for calibration and 0.75 and 0.13 for validation, respectively. All in all, although the hybrid M5-RF estimated a drop in average monthly precipitation in most months, it demonstrated a good performance in simulating the precipitation of Mahabad for the historical period.

References

1. Xu R, Chen N, Chen Y, Chen Z (2020) Downscaling and projection of multi-CMIP5 precipitation using machine learning methods in the Upper Han River basin. Adv Meteorol
2. Genuer R, Poggi JM, Tuleau-Malot C (2010) Variable selection using random forests. Pattern Recognit Lett 31:2225–2236
3. Cateni S, Vannucci M, Vannocci M, Coll V (2013) Variable selection and feature extraction through artificial intelligence techniques. Multivar Anal Manag Eng Sci
4. Campozano L, Tenelanda D, Sanchez E, Samaniego E, Feyen J (2016) Comparison of statistical downscaling methods for monthly total precipitation: case study for the Paute River basin in Southern Ecuador. Adv Meteorol
5. Nourani V, Rouzegari N, Molajou A, Hosseini Baghanam A (2020) An integrated simulation-optimization framework to optimize the reservoir operation adapted to climate change scenarios J Hydrol 587:125018
6. Quinlan JR (1992) Learning with continuous classes. In: Australian joint conference on artificial intelligence, vol 92, pp 343–348
7. Witten IH, Frank E (1999) Data mining: practical machine learning tools and techniques with java implementations. Morgan Kaufmann, San Fransisco
8. Breiman L (2001) Random forest. Mach Learn 45:5–32
9. Breiman L, Friedman J, Charles JS, Olshen RA (1984) Classification and regression trees
10. He X, Chaney NW, Schleiss M, Sheffield J (2016) Spatial downscaling of precipitation using adaptable random forests. Water Resour Res 52:8217–8237
11. Jato-Espino D, Sillanpää N, Charlesworth SM, Rodriguez-Hernandez J (2019) A simulation-optimization methodology to model urban catchments under non-stationary extreme rainfall events. Environ Model Softw 122

Multi-Step-Ahead Forecasting of Groundwater Level Using Model Ensemble Technique

Vahid Nourani, Parnian Ghaneei, and Elnaz Sharghi

Abstract Multi-step-ahead forecasting of groundwater level (GWL) is an important object that can be utilized for long-term policies and effective implementation of mitigation measures in the future. In this study, first three single artificial intelligence (AI)-based models including feed-forward neural network (FFNN), adaptive neural fuzzy inference system (ANFIS), and the group method of data handling (GMDH) network were used for predicting the multi-step-ahead GWL of Ghorveh–Dehgolan plain (GDP). Then, as a post-processing step to enhance the outcomes of the single models, their results were combined via two linear (simple and weighted) and one nonlinear (neural network) ensemble techniques. The results showed the superiority of the neural averaging ensemble technique because of its ability to cope with complex data. The neural model ensemble could improve the accuracy of the single models up to 23% in the testing phase. It could be concluded that the model ensemble techniques could enhance the performance of the single models for reliable forecasting of the future GWL condition.

Keywords Groundwater level (GWL) · Multi-step-ahead forecasting · Model ensemble · Ghorveh–Dehgolan plain (GDP)

1 Introduction

Groundwater is the most precious resource of the freshwater supply in arid and semi-arid countries, including Iran, where the demands for drinking and irrigation are satisfied by groundwater. Groundwater quantity and quality have undergone deterioration due to climate change and anthropogenic activities. The comprehensive

V. Nourani · P. Ghaneei (✉) · E. Sharghi
Center of Excellence in Hydroinformatics, Faculty of Civil Engineering, University of Tabriz, 29 Bahman Ave, Tabriz, Iran
e-mail: Parnian.Ghaneei@yahoo.com; p.ganei99@ms.tabrizu.ac.ir

V. Nourani
Faculty of Civil and Environmental Engineering, Near East University, Near East Boulevard via Mersin 10, 99138 Nicosia, Northern Cyprus, Turkey

J. H. Kim et al. (eds.), *Proceedings of 7th International Conference on Harmony Search, Soft Computing and Applications*, Lecture Notes on Data Engineering and Communications Technologies 140, https://doi.org/10.1007/978-981-19-2948-9_24

assessment of past, current, and future conditions of groundwater is a fundamental aspect of developing water resource strategies. Hot spots of groundwater depletion were reported in other countries including the northwest of India, northeast of China, northeast of Pakistan, Iran, central Yemen, and southern Spain, and in the High Plains and California Central Valley aquifers in the United States (e.g., see [1–3]).

Water resources managers need to know the future condition of groundwater level (GWL) to make appropriate decisions. One of the best ways to forecast GWL is to utilize modeling techniques. GWL modeling process includes nonlinear and complex features associated with natural factors and/or anthropogenic activities, affecting the dynamic predictions. Although there are different types of models for predicting groundwater quantity and quality parameters, the artificial intelligence (AI) models have gained considerable attention over the last decades because of their superiority in predicting complex hydrologic systems [4]. Due to the ability of recognizing complex and nonlinear relationships between inputs and outputs that are difficult to find by classic parametric methods, data-driven methods like ANNs (i.e., black-box models) have been widely utilized to forecast time series data of groundwater quantity and quality [5].

As another type of A.I. model, the adaptive neural fuzzy inference system (ANFIS) is a Takagi–Sugeno–Kang (TSK) fuzzy-based mapping algorithm providing minimal training time and error, less overshoot, and oscillation. ANFIS model benefits from both the ANN strength and fuzzy logic features representing accurate outcomes for groundwater modeling due to the ability of fuzzy concept in dealing with the involved uncertainties in the hydrological phenomena. Some previous investigations indicated that ANFIS is an efficient method for groundwater modeling (e.g., see [6]).

In addition to the ANN and ANFIS methods, the group method of data handling (GMDH) network is one of the feed-forward neural network (FFNN) models based on the Kolmogorov–Gabor (K–G) polynomial model. GMDH is an inductive learning algorithm that can recognize a relationship between inputs and outputs, finding an optimal model structure through quadratic regression polynomials. GMDH provides an objective network of a high order polynomial in the inputs to solve control, identification, and prediction problems. During the self-organized process of the model, GMDH creates, validates, and selects many alternative networks of growing complexity until to reach an optimized structure [7].

There is no agreement between researchers on which modeling method can present the highest accuracy and performance. The ensemble technique can serve the advantages of every single model; therefore, the model ensemble technique can be an efficient solution to overcome this problem and represent an enhanced outcome. Such model ensemble techniques have already been applied in hydrological modeling (e.g., see [8]). This study presents an attempt to use the model ensemble techniques to multi-step-ahead forecasting of GWL over the Ghorveh–Dehgolan plain (GDP) where has undergone several problems due to sharp GWL depletions. By integrating different robust forecasting models, the difficulty of opting for the proper approach was overcome, and the unique ability and feature of each method were employed in a more accurate process.

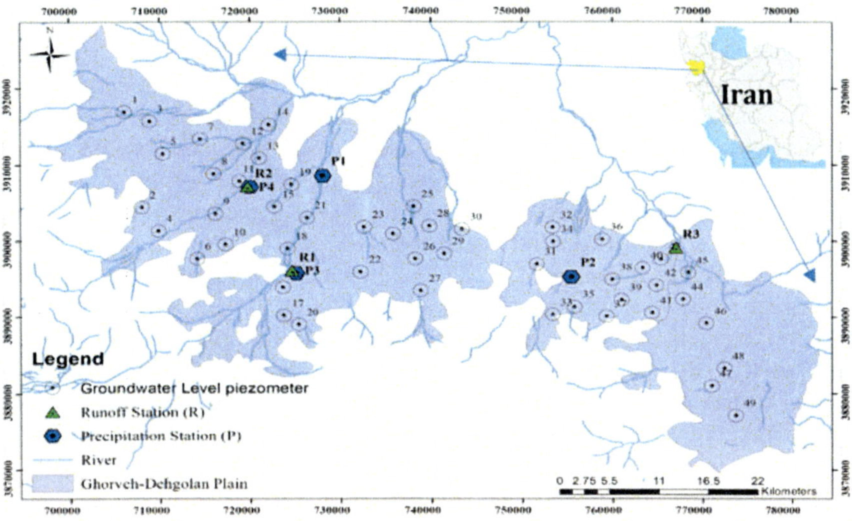

Fig. 1 Location of GDP and observation stations

2 Materials and Methods

2.1 Case Study

The GDP is located in western Iran within the longitude from 47° 38′ 52″ to 48° 06′ 03″ east and latitude from 35° 02′ 22″ to 35° 30′ 54″ north. This plain covers an area of approximately 1270 km^2 in Kurdistan Province. The climate of the area is semi-arid, while during winter, the average minimum temperature is 5.5 °C, and the region experiences a daily maximum temperature of 36 °C in the summer. Moreover, the average annual precipitation in this area is 345 mm. In this area, the significant factors influencing groundwater quantity and quality include overpumping for irrigation and drinking purposes, cultivation patterns changes, and lack of standard control with regard to waste disposal. The monthly GWL time series (1989–2018) from four piezometers (P15, P42, P30, and P35), four rainfall, and three runoff stations were gathered and used in this study. The location map, positions of piezometers, rainfall stations (P1, P2, P3, and P4), and runoff stations (R1, R2, and R3) are shown in Fig. 1.

2.2 Input Selection and Single AI-Based Models

Generally, as a conventional method, linear correlation coefficients between inputs and output are employed to determine the dominant inputs of A.I. models, but as criticized by [9], it is more appropriate to use a nonlinear measure such as mutual information (MI) in a nonlinear modeling framework, since sometimes there may be a weak linear but strong nonlinear relationship between input and output. The selection process of the dominant inputs for a nonlinear modeling framework requires nonlinear measures such as MI because of nonlinear relationships between inputs and outputs in environmental and hydrological parameters.

ANN is a mathematical model and one of the AI-based models for handling the nonlinear connections between inputs and output datasets. FFNN is one of the ANN algorithms with B.P. training used widely in different fields of hydroclimatological research and is the most common class of ANNs [8].

Neuro-fuzzy simulation employs various learning algorithms for fuzzy modeling in the neural network or fuzzy inference system. Every fuzzy system contains three essential parts; fuzzy database, fuzzifier, and defuzzifier. The fuzzy database includes the inference engine and fuzzy rule base involving rules related to fuzzy propositions. The combination of ANN and fuzzy modeling is used to consider the existing uncertainty of hydrological processes. There are various types of fuzzy inference engines to simulate the ANFIS model in which the Sugeno FIS was selected to be used in this study [8].

The GMDH network is an enhanced type of FFNN. GMDH to simulate the complex relationships passes the inputs through the input layer and enters the hidden layers. Unlike the other models, in GMDH, inputs of the hidden layer nodes may be from any layers before the current layer, and each node of hidden layers has two inputs. The selection of these two inputs is conducted randomly among all nodes before the current node and then forms an input group. The best input group is found by the least square optimization of the objective function in the GMDH network [10].

2.3 *Model Ensemble Unit*

Since different models can have different results at different conditions, intervals, or simulations of peak values, to improve the final accuracy and predicted time series ensembling (combining) the outputs of different prediction models is suggested. Another advantage of the model ensemble technique is that the final outcome will not depend on selecting the best method. This combining method aims to create an ensemble of the single methods that are diverse and yet accurate.

In this study, to enhance the overall efficiency of the predictions, two ensemble techniques were conducted to combine the results of single models, including linear ensemble technique (simple averaging and weighted averaging) and the neural ensemble technique as FFNN [8].

The first linear ensemble technique (simple averaging) is formulated as

$$\overline{\mathrm{GWL}}(t) = \frac{1}{N}\sum_{i=1}^{N}\mathrm{GWL}_i(t) \tag{1}$$

where N indicates the number of single models (here $N = 3$), $\text{GWL}_i(t)$ is the outcome of the ith single model (i.e., FFNN, ANFIS, and GMDH), and $\overline{GWL}(t)$ is the final outcome for the simple averaging ensemble technique.

The second linear ensemble technique (weighted averaging) is conducted using

$$\overline{\text{GWL}}(t) = \sum_{i=1}^{N} w_i \text{GWL}_i(t) \tag{2}$$

where w_i is the weight imposed on the ith single method output that may be assumed based on the performance criteria of the single ith model as

$$w_i = \frac{R_i^2}{\sum_{i=1}^{N} R_i^2} \tag{3}$$

where R_i^2 is the accuracy measure of the ith single model.

In the neural ensemble technique, the outputs of single AI models are used as inputs to the neurons of the input layer of FFNN models. Although other types of AI-based models (e.g., GMDH and ANFIS) can be used as a kernel for the neural ensemble technique, in this study, the FFNN method was utilized as a nonlinear technique to combine the outputs of the single AI-based models.

2.4 Forecasting Performance Criteria

The assessments and comparison procedure of the proposed modeling performance were conducted by the coefficient of determination (R^2) and root mean square error (RMSE). The equations for R^2 and RMSE are given by

$$\text{RMSE} = \sqrt{\frac{1}{N}\sum_{i=1}^{N}\left(Q_i - \hat{Y}_i\right)^2} \tag{4}$$

$$R^2 = 1 - \frac{\sum_{i=1}^{N}\left(Q_i - \hat{Y}_i\right)^2}{\sum_{i=1}^{N}\left(Q_i - \overline{Q_i}\right)^2} \tag{5}$$

where N, Q_i, $\hat{Y}_i$, and $\overline{Q_i}$ are the number of observations, observed, computed values, and mean of observed data, respectively [8].

3 Results and Discussions

3.1 *Results of GWL Forecasting via Single Models*

Multi-step-ahead forecasting of GWL is an important object that can be utilized for long-term policies and effective implementation of mitigation measures in the future. In this regard, the forecasting models were applied to the data of four piezometers (P15, P42, P30, and P35) of GDP to forecast one, two, and three months ahead GWLs. For this aim, all proposed models utilized monthly GWL, rainfall, and runoff data of the GDP in training (70% of the dataset) and testing (30% of the dataset) phases. The MI measure was computed for the data set of parameters to find the best combination of inputs and lag times between hydrological processes (rainfall (P), runoff (R), and GWL). The selected dominant inputs used in all models are shown in Table 1.

To predict the future conditions of GWL of GDP, first, three AI-based models (FFNN, ANFIS, and GMDH) were separately trained for all selected inputs. The GWL forecasting results using the FFNN, ANFIS, and GMDH methods are tabulated in Table 2.

Table 1 Information of developed models for all clusters

Piezometer	Inputs	Output
15	GWL15 (t), GWL15 ($t-1$), GWL15 ($t-4$), P1 ($t-4$), R2 ($t-5$)	GWL15 (t+1)
	GWL15 (t), GWL15 ($t-1$), GWL15 ($t-3$), P1 ($t-3$), R2 ($t-4$)	GWL15 (t+2)
	GWL15 (t), GWL15 ($t-1$), GWL15 ($t-2$), P1 ($t-2$), R2 ($t-3$)	GWL15 (t+3)
42	GWL42 (t), GWL42 ($t-2$), GWL42 ($t-7$), P2 ($t-2$), R3 ($t-1$)	GWL42 (t+1)
	GWL42 (t), GWL42 ($t-2$), GWL42 ($t-7$), P2 ($t-1$), R3 ($t-2$)	GWL42 (t+2)
	GWL42 (t), GWL42 ($t-2$), GWL42 ($t-6$), P2 (t), R3 (t)	GWL42 (t+3)
30	GWL30 (t), GWL30 ($t-1$), GWL30 ($t-9$), P3 ($t-2$), R1 (t)	GWL30 (t+1)
	GWL30 (t), GWL30 ($t-1$), GWL30 ($t-8$), P3 ($t-1$), R1 ($t-1$)	GWL30 (t+2)
	GWL30 (t), GWL30 ($t-1$), GWL30 ($t-8$), P3 ($t-1$), R1 ($t-1$)	GWL30 (t+3)
35	GWL35 (t), GWL35 ($t-1$), GWL35 ($t-3$), GWL35 ($t-9$), P2 ($t-2$)	GWL35 (t+1)
	GWL35 (t), GWL35 ($t-1$), GWL35 ($t-3$), GWL35 ($t-8$), P2 (t)	GWL35 (t+2)
	GWL35 (t), GWL35 ($t-1$), GWL35 ($t-2$), GWL35 ($t-8$), P2 (t)	GWL35 (t+3)

Table 2 Results of single-developed models for multi-step-ahead GWL forecasting

Piezometer	Model	Testing phase					
		GWL ($t + 1$)		GWL ($t + 2$)		GW L($t + 3$)	
		R^2	RMSE	R^2	RMSE	R^2	RMSE
15	FFNN	0.80	0.44	0.79	0.44	0.77	0.46
	ANFIS	0.81	0.43	0.82	0.43	0.78	0.46
	GMDH	0.85	0.40	0.84	0.41	0.82	0.45
42	FFNN	0.79	0.46	0.67	0.48	0.72	0.47
	ANFIS	0.80	0.45	0.68	0.47	0.71	0.46
	GMDH	0.83	0.44	0.76	0.42	0.75	0.42
30	FFNN	0.76	0.44	0.74	0.46	0.72	0.47
	ANFIS	0.79	0.44	0.78	0.45	0.74	0.46
	GMDH	0.81	0.39	0.80	0.40	0.78	0.43
35	FFNN	0.70	0.58	0.68	0.59	0.69	0.49
	ANFIS	0.74	0.56	0.72	0.57	0.71	0.48
	GMDH	0.77	0.55	0.75	0.56	0.78	0.45

As shown in Table 2, in one, two, and three time step-ahead forecasting, the R^2 and RMSE values indicate better performance of the GMDH method compared to other single AI-based models. The main reason for the superiority of GMDH is that just neurons with better external criterion values are kept, and others are removed. The mathematical prediction function of the GMDH is found by the polynomial expression of the best neuron of the last layer. On the other hand, both ANN and ANFIS models use all of the neurons' information. Too many neurons may lead to overfitting, while insufficient neurons may capture unsatisfactory information, decline the models' performance, and reduce the efficiency.

It is clear from Fig. 2a that piezometer 15, which is located in the western part of the GDP, experienced a sharp GWL drop (46 m in total period). One of the significant reasons for the over pumping of groundwater is that Dehgolan city, with a population of 26,000 people. The main occupation of the inhabitants of Dehgolan is agriculture, and the predominant crop cultivated is potatoes, which is a product with high irrigation water requirements. Also, the largest industrial town in Kurdistan province is located in Dehgolan city, which has increased the water stress [9]. Excessive groundwater withdrawals to supply all these factors have faced this area with experiencing the lowest GWL.

The single AI-based models' outputs of piezometers 42 and for the test phase are shown in Fig. 2b, c, respectively. These members are located in the eastern and central parts of GDP, experiencing almost 53 and 34 m GWL depletions in the last 30 years, respectively. The main reason for GWL depletion in these zones is an increasing tendency among local farmers to cultivate irrigated crops. The cropping pattern shifted from wheat and barley to cultivating water-intensive crops such as potato and forage crops in recent years [9].

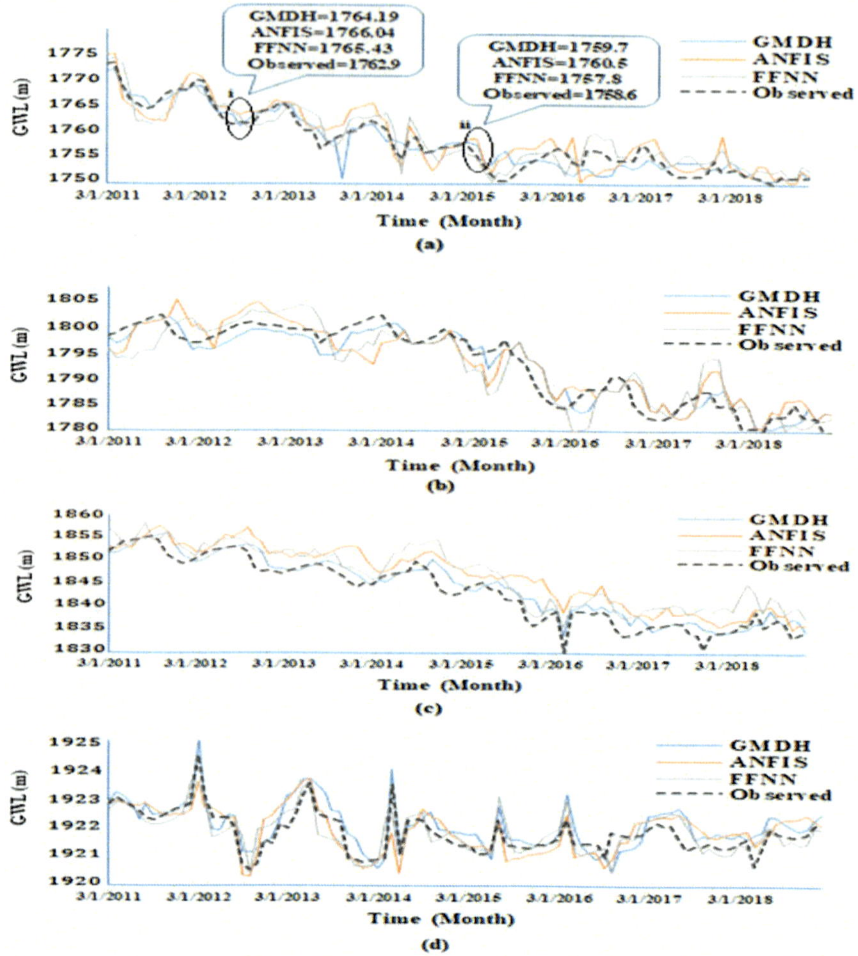

Fig. 2 The best results of single AI-based models for one step-ahead GWL forecasting in testing phase for **a** piezometer 15, **b** piezometer 42, **c** piezometer 30, and **d** piezometer 35

Piezometer 35, with just 3 m GWL depletion during the total period, is the representative zone of a place with the satisfactory condition of GWL located in the southern part of the GDP. Notably, this area experienced the most considerable GWL fluctuations (Fig. 2d). The existence of more extremum values in the GWL time series of Piezometer 35 made the forecasting procedure a bit complicated. Consequently, the performance of the single models for multi-step-ahead forecasting of GWL for this cluster decreased and led to lower accuracy compared to others (see Table 2).

Overall comparison of plots in Fig. 2 and information in Table 2 revealed that forecasting of piezometers with high amounts of groundwater depletion led to better performance than members with the low level of GWL depletion. This result implies that excessive water abstraction for anthropogenic activities and consequently experiencing a sharp decrease in GWL may cause the aquifer to lose its ability to react to natural events such as rainfall and seasonal changes, which has led to a decrease in natural distributions of changes and fluctuations of the GWL time series in recent years over the plain that consequently decreased the complexity of time series and increased the accuracy of the modeling [11, 12].

The details of Fig. 2a show the values of two points (i and ii), representing the different performances of single models in various time spans. Similarly, in some points, one of the single models was superior and in other points, other models presented better results. These differences make the comparison of single models a challenging issue. In order to overcome this problem, the ensemble approaches could be the logical solution to achieve more integrated and accurate outcomes.

3.2 *Results of GWL Forecasting via Model Ensemble Techniques*

The integration of the outcomes of FFNN, ANFIS, and GMDH models was carried out based on the three proposed ensemble techniques (simple, weighted, and neural averaging). Although other types of AI-based models (e.g., GMDH and ANFIS) can also be used as a kernel for the neural ensemble, FFNN was selected as a commonly used model in dealing with complex data for the neural ensemble step of this study. Table 3 shows the results of model ensemble techniques obtained in this study.

The final results of model ensemble techniques indicated that the simple averaging, weighted averaging, and neural ensemble method by GMDH, increased the testing phase accuracy of single models up to 13, 13, and 18% for GWL $(t + 3)$ forecasting of piezometer 15; 9, 9, and 23% for GWL $(t + 2)$ forecasting of piezometer 42; 7, 8, and 15% for GWL $(t + 2)$ forecasting of piezometer 30; and 11, 12, and 17% for GWL $(t + 2)$ forecasting of piezometer 35, respectively. It can be concluded that the neural ensemble technique could be a practical alternative to design a more accurate model for forecasting with good agreement between the observed and computed values.

4 Conclusion

In the present study, predictive modeling's were first conducted by three AI models (FFNN, ANFIS, and GMDH) for one, two, and three time step-ahead forecasting. Results indicated that the GMDH model outperforms other AI-based models. Then, as a post-processing step to improve the accuracy of single modeling, the outcomes of single models were combined by two linear (simple and weighted) and one nonlinear neural ensemble technique. Up to 23% increase in the performance of

Table 3 Results of the proposed ensemble techniques for multi-step-ahead forecasting of GWL

piezometer	Method	Testing phase					
		GWL ($t+1$)		GWL ($t+2$)		GWL ($t+3$)	
		R^2	RMSE	R^2	RMSE	R^2	RMSE
15	Simple averaging ensemble	0.85	0.38	0.86	0.39	0.90	0.32
	Weighted averaging ensemble	0.87	0.37	0.90	0.37	0.90	0.31
	Neural averaging ensemble	0.95	0.33	0.95	0.34	0.95	0.34
42	Simple averaging ensemble	0.86	0.44	0.76	0.47	0.73	0.43
	Weighted averaging ensemble	0.88	0.43	0.77	0.45	0.74	0.40
	Neural averaging ensemble	0.90	0.40	0.90	0.39	0.87	0.39
30	Simple averaging ensemble	0.82	0.42	0.80	0.41	0.79	0.44
	Weighted averaging ensemble	0.85	0.41	0.82	0.38	0.80	0.43
	Neural averaging ensemble	0.90	0.37	0.89	0.36	0.85	0.39
35	Simple averaging ensemble	0.74	0.44	0.78	0.49	0.74	0.46
	Weighted averaging ensemble	0.79	0.49	0.79	0.51	0.78	0.45
	Neural averaging ensemble	0.84	0.41	0.83	0.48	0.84	0.42

single models implied the superiority of the nonlinear neural ensemble technique over other ensemble techniques. Taken together, the neural model ensemble technique could enhance the final outcomes to fulfill the aim of obtaining the most accurate and reliable information and knowledge about the past, current, and future changes of groundwater depletion.

References

1. Chen W, Panahi M, Khosravi K, Reza H, Rezaie F (2019) Spatial prediction of groundwater potentiality using ANFIS ensembled with teaching-learning-based and biogeography-based optimization. J Hydrol 572:435–448. https://doi.org/10.1016/j.jhydrol.2019.03.013
2. Gong Y, Zhang Y, Lan S, Wang H (2016) A Comparative study of artificial neural networks, support vector machines and adaptive neuro fuzzy inference system for forecasting groundwater levels near Lake Okeechobee, Florida. Water Resour Manage 30:375–391. https://doi.org/10.1007/s11269-015-1167-8

3. Rajaee T, Ebrahimi H, Nourani V (2019) A review of the artificial intelligence methods in groundwater level modeling. J Hydrol 572:336–351. https://doi.org/10.1016/J.JHYDROL.2018.12.037
4. Tao H, Al-Khafaji ZS, Qi C, Zounemat-Kermani M, Kisi O, Tiyasha T, Chau K-W, Nourani V, Melesse AM, Elhakeem M, Farooque AA, Pouyan Nejadhashemi A, Khedher KM, Alawi OA, Deo RC, Shahid S, Singh VP, Yaseen ZM (2021) Artificial intelligence models for suspended river sediment prediction: state-of-the art, modeling framework appraisal, and proposed future research directions. Eng Appl Comput Fluid Mech 15:1585–1612. https://doi.org/10.1080/19942060.2021.1984992
5. Lee S, Lee K-K, Yoon H (2019) Using artificial neural network models for groundwater level forecasting and assessment of the relative impacts of influencing factors. Hydrogeol J 27:567–579. https://doi.org/10.1007/s10040-018-1866-3
6. Azad A, Karami H, Farzin S, Mousavi SF, Kisi O (2019) Modeling river water quality parameters using modified adaptive neuro fuzzy inference system. Water Sci Eng 12:45–54. https://doi.org/10.1016/J.WSE.2018.11.001
7. Lambert RSC, Lemke F, Kucherenko SS, Song S, Shah N (2016) Global sensitivity analysis using sparse high dimensional model representations generated by the group method of data handling. Math Comput Simul 128:42–54. https://doi.org/10.1016/j.matcom.2016.04.005
8. Elkiran G, Nourani V, Abba SI (2019) Multi-step ahead modelling of river water quality parameters using ensemble artificial intelligence-based approach. J Hydrol 577:123962. https://doi.org/10.1016/J.JHYDROL.2019.123962
9. Nourani V, Ghaneei P, Kantoush SA (2021) Robust clustering for assessing the spatiotemporal variability of groundwater quantity and quality. J Hydrol 604:127272. https://doi.org/10.1016/J.JHYDROL.2021.127272
10. Tsai TM, Yen PH (2017) GMDH algorithms applied to turbidity forecasting. Appl Water Sci 7–1151–1160. https://doi.org/10.1007/s13201-016-0458-4
11. Foroumandi E, Nourani V, Dabrowska D, Kantoush SA (2022) Linking spatial-temporal changes of vegetation cover with hydroclimatological variables in terrestrial environments with a focus on the Lake Urmia basin 11. https://doi.org/10.3390/land11010115
12. Sharghi E, Nourani V, Zhang Y, Ghaneei P (2022) Conjunction of cluster ensemble-model ensemble techniques for spatiotemporal assessment of groundwater depletion in semi-arid plains. J Hydrol 610:127984. https://doi.org/10.1016/j.jhydrol.2022.127984

AMHS: Archive-Based Multi-objective Harmony Search Algorithm

Nima Khodadadi, Farhad Soleimanian Gharehchopogh, Benyamın Abdollahzadeh, and Seyedali Mirjalili

Abstract Meta-heuristics have been widely used in both science and industry as reliable alternatives to conventional optimization algorithms to solve challenging, real-world problems. Despite being general-purpose and having a black-box nature, they require changes to solve multi-objective optimization problems. This paper proposes a multi-objective version of harmony search based on the archive. Archive, grid, and leader selection mechanisms are applied in multi-objectives of HS. Five real-engineering problems are evaluated with the results of three indexes. Based on the results, the AMHS is capable of providing acceptable results than other alternatives.

Keywords Multi-objective harmony search · Harmony search · Real-engineering problems

1 Introduction

Optimization problems can be found everywhere. All engineering design problems can be considered optimization as the goal is to find the best design parameters to improve a set of objectives. Data analysis, which is very popular in the machine learning space, is also optimization as the goal is to find the best configuration and structural parameters for a machine learning model to reduce error and/or improve

N. Khodadadi
Department of Civil and Environmental Engineering, Florida International University, Miami, FL, USA

F. S. Gharehchopogh · B. Abdollahzadeh
Department of Computer Engineering, Urmia Branch, Islamic Azad University, Urmia, Iran

S. Mirjalili (✉)
Centre for Artificial Intelligence Research and Optimisation, Torrens University Australia, Brisbane, QLD, Australia
e-mail: ali.mirjalili@gmail.com

Yonsei Frontier Lab, Yonsei University, Seoul, South Korea

J. H. Kim et al. (eds.), *Proceedings of 7th International Conference on Harmony Search, Soft Computing and Applications*, Lecture Notes on Data Engineering and Communications Technologies 140, https://doi.org/10.1007/978-981-19-2948-9_25

accuracy. Most decision-making processes are optimized in businesses and organizations as several decision parameters are changed to reach a goal (e.g., increasing revenue, reducing cost, etc.).

To understand the key concepts in this area, the following subsections discussed preliminaries and basic definitions about optimization problems and optimization algorithms. We start with problems because finding a suitable algorithm without knowing the characteristic of an optimization problem is meaningless. In other words, optimization algorithms are very diverse and fit specific types of problems. The following sub-section first covers the types of optimization problems investigated in this study. Then, the suitable class of optimization algorithms to solve them is discussed.

1.1 Optimization Problems

Despite the differences of such optimization problems, they can be formulated as follows:

$$P\text{: Minimize } f(x) \tag{1}$$

$$\text{Subject to: } x \in X \tag{2}$$

$$\text{where: } f\text{: } R^n \rightarrow R \text{ and } X \text{ is a subset of } R^n \tag{3}$$

In these equations, f represents an objective function, which is often called cost, merit, or fitness function, x shows a solution, R^n includes all the possible solutions for the problem, X is a subset of R^n defined by constraints and includes only feasible solutions that do not violate constraints, and the problem is formulated as a minimization problem without the loss of generality [1].

The above formulation is for a single-objective minimization problem. Therefore, one of the solutions in set X is considered the global minimum and considered the best solution for this problem. The mathematical definition of this solution is as follows:

$$\forall x \in X\text{: } f\left(x^*\right) \leq f(x) \tag{4}$$

$$\text{where}\left(x^* \in X\right) \tag{5}$$

Note that the above formulation is for a single-objective problem. Multi-objective problems have multiple objective functions that provide multiple comparison measures for the same solution x. The formulation is as follows:

$$P\colon \text{Minimize } f_1(x),\ f_2(x), \ldots,\ f_o(x), \tag{6}$$

$$\text{Subject to:}\ x \in X \tag{7}$$

$$\text{where: } f\colon R^n \rightarrow R \text{ and } X \text{ is a subset of } R^n \tag{8}$$

When dealing with such problems, the challenge is finding a solution that minimizes all objectives simultaneously, which is often difficult to achieve due to the conflicting nature of objectives. Also, there is no longer a single solution for such a problem. A set of a solution called Pareto optimal solution set includes the best solution that represents the best trade-offs between the objectives. Pareto optimality is that defines the superiority of a solution over another as follows [2]:

$$\forall i \in \{1, 2, 3, \ldots, o\}\colon f_i(x_1) \leq f_i(x_2) \wedge \exists i \in \{1, 2, 3, \ldots, o\}\colon f_i(x_1) < f_i(x_2) \tag{9}$$

In this formulation, x_1 is better than x (denoted as $x_1 \prec x_2$). A solution (x^*) to a multi-objective optimization problem is called Pareto optimal if:

$$\nexists x \in X \middle| x \prec x^* \tag{10}$$

1.2 Optimization Algorithms

After the formulation of an optimization problem, we are ready to choose a suitable algorithm. Of course, finding the best algorithm requires extensive comparison, but knowing the problem and its difficulties allows us to narrow the search to a particular class of optimization algorithms. There is a different classification of optimization algorithms in the literature, but one that fits well the scope of this work is based on the number of solutions they use and improve to find the best solution for the problem. An algorithm can start with one solution, which is called solitary optimizer. Two of the most popular algorithms in this class are gradient descent [3] and hill climbing [4]. In the former case, the algorithm follows the negative gradient iteratively, which leads to finding the best locally optimal solution in the vicinity of the solution. In the latter case, the algorithm creates a possible solution at each optimization step and chooses the best one, which is quite similar to a greedy search [5].

The second class of optimization algorithm includes collective algorithms, in which more than one solution work together collectively to find the best solution for a given optimization problem. Some of the best examples are the genetic algorithm (GA) [6] or particle swarm optimization (PSO) [7]. In both algorithms, the optimization process is commenced with a random set of solutions, often called population. This population is then iteratively evaluated using an objective function(s) and improved using specific mechanisms. Due to information sharing between the solutions and their collective behavior tends to outperform solitary optimization algorithms.

Both classes presented in this sub-section have their advantages and drawbacks. The benefits of solitary optimization algorithms are low-computational cost and high-convergence speed. However, they suffer from finding locally optimal solutions as there is usually little mechanism and capability to search globally in a search space of an optimization problem. On the other hand, collective algorithms benefit from a high chance of locally optimal solution avoidance due to multiple solutions in each step of optimization. However, they are computationally more expensive as each solution requires calculating the objective value and the need for complicated operators to merge them and maintain their diversity/quality.

Real-world problems tend to have many locally optimal solutions, so it makes sense to use collective optimization algorithms for them. This is why we used collective meta-heuristics, which are the most reliable and efficient optimization algorithms in the current century. Harmony search algorithm [8], stochastic paint optimizer (SPO) [9], ant colony optimization (ACO) [10], hybrid invasive weed optimization-shuffled frog-leaping algorithm (IWOSFLA) [11], bat algorithm (BA) [12], dynamic water strider algorithm (DWSA) [13], flow direction algorithm (FDA) [14], and advanced charged system search (ACSS) [14] are some of the well-formulated meta-heuristic optimization algorithms alongside the application of some algorithms in different fields [15–17]. The rest of the paper is organized as follows:

Section 2 briefly presents HS algorithms. Multi-objective harmony searched based on the archive is proposed in Sect. 3. The fundamental engineering problems design is discussed and provided an extensive comparative study between multi-objective particle swarm optimization (MOPSO) [18], multi-objective ant lion optimizer (MOALO) [19], and multi-objective multi-verse optimizer (MOMVO) [20] algorithms in Sect. 4. Finally, Sect. 5 concludes the work and suggests future directions.

2 Harmony Search Algorithm (HS)

Geem et al. [8] proposed the harmony search method as a population-based meta-heuristic optimization technique. HS is a music-inspired algorithm and mimic the process of improvising a piece of music by a musician by tuning pitch of an instrument to achieve a desirable harmony. A special relationship between several sound waves of different frequencies in nature is presented by harmony. Harmony is the name given to a feasible solution in the HS algorithm, and each decision variable corresponds to a note. The harmony memory (HM) in HS stores a predetermined number of harmonies (N) with d decision variables and tune them to optimize an objective function (f). This algorithm is made of the following steps:

Step 1. *Initializing the harmony memory.*
Step 2. *Create a new harmony.*
Step 3. *Accept the new harmony to HM* if better than any in the harmony memory.
Step 4. *Repeat steps 2 and 3 until an end condition is met.*
Step 5. *The found optimum solution is the best harmony stored in HM*.

3 Multi-objective Harmony Search Algorithm (AMHS)

There are some multi-objective versions of harmony search with different mechanisms. Objective harmony search (MOHS) algorithm is proposed.

It is possible to formulate the optimal power flow (OPF) problem as a non-linear constrained multi-objective optimization problem, in which different objectives and constraints have been taken into consideration during the formulation process. Sivasubramani and Swarup [21] used fast, non-dominated, elitist sorting, and crowding distance to find the Pareto optimal front and manage it. Finally, a fuzzy-based mechanism was used to select a compromise solution from the Pareto set.

The peak load management problem in modern power systems necessitates efficient methods like demand side management (DSM). A model of multi-objective dynamic optimal power flow (MODOPF) is combined with a DSM technique based on game theory by Lokeshgupta and Sivasubramani [22]. The DSM method, which is based on game theory, is applied to a single utility company and a large number of individual residential energy customers.

Four variations of the harmony search algorithm were implemented in the framework of NSGA-II by Pavelski et al. [23]. A set of benchmark instances proposed in CEC 2009 is used to compare the four methods. The result of this method shows good performance.

An improved metaheuristic searching algorithm that combines harmony search and fast non-dominated sorting to solve the multi-objective optimization problem is proposed by Sheng et al. [24]. For the purpose of finding multi-objective harmony, they have developed a new intelligent optimization algorithm (MOHS). Modified IEEE 33-bus test system simulation results and comparison with NSGA-II algorithm have shown that proposed MOHS can achieve promising results for engineering applications.

To handle multimodal multi-objective optimization issues, a modified multi-objective harmony search algorithm named niching multi-objective harmony search algorithm (NMOHSA) is developed by Qu et al. [25]. It makes use of neighborhood information to create dynamic harmony memories, which are essential for preserving population diversity. To avoid being stuck in a local optimal solution, a new memory consideration rule is implemented. There are a number of variables that can be dynamically modified, including the harmony memory consideration rate (HMCR) and pitch adjustment rate (PAR) values. Experimental results demonstrate that the suggested method outperforms the other multimodal multi-objective algorithms in terms of solution quality when compared to the other existing algorithms.

According to the No Free Lunch (NFL) [26] theorem, none of these algorithms will solve all optimization problems. This means that one or more algorithms may perform well on one set of problems while poorly on others. This emphasizes the importance of conducting comparative research on various problem areas. This type of comparative analysis also provides new insights into the performance of meta-heuristics and potential improvements. The present paper introduces and applies an archive-based multi-objective harmony search algorithm (AMHS) to several engineering examples. HS was designed to solve single-objective optimization problems and cannot be simply used to multi-objective challenges. As a result, a multi-objective variant of HS is presented in this study for tackling multi-criterion optimization issues. AMHS contains three multi-objective optimization mechanisms as following subsections [27]:

3.1 Archive Mechanism

It acts as a repository for the derived Pareto optimal solutions, which can be stored or restored. A single controller manages the archive, determining which solutions are added to it and when it is complete. There is a limit to the number of solutions that can be saved in the archive. The occupants of the archive are compared to the non-dominated solutions developed thus far during iteration. If at least one archive member dominates the new solution, it is not permitted to enter the archive. The new solution may be included in the archive if it outperforms at least one existing solution by omitting the existing one. The new solution is added to the archive if the new and archive solutions do not dominate each other.

3.2 Grid Mechanism

It is an excellent way to improve non-dominated solutions in the archive. If the archive fills, the grid technique will reorganize the object space's division and identify the most occupied area, allowing one of the solutions to be eliminated. The additional member should be added to the least crowded segment to increase the diversity of the final approximated Pareto optimal front. As the hypercube's number of possible solutions grows, so does the possibility of removing a solution. If the archive is complete, the busiest sections are selected first, and a solution from one of them is randomly drawn to make space for the new one. A particular case occurs when a solution is placed outside of the hypercubes. This scenario's sections have all been expanded to include the most recent solutions. As a result, the sections of alternative solutions can be modified.

3.3 Leader Selection Mechanism

Using this mechanism, the results of a search with multiple objectives can be compared. As a result, the top search candidates are guided toward areas of the search space to find a solution as close to global optimal as feasible. There are no dominant solutions in the archive, and the leader selection mechanism selects the most minor crowded areas of search space and displays the best non-dominated answers. Each hypercube is chosen using a roulette wheel with the following probability [27]:

$$P_i = \frac{C}{N_i} \tag{11}$$

N is the variety of obtained Pareto optimal answers in the ith section, where c is a constant number greater than one. Equation (11) shows that less crowded hypercubes are more likely to suggest new leaders. The probability of selecting leaders from a hypercube increases when the number of solutions is reduced. The AMHS algorithm, of course, relies on HS for its convergence. AMHS's already high level of reliability is likely to increase if we select one of the solutions from the archive. Finding Pareto optimal solutions when there is much variation is challenging.

4 Results and Discussion

The proposed method was determined to be efficient using performance measures and case studies, which included real-world engineering design problems. Inverted generational distance (IGD), maximum spread (MS), and spacing (S) are used to assess the algorithms' performance. The details of mentioned metrics can be found in [28]. Algorithms were written in MATLAB R2021a, and datasets were tested in 30 separate iterations with a sample size of 50 on an Intel Core i9 computer running at 2.3 GHz with 16 GB of RAM running macOS Big Sur. The initial parameters of all described algorithms are summarized in Table 1.

Table 1 The parameters settings of all algorithms

Algorithm	Parameters and values
MOPSO	Probability of mutation = 0.5, size of population = 100, archive capacity = 100, archive grid index = 30, $c_1 = 1$, $c_2 = 2$, w = 0.4, beta = 4, gamma = 4
MOALO	Size of population = 100, archive capacity = 100, archive grid index = 30, beta = 4, gamma = 4
MOMVO	Size of population = 100, archive capacity = 100, archive grid index = 30, beta = 4, gamma = 4
AMHS	Size of population = 100, archive capacity = 100, archive grid index = 30, beta = 4, gamma = 4

The IGD and S performance metrics results are presented in Tables 2 and 3 for engineering design problems. When it comes to IGD metrics, the AMHS algorithm outranks the other algorithms in 4 of the 5 cases. In terms of the S metric, the AMHS surpasses most other algorithms except WELDED BEAM. Moreover, the SD values confirmed the proposed method's ability to get similar outcomes over various executions. The accurate and obtained Pareto fronts for the considered engineering design problems are shown in Fig. 1, demonstrating the algorithm's capability to provide better solutions closer to the Pareto front. As shown in this figure, the proposed AMHS presents a perfect convergence approaching all actual Pareto optimal fronts.

Regarding MS performance metric in dealing with engineering design problems (Table 4), the AMHS is capable of calculating better results than MOPSO, MOMVO, and MOALO in 4 of the five problems in terms of average, while for the DISK BRAKE, the results of MOMVO are better than other algorithms. In terms of standard deviation, AMHS obtained the first rank. This paper demonstrates that the proposed

Table 2 The results of the IGD metric

Problem	Metric	MOPSO	MOALO	MOMVO	AMHS
P1: BNH	AVG	9.8964E–04	2.3879E–03	2.4514E–03	**8.4345E–04**
	SD	2.7342E–04	5.4986E–03	4.3865E–04	**1.6761E–04**
P2: CONSTR	AVG	6.3478E–04	1.2840E–03	2.3648E–03	**4.3489E–04**
	SD	**4.4356E–05**	8.3473E–04	4.3495E–04	7.3545E–05
P3: DISK BRAKE	AVG	5.7489E–04	1.3489E–03	1.9040E–03	**4.9151E–04**
	SD	6.3384E–05	8.3439E–04	2.4738E–04	**4.9967E–05**
P4: WELDED BEAM	AVG	**5.6520E–04**	7.1434E–03	2.3740E–03	6.3476E–04
	SD	5.6341E–05	3.4559E–03	3.9043E–04	**3.3397E–05**
P5: SRN	AVG	5.0046E–04	7.2007E–03	2.1372E–03	**3.9087E–04**
	SD	1.3475E–04	2.0310E–03	3.5346E–04	**8.5663E–05**

Table 3 The results of the S metric

Problem	Metric	MOPSO	MOALO	MOMVO	AMHS
P1: BNH	AVG	1.1256E+00	9.3402E–01	1.3457E+00	**7.9834E–01**
	SD	1.7681E–01	5.4898E–01	6.7653E–01	**1.0085E–01**
P2: CONSTR	AVG	6.8740E–02	6.9981E–02	5.9139E–02	**5.6579E–02**
	SD	7.9088E–03	1.9823E–02	2.3327E–02	**6.7690E–03**
P3: DISK BRAKE	AVG	1.1487E–01	1.4877E–01	2.9820E–01	**1.1279E–01**
	SD	1.9022E–02	2.1759E–02	2.8742E–01	**9.7786E–03**
P4: WELDED BEAM	AVG	**2.3432E–01**	4.6570E–01	5.8977E–01	3.6759E–01
	SD	2.5982E–02	2.9895E–01	**1.0061E–02**	1.2009E–02
P5: SRN	AVG	2.7896E+00	2.7386E+00	2.9018E+00	**2.0154E+00**
	SD	5.7994E–01	7.6072E–01	1.0095E+00	**3.4347E–01**

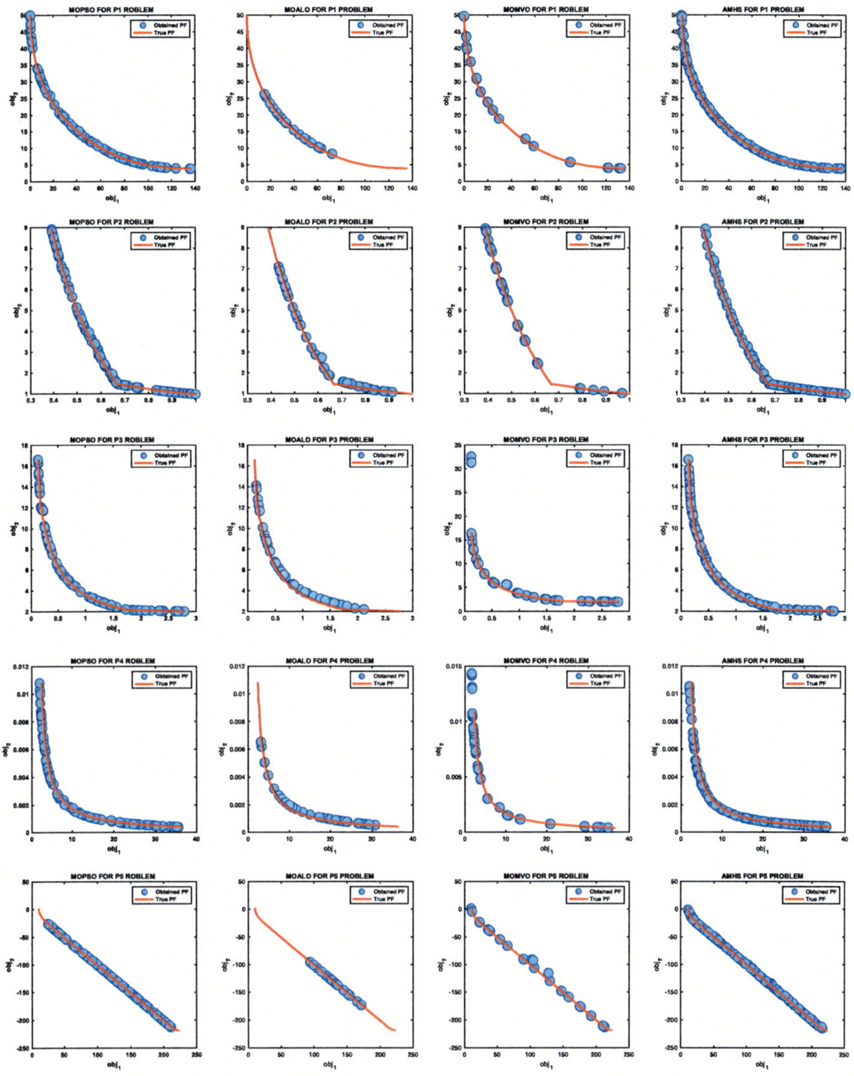

Fig. 1 Estimated Pareto optimal solutions using MOPSO, MOALO, MOMVO, and AMHS

AMHS can outperform existing approaches in the IGD, S, and MS indices and may produce highly competitive results. When the true and achieved Pareto fronts are considered, it is concluded that the proposed AMHS technique may deliver better solutions with a closer distance to the Pareto front for the engineering issues.

Table 4 The results of the MS metric

Problem	Metric	MOPSO	MOALO	MOMVO	AMHS
P1: BNH	AVG	**1.0000E+00**	6.7680E–01	8.9989E–01	**1.0000E+00**
	SD	**0.0000E+00**	1.5647E–01	3.7988E–02	**0.0000E+00**
P2: CONSTR	AVG	9.96758E–01	6.7786E–01	9.7857E–01	**9.9978E–01**
	SD	**6.6578E–03**	6.8879E–02	1.7783E–02	3.7869E–02
P3: DISK BRAKE	AVG	9.9956E–01	9.5469E–01	**1.8730E+00**	9.9987E–01
	SD	1.3425E–03	3.7685E–01	3.8790E–01	**1.0902E–03**
P4: WELDED BEAM	AVG	1.0341E+00	6.8796E–01	1.0345E+00	**1.1991E+00**
	SD	6.1053E–02	6.9879E–02	9.3781E–02	**6.0879E–02**
P5: SRN	AVG	9.3980E–01	3.8977E–01	9.3897E–01	**9.4380E–01**
	SD	4.7860E–02	8.8789E–02	4.8799E–02	**3.7685E–02**

5 Conclusion and Future Works

An archive-based multi-objective version of the well-known proposed innovative metaheuristic algorithms, the harmony search algorithm (HS), is the subject of this research. For result confirmation, the proposed method was compared to well-known algorithms such as MOPSO, MOMVO, and MOALO. Thus, when compared to the previous method, the results are highly competitive. The engineering problems are used to evaluate the performance of multi-objective HS. The proposed AMHS algorithm can outrank the other methods by considering the IGD, S, and MS performance metrics for these problems. Engineering design challenges, such as the design and development of the structural health evaluation, will be addressed as part of the AMHS's future development efforts.

References

1. Yang X-S (2010) Nature-inspired metaheuristic algorithms. Luniver Press
2. Khodadadi N, Azizi M, Talatahari S, Sareh P (2021) Multi-objective crystal structure algorithm (MOCryStAl): introduction and performance evaluation. IEEE Access
3. Mandic DP (2004) A generalized normalized gradient descent algorithm. IEEE Sig Process Lett 11(2):115–118
4. Selman B, Gomes CP (2006) Hill-climbing search. Encycl Cogn Sci 81:82
5. Ahmadianfar I, Bozorg-Haddad O, Chu X (2020) Gradient-based optimizer: a new metaheuristic optimization algorithm. Inf Sci (Ny) 540:131–159
6. Whitley D (1994) A genetic algorithm tutorial. Stat Comput 4(2):65–85. https://doi.org/10.1007/BF00175354
7. Kennedy J, Eberhart R (1995) Particle swarm optimization. In: Proceedings of ICNN'95-international conference on neural networks, vol 4, pp 1942–1948
8. Geem ZW, Kim JH, Loganathan GV (2001) A new heuristic optimization algorithm: harmony search. SIMULATION 76(2):60–68

9. Kaveh A, Talatahari S, Khodadadi N (2020) Stochastic paint optimizer: theory and application in civil engineering. Eng Comput, 1–32
10. Dorigo M, Blum C (2005) Ant colony optimization theory: a survey. Theor Comput Sci 344(2–3):243–278
11. Kaveh A, Talatahari S, Khodadadi N (2019) Hybrid invasive weed optimization-shuffled frog-leaping algorithm for optimal design of truss structures. Iran J Sci Technol Trans Civ Eng 44(2):405–420
12. Yang X-S (2010) A new metaheuristic bat-inspired algorithm. Nature inspired cooperative strategies for optimization (NICSO 2010). Springer, pp 65–74
13. Kaveh A, Eslamlou AD, Khodadadi N (2020) Dynamic water strider algorithm for optimal design of skeletal structures. Period Polytech Civ Eng 64(3):904–916
14. Karami H, Anaraki MV, Farzin S, Mirjalili S (2021) Flow direction algorithm (FDA): a novel optimization approach for solving optimization problems. Comput Ind Eng 156:107224
15. Kaveh A, Khodadadi N, Talatahari S (2021) A comparative study for the optimal design of steel structures using CSS and ACSS algorithms. Iran Univ Sci Technol 11(1):31–54
16. Arora S, Anand P (2019) Binary butterfly optimization approaches for feature selection. Expert Syst Appl 116:147–160
17. Kaveh A, Talatahari S, Khodadadi N (2019) The hybrid invasive weed optimization-shuffled frog-leaping algorithm applied to optimal design of frame structures. Period Polytech Civ Eng 63(3):882–897
18. Coello CAC, Lechuga MS (2002) MOPSO: a proposal for multiple objective particle swarm optimization. In: Proceedings of the 2002 congress on evolutionary computation. CEC'02 (Cat. No.02TH8600), vol 2, pp 1051–1056. https://doi.org/10.1109/CEC.2002.1004388.
19. Mirjalili S, Jangir P, Saremi S (2017) Multi-objective ant lion optimizer: a multi-objective optimization algorithm for solving engineering problems. Appl Intell 46(1):79–95
20. Mirjalili S, Mirjalili SM, Hatamlou A (2016) Multi-verse optimizer: a nature-inspired algorithm for global optimization. Neural Comput Appl 27(2):495–513
21. Sivasubramani S, Swarup KS (2011) Multi-objective harmony search algorithm for optimal power flow problem. Int J Electr Power Energy Syst 33(3):745–752
22. Bhamidi L, Shanmugavelu S (2019) Multi-objective harmony search algorithm for dynamic optimal power flow with demand side management. Electr Power Compon Syst 47(8):692–702
23. Pavelski LM, Almeida CP, Goncalves RA (2012) Harmony search for multi-objective optimization. In: 2012 Brazilian symposium on neural networks, pp 220–225
24. Sheng W, Liu K, Li Y, Liu Y, Meng X (2014) Improved multiobjective harmony search algorithm with application to placement and sizing of distributed generation. Math Probl Eng 2014
25. Qu B-Y, Li GS, Guo QQ, Yan L, Chai XZ, Guo ZQ (2019) A niching multi-objective harmony search algorithm for multimodal multi-objective problems. In: 2019 IEEE congress on evolutionary computation (CEC), pp 1267–1274
26. Wolpert DH, Macready WG (1997) No free lunch theorems for optimization. IEEE Trans Evol Comput 1(1):67–82
27. Zapotecas-Martinez S, Garcia-Najera A, Lopez-Jaimes A (2019) Multi-objective grey wolf optimizer based on decomposition. Expert Syst Appl 120:357–371
28. Coello CAC, Sierra MR (2004) A study of the parallelization of a coevolutionary multi-objective evolutionary algorithm. In: Mexican international conference on artificial intelligence, pp 688–697

Control of Reinforced Concrete Frame Structures via Active Tuned Mass Dampers

Aylin Ece Kayabekir, Gebrail Bekdaş, and Sinan Melih Nigdeli

Abstract In this paper, an active structural control method called active tuned mass dampers (ATMDs) was applied to reinforced concrete frame structures. ATMD is positioned on the top of the structure and uses a proportional integral derivative (PID) controller. Both damper and PID controller properties are optimized via an improved harmony search algorithm. During the optimization, the limit of control force, stroke capacity of ATMD and time-delay of the control system is considered. As the numerical example, a 15-story frame building is presented and the optimum ATMD is effective to reduce maximum displacement by 37.5% for critical excitation.

Keywords Active tuned mass damper · Metaheuristic · Harmony search · Control

1 Introduction

To reduce structural vibrations resulting from distributing factors such as wind, traffic and ground motion, control systems that are passive, active and semi-active or hybrid are used. Tuned mass dampers (TMDs) have been used in several structures including the TV tower in Berlin, Citicorp Center in New York, Taipei 101 in Taipei, etc. TMDs can be designed as passive, active, semi-active or hybrid. For all types of control system and TMD, it is essential to tune the system perfectly for efficiency in vibration reduction. In that case, optimization is a must in structural control and TMDs.

A. E. Kayabekir (✉)
Department of Civil Engineering, Istanbul Gelişim University, 34310 Avcılar, Istanbul, Turkey
e-mail: aekayabekir@gelisim.edu.tr

G. Bekdaş · S. M. Nigdeli
Department of Civil Engineering, Istanbul University-Cerrahpaşa, 34320 Avcılar, Istanbul, Turkey
e-mail: bekdas@iuc.edu.tr

S. M. Nigdeli
e-mail: melihnig@iuc.edu.tr

J. H. Kim et al. (eds.), *Proceedings of 7th International Conference on Harmony Search, Soft Computing and Applications*, Lecture Notes on Data Engineering and Communications Technologies 140, https://doi.org/10.1007/978-981-19-2948-9_26

After the invention of Frahm [1], TMDs are designed to reduce vibration in structures [2] and this situation created the need for optimum design. Then, basic expressions for simplified main systems are proposed [3–8]. For optimization of TMD parameters by considering detailed structure properties, metaheuristic-based methods have been developed [9–12]. For active TMDs (ATMDs), optimization is more important since controller parameters are also needed to optimize in addition to damper properties. For that reason, metaheuristics have been also used in the optimization of ATMDs [13–17].

In this study, an optimization method is presented for ATMDs positioned on the top of the structures for reducing seismic effects. During the optimization, an improved harmony search (HS) algorithm was used by considering the time-delay of the control signal and force limitation for control force for practical use. The efficiency of the method is presented on a reinforced concrete (RC) frame structure that is subjected to various earthquake records in FEMA P-695 [18] during the optimization process.

2 Methodology

HS is a metaheuristic algorithm that uses musical performances for inspiration [19]. An improved HS is used in this study. One improvement is the usage of the best solution in local search with a probability called best solution considering rate (BSCR). Also, adaptive changing of harmony memory considering rate (HMCR) and fret width (FW) was used. The pseudo-code of the algorithm is shown in Fig. 1.

During the optimization, the dynamic analysis is done via Matlab and Simulink [20]. The equation of motion is given as Eq. (1). The matrices and vectors are shown in Fig. 2 with the shear building structure model with ATMD.

Define design constants, algorithm parameters and design variable ranges
Generate an initial harmony memory matrix
if HMCR $\leq$ rand(1)
  Global optimization
  $X_i^{j,t+1}=X_{i,min}+rand(1)\,(X_{i,max}-X_{i,min})$
else
  Local optimization
  if BMCR $\leq$ rand(1)
  Chose best solution as mth one
  else
  Chose a random solution as mth one
  $X_i^{j,t+1}= X_i^{m,t}+rand(1)\,FW(X_{i,max}-X_{i,min})$
Update harmony memory matrix
Update FW and HMCR
Generate a new solution for the maximum number of iterations
end

Fig. 1 Pseudo-code of the algorithm

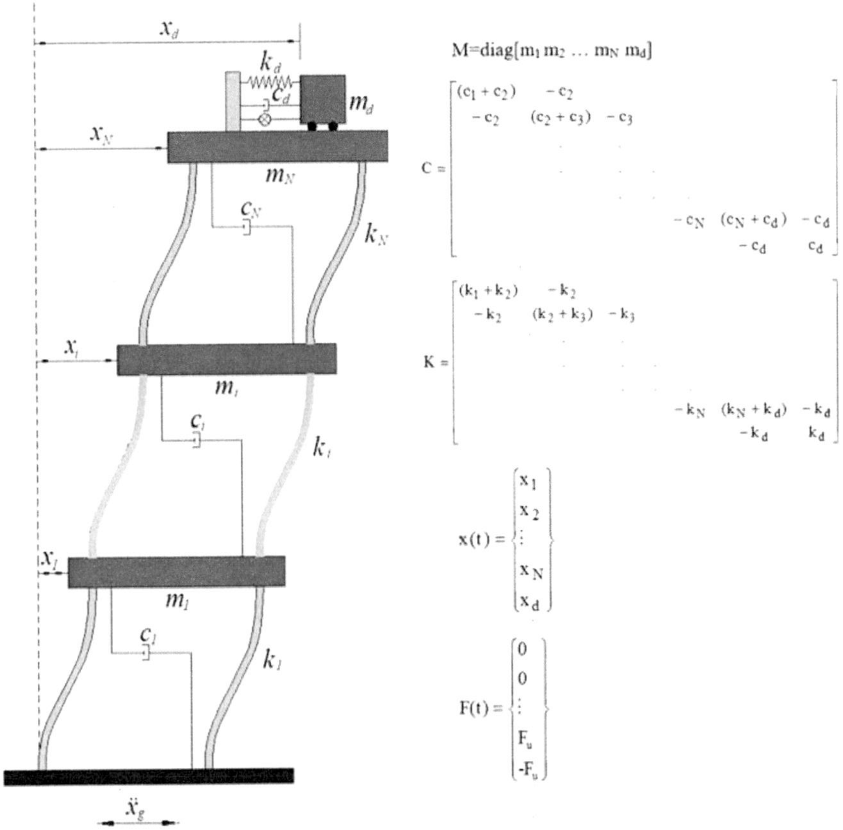

Fig. 2 The structure with ATMD

$$M\ddot{x}(t) + C\dot{x}(t) + Kx(t) = -M\{1\}\ddot{x}_g(t) + F(t) \tag{1}$$

The design variables are shown in Table 1 with chosen ranges in the numerical example.

Table 1 The design variables

Symbol	Definition	Range	Unit
T_{atmd}	Period of ATMD	0.5–1.5 times of period of structure	s
ξ_{d}	Damping ratio of ATMD	1–50	%
K_{p}	Proportional gain	(−10,000)–(10,000)	Vs/m
T_{d}	Derivative time	(−10,000)–(10,000)	s
T_{i}	Integral time	(−10,000)–(10,000)	s

Proportional Integral Derivative (PID) type controllers are used to generate control force (F_u) via control signal (u) as given Eqs. (2) and (3). In PID control, Eq. (4) is used to transform error signal ($e(t)$) to u. As the error, the velocity of the top story of the structure is taken. The period and damping ratio formulations for ATMD are shown as Eqs. (5) and (6), respectively.

$$F_u = K_f i_{ATMD} \tag{2}$$

$$R i_{ATMD} + K_e(\dot{x}_d - \dot{x}_N) = u \tag{3}$$

$$u = K_p\left[e(t) + T_d\frac{de(t)}{dt} + \frac{1}{T_i}\int e(t)dt\right] \tag{4}$$

$$T_{atmd} = 2\pi\sqrt{\frac{m_d}{k_d}} \tag{5}$$

$$\xi_d = \frac{c_d}{2m_d\sqrt{\frac{k_d}{m_d}}} \tag{6}$$

The design constants are shown in Table 2 with the numerical values. The problem has objectives (f_1 and f_2) and one constraint related to the control force.

The objective function (f_1) shown as Eq. (7) is penalized with a pen value (Eq. 8) if the maximum control force is higher than 10% of the total weight of the structure. The second objective (Eq. 9) must be smaller than the defined stmax value and it is firstly considered in the optimization until the set of design variables provides the requirement and then, the first objective is considered. The usual and specific parameters used in the optimization are reported in Table 3.

$$f_1 = \max|x_N| + \text{pen} \tag{7}$$

Table 2 The design constants

Symbol	Definition	Value	Unit
m_i	Mass of the story	3590	ton
k_i	Rigidity coefficient of the story	5520	MN/m
c_i	Damping coefficient of the story	According to 5% Rayleigh damping	MNs/m
m_d	Mass of ATMD	2692.5	ton
stmax	Stroke limit of ATMD	3	–
t_d	Time-delay	20	ms
R	Resistance value	4.2	Ω
K_f	Trust constant	2	N/A
K_e	Induced voltage constant of armature coil	2	V

Table 3 The algorithm parameters

Symbol	Definition	Value
pn	Population number	10
mt	Maximum iteration number	5000
$HMCR_{in}$	Initial harmony memory considering rate	0.5
PAR_{in}	Initial pitch adjusting rate	0.05
BSCR	Best solution considering rate	0.3

$$\text{pen} = \max |F_{\text{u}}| \tag{8}$$

$$f_2 = \frac{\max(|x_{\text{d}} - x_{\text{N}}|)_{\text{with ATMD}}}{\max(|x_{\text{N}}|)_{\text{without ATMD}}} \tag{9}$$

3 The Numerical Example

A case of a 3-d structure is presented to verify the method on real-size buildings. For that reason, a 15-story reinforced concrete structure with the story plan given as Fig. 3 is controlled with an ATMD. The damping of the main structure was taken according to Rayleigh damping and it is assumed as 5% for reinforced concrete structures.

The 3-d structure has 9 axes with equal distances between them in both directions. The distance between axes is 8 m and the height of a story is 3.5 m. The rigidity of the structure in both translational directions was calculated as 5520 MN/m. The structure has 3590 tons of story masses.

4 Application Results

According to these values, the optimum ATMD parameters such as T_{atmd}, ξ_{d}, K_{p}, T_{d} and T_{i} were found as 1.6027 s, 0.1828, −1159.8 Ns/m, 5312.1 s and 5950.5 s, respectively. The maximum values of top story displacement of the 15-story structure are between 0.0425 and 0.4638 m for the far-fault records taken from FEMA P-695 [18]. These displacements are reduced to 0.0405–0.2900 m by using the optimum ATMD.

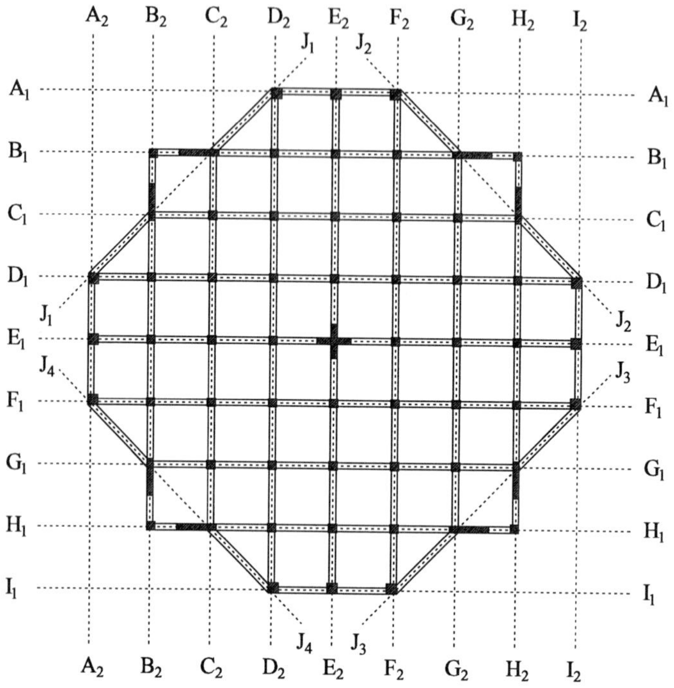

Fig. 3 The story plan of 3D reinforced concrete structure

5 Conclusions

A 3-d real-size reinforced concrete structure was considered for the employed method to add an ATMD on the top. According to the results, the proposal is feasible and effective on 3-d real structures. The maximum effect occurs under the MUL009 component of the Northridge earthquake. For this excitation, the reduction percentage for ATMD controlled structure is 37.5%. The ATMD controlled structure values for the maximum top story displacement are between 0.0405 and 0.2900 m with a maximum 2.2487 stroke value and 51.893 MN control force.

References

1. Frahm H (1911) Device for damping of bodies. U.S. Patent No: 989,958
2. Ormondroyd J, Den Hartog JP (1928) The theory of dynamic vibration absorber. T ASME 50:922
3. Den Hartog JP (1947) Mechanical vibrations. McGraw-Hill, New York
4. Bishop RED, Welbourn DB (1952) The problem of the dynamic vibration absorber. Engineering (London) 174:769

5. Snowdon JC (1959) Steady-state behavior of the dynamic absorber. J Acoust Soc Am 31:10961103
6. Ioi T, Ikeda K (1978) On the dynamic vibration damped absorber of the vibration system. Bull JSME 21:6471
7. Warburton GB (1982) Optimum absorber parameters for various combination of response and excitation parameters. Earthq Eng Struct Dynam 10:381401
8. Sadek F, Mohraz B, Taylor AW, Chung RM (1997) A method of estimating the parameters of tuned mass dampers for seismic applications. Earthq Eng Struct Dynam 26:617635
9. Bekdaş G, Nigdeli SM, Yang XS (2018) A novel bat algorithm based optimum tuning of mass dampers for improving the seismic safety of structures. Eng Struct 159:89–98
10. Yucel M, Bekdaş G, Nigdeli SM, Sevgen S (2019) Estimation of optimum tuned mass damper parameters via machine learning. J Build Eng 26:100847
11. Farshidianfar A, Soheili S (2013) ABC optimization of TMD parameters for tall buildings with soil structure interaction. Interact Multiscale Mech 6:339–356
12. Bekdaş G, Kayabekir AE, Nigdeli SM, Toklu YC (2019) Transfer function amplitude minimization for structures with tuned mass dampers considering soil-structure interaction. Soil Dyn Earthq Eng 116:552–562
13. Pourzeynali S, Lavasani HH, Modarayi AH (2007) Active control of high rise building structures using fuzzy logic and genetic algorithms. Eng Struct 29(3):346–357
14. Kayabekir AE (2021) Control of structures by active tuned mass dampers optimized via metaheuristic algorithms. Ph.D. Thesis, Istanbul University-Cerrahpaşa Institute of Graduate Studies
15. Kayabekir AE, Bekdaş G, Nigdeli SM, Geem ZW (2020) Optimum design of PID controlled active tuned mass damper via modified harmony search. Appl Sci 10(8):2976
16. Kayabekir AE, Nigdeli SM, Bekdaş G (2020) Robustness of structures with active tuned mass dampers optimized via modified harmony search for time delay. In: International conference on harmony search algorithm. Springer, Singapore, pp 53–60
17. Kayabekir AE, Nigdeli SM, Bekdaş G (2021) A hybrid metaheuristic method for optimization of active tuned mass dampers. Comput-Aided Civ Infrastruct Eng. https://doi.org/10.1111/mice.12790
18. FEMA P-695. Quantification of building seismic performance factors. Washington
19. Geem ZW, Kim JH, Loganathan GV (2001) A new heuristic optimization algorithm: harmony search. Simulation 76(2):60–68
20. The MathWorks, Matlab R2018a (2018) Natick, MA

Table 1 Search space of the design variables

Variables	Parameter[a]	Search space
y_1	Number of layers	1, 2, 3, 4, 5
y_2	Number of starting kernels	2, 4, 8, 16, 32, 64
y_3	Size of kernel	1, 2, 3, 4, 5
y_4	Number of dense layers	1, 2, 3, 4, 5
y_5	Number of nodes	16, 32, 64, 128, 256

[a]Is important to remark that there are many other hyperparameters on a neural network, but this work limits itself only to try to solve the topological search problem. Future works might try adding other hyperparameters as part of the search space

2 Methodology

To be able to use HS on the problem, three things have to be accomplished: The topological problem has to be describe in a finite and fixed number of variables; A objective function has to be defined and has to be able to be computed in function of the variables defined [9].

Describing a CNNs in terms of metaheuristic parameters has been done before [12]. The variables can be defined as an array y:

$$y = \begin{bmatrix} y_1 & y_2 & y_3 & y_4 & y_5 \end{bmatrix} \tag{4}$$

where y_1 is the number of layer, y_2 is the number of starting kernels, y_3 is the size of the kernel, y_4 is the number of dense layers, and y_5 is the number of nodes.

A generator code can be written to assemble a network topology with the characteristics given by y and passing that topology for training and evaluation. Encapsulating both the generation of the network as well as training and validation into a function which yields our defined metric as output will give us our optimization function.

The bounds of our search space is mostly defined by the code implementation of the network assembly, which in our case uses a Python package for neural networks, Keras. The search space for each variable is describe in Table 1.

Once both the optimization function and the search space have been defined, a HS implementation has to be selected. In this work, the library pyHarmonySeach [25] was used.

The dataset [26] contains a total of 27,558 cell images with equal instances of parasitized and uninfected cells. A 80–20 split was made for train-test. To calculate our metrics, the topology is assembled for the parameters in y and then used in the train-test. The test dataset is then used to calculate the optimization metric (either F1CCC or $\text{F1CCC}_{\log}$). This feeds the Harmony Search algorithm as the performance metric. The parameters used in the Harmony Search were: HMS = 10; HMCR = 0.75; PAR = 0.5; MPAI = 1; HMS and MPAI were mainly chosen due to the relatively small search space, while MCR and PAR were chosen as non-committal values, since tuning for these parameters would be computationally

Table 2 Summary of the classification results

Opt. metric[a]	Obs. metric[a]	Mean	Best	Worst	Median
F1CCC	$F1$	0.9559	0.9592	0.9508	0.9570
$F1CCC_{log}$	$F1$	0.9546	0.9585	0.9512	0.9556
Random	$F1$	0.5714	0.9406	0.3290	0.3376
F1CCC	F1CCC	4.8e−05	8.6e−05	2.8650e−05	3.2320e−05
$F1CCC_{log}$	F1CCC	4.0e−05	7.5e−05	8.9900e−06	3.1550e−05
Random	F1CCC	1.0e−06	3.0e−06	1.1830e−07	6.2520e−07
F1CCC	$F1CCClog$	0.2207	0.2369	0.2120	0.2140
$F1CCC_{log}$	$F1CCC_{log}$	0.2139	0.2319	0.1906	0.2132
Random	$F1CCC_{log}$	0.0972	0.1730	5.2290e−02	6.4260e−02

[a]Obs. metric being which values are we observing in the cells of the table, while opt. metric is the metric by which the search algorithm tried to optimize

expensive. The code is developed in python and can be found here. The deployment was made on a Colab notebook using Python 3 Google Compute Engine backend. A cache was implemented to return the same results for multiple runs of the same topology in a single run, such that search space remained constant during a single run. Otherwise for the same y, two different results would occur, and the search would converge poorly. It was discussed to use a average of many trains on the same topology for calculation of the F1CCC, but it incurred in significant computational overhead and the idea was dropped, but that is a point of concern for future works.

3 Results and Discussion

Five independent runs of the Harmony Search were calculated, for both the F1CCC and $F1CCC_{log}$, each with 50 iterations ($N_{\text{improvisations}}$). Table 2 has the summary of the best results of each run.

It can be seen that while the F1-scores of both optimization metrics are very similar, they clearly diverge from the random samples. A Mann–Whitney U test can be performed to verify this, comparing the random samples with the best results of the five runs(for each objective function separately). Doing so shows that both objective functions, $p \approx 0.006$, for every metric (F1-score, F1CCC and $F1CCC_{log}$). Meaning that is very likely that the results are indeed different. Meanwhile perform the same test between the best results of the two objective functions and $p \geq 0.26$, which implies in no statistical difference between the outcomes, implying that both metrics achieve the same results. Therefore, it can be concluded that the optimization searches in fact work, increasing the desired metric while maintaining a reasonable F1-score, however, that there is no reason to opt for one objective function (F1CCC) over another ($F1CCC_{log}$).

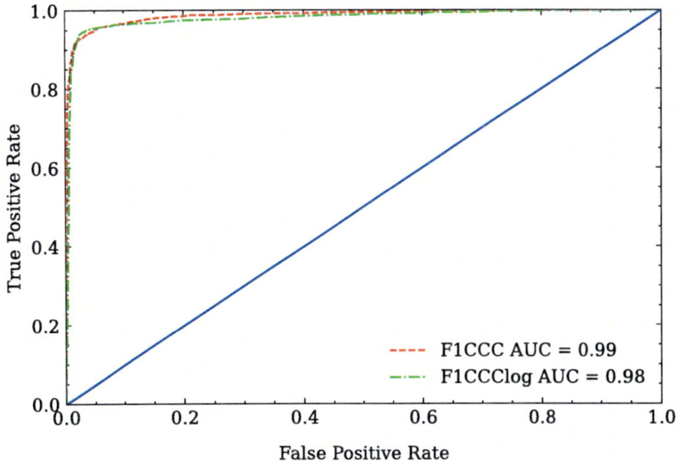

Fig. 1 ROC-AUC curve for the best model for each of the strategies

Table 3 Optimization function search space

References	F1-score	F1CCC	$F1CCC_{log}$	N_{params}	Size (KB)
[4]	0.97	2.37E−6	54.43E−2	409146	–
[7]	0.99	–	–		73.70
[8]	0.98	–	–	–	–
Best F1CCC	0.95	19.4E−5	0.2584	4909	246
Best F1CCClog	0.95	7.5E−5	0.2319	12637	215

*Its important to note, this work did not perform a systematic review of the literature, and thus, the result comparison should be taken with a grain of salt
**Other important thing to take notice is the absence of the number of parameters in other works. It becomes complicated to compare outcomes when one of the results used 400 thousand parameters and another uses 10 million. Future works should strive to always disclose the number of parameters of the model, since it is highly relevant for the application

Sadly, it was posteriorly noted that the library used, PyHarmonySearch had a poor historic retention implementation not allowing us to plot a convergence curve, and due to computational and time constraints, another series of runs was not made. In Fig. 1 the ROC-AUC can be seen for the best model for each of the strategies. The AUC is import because it allows us to visualize how can we change the true positive rate in exchange of false positive rate, which is very valuable to know when dealing with medical diagnosis. Larger AUC implies a better model. Therefore, from the AUC perspective, the models are again, highly performant.

The results of this models can be compared with other state-of-the-art models, and can be seen in Table 3.

From this table, we can conclude the F1CCC achieved remarkable results, obtaining a model a hundred times smaller than the one reported, while maintaining reasonable F1-scores. Meanwhile $F1CCC_{log}$ seems to have provided worse reductions of the N_{params}, nevertheless producing a 32-fold reduction. This worse performance is expected, since by applying a logarithm to the N_{params}, such factor has its impact dampened in the objective function. However, it seems that the initial concern of the objective function prioritizing reducing the number of parameters instead of accuracy was misguided, and for both objective functions, F1-score remained relatively constant.

4 Conclusion and Future Works

This worked explored the utilization of a HS algorithm in order to obtain a CNN topology for malaria parasite detection. To do so it suggested function to be optimized that took parameters related to the CNN topology and assessed the resulting network by using two suggested metrics: F1CCC and $F1CCC_{log}$, yielding remarkable results, reducing the number of parameters hundred-fold, while losing only a couple percentage points, definitely proving the potential of using HS to find better topologies in the CNN for malaria space. There is much to be done regarding future works, however. Better optimization functions may be suggested. Using a larger search space or a search space with more variables might be attempted or even other topology structures. Different HS hyperparameters or a auto-tuning approach can also be tested. At last, other search algorithms may be attempted, such as particle swarm optimization or differential evolution.

References

1. Bronzan RN, McMorrow ML, Kachur SP (2008) Diagnosis of malaria. Mol Diagn Ther 12(5):299–306
2. Araújo T et al (2017) Classification of breast cancer histology images using convolutional neural networks. PLoS ONE 12(6):e0177544
3. Alakwaa W, Nassef M, Badr A (2017) Lung cancer detection and classification with 3D convolutional neural network (3D-CNN). Lung Cancer 8(8):409
4. Pattanaik PA et al (2020) Malaria detection using deep residual networks with mobile microscopy. J King Saud Univ-Comput Inform Sci. https://doi.org/10.1016/j.jksuci.2020.07.003
5. Shekar G, Revathy S, Goud EK (2020) Malaria detection using deep learning. In: 2020 4th international conference on trends in electronics and informatics (ICOEI)(48184), IEEE. https://doi.org/10.1109/ICOEI48184.2020.9143023
6. Manescu P et al (2020) Expert-level automated malaria diagnosis on routine blood films with deep neural networks. Am J Hematol 95(8):883–891
7. Masud M et al (2020) Leveraging deep learning techniques for malaria parasite detection using mobile application. Wirel Commun Mob Comput. https://doi.org/10.1155/2020/8895429
8. Fuhad KM et al (2020) Deep learning based automatic malaria parasite detection from blood smear and its smartphone based application. Diagnostics 10(5):329

9. Osaba E et al (2021) A tutorial on the design, experimentation and application of metaheuristic algorithms to real-world optimization problems. Swarm Evol Comput 64:100888. https://doi.org/10.1016/j.swevo.2021.100888
10. Dokeroglu T et al (2019) A survey on new generation metaheuristic algorithms. Comput Ind Eng 137:106040
11. Talbi E-G (2021) Machine learning into metaheuristics: a survey and taxonomy. ACM Comput Surv (CSUR) 54(6):1–32
12. Lee W-Y, Park S-M, Sim K-B (2018) Optimal hyperparameter tuning of convolutional neural networks based on the parameter-setting-free harmony search algorithm. Optik 172:359–367
13. Ayumi V et al (2016) Optimization of convolutional neural network using microcanonical annealing algorithm. In: 2016 international conference on advanced computer science and information systems (ICACSIS). IEEE
14. de Souza RCT et al (2020) Binary coyote optimization algorithm for feature selection. Pattern Recogn 107:107470
15. Xue B et al (2015) A survey on evolutionary computation approaches to feature selection. IEEE Trans Evol Comput 20(4):606–626
16. Lee W-Y, Park S-M, Sim K-B (2018) Optimal hyperparameter tuning of convolutional neural networks based on the parameter-setting-free harmony search algorithm. Optik 172:359–367. https://doi.org/10.1016/j.ijleo.2018.07.044
17. Kim S-H, Zong WG, Han G-T (2020) Hyperparameter optimization method based on harmony search algorithm to improve performance of 1D CNN human respiration pattern recognition system. Sensors 20(13):3697
18. Awad N, Mallik N, Hutter F (2020) Differential evolution for neural architecture search. arXiv preprint arXiv:2012.06400
19. Wei C et al (2020) Npenas: neural predictor guided evolution for neural architecture search. arXiv preprint arXiv:2003.12857
20. Wei C et al (2021) Self-supervised representation learning for evolutionary neural architecture search. IEEE Comput Intell Mag 16(3):33–49
21. Yang X-S (2009) Harmony search as a metaheuristic algorithm. In: Music-inspired harmony search algorithm. Springer, Berlin, Heidelberg, pp 1–14
22. Nasir M et al (2021) Harmony search algorithm and fuzzy logic theory: an extensive review from theory to applications. Mathematics 9(21):2665
23. Goutte C, Gaussier E (2005) A probabilistic interpretation of precision, recall and F-score, with implication for evaluation. In: European conference on information retrieval. Springer, Berlin, Heidelberg
24. Handelman GS et al (2019) Peering into the black box of artificial intelligence: evaluation metrics of machine learning methods. Am J Roentgenol 212(1):38–43
25. https://github.com/gfairchild/pyHarmonySearch
26. Rajaraman S et al (2018) Pre-trained convolutional neural networks as feature extractors toward improved malaria parasite detection in thin blood smear images. PeerJ 6:e4568

Net-Zero Energy Building Using Metaheuristics in Melbourne City

Seyed Mohammad Ardehali Zadeh and Ali Sadollah

Abstract In this paper, a net-zero energy building setup over an existing building in the suburbs of Melbourne city is proposed. Based on the power consumption, a solar panel system considering minimum size is done to determine the optimum number of solar panels required for the building. An anisotropic method is utilized to model the incident radiation. For optimization purpose, various metaheuristic optimization methods (i.e., particle swarm optimization, water cycle algorithm, and neural network algorithm) are used to maximize the incident radiation on solar panels by optimizing tilt angle of solar modules to reduce the number of solar panels required for the system. Statistical optimization results prove the efficiency of applied optimizers and show that seasonal adjustment can increase the incident radiation per meter squared significantly and simultaneously decrease the required number of solar panels (reduction in equipment costs).

Keywords Net-zero · Solar panels · Tilt angle · Neural network algorithm · Water cycle algorithm · Particle swarm optimization

1 Introduction

Net-zero energy building is a building which consumed electricity is equal to electricity which is generated annually by renewable energies both onsite or nearby site [1]. In this paper, studied building is located on the suburbs of Melbourne city near Toorak. In order to choose the most appropriate solar panel system in the studied case, we need to calculate the annual electricity which is produced by 1 m^2 solar panels, and then decide on number of solar panels.

S. M. A. Zadeh
Faculty of Mechanical Engineering, Monash University, Melbourne 3800, Australia

A. Sadollah (✉)
Department of Mechanical Engineering, University of Science and Culture, Tehran, Iran
e-mail: sadollah@usc.ac.ir

J. H. Kim et al. (eds.), *Proceedings of 7th International Conference on Harmony Search, Soft Computing and Applications*, Lecture Notes on Data Engineering and Communications Technologies 140, https://doi.org/10.1007/978-981-19-2948-9_28

Most of the studies done in the past showed that tilt angle has a great connection with latitude and directly affect the annual electricity. Namely, Heywood [2] concluded that the annual optimum angle is $\beta_{opt} = \varphi - 10$. Lunde [3] assumed that the tilt angle $\varphi \pm 15$, which means in summer months the angle is 15° less than latitude and in winter months 15° more than latitude. In the northern hemisphere, Abdulaziz [4] measured the latitude tilt angle from 10° to 50°. The tilt angle measurement equations have been improved by Nijegorodov et al. [5], and the results seem to be very useful. Later, Duffie and Beckman [6] achieved tilt angle equal to $(\varphi + 15) \pm 15$, as it is obvious latitude is specific for each coordinate. Darhmaoui and Lahjouji [7] obtained the optimum tilt angle by developing an isotropic method.

Modeling of solar radiation have been a topic of much thought and discussion. In this regard, Chang [8] calculated optimum tilt angle by various models in six locations. The ANNs was used by Shaddel et al. [9] in Mashhad. Bulut and Buyukalaca [10] modeled the prediction of monthly average daily global solar radiation on a horizontal surface in different regions in Turkey. Comparison between isotropic and anisotropic methods was proposed by Noorian et al. [11] and twelve methods are compared together in their paper. Stanciu and Stanciu [12] created a model for isotropic sky which optimization of solar panels is available in their method.

For optimal solving of tilt angle, metaheuristic optimization methods are widely used in the literature. Metaheuristic optimization algorithms are typically inspired and designed based on to natural phenomenon [13]. Complex problems are not far from finding solutions with the advent of metaheuristic optimization algorithms. Metaheuristics have proved their capacity to find near-optimal solutions to digital real-world questions that cannot be answered by reliable and analytical methods in reasonable computational time [14].

In the current state of the art, the solar systems produce needed electricity throughout the year partially, also panels are just installed based on the annual optimum tilt angle and other adjustments are neglected. In this paper, using an anisotropic prediction model with the aid of metaheuristics for a real case study located in Melbourne city (a net-zero house including 4 persons) have been optimized by considering the minimum cost of solar panel equipment and maximum incident energy per meter squared. The tilt angle is considered as design variable and different adjustments and scenarios have been considered.

The rest of this paper is organized as follows. Next section provides mathematical formulation for used predicted model of solar energy. Afterward, considered optimization methods are briefly introduced. In Sect. 4, obtained statistical optimization results by applied optimizers are compared and discussions over the results quality are conducted. Finally, conclusions and future directions ends this paper.

2 Mathematical Modeling

2.1 *Model of Incident Radiation on Solar Modules*

Total monthly average daily radiation on a tilted surface ($\overline{H_T}$) is obtained by summations of beam ($\overline{H_B}$), diffuse ($\overline{H_D}$), and reflected ($\overline{H_R}$) solar radiation, and it is given as follow [15]:

$$\overline{H_T} = \overline{H_B} + \overline{H_D} + \overline{H_R}, \tag{1}$$

In this paper, in order to calculate $\overline{H_T}$, an anisotropic approach proposed by Reindle et al. [16], considering the diffuse radiation, is influenced by cloudiness. The average monthly radiation of the regular beam is as follows [15]:

$$\overline{H_B} = \left(\overline{H_g} - \overline{H_b}\right)\overline{R_b}, \tag{2}$$

where $\overline{H_g}$ is normal global monthly radiation (in MJ/m^2/day) and $\overline{H_b}$ is Horizontal diffuse radiation (in MJ/m^2/day). $\overline{R_b}$ seen in Eq. (2) can be calculated in accordance with the following horizontal surface:

$$\overline{R_b} = \frac{\cos(\phi - \beta)\cos(\delta)\sin\left(\omega_s'\right) + \left(\frac{\Pi}{180}\right)\omega_s' \sin(\phi - \beta)\sin(\delta)}{\cos(\phi)\cos(\delta)\sin(\omega_s) + \left(\frac{\Pi}{180}\right)\omega_s \sin(\phi)\sin(\delta)}, \tag{3}$$

where the solar declination angle is given by Duffie and Beckman [6]:

$$\delta = 23.45 \sin\left[360 * \left(\frac{284 + n}{365}\right)\right], \tag{4}$$

The sunset and sunshine hour angles for tilted surface (in degree) is recommended:

$$\omega_s' = \min\left\{\cos^{-1}(-\tan(\phi)\tan(\delta)), \cos^{-1}(-\tan(\phi - \beta)\tan(\delta))\right\}, \tag{5}$$

$$\omega_s = \cos^{-1}(-\tan(\phi)\tan(\delta)), \tag{6}$$

where ϕ is the latitude of the site (in degree) and β is the horizontal (in degree) tilt angle of the collector. The average monthly clearness index in anisotropic system $\overline{K_T}$ which is the monthly average ratio of global horizontal radiation per month to the daily average alien radiation per month is used to calculate $\overline{H_d}$ given as follows:

$$\overline{K_T} = \frac{\overline{H_g}}{\overline{H_0}}, \tag{7}$$

where $\overline{H_0}$ is suggested as given follows:

$$\overline{H_0} = \left(\frac{24 * 3600}{\Pi * 10^6}\right) G_{\text{on}} \left(\cos(\phi)\cos(\delta)\sin(\omega_s) + \frac{\Pi}{180}\omega_s \sin(\phi)\sin(\delta)\right), \quad (8)$$

and

$$G_{\text{on}} = G_{\text{sc}}\left(1 + 0.033\cos\left(\frac{360 * n}{365}\right)\right), \quad (9)$$

The levels of monthly average daily diffuse radiation on horizontal surface depends on three different ranges of $\overline{K}_T$ reported in the literature [17]. The monthly average ground reflected radiation incident on an inclined surface can be calculated as follows [15]:

$$\overline{H_R} = \overline{H_g}\,\rho \frac{1 - \cos(\beta)}{2}, \quad (10)$$

where ρ is the ground reflectivity coefficient and it is assumed to be 0.2 [6]. The monthly average daily sky diffuse radiation incident on inclined surface can be assumed by Reindle et al. model as given follows [16]:

$$\overline{H_D} = \overline{H_d}\,\overline{R_d} \quad (11)$$

where

$$\overline{R_d} = \frac{\overline{H_b}}{\overline{H_0}}\overline{R_b} + \left(1 - \frac{\overline{H_b}}{\overline{H_0}}\right)\left(\frac{1 + \cos(\beta)}{2}\right)\left[1 + \sqrt{\frac{\overline{H_b}}{\overline{H_0}}}\sin\left(\frac{\beta}{2}\right)^3\right] \quad (12)$$

2.2 *Net-Zero Solar Panel Systems*

For having a net-zero house, we need to generate the amount of energy which is consumed annually. In this regard, Fig. 1 shows the energy demand for a household including 4 persons in each season and annually.

Therefore, the total number of solar panels required to the obtained annual energy can be calculated using the following equation:

$$N = \frac{\text{Annual Energy Demand(KWh)}}{\text{Incident Solar Energy}\left(\frac{\text{KWh}}{\text{m}^2}\right) \times A \times \text{EP} \times \text{EI}} \quad (13)$$

where N is the number of panel, A is the area of solar panel, EP is efficiency of solar panels, and EI is efficiency of inverter. Besides, system size for a house with having

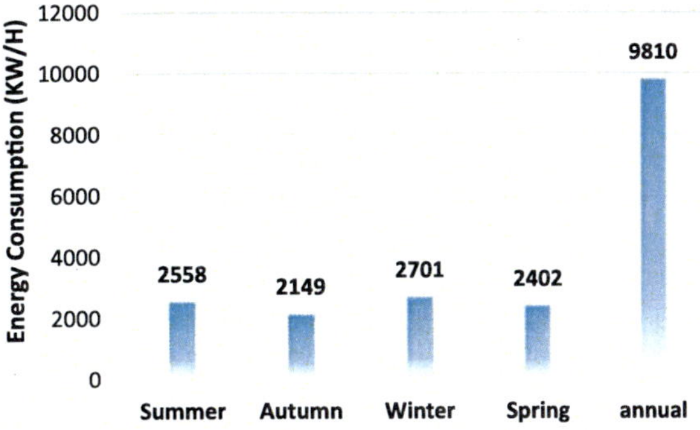

Fig. 1 Energy consumption in Melbourne city on average, seasonally and annually (Australian Energy Regulator, Available at https://www.aer.gov.au/wholesale-markets/wholesale-statistics/annual-electricity-consumption-nem)

4 households can be attained as given follows:

$$\text{Size of System} = N \times \text{Watt(power of panel)} \tag{14}$$

3 Applied Optimization Methods

3.1 *Particle Swarm Optimization*

Particle swarm optimization (PSO) is a swarm intelligence technique proposed by Kennedy's and Eberhart's [18] for tackling global optimization problems. The PSO inspired by nature phenomenon performs by improving and cooperating with individuals (i.e., particles). The PSO is based on simplified social models like bird flocking, fish education, and swarm theory.

The PSO strategy is given as follows. The problem space is flown through potential solutions, called birds or particles, through the current optimal particles. The co-ordinates in the space where the problem occur are tracked by each particle and are linked to the best possible solution (i.e., fitness) which is known as *p*Best (best personal experience in the current population).

The overall best value is another best value that is tracked by the PSO and is the best position that any particle in the population has obtained so far which is known as g_{Best} (best global experience in the current population). The PSO concept involves changing the speed of each particle at each step toward its (i.e., acceleration) p_{Best} and g_{Best} locations. A random term with separate random numbers is used to speed up p_{Best} and g_{Best} locations. As a results, the position and the speed of an individual candidate are updated using the following equations [18]:

$$\vec{V}(t+1) = w\vec{V}(t) + C_1 r_1\left(p\vec{\text{Best}}(t) - \vec{X}(t)\right) + C_2 r_2\left(g\vec{\text{Best}}(t) - \vec{X}(t)\right), \tag{15}$$

$$\vec{X}(t+1) = \vec{X}(t) + \vec{V}(t+1), \tag{16}$$

where w is the speed inertia weight, $X(t)$ is the current particle location ith, $V(t)$ is the up to dated particle's velocity ith, pBest(t) is the so far best location found by particle ith, gBest(t) is the best place the swarm has found, r_1 and r_2 are uniform random numbers between zero and one, C_1 is the cognitive component, and C_2 is the social component. Here, C_1 and C_2 are constants that influence how each particle is directed to good positions in the best and best personal information.

3.2 Water Cycle Algorithm

Inspired by nature, the idea of water cycle algorithm (WCA) is based on observing the process of the water cycle and on how rivers and rivers flow downwards to sea. The WCA starts with an initial population known as streams. First of all, presume we have snow or raining. The best individual (i.e., the best stream) is chosen as a sea. Also, there are some streams that run to the rivers and the sea, while the rivers flow toward to the sea. Rivers and sea collect the water from the lakes based on the intensity of the flow [19]. Based on surface runoff as can be seen in nature, streams flow to the rivers and rivers flow to the see. Also, there may be streams flowing directly to the sea. The best equilibrium location is the sea shore with gathering both the river and streams. Therefore, it is possible to create new locations for streams and rivers as flow during surface runoff given as follows [19]:

$$\vec{X}^{i+1}_{\text{Stream}} = \vec{X}^{i}_{\text{Stream}} + \text{rand}(0, 1) \times 2 \times \left(\vec{X}^{i}_{\text{Sea}} - \vec{X}^{i}_{\text{Stream}}\right), \tag{17}$$

$$\vec{X}^{i+1}_{\text{Stream}} = \vec{X}^{i}_{\text{Stream}} + \text{rand}(0, 1) \times 2 \times \left(\vec{X}^{i}_{\text{River}} - \vec{X}^{i}_{\text{Stream}}\right), \tag{18}$$

$$\vec{X}^{i+1}_{\text{River}} = \vec{X}^{i}_{\text{River}} + \text{rand}(0, 1) \times 2 \times \left(\vec{X}^{i}_{\text{Sea}} - \vec{X}^{i}_{\text{River}}\right), \tag{19}$$

If a new position of a stream is better than that of its corresponding river (sea), the position of the river (sea) and the stream is changed. Similarly, such exchanges can occur for movements of rivers and sea. Evaporation is one of the key factors in introducing another operator which can prevent an immature convergence of the studied algorithm. In the WCA, the process of evaporation causes sea water to evaporate as the water flows to the sea. In order to not trap in local optima in the WCA, the following pseudo-code indicates whether or not the evaporation is satisfied [20]:

$$\begin{aligned} &\text{if } \left\| \vec{X}^{i}_{\text{Sea}} - \vec{X}^{i}_{\text{River}} \right\| < d_{\max} \text{ or rand} < 0.1 \qquad i = 1, 2, 3, \ldots, N_{sr} - 1 \\ &\quad \text{Perform raining process} \\ &\text{end} \end{aligned} \tag{20}$$

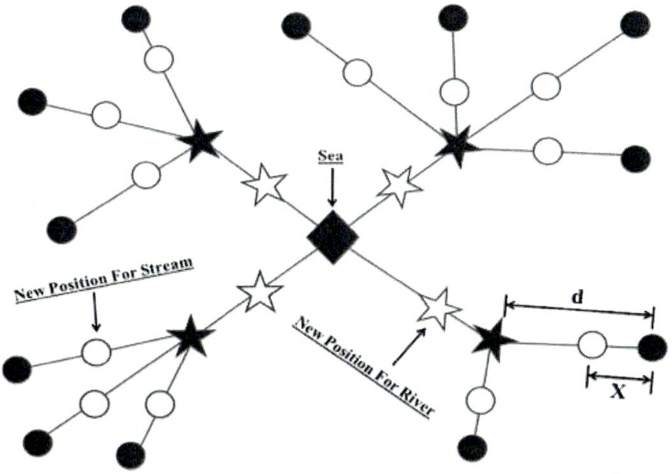

Fig. 2 Schematic view of the WCA's movement toward the best optimum point (sea) [19]

where d_{max} is a small (close to zero) value. The raining cycle is implemented after satisfying the evaporation mechanism. New streams are emerging in the raining cycle at various locations. In fact, the evaporation is the reason of the exploration in the WCA. The schematic view of WCA is shown in Fig. 2 where circles, stars, and the diamond correspond to streams, rivers, and sea, respectively. From Fig. 2. The new positions of rivers and streams are referred to by the white (empty) forms.

3.3 Neural Network Algorithm

Neural network algorithm (NNA) is motivated by unique structure of ANNs and biological nervous system [21]. The NNA starts with an initial population called population of pattern solution. Inspired by the ANNs, each iteration in the NNA assumes the best solution is achieved as target data and the aim is error minimization between target data and other predicted pattern solutions.

The target weight (WTarget) corresponding to the target solution must be selected from the weight population (weight matrix) after setting the target solution (XTarget) among the other pattern solutions. The original ANNs weights are random numbers, which are modified with the estimated error of the network as the iteration number increases. Back to the NNA, the initial weights are defined as random numbers uniformly from zero to one.

However, there is a limitation for the weight matrix. The restriction that is imposed should not exceed one. The following equation inspired by the weight summation method used in ANNs is used to quantify the new location (XNews) after generating the weight matrix (W) and finally updating the new locations as given follows [21]:

$$\vec{X}_j^{\text{New}}(t+1) = \sum_{i=1}^{N_{\text{pop}}} w_{ij}(t) \times \vec{X}_i(t), \quad j = 1, 2, 3, ..., N_{\text{pop}}, \tag{21}$$

$$\vec{X}_i(t+1) = \vec{X}_i(t) + \vec{X}_i^{\text{New}}(t+1), \quad i = 1, 2, 3, \ldots, N_{\text{pop}}, \tag{22}$$

For the sake of conducting exploration phase in the NNA, the existence of bias operator plays a crucial role. In the current population of pattern solutions and an updated weights matrix (acting as noise), the biased operator modifies a certain portion of pattern solutions in the NNA.

As an another modification factor in the NNA, transfer function operator, unlike the ANNs, is moving from existing search positions to the new positions of the population so that higher performing solutions are revised and implemented against the target solution. Therefore, for the proposed method, the following equation is defined as an operator of the transmission function [21]:

$$\vec{X}_i^*(t+1) = \text{TF}\left(\vec{X}_i(t+1)\right) = \vec{X}_i(t+1) + 2 \times \text{rand} \times \left(\vec{X}^{\text{Target}}(t) - \vec{X}_i(t+1)\right), \quad i = 1, 2, 3, \ldots, N_{\text{pop}} \tag{23}$$

Using the aforementioned steps, the objective function value of an updating pattern solution (individual) is compared to the target solution (i.e., temporal best solution) and modified the target solution and its corresponding target weights if the created pattern solution is better than the target solution.

4 Results and Discussion

In this paper, information about horizontal surface of the area are barrowed from NASA surface meteorology based on a 22-year average data [22]. Calculations and optimization task for obtaining optimum tilt angle of the PV modules are carried out for 20 independent runs via MATLAB in a way that incident energy is maximized by finding the optimum tilt angle in different adjustments. Concerning user parameters of applied optimizer population size of 50 and maximum iteration of 1000 have been used. Regarding the initial parameters of each optimizer, suggested values in the literature [18–21] have been utilized.

In order to find optimum tilt angle, it is vital to know annual average daily solar radiation on a horizontal surface in Melbourne city which is 15.78 MJ/m^2 [4]. At first, the annual incident energy is provided by optimization methods based on annual adjustment of the solar panel modules provided in Table 1. By observing Table 1, fixed tilt angel of 47.7° is suggested for the whole year. Besides, the results for different adjustment can be seen in Table 2.

According to the results obtained from Tables 1 and 2, details of desired equipment are reported in Table 3. Based on Eq. (12) and obtained optimization results reported

Table 1 Statistical optimization results based on annual adjustment using three optimization methods for Reindle's model. "SD" stands for standard deviation

Annual data	Average radiation (MJ/m^2)	Optimum tilt angle (°)	SD
PSO	7615.68	47.7	1.81e−12
NNA	7615.68	47.7	1.52e−12
WCA	7615.68	47.7	1.90e−03

Table 2 Comparison of different adjustment using Reindle's model

Adjusting period	H_T (MJ/m^2)	Ratio to radiation on horizontal surface
Fixed horizontally	6570	1.00
Fixed vertically	3193	0.48
Fixed annually	7615	1.15
Seasonally	9710	1.47

Table 3 Information of designated tools for solar panel system

Power of panels (Wp)	Area of panel (m^2)	EP	EI
220	1.45	20%	90%

in Tables 1 and 2, the required number of panels, if we use seasonal adjustment will be 15, while if annual adjustment is considered, we will need 19 solar panels. Indeed, using seasonal adjustment, we can decrease equipment costs. Looking at Fig. 3, the generated electricity (obtained by the predicted model in this paper) and electricity demand (see Fig. 1) in a year are shown. As can be seen in Fig. 3, in annual consumption, both energies met each other at the same energy level which resembles a net-zero energy building.

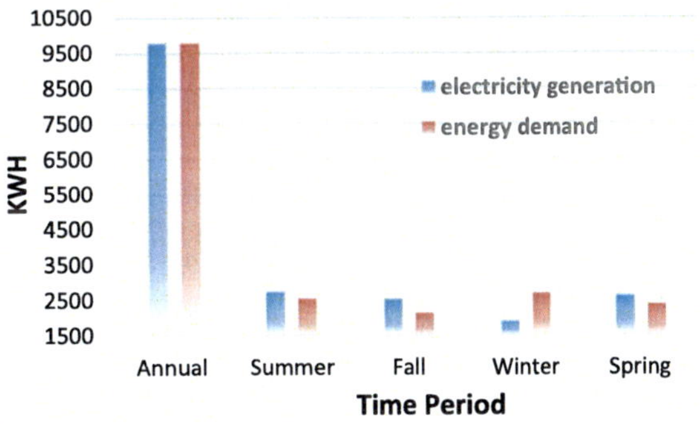

Fig. 3 Demanded and generated electricity in a year

5 Conclusions and Further Directions

In this paper, a net-zero house in Melbourne city has been considered. In this regard, an anisotropic model (i.e., Reindle) was utilized to calculate total incident radiation in the studied region. Metaheuristic optimization methods helped us to maximize the generated electricity by optimizing the tilt angle. Different adjustments (i.e., annual and seasonal adjustments) were also considered in this paper, and it is concluded that seasonal adjustment could increase the incident radiation per meter squared significantly and simultaneously could decrease the required number of solar panels. Besides, the solar panel system based on the electricity demand was optimized with respect to its number for a house including 4 households. The required sizes of the obtained solar panel system were 3300 W and 4400 W for seasonal and annual adjustments, respectively.

Further researches can be focused on using solar trackers and comparing it with the current model in terms of generated electricity and economics. Also, C-bus system or other atomized system for controlling the electrical services such as controlling lighting, under floor heating, and motorized blinds can be considered. These systems can reduce the electricity demand for the house and can affect the payback period of the system.

References

1. Kumar S (2016) Towards net zero energy solar building, system, and concepts. In: 2016 first international conference on sustainable green buildings and communities (SGBC), Chennai, India
2. Heywood H (1971) Operating experience with solar water heating. J Ind Heat 39:63–99
3. Lunde PJ (1980) Space heating and hot water systems. Solar thermal engineering. Wiley, New York, NY
4. Mujahid AM (1994) Optimum tilt angle for solar collection systems. Int J Solar Energy 14(4):191–202
5. Nijegorodov N, Devan KRS, Jain PK, Carlsson S (1994) Atmospheric transmittance models and an analytical method to predict the optimum slope of an absorber plate, variously oriented at any latitude. Renew Energy 4(5):529–543
6. Duffie J, Beckman WA (2006) Solar engineering of thermal processes, 4th edn. Wiley, New York
7. Darhmaoui H, Lahjouji D (2013) Latitude based model for tilt angle optimization for solar collectors in the Mediterranean region. Energy Procedia 42:426–435
8. Chang TP (2008) Study on the optimal tilt angle of solar collector according to different radiation types. Int J Appl Sci Eng 6(2):151–161
9. Shaddel M, Javan SD, Baghernia P (2016) Estimation of hourly global solar irradiation on tilted absorbers from horizontal one using artificial neural network for case study of Mashhad. Renew Sustain Energy Rev 53:59–67
10. Bulut H, Büyükalaca O (2007) Simple model for the generation of daily global solar-radiation data in turkey. Appl Energy 84(5):477–491
11. Noorian AM, Moradi I, Kamali GA (2008) Evaluation of 12 models to estimate hourly diffuse irradiation on inclined surfaces. Renew Energy 33(6):1406–1412

12. Stanciu C, Stanciu D (2014) Optimum tilt angle for flat plate collectors all over the world—a declination dependence formula and comparisons of three solar radiation models. Energy Convers Manage 81:133–143
13. Yang X (2010) Nature-inspired metaheuristic algorithms, 2nd edn. Luniver Press
14. Yang X (2010) Engineering optimization: an introduction with metaheuristic applications. Wiley
15. Hagen K (2015) Introduction to renewable energy for engineers, 1st edn. Pearson
16. Reindl DT, Beckman WA, Duffie JA (1990) Diffuse fraction correlations. Sol Energy 45(1):1–7
17. Talebizadeh P, Mehrabian MA, Abdolzadehb M (2011) Prediction of the optimum slope and surface azimuth angles using the genetic algorithm. Energy Build 43(11):2998–3005
18. Kennedy J, Eberhart R (1995) Particle swarm optimization. In: Proceedings of ICNN'95-international conference on neural networks, Perth, WA, Australia
19. Eskandar H, Sadollah A, Bahreininejad A, Hamdi M (2012) Water cycle algorithm—a novel metaheuristic optimization method for solving constrained engineering optimization problems. Comput Struct 110–111:151–166
20. Sadollah A, Eskandar H, Kim JH (2015) Water cycle algorithm for solving constrained multi-objective optimization problems. Appl Soft Comput 27:279–298
21. Sadollah A, Sayyadi H, Yadav A (2018) A dynamic metaheuristic optimization model inspired by biological nervous systems: neural network algorithm. Appl Soft Comput 71:747–782
22. NASA: Surface meteorology and solar energy. A renewable energy resource web site. NASA LaRC POWER Project: NASA Langley Research Center Atmospheric Science Data Center Surface meteorological and Solar Energy (2015)

A Hybrid Firefly Algorithm and Particle Swarm Optimization Algorithm for Mesh Routers Placement Problem in Wireless Mesh Networks

Sylia Mekhmoukh Taleb, Yassine Meraihi, Asma Benmessaoud Gabis, and Seyedali Mirjalili

Abstract This paper proposes the application of the recently proposed hybrid particle swarm optimization (PSO) and firefly algorithm (FA), called HFPSO, for solving the mesh routers placement problem in wireless mesh networks (WMNs). HFPSO combines the local search ability of FA and the fast convergence ability of PSO algorithm. The effectiveness of HFPSO was demonstrated using many generated instances in comparison with FA and PSO algorithms taking into account the metrics of user coverage and network connectivity. The results showed that HFPSO is more effective than FA and PSO in finding optimal mesh routers locations.

Keywords Mesh router nodes placement · Wireless mesh networks · Firefly algorithm · Particle swarm optimization · Network design

1 Introduction

Wireless mesh network (WMN) has seen accelerated development in both academia and industry. This is due to its easy implementation at a low cost, scalability, connectivity, and dynamic self-organization. This attractive communication technology was applied in different fields such as broadband home networking, education, healthcare, military, intelligent transportation systems, disaster management, and emer-

S. M. Taleb · Y. Meraihi
LIST Laboratory, University of M'Hamed Bougara Boumerdes, Avenue of Independence, 35000 Boumerdes, Algeria
e-mail: s.mekhmoukh@univ-boumerdes.dz

A. B. Gabis
Laboratoire des Méthodes de Conception des Systèmes, Ecole Nationale Supérieure d'Informatique, BP 68M, 16309 Oued-Smar, Alger, Algeria

S. Mirjalili (✉)
Centre for Artificial Intelligence Research and Optimisation, Torrens University Australia, Fortitude Valley, Brisbane, QLD 4006, Australia
e-mail: ali.mirjalili@gmail.com

Yonsei Frontier Lab., Yonsei University, Seoul, South Korea

J. H. Kim et al. (eds.), *Proceedings of 7th International Conference on Harmony Search, Soft Computing and Applications*, Lecture Notes on Data Engineering and Communications Technologies 140, https://doi.org/10.1007/978-981-19-2948-9_29

gency operations [1]. WMN consists of three kinds of nodes such as mesh routers (MRs), mesh gateways (MGs), and mesh clients (MCs). Its performance is mainly affected by the poor and bad positioning of mesh nodes (MR and/or MG) [2]. As a result, there is a lot of interference and congestion, leading to significant packet loss, high delay, and low throughput. To overcome these drawbacks, network operators need to apply efficient optimization techniques for WMN nodes placement. So meta-heuristics have been presented as successful optimization algorithms for solving it by providing acceptable solutions in a reasonable execution time. In this sense, several meta-heuristic algorithms have been proposed in the literature to solve this problem such as genetic algorithm (GA) [3], simulated annealing (SA) [4], Tabu search (TS) [5, 6], firefly optimization algorithm (FA) [7], African vulture optimization algorithm (AVOA) [8], artificial hummingbird algorithm (AHA) [9], and many others. Particle swarm optimization (PSO) [10–12] is the most commonly used algorithm for solving the node placement problem in WMNs. This is due to its simplicity, adaptability, and low memory requirement. To obtain an effective solution for the mesh routers placement problem, PSO has been combined with other meta-heuristics such as GA [13], SA [14], and HC [13].

In this paper, we apply the recently proposed hybrid algorithm [15], called HFPSO, for solving the mesh router nodes placement problem. HFPSO is based on the combination of PSO and FA. The performance of HFPSO was assessed taking into account the coverage and the connectivity metrics in comparison with the original FA and PSO algorithms.

The rest of this paper is organized as follows. The formulation of the mesh routers node placement problem is given in Sect. 2. Section 3 presents the FA and PSO algorithms. Section 4 describes the hybrid HFPSO for solving the formulated problem. Moreover, we evaluate the performance of HFPSO and provide the obtained results in Sect. 5. Finally, we conclude the paper in Sect. 6.

2 Mesh Router Nodes Placement Problem Formulation

WMN can be formulated mathematically as an undirected graph $G = (N, L)$ where N is the set of network nodes and L is the set of links between these nodes. The network G is formed by a set of disjoint sub-nets. In this work, we consider the WMN with two types of nodes including mesh clients and mesh routers. Thus, $N = R \cup C$ where MR represents the set of m mesh routers ($\text{MR} = \{r_1, r_2, \ldots, r_m\}$) and C represents the set of n mesh clients ($C = \{c_1, c_2, \ldots, c_n\}$).

In this paper, the static continuous version of mesh routers nodes placement problem was considered. Therefore, the main goal is to find the optimal placement of m mesh routers in a $2D$ area of dimensions $W \times H$, depending on the location of n mesh clients.

The problem studied in this work considers two main objectives that need to be optimized such as user coverage $\Psi(G)$ and network connectivity $\Phi(G)$. The user coverage defines the number of covered users by at least one mesh router, and it is given as:

$$\Psi(G) = \sum_{i=1}^{n} \left(\max_{j \in \{1,\ldots,m\} \sigma_{ij}} \right) \quad (1)$$

where σ_{ij} represents the coverage variable given as follows:

$$\sigma_{ij} = \begin{cases} 1 \text{ if mesh client } c_i \text{ is covered by mesh router } r_j, \\ 0 \text{ Otherwise.} \end{cases} \quad (2)$$

The network connectivity is represented as the geant sub-net among k formed sub-net with regard to the number of mesh nodes (mesh routers and mesh clients). It is calculated as follows:

$$\Phi(G) = \text{Max}_{i \in \{1,\ldots,k\}} |G| \quad (3)$$

where $|G_i|, i \in \{1, k\}$ is the size of ith sub-network and $G = G_1 \cup G_2 \cup \cdots \cup G_k$.

3 Firefly and Particle Swarm Optimization Algorithms

3.1 Firefly Algorithm

Firefly algorithm (FA) [16] is a population-based meta-heuristic inspired by the blinking behavior of fireflies. Its concept is based on the bio-luminescence process. FA assumes the following idealized rules:

- All fireflies are unisex.
- The attractiveness of each firefly is proportional to its brightness.
- The brightness decreases as the distance increases.
- If there is no more attractive firefly than a particular one, it will move randomly.

Based on the mentioned rules, the basic steps of FA are described below:

1. A set of initial solutions are randomly generated.
2. Let i and j be two fireflies and their positions be $X_i(x_i, y_i)$ and $X_j(x_j, y_j)$, respectively. The distance r_{ij} between fireflies i and j is calculated as follows:

$$\mathrm{r}_{ij} = \left\| X_j - X_i \right\| = \sqrt{(y_j - y_i)^2 - (x_j - x_i)^2} \quad (4)$$

3. As mentioned earlier, the attractiveness of each firefly is proportional to its brightness and inversely proportional to the distance between two fireflies. It is calculated as follows:

$$B(r) = \beta_0 \mathrm{e}^{-\gamma r^2} \quad (5)$$

where $\beta_0(r)$ is the attractiveness of firefly at distance r. β_0 is the attractiveness of firefly at distance 0. γ is the light absorption coefficient.

4. In this step, FA performs the exploration of search space. More precisely, fireflies move according to their attractiveness and the attractiveness of other fireflies. When a firefly i is attracted by another firefly j, it moves according to the following equation:

$$X_i = X_i + \beta_0 e^{-\gamma r_{ij}^2}(X_j - X_i) + \alpha\left(\text{rand} - \frac{1}{2}\right) \tag{6}$$

where rand is a random number in [0, 1]. α is a randomized parameter.

3.2 Particle Swarm Optimization Algorithm

Particle swarm optimization algorithm is a population-based meta-heuristic suggested by [17] in 1995. PSO concept is based on mimicking birds around food or movement of fishes. In PSO algorithm, the possible solution of the problem is represented by a particle. Each particle has a position X_i and a velocity V_i. Position and velocity are updated according to Eqs. (8) and (7), respectively. When updating the position and velocity of each particle, the personal best (*Pbest*) and global best (*Gbest*) results are considered. *Pbest* stores the best results for each particle. *Gbest* records the best result among all particles.

$$V_i(t+1) = wV_i(t) + c_1r_1(Pbest_i(t) - X_i(t)) + c_2r_2(Gbest_i(t) - X_i(t)) \tag{7}$$

$$X_i(t+1) = X_i(t) + V_i(t+1) \tag{8}$$

where w is the inertia weight parameter. r_1 and r_2 are random numbers in the range of [0, 1]. c_1 and c_2 are acceleration coefficients. t and $t+1$ denote the current and next iteration of the algorithm, respectively.

4 HFPSO Applied for Mesh Routers Placement Problem

The HFPSO algorithm was first proposed by Aydilek in 2018 [15]. As of 2021, this algorithm has been cited over 200 times in Google scholar. It is a very efficient hybrid algorithm on challenging optimization problems, which motivates our attempt to test it for the mesh routers placement problem, which is quite challenging. HFPSO combines the local search capabilities of FA with the global search capabilities of PSO algorithm.

4.1 Algorithm

The pseudo-code of HFPSO is shown in Algorithm 1. It defines five steps. The first step consists of generating initial solutions randomly in the defined search and velocity ranges. *Gbest* and *Pbest* are calculated and assigned. The inertia weight is dynamically adjusted according to (Eq. 9). The current position is saved in a temporary variable ($X_{i_{\text{temp}}}$). Then, FA is used to determine position and velocity according to (Eq. 11) and (Eq. 12), respectively. Otherwise, it continues to search with PSO according to (Eq. 8) and (Eq. 7), respectively.

$$w = w_i - \left(\frac{(w_i - w_f)}{T}\right) * t \tag{9}$$

$$h(i,t) = \begin{cases} \text{true, if } g(X_i^t) \leq Gbest^{t-1} \\ \text{false, if } g(X_i^t) > Gbest^{t-1}. \end{cases} \tag{10}$$

$$X_i(t+1) = X_i(t) + \beta_0 \mathrm{e}^{-\gamma r_{ij}^2}(X_i(t) - Gbest^{t-1}) + \alpha\left(\text{rand} - \frac{1}{2}\right) \tag{11}$$

$$V_i(t+1) = X_i(t+1) + X_{i\text{temp}} \tag{12}$$

Algorithm 1 The proposed HFPSO for the mesh routers placement problem

Input:
m: Number of mesh routers (dimension of the problem)
n: Number of mesh clients
CR: Coverage radius
pop: Population size
T: Maximum number of iterations
Output:
$Gbest$ particle and its fitness value

for i= 1 to pop **do**
    for j= 1 to m **do**
        Randomly place the *j-th* router of the particle X_i in the deployment area
        Randomly initialize the velocity of the *j-th* router of the particle X_i in search range (V_max,V_min)
    end for
end for
for i= 1 to n **do**
    Randomly place the *i-th* mesh client in the deployment rea
end for
Calculate the fitness (Eq. 13), $Pbest$, and $Gbest$
for t= 1 to T **do**
    for p= 1 to pop **do**
        if particle has an improvement in the last iteration (Eq. 10) **then**
            Save the current particle position in a temp variable
            Update particle position (Eq. 11)
            Update particle velocity (Eq. 12)
        else
            Update inertia weight (Eq. 9)
            Update particle velocity (Eq. 7)
            Update particle position (Eq. 8)
        end if
        Calculate the new fitness value (Eq. 13)
        Do adaptation
    end for
end for
Return the $Gbest$ particle and its fitness value

4.2 Objective Function

The aim of our approach determines the optimal placement of a given number of mesh routers in order to maximize the user coverage and network connectivity parameters. These parameters are used to define the fitness function f in order to evaluate the quality of solutions. The objective function is given as:

$$f(X_i) = \lambda.\frac{\Psi(G)}{n} + (1-\lambda).\frac{\Phi(G)}{m+n} \tag{13}$$

where G is the graph corresponding to the solution X_i and λ is a parameter used to control the importance of metrics.

The HFPSO suggested in [15] was developed for a minimization problem. Thus, we need to convert our objective function into a minimization function. Therefore, we have defined another function g as:

$$g = 1 - f(X_i) \tag{14}$$

4.3 Solution Representation

A potential solution of a mesh routers placement problem consists of defining the locations of all mesh routers. In this work, the solution of the problem is represented by an array $X_i = \{x_i^{r1}, y_i^{r1}, x_i^{r2}, y_i^{r2}, \ldots, x_i^{rm}, y_i^{rm}\}$, where (x_i^{rk}, y_i^{rk}) are the (x, y) coordinates of the mesh router r^k, with $0 \leq x_i^{rk} \leq W$ and $0 \leq y_i^{rk} \leq H$; $\forall$ $m \in \{1, 2, \ldots, m\}$.

5 Simulation Results and Discussion

In this section, we investigate the performance of HFPSO algorithm in solving the mesh routers placement problem in WMNs. The obtained results are compared with the original PSO and FA algorithms considering the metrics of user coverage, network connectivity, and objective function value. Therefore, all experiments were performed on a 2.5 GHz core i7 PC with 8 GB RAM under the MATLAB environment. In all simulations, the number of mesh routers is varied from 5 to 40 for covering 50–400 mesh clients in a rectangular deployment area of 2000 m × 2000 m. The maximum number of iterations is 1000, and each result presented in this section is an average of 10 runs. The parameter values used during the simulation are summarized in Table 1. The performance of HFPSO is evaluated and analyzed by investigating the impact of varying the number of mesh clients, number of mesh routers, and coverage

Table 1 Parameters values considered in our simulations

Parameter	Description	Value	Default value
n	No. of mesh clients	[50 400]	100
m	No. of mesh routers	[5 40]	20
CR	Coverage radius	[50 400]	200 m
W	Width	2000	2000 m
H	Height	2000	2000 m
λ	Control parameter	[0 1]	0.5
Pop	Population size	25	25
T	No. of iteration	1000	1000

Table 2 Coverage, connectivity, fitness under various number of mesh clients

	n	50	100	150	200	250	300	350	400
Coverage	HFPSO	42.2	82.6	113.1	150.8	188.6	219.7	256.4	321.7
	FA	43.4	79.6	115.3	148.5	184.5	218	252.5	284.5
	PSO	38.3	76.2	113.5	148.5	183.3	215.5	242.4	273.2
Connectivity	HFPSO	62.2	102.2	130.4	169.5	208.6	239.7	275.4	301.4
	FA	62.4	98.6	135.3	165.1	201	236.3	271	302.4
	PSO	58.3	96.2	133.5	168.5	203.3	233.4	259.1	293.2
Fitness	HFPSO	0.86	0.83	0.76	0.76	0.76	0.74	0.73	0.76
	FA	0.88	0.80	0.78	0.73	0.74	0.73	0.72	0.71
	PSO	0.79	0.78	0.77	0.75	0.74	0.72	0.69	0.69

radius values. The results are given in Tables 2, 3, and 4, respectively. Their graphical representations are shown in Figs. 1, 2, and 3.

Table 2 and Fig. 1 show the user coverage, network connectivity, and fitness value when varying the number of mesh clients (from 50 to 400). It is observed that when increasing the number of mesh clients, the fitness value is slightly decreased even though the values of coverage and connectivity were increased. The obtained results confirmed the effectiveness of HFPSO when compared to the original FA and PSO algorithms for most of the cases.

Table 3 and Fig. 2 illustrate the impact of varying the number of mesh routers (varies from 5 to 40) on user coverage, network connectivity, and fitness, respectively. It is demonstrated that the user coverage, network connectivity, and fitness value increase every time we increase the number of mesh routers. As more routers are added in the network, mesh routers have the chance to cover more clients and to connect to other mesh routers, until covering and including all mesh nodes when the number of mesh routers exceeds 30. Again, when the number of mesh routers exceeds 5, HFPSO gives better results in terms of coverage, connectivity, and fitness when compared to FA and PSO.

The impact of varying coverage radius values from 50 to 400 on user coverage, network connectivity, and fitness value is illustrated in Table 4 and Fig. 3. The obtained

Table 3 Coverage, connectivity, fitness under various number of mesh routers

	m	5	10	15	20	25	30	35	40
Coverage	HFPSO	31.7	55	69.3	83	89.9	96.6	98.7	99.2
	FA	31.5	53.1	64.7	73.6	86.3	92.8	94.8	97
	PSO	33.9	45.9	65.6	81	90.1	93.2	94.8	98.8
Connectivity	HFPSO	27	61.6	83.2	103	114.4	126.3	133.7	139.2
	FA	30.2	48	79.7	88.2	109.8	122.7	129.8	136.9
	PSO	36.6	53.1	76.8	101	113.5	122.4	129.8	138.8
Fitness	HFPSO	0.28	0.55	0.70	0.84	0.9	0.96	0.98	0.99
	FA	0.3	0.48	0.67	0.73	0.87	0.93	0.95	0.97
	PSO	0.34	0.47	0.66	0.82	0.9	0.93	0.95	0.98

Table 4 Coverage, connectivity, fitness under various coverage radius values

	CR	50	100	150	200	250	300	350	400
Coverage	HFPSO	26	39.7	58.8	82	96.4	99.8	100	100
	FA	25.2	41.5	54.1	78.5	95.1	99	100	100
	PSO	22.5	27.6	50.7	76.9	94.6	99.2	100	100
Connectivity	HFPSO	8.7	31.8	74	100.8	116.4	119.8	120	120
	FA	4.9	26.2	66.7	95.7	115.1	119	120	120
	PSO	9.5	37.7	70	96	114.6	119.2	120	120
Fitness	HFPSO	0.16	0.33	0.59	0.83	0.96	0.99	1	1
	FA	0.14	0.31	0.54	0.79	0.95	0.99	1	1
	PSO	0.15	0.29	0.54	0.78	0.95	0.99	1	1

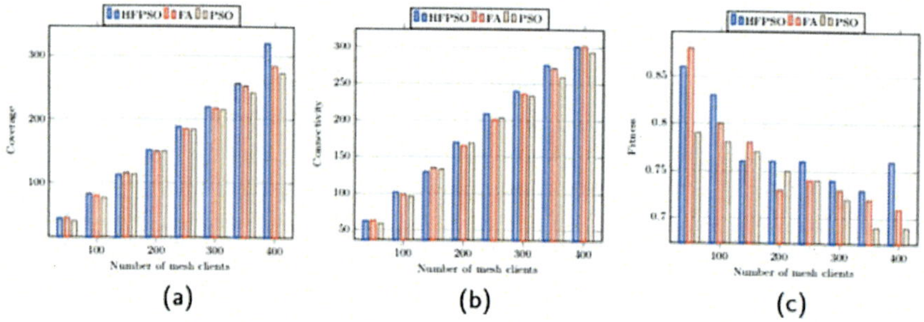

Fig. 1 Impact of varying the number of mesh clients on: **a** Coverage **b** connectivity **c** fitness

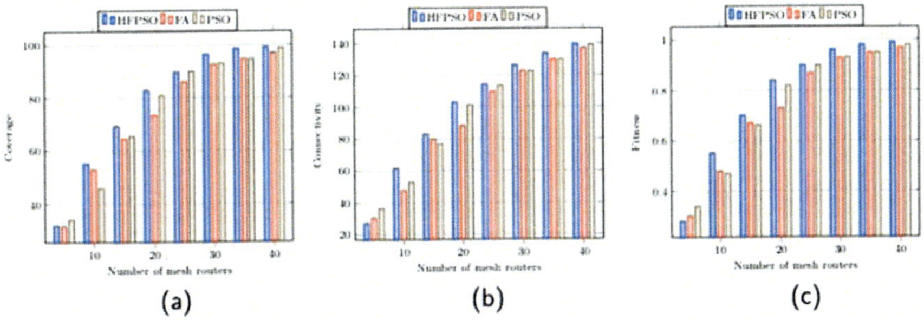

Fig. 2 Impact of varying the number of mesh routers on: **a** Coverage **b** connectivity **c** fitness

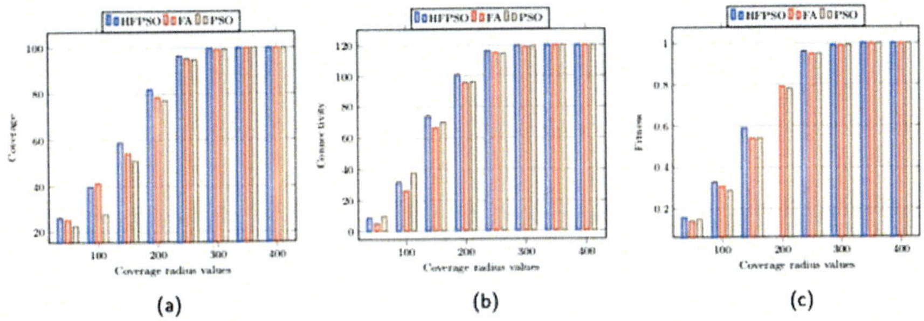

Fig. 3 Impact of varying coverage radius values on: **a** Coverage **b** connectivity **c** fitness

results show that when increasing the radius coverage values, the user coverage, network connectivity, and fitness value will increase too. In fact, when increasing the radius coverage, mesh routers cover a large area until covering and connecting approximately all mesh nodes when the coverage radius exceeds 300. Moreover, when the coverage radius exceeds 100, HFPSO outperforms FA and PSO algorithms.

6 Conclusion

This paper considers the mesh routers placement problem in WMNs. To solve this issue, which is considered as an NP-hard problem, a hybrid approach (HFPSO) based on the combination of PSO and FA was employed. HFPSO was assessed in comparison with PSO and FA using many generated instances. The simulation results demonstrated the performance of HFPSO in terms of user coverage and network connectivity. In future works, we will apply HFPSO for solving joint design including gateway selection, routing, antenna placement, and channel assignment.

References

1. Karthika KC (2016) Wireless mesh network: a survey. In: 2016 international conference on wireless communications, signal processing and networking (WiSPNET). IEEE, pp 1966–1970
2. Qiu L, Bahl P, Rao A, Zhou L (2006) Troubleshooting wireless mesh networks. ACM SIGCOMM Comput Commun Rev 36(5):17–28
3. Xhafa F, Sanchez C, Barolli L, Spaho E (2010) Evaluation of genetic algorithms for mesh router nodes placement in wireless mesh networks. J Ambient Intell Hum Comput 1(4):271–282
4. Xhafa F, Barolli A, Sánchez C, Barolli L (2011) A simulated annealing algorithm for router nodes placement problem in wireless mesh networks. Simul Modell Pract Theory 19(10):2276–2284
5. Xhafa F, Sánchez C, Barolli A, Takizawa M (2015) Solving mesh router nodes placement problem in wireless mesh networks by Tabu search algorithm. J Comput Syst Sci 81(8):1417–1428
6. Sayad L, Bouallouche-Medjkoune L, Aissani D (2017) A chemical reaction algorithm to solve the router node placement in wireless mesh networks. In: Mobile networks and applications, pp 1–14
7. Sayad L, Aissani D, Bouallouche-Medjkoune L (2018) Placement optimization of wireless mesh routers using firefly optimization algorithm. In: 2018 international conference on smart communications in network technologies (SaCoNeT). IEEE, pp 144–148
8. Abdollahzadeh B, Gharehchopogh FS, Mirjalili S (2021) African vultures optimization algorithm: a new nature-inspired metaheuristic algorithm for global optimization problems. Comput Ind Eng 158:107408
9. Zhao W, Wang L, Mirjalili S (2022) Artificial hummingbird algorithm: a new bio-inspired optimizer with its engineering applications. Comput Methods Appl Mech Eng 388:114194
10. Nouri NA, Aliouat Z, Naouri A, Hassak SA (2021) Accelerated PSO algorithm applied to clients coverage and routers connectivity in wireless mesh networks. J Ambient Intell Hum Comput, 1–15
11. Lin C-C (2013) Dynamic router node placement in wireless mesh networks: a PSO approach with constriction coefficient and its convergence analysis. Inform Sci 232:294–308
12. Lin C-C, Tseng P-T, Wu TY, Deng D-J (2016) Social-aware dynamic router node placement in wireless mesh networks. Wireless Netw 22(4):1235–1250
13. Sakamoto S, Barolli L, Okamoto S (2020) Performance comparison of CM and RDVM router replacement methods for WMNs by WMN-PSOHC hybrid simulation system considering normal distribution of mesh clients. In: International conference on P2P, parallel, grid, cloud and internet computing. Springer, pp 9–17
14. Sakamoto S, Ozera K, Barolli A, Ikeda M, Barolli L, Takizawa M (2019) Implementation of an intelligent hybrid simulation systems for WMNs based on particle swarm optimization and simulated annealing: performance evaluation for different replacement methods. Soft Comput 23(9):3029–3035
15. Aydilek IB (2018) A hybrid firefly and particle swarm optimization algorithm for computationally expensive numerical problems. Appl Soft Comput 66:232–249
16. Yang X-S (2009) Firefly algorithms for multimodal optimization. In: International symposium on stochastic algorithms. Springer, pp 169–178
17. Kennedy J, Eberhart R (1995) Particle swarm optimization. In: Proceedings of ICNN'95-international conference on neural networks, vol 4. IEEE, pp 1942–1948

Spur Gear Optimization Using Metaheuristics and Computer Aided Engineering Design

Muhammad Sadeqi, Ali Sadollah, and Seyed Morteza Razavi

Abstract Gearing is one of the most efficient methods of transmitting power from a source to its application with or without change of speed or direction. In this paper, a spur gear model is optimized aiming to maximize its transmission power and minimize its weight. Several design variables named as transmitted power, number of pinion teeth, modules, and thickness of gears have been considered during optimization process. For the sake of optimization, two developed metaheuristics named as water cycle and neural network algorithms have been examined using MATLAB programming language platform. Besides, obtained optimization results have been validated and analyzed using well-known commercial computer aided engineering software ANSYS. Based on the obtained optimization results, optimum design has been found using optimizers and in terms of engineering analysis good agreement has been observed between the applied finite element approach.

Keywords Spur gear · Metaheuristics · Neural network algorithm · Water cycle algorithm · Finite element method

1 Introduction

The development of gear optimization has been on the rise for several years. Gears are defined as the friction wheels, which transmits rotational power among the shafts by increasing or decreasing the speed and torque. To increase the friction and transmission effectiveness, teethes are formed over the surface of the gear wheel. The straight teethed gears are normally called as spur gears [1].

Generally one gear receives power and transmitted to the other. The former is called driver and the latter is called driven. Spur gears are easy to manufacture and commonly used gears in all type of machine tools and automobiles to transmit power between the parallel shafts [2, 3]. Hence, the design of spur gear pair is taken in this research for optimization of the parameters.

M. Sadeqi · A. Sadollah (✉) · S. M. Razavi
Department of Mechanical Engineering, University of Science and Culture, Tehran, Iran
e-mail: sadollah@usc.ac.ir

J. H. Kim et al. (eds.), *Proceedings of 7th International Conference on Harmony Search, Soft Computing and Applications*, Lecture Notes on Data Engineering and Communications Technologies 140, https://doi.org/10.1007/978-981-19-2948-9_30

Optimization is the process of identifying the better solution from the predefined solution space by considering the constraints. All suitable parameter values influencing the problem are possible solutions, which used to form the solution space and the best value in the solution space is the optimum solution [4].

Often optimization problems are solved through deterministic algorithms and stochastic algorithms. As the name implies, the former algorithm follows a rigorous procedure to identify the optimal solution, but it might not be best suitable for all kinds of optimization problems. The latter algorithm is further sub-divided into heuristic and metaheuristic algorithms. Heuristic algorithms are the more efficient and effective algorithm for solving a particular category of the optimization problems and cannot be the best suitable for all category of the problems [5].

On the other hand, nature-inspired metaheuristic optimization algorithms can be used to solve global optimization problems. Because these algorithms use random values to identify the best suitable path and to avoid the stagnation at local optimal points, similar to the nature. In contradict, the random values can also deviate the convergence path away from the solution. Thus, there exists a tradeoff between randomization and local search. Hence, the metaheuristic optimization algorithms possess a deterministic component and random component for randomly sampling the search space and for random walk along the path [6].

Zolfaghari et al. [7] proposed an optimal design of straight bevel gear considering the minimum volume of used material. They used simulated annealing and genetic algorithms (GAs) for the optimization task. The obtained optimization results proved better efficiency compared with the existing design approach. Yao [8] proposed a multi-objective model for a spur design using non-dominated sorting GAs involving different decision making approaches. Savsani et al. [9] proposed an optimization model for simple and multi-stage spur gear trains using particle swarm optimization (PSO) and SA. Their result showed better design compared with the previously published results. Hoseiniasl and Jafari Fesharaki [10] optimized the volume/weight of the gearbox in 3 directions (i.e., height, length, and width of the gearbox) using PSO under different input conditions.

Prabhakaran et al. [11] constructed a multi-objective optimization model considering maximizing transmitted power, efficiency and minimizing overall weight for a paired gear system. Rajesh Kanna et al. [12] proposed a single objective function using weighted summation method to maximize transmitted power, efficiency, while minimizing overall weight and center distance (overall dimension of a paired gear). Furthermore, a comprehensive review has been carried out over gear design optimization using various evolutionary optimization algorithms [13].

This paper aims to identify the optimal parameters for the spur gear pair by considering the design and manufacturing constraints in a non-lubricated condition. Two modern metaheuristic optimization methods named as water cycle and neural network algorithms have been used. Besides, the obtained optimization results using considered optimizers have been analyzed and verified in finite element method approaches.

The rest of this paper is organized as follows: Next section describes the detail mathematical formulation (i.e., objective function and imposed constraints).

Section 3 in brief explains the searching operators of studied optimizers and their introductions. Results and discussion along with finite element method (FEM) results have been provided in Sect. 4. Finally, Sect. 5 concludes the finding of this research.

2 Mathematical Formulation

In this section, components of optimization process have been listed including objective function, design variables, and subjected constraints as follows.

2.1 *Objective Function*

The combined objective function (COF) for optimizing the spur gear is given in the following equation:

$$\text{COF} = 0.6 \times F_1 + 0.44 \times F_2 \tag{1}$$

where F_1 is power transmitted between gear and pinion equals to P and F_2 is weights of gear and pinion given as follows:

$$F_2 = \left(\frac{\pi}{2} \times W \times d_{\text{p}}^2 \times \rho\right) + \left(\frac{\pi}{4} \times W \times d_{\text{G}}^2 \times \rho\right) \tag{2}$$

where d_{p} and d_{G} are pitch circle diameters of pinion and gear (mm), respectively. W is the thickness of gear and pinion (mm), and ρ is the density of used material (kg/mm^3).

2.2 *Imposed Constraints*

In general, design of all mechanical components should be within the ultimate strength and the allowable stress values called as design constraints. Also, the gear parameters should align with the standard design parameters for ease of manufacturing and are generally called as manufacturing constraints. Major constraints considered in this research are given as follows [6]:

Bending Stress Constraint

While transmitting the power between the shafts, the top portion of the gear teeth will subject to bending stress and the condition for the bending stress constraint is given as follows:

$$\sigma_b - \sigma_{b_{\text{all}}} < 0 \tag{3}$$

where $\sigma_{b_{\text{all}}}$ and σ_b are allowable and applied bending stresses (N/mm^2), respectively. σ_b can be calculated based on the handbooks in the literature [1–3].

Contact Stress Constraint

Contact stress analysis between two spur gear teeth was considered in different contact positions, representing a pair of mating gears during rotation [14] and the condition for the contact stress constraint is given in Eq. 5.

$$\sigma_c - \sigma_{c_{\text{all}}} < 0 \tag{4}$$

where $\sigma_{\mathrm{c_{all}}}$ and σ_{c} are allowable and applied compressive (contact) stresses (N/mm^2), respectively. σ_{c} can be calculated based on the handbooks in the literature [1–3].

Module Constraint

The module of pinion and gear should be the same and it should be higher than the following equation [12]:

$$m \geq 1.26\sqrt[3]{\frac{M_{\mathrm{t}}}{y\sigma_{\mathrm{b}}\psi_{\mathrm{m}}N_{\mathrm{P}}}} \tag{5}$$

where ψ_{m} is ratio between the gear pair thickness and module, N_{p} is the number of pinion teeth, y is the Lewis form factor and M_{t} is normal twisting moment transmitted by the pinion in N/mm.

Center of Distance Constraint

Center of distance between pinion and gear shafts should be higher than the suggested following value [12]:

$$C \geq (i+1)\sqrt[3]{\left(\frac{0.74}{\sigma_{\mathrm{c}}}\right)^2 \times \left(\frac{EM_{\mathrm{t}}}{i\psi}\right)} \tag{6}$$

where i is gear ratio, E is young' s module, and ψ is ratio between the gear pair thickness and center of distance.

Safety Factor for Bending and Contact Stress Constraints

In this problem, safety factor of bending and contact stresses were selected as 1.2 and therefore, the following constraints must be met:

$$\mathrm{SF_b} = \frac{\sigma_{\mathrm{b_{all}}}}{\sigma_{\mathrm{b}}} \geq 1.2 \tag{7}$$

$$SF_c = \frac{\sigma_{c_{all}}}{\sigma_c} \geq 1.2 \tag{8}$$

Efficiency Constraint

Engraining gear should have efficiency higher than 98% as can be seen follows:

$$\eta = 100 - P_L \geq 98 \tag{9}$$

where P_L is power loss and can be calculated based on the handbooks [1–3]. Table 1 tabulates used constant parameters for mechanical calculation in this paper. Also, Table 2 lists the design variables considered in this paper along with their upper and lower bounds.

Table 1 Used constant parameters in the proposed optimization model

Constant parameters	Value
i (Gear ratio)	2
E (Young's Module)	1.5×10^5 N/mm^2
Q_v(Quality class)	6
n_1 (Angular velocity of pinion)	500 rpm
K (Load concentration factor)	1.4
ν (Poisson's ratio)	0.3
φ (Pressure angle)	20°
f (Friction coefficient)	0.05
ρ (Density)	7×10^{-6} kg/mm^3
$\sigma_{b_{all}}$ (Allowable bending stress)	35 N/mm^2
$\sigma_{c_{all}}$ (Allowable compressive stress)	350 N/mm^2
Material of gear and pinion	40 Ni 2 Cr 1 Mo 28

Table 2 Side constraints and decision variables of the proposed optimization model

Design variables	Definition	Lower bound	Upper bound
N_P	Number of pinion teeth	15	60
P	Transmitted power (kW)	20	100
F	Thickness of gears (mm)	10	100
m	Module (mm)	1	24

3 Applied Optimization Methods

3.1 Water Cycle Algorithm

Inspired by nature, the idea of water cycle algorithm (WCA) is based on observing the process of the water cycle and on how rivers and rivers flow downwards to sea. The WCA starts with an initial population known as streams. First of all, presume we have snow or raining. The best individual (i.e., the best stream) is chosen as a sea. Also, there are some streams that run to the rivers and the sea, while the rivers flow toward to the sea. Rivers and sea collect the water from the lakes based on the intensity of the flow [15]. Based on surface runoff as can be seen in nature, streams flow to the rivers and rivers flow to the sea. Also, there may be streams flowing directly to the sea. The best equilibrium location is the sea shore with gathering both the river and streams. Therefore, it is possible to create new locations for streams and rivers as flow during surface runoff given as follows [15]:

$$\vec{X}^{i+1}_{\text{Stream}} = \vec{X}^{i}_{\text{Stream}} + \text{rand}(0, 1) \times 2 \times \left(\vec{X}^{i}_{\text{Sea}} - \vec{X}^{i}_{\text{Stream}}\right), \tag{10}$$

$$\vec{X}^{i+1}_{\text{Stream}} = \vec{X}^{i}_{\text{Stream}} + \text{rand}(0, 1) \times 2 \times \left(\vec{X}^{i}_{\text{River}} - \vec{X}^{i}_{\text{Stream}}\right), \tag{11}$$

$$\vec{X}^{i+1}_{\text{River}} = \vec{X}^{i}_{\text{River}} + \text{rand}(0, 1) \times 2 \times \left(\vec{X}^{i}_{\text{Sea}} - \vec{X}^{i}_{\text{River}}\right), \tag{12}$$

If a new position of a stream is better than that of its corresponding river (sea), the position of the river (sea) and the stream is changed. Similarly, such exchanges can occur for movements of rivers and sea. Evaporation is one of the key factors in introducing another operator which can prevent an immature convergence of the studied algorithm. In the WCA, the process of evaporation causes sea water to evaporate as the water flows to the sea. In order to not trap in local optima in the WCA, the following pseudo-code indicates whether or not the evaporation is satisfied [16]:

$$\begin{aligned}
&\text{if } \left\| \vec{X}^{i}_{\text{Sea}} - \vec{X}^{i}_{\text{River}} \right\| < d_{\max} \text{ or rand} < 0.1 \qquad i = 1, 2, 3, \ldots, N_{sr} - 1\\
&\quad \text{Perform raining process}\\
&\text{end}
\end{aligned} \tag{13}$$

where $d_{\max}$ is a small (close to zero) value. The raining cycle is implemented after satisfying the evaporation mechanism. New streams are emerging in the raining cycle at various locations. In fact, the evaporation is the reason of the exploration in the WCA. The schematic view of WCA is shown in Fig. 1 where circles, stars, and the diamond correspond to streams, rivers, and sea, respectively. From Fig. 1. The new positions of rivers and streams are referred to by the white (empty) forms.

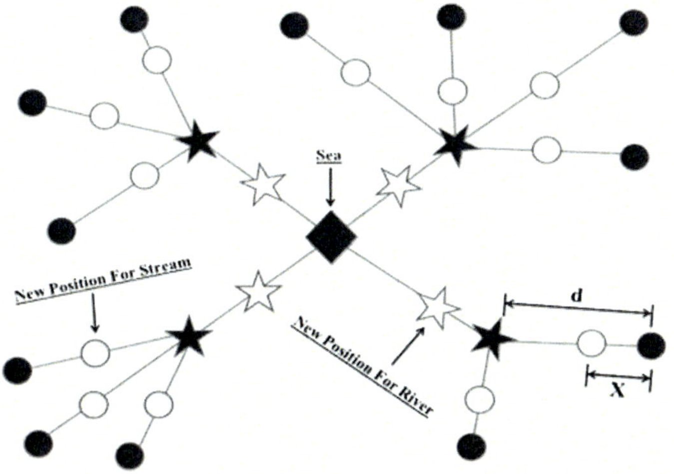

Fig. 1 Schematic view of the WCA's movement toward the best optimum point (sea) [15]

3.2 *Neural Network Algorithm*

Neural network algorithm (NNA) is motivated by unique structure of ANNs and biological nervous system [17]. The NNA starts with an initial population called population of pattern solution. Inspired by the ANNs, each iteration in the NNA assumes the best solution is achieved as target data and the aim is error minimization between target data and other predicted pattern solutions.

The target weight (W_{Target}) corresponding to the target solution must be selected from the weight population (weight matrix) after setting the target solution (X_{Target}) among the other pattern solutions. The original ANNs weights are random numbers, which are modified with the estimated error of the network as the iteration number increases. Back to the NNA, the initial weights are defined as random numbers uniformly from zero to one.

However, there is a limitation for the weight matrix. The restriction that is imposed should not exceed one. The following equation inspired by the weight summation method used in ANNs is used to quantify the new location (X_{News}) after generating the weight matrix (W) and finally updating the new locations as given follows [17]:

$$\vec{X}_j^{\text{New}}(t+1) = \sum_{i=1}^{N_{\text{pop}}} w_{ij}(t) \times \vec{X}_i(t), \quad j = 1, 2, 3, ..., N_{\text{pop}}, \tag{14}$$

$$\vec{X}_i(t+1) = \vec{X}_i(t) + \vec{X}_i^{\text{New}}(t+1), \qquad i = 1, 2, 3, ..., N_{\text{pop}}, \tag{15}$$

For the sake of conducting exploration phase in the NNA, the existence of bias operator plays a crucial role. In the current population of pattern solutions and an updated weights matrix (acting as noise), the biased operator modifies a certain portion of pattern solutions in the NNA.

As an another modification factor in the NNA, transfer function operator, unlike the ANNs, is moving from existing search positions to the new positions of the population so that higher performing solutions are revised and implemented against the target solution. Therefore, for the proposed method, the following equation is defined as an operator of the transmission function [17]:

$$\vec{X}_i^*(t+1) = \text{TF}\left(\vec{X}_i(t+1)\right) = \vec{X}_i(t+1) + 2 \times \text{rand} \times \left(\vec{X}^{\text{Target}}(t) - \vec{X}_i(t+1)\right), \quad i = 1, 2, 3, \ldots, N_{\text{pop}} \tag{16}$$

Using the aforementioned steps, the objective function value of an updating pattern solution (individual) is compared to the target solution (i.e., temporal best solution) and modified the target solution and its corresponding target weights if the created pattern solution is better than the target solution.

4 Results and Discussion

In this section, statistical optimization results have been provided. MATLAB programming language has been utilized. Table 3 shows optimal design parameters

Table 3 Obtained statistical optimization results using two considered optimizers

Optimization	Parameters	WCA	NNA
Design variables	N_p (Number of teeth)	38	38
	Power (kW)	67.238	66.867
	Face thickness (mm)	100	100
	Module (mm)	2.774	2.755
Imposed constraints	SF (Safety factor of bending)	1.2	1.2
	SH (Safety factor of contact)	1.302	1.302
	Weight (Kg)	17.192	16.948
	η (Efficiency)	98.025	98.025
	C (Center of distance)	158.12	157.03
	σ_b (Applied bending stress)	29.17	29.16
	σ_c (Applied contact stress)	268.68	268.65
Objective function	Best COF	−16.580	−16.577
	Worst COF	−16.572	−15.426
	Average COF	−16.579	−16.396
	Standard deviation COF	0.001	0.264

Table 4 Practical and evaluated parameters of pinion and gear for simulation in ANSYS

Parameters (in mm)	Pinion	Gear
Addendum circle diameter	110.96	216.372
Dedendum circle diameter	98.478	203.89
Clearance circle diameter	99.864	205.276
Pitch circle diameter	105.412	210.824
Addendum	2.774	2.774
Dedendum	3.467	3.467
Clearance	0.693	0.693

along with imposed constraints of the proposed model. As can be seen in Table 3, both applied optimizers could achieve almost similar results. Furthermore, based on obtained optimization results, Table 4 gives kinematics of each gearing (i.e., pinion and gear) in terms of their dimensions.

Concerning FEM analysis of studied problem, ANSYS has been utilized to check the quality of obtained solution using considered optimizers in this study. Figure 2 demonstrates stress distribution (von-mises) of a paired spur gear. Looking at Fig. 2, maximum stress has been occurred between engaged gears (see red color). To validate the obtained results using MATLAB, Table 5 compares the optimization and FEM results with each other. By observing Table 5, studied optimizers could attain very close solution with respect to the FEM results (i.e., ANSYS) with less than 1% error (deviation). Therefore, it can be inferred using metaheuristic optimization methods, design of mechanical elements such as gearing can be beneficial.

5 Conclusions and Future Directions

In this paper, metaheuristic optimization methods along with computer aided engineering (CAE) approach have been implemented to identify optimal spur gear parameters by considering the design and manufacturing constraints in a non-lubricated environment. ANSYS software as one of the well-used CAE method, and water cycle and neural network algorithms as optimization methods have been utilized for the sake of obtaining numerical and CAE results. Optimal design parameters of the paired spur gear using metaheuristic optimizers were proposed and the attained results compared with the FEM results by ANSYS. Comparison between two approaches verified good conformity between the results with lower than 1% deviation.

As future research, different optimization techniques can be tested to verify the quality of the obtained results. Conducting optimization using multi-objective optimization algorithms such as NSGA-II may provide us with the wider range of optimal solution. Besides, radial and backlash clearances can be considered as design variables in which their values can affect the performance and efficiency of a paired spur

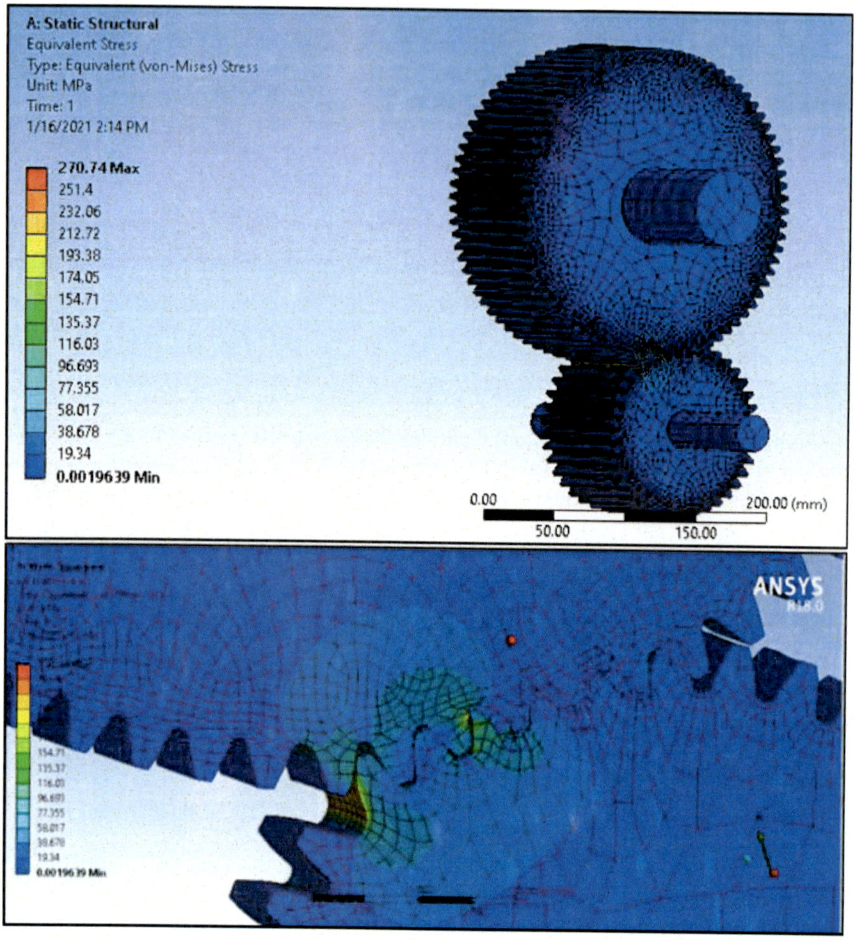

Fig. 2 Equivalent Von-Mises Stress analysis conducted using ANSYS for the paired spur gear

Table 5 Comparison of obtained optimization results and simulated model in CAE software

Methods	Equivalent stress (MPa)	Error (%)
MATAB	268.68	−0.76
ANSYS	270.74	

gear. Furthermore, investigating different materials for a paired spur gear can modify the stress distribution in the studied system.

References

1. Budynas RG, Nisbett JK (2019) Shigley's mechanical engineering design, 11th edn
2. Norton RL (2004) Design of machinery, 3rd edn. McGraw-Hill Professional
3. Townsend DP (1992) Dudley's gear handbook, 2nd edn. McGraw-Hill
4. Yang XS (2010) Nature-inspired metaheuristic algorithms, 2nd edn. Luniver Press
5. Yang XS (2010) Engineering optimization: an introduction with metaheuristic applications. Wiley
6. Baeck T, Fogel DB, Michalewicz Z (1997) Handbook of evolutionary computation. Taylor & Francis
7. Zolfaghari A, Goharimanesh M, Akbari AA (2017) Optimum design of straight bevel gears pair using evolutionary algorithms. J Braz Soc Mech Sci Eng 39:2121–2129
8. Yao Q (2019) Multi-objective optimization design of spur gear based on NSGA-II and decision making. Adv Mech Eng 11(3):1–8
9. Savsani V, Rao RV, VakhariaM DP (2010) Optimal weight design of a gear train using particle swarm optimization and simulated annealing algorithms. Mech Mach Theory 45:531–541
10. Hoseiniasl M, Jafari Fesharaki J (2020) 3D optimization of gear train layout using particle swarm optimization algorithm. J Appl Comput Mech 6(4):823–840
11. Prabhakaran S, Padmanabhan S, Guhan CP, Mugilvalavan M (2016) Optimization of gear drive using genetic algorithm. Int J Innovation Sci Res 25:536–539
12. Rajesh Kanna SK, Sethuramalinmga P, Kanagaraja Sree JS (2015) A firefly algorithm for optimizing spur gear parameters under non-lubricated condition. Int Refereed J Eng Sci (IRJES) 4(7):31–37
13. Usman YO, Odion PO, Onibere, EO, Egwoh AY (2020) Gear design optimization algorithms: a review. J Comput Sci Appl 27(1)
14. Hassan AR (2009) Contact stress analysis of spur gear teeth pair. World Acad Sci Eng Technol 58:1–6
15. Eskandar H, Sadollah A, Bahreininejad A, Hamdi M (2012) Water cycle algorithm—a novel metaheuristic optimization method for solving constrained engineering optimization problems. Comput Struct 110–111:151–166
16. Sadollah A, Eskandar H, Kim JH (2015) Water cycle algorithm for solving constrained multi-objective optimization problems. Appl Soft Comput 27:279–298
17. Sadollah A, Sayyadi H, Yadav A (2018) A dynamic metaheuristic optimization model inspired by biological nervous systems: neural network algorithm. Appl Soft Comput 71:747–782

Analysis of COVID-19 Epidemic Disease Dynamics Using Deep Learning

K. Nirmala Devi, S. Shanthi, K. Hemanandhini, S. Haritha, and S. Aarthy

Abstract COVID-19 has induced anxiety, depression, and fear among people around the world with its cases. During this period, people undergo mixed emotion. Social media is a tool that affected human life during this time in a dominant manner. Twitter is a trending social media platform. Analyzing sentiment of tweets related to COVID-19 can help to analyze the sentiments around the world. In this system, we have taken the dataset which contains tweets related to COVID-19 from IEEE dataport. SVM and LSTM models are built which classifies the tweets as positive, negative, and neutral accordingly. The performance of LSTM model is further analyzed by using hyperparameter tuning method. LSTM gave better results than SVM. It gave an accuracy of 94.58%.

Keywords Sentiment analysis · COVID-19 · Machine learning · Deep learning · TextBlob · Term frequency-inverse document frequency · Twitter

1 Introduction

Because of the spread of COVID, world is facing a very difficult situation. Thus, the government-imposed lockdowns for keeping people safe and stopping the spread of coronavirus. People exposed themselves to the Internet world during the lockdown period. Social media became an important source for communication and getting information about the condition lying outside the houses. The updates about the cases which are positive, got recovered, and the death rate are given by social medias. So, the people totally depend on them during that period. One of the trending social media is Twitter. People tweeted with trending hashtags related to COVID-19.

K. Nirmala Devi (✉) · S. Shanthi · K. Hemanandhini · S. Haritha · S. Aarthy
Department of Computer Science and Engineering, Kongu Engineering College, Erode, Tamil Nadu, India
e-mail: k_nirmal.cse@kongu.edu

J. H. Kim et al. (eds.), *Proceedings of 7th International Conference on Harmony Search, Soft Computing and Applications*, Lecture Notes on Data Engineering and Communications Technologies 140, https://doi.org/10.1007/978-981-19-2948-9_31

Analyzing sentiments of such hash tagged tweets can help to predict the overall sentiment around the world during a specific period [1]. Through the use of inbuilt methods, machine learning techniques are trained, so that they are useful for classification or prediction purpose. It uses time-series, historical data, or labeled data as input for training. Deep learning differs from machine learning algorithm by the data and methods involved in training them. The deep learning algorithms have multiple layers such as input, output, and hidden layers. It can have more than one hidden layers based on its purpose. Our objective is to analyze the feelings conveyed by the public around the globe during the COVID period and classifying them as positive, negative, or neutral.

2 Literature Survey

Chakraborty and Bhatia [2] analyzed the truth and facts of comments related to COVID-19 pandemic. It shows two kinds of tweets gathered in pandemic times. The algorithm used in this sentiment analysis of deep learning classifiers is Naïve Bayes, classification algorithm, and AdaBoost classifier. The maximum accuracy obtained in this result is 81%.

Basiri and Nemati [3] analyzed the people expressed their feelings in their social media comments. Sentiment patterns are observed commonly in different countries during this COVID-19 period. The algorithm used in this sentiment analysis for deep learning models is FastText, BIGRU, NBSVM, DistilBERT, and CNN. The highest accuracy acquired is 85.8%.

Imran and Daudpota [4] analysis the citizens reactions from different cultures to novel. Social media was attacked with the sample post of both negative and positive messages in the COVID-19 lockdown, hashtag, and pandemic. The ML algorithm used in this sentiment analysis is logistic regression and Naïve Bayes. The deep learning technique utilized in this sentiment analysis is LSTM. The greatest accuracy acquired is 90%.

Kaur and Ahsaan [5] approached the analysis the sentiment of Twitter data had been completed by R programming language. The social media has lots of data which is related with our life events like COVID-19. The technique used in the deep learning approach is recurrent neural network (RNN), support vector machine (SVM), and hybrid support vector machine (H-SVM). The maximum accuracy obtained in this result is 80%.

Nirmala Devi and Shanthi [6] examined that the destructive diseases in the globe are the Diabetes Mellitus. In order to obtain beneficial information for exploring the data, machine learning algorithms are employed from various perspectives. Naive Bayes, decision tree, and KNN are the algorithms. 81.33% is the maximum accuracy obtained.

Shanthi and Nirmaladevi [7] examined that Keratoconus is type of an eye illness and it have to be identified in the initial phase, for making refractive surgical process possible which will expand the delightfulness. Support vector machine technique is used. This model obtains the maximum accuracy of 100%.

3 Proposed Work

In this proposed system, the results obtained from the classifiers using deep learning and machine learning algorithms are analyzed, and the finest algorithm among them is identified [8]. The first step in this proposed system is the collection of dataset from the IEEE port. The dataset used in this proposed system contains tweets related to COVID-19. After the collection of dataset, preprocessing takes place. Preprocessing is done to remove the URLs, stopwords, abbreviation to improve the classifiers performance. Then, the datasets are labeled using TextBlob based on the polarity value of the tweets into negative (weakly bad, bad, and strongly bad), positive (strongly positive, positive, and weekly positive), and neutral. After preprocessing, we cannot create models directly, so the preprocessed tweets need to be vectorized. The vectorization process converts the text to number, and it can be fed into the classifier. The vectorizer used here is CountVectorizer. CountVectorizer transforms the tweets into vectors based on count of each word in the given dataset. Figure 1 shows the architecture of this proposed system.

For deep learning algorithms after vectorization process, padding the sequence process will take place. Here, padding the sequence is done to ensure that all the sequences have the same length. The models are built using the classifiers SVM and LSTM. LSTM model is further analyzed by hyperparameter tuning. The results obtained using SVM and LSTM are evaluated and compared. The metrics used for evaluation are accuracy, precision, and F1-score. In this proposed system, 80% of data is used for training the given data, and 20% data is used for testing the given data, that is, 80% data is used for fitting in the machine learning model, and 20% data is used for evaluating the fitted machine learning model.

3.1 Dataset Description

In this proposed system, dataset is chosen from dataport of IEEE, and this dataset contains COVID-19-related tweets. Our proposed work considers 88,767 tweets. The specification of the dataset is revealed in Table 1.

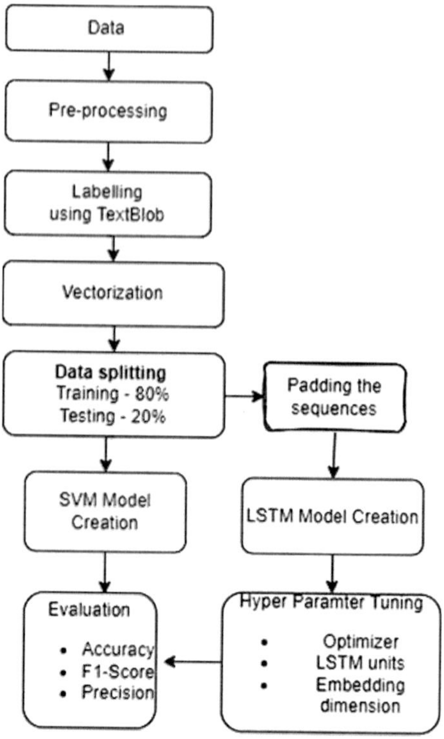

Fig. 1 This figure shows the proposed system architecture

3.2 Data Preprocessing

The tweet in the given dataset contains noise like slang words, abbreviated words, URLs. This noise will disturb the overall performance of the algorithms and sentiment analysis process. Thus, we are preprocessing the data before feature extraction. Initially, URLs are removed using regular expression in Python. Then, all the tweets are converted to lowercase, tokenized in which the sentence is divided into words and removed the stopwords in the tweets and finally lemmatized. Lemmatization is used for grouping together a word which is in different forms since they can be considered and analyzed as a single word. Tokenization and lemmatization process take place using NLTK. The mechanism that is applied for preprocessing the data. Two algorithms are used here since the results are to be compared.

Table 1 Description of dataset

Duration	No. of tweets	Keywords used while collecting	Source of the dataset
1 January, 2021	88,676	#2019-ncov,#ncov,#covid_19,#sarscov2,#covid-19,#ncov2019,#corona,#covid,#lockdown	Taken from IEEE dataport (https://ieee-dataport.org/open-access/coronavirus-covid-19-tweets-dataset/corona_tweets_289.zip)

3.3 Labeling

The tweet in the dataset contains noise like slang words, abbreviated words, URLs. This noise will disturb the overall performance of the algorithms and sentiment analysis process. Thus, we are preprocessing the data before feature extraction. Initially, URLs are removed using regular expression in Python. Then, all the tweets are converted to lowercase, tokenized in which the sentence is divided into words and removed the stopwords in the tweets and finally lemmatized. Lemmatization is used for grouping together a word which is in different forms since they can be considered and analyzed as a single word. Tokenization and lemmatization process take place using NLTK. The mechanism that is applied for preprocessing the data. Two algorithms are used here since the results are to be compared.

3.4 Vectorization

The features need to be extracted from the labeled tweets in order to fed input to the classifier. The tweets need to be converted into number in order to build a model [9]. In this proposed system, CountVectorizer is employed to build a bag of words for converting words to numbers. CountVectorizer is used to transform the tweets into vector based on the count or frequency of every word in the dataset.

3.5 Machine Learning Model

In this proposed system, SVM is used to build model and then compared and evaluated with the deep learning model. The support vector machine algorithm distinctly classifies the data points in an N-dimensional space into two different classes using a straight line known as hyperplane, in which N is referred to the number of features. At first, every data item is marked as a point in a n-dimensional space. Then, the classification is achieved by identifying the hyperplane for differentiating the data points into two different classes. Hyperplane finds the best boundary which helps in classifying the data points.

SVM algorithm uses a kernel trick for getting optimal decision boundary for classifying the data into two classes. When the data is linearly separable data, that is, it is used when the data can be separated by using a single line. It is one of the most common kernels to be used. For text classification, linear kernel is mostly used. Training a SVM using a linear kernel is faster than using other kernels that are used for SVM.

3.6 Deep Learning Model

LSTM is a type of RNN, designed to avoid long-term dependency problems in RNN. During back propagation, vanishing gradient problem exists in RNN. If the value of the gradient is utmost small, it does not commit much learning. RNN, thus having a short-term memory, forget what it seen in longer sequences.

LSTM has internal mechanisms called gates which is used for regulating the flow of information. These gates learn that which data in a series to throw away or store for passing the suitable information to the sequences to make prediction. The three major mechanisms used here are input gate, forget gate, and output gate. A LSTM network is composed of distinct memory blocks known as cells. The states that are being transferred to next cell consist of two states: the cell state and the hidden state.

The pad_sequence ( ) function, Keras deep learning library used for padding the variable length sequences. Padding the sequence is done to assure that all the sequences of tweets in the list have the same length. The pad_sequence ( ) function used in padding the sequence to a desired length which may be longer than observed sequences. After vectorizing the preprocessed tweets [10], all the tweets are padded to a length of 20.

Hyperparameter Tuning: Optimizers are methods or algorithms used for reducing the loses by changing the feature of the neural network like weights and learning rate. It minimizes the function for solving optimization problems. Optimizers used for RNN are Adam, Nadam, RMSprp, SGD, Ftrl, Adamax, Adagrad, and Adadelta.

The value in the hyperparameter is used in controlling the learning process. Hyperparameter is tuned for making the model to solve the learning problem optimally. Hyperparameter optimization identifies a tuple of hyperparameters which yields an optimal model for minimizing a loss function which is predefined on given independent data.

Embedding layer specifies arguments like input dim, output dim, input length. Input dim describes the dimension of the words in the given dataset. Output dim describes the dimensions of the output vectors for each word. Input length expresses the length of input sequences, and it is described for any input layer of a Keras model. Embedding layer converts each and every word into a certain length of vector with defined size. The output of the vector has real values instead of 0's and 1's. The number of layers used in this proposed system are 20 and 150.

LSTM units, also called as latent dimensions, define the length of hidden state or length of activation vector. It defines the shape of the tensor, produced by the layer and it will be the input to the next layer. In this proposed system, the LSTM units used are 15 and 20 unit.

4 Results and Discussion

The goal of this study is to construct ML and DL models that can analyze the emotion of tweets. Here, tweets related to COVID-19 is the dataset which was extracted from dataport of IEEE is employed to create various ML model such as SVM and DL model such as LSTM and the performance of classifiers are compared. The experiments were done using open-source libraries such as scikit-learn and TensorFlow.

In this experiment, the tweets are labeled as positive (weakly, normal, strongly), negative (weakly, normal, strongly), and neutral using TextBlob sentiment values. CountVectorizer and vectorizer methods are used for vectorization purpose. The performance of the models created is compared using accuracy, precision, and f1-score. In LSTM model, the influence of the optimizers and the other hyperparameters such as embedding dimension and LSTM units is analyzed. Among the models created, LSTM gives a maximum accuracy, precision, and F1-score of 94.58, 94.67, and 94.61%. The linear SVM model gives an accuracy of 92.94%, F1-score of 92.92%, and a precision of 92.91%. These details are shown in Fig. 2.

4.1 Optimizer Tuning

Optimizers are techniques employed for changing the feature of the neural network like weights and learning rate for reducing the losses. In order to solve the optimization problems by minimizing the function, optimizers are used. Here, we have used six optimizers such as Nadam, Adam, RMSprop, Adagrad, Adamax, Adadelta, Ftrl, and SGD. All optimizers are different in their mechanism, thus are different in their optimization performance. The performance of LSTM model with the listed

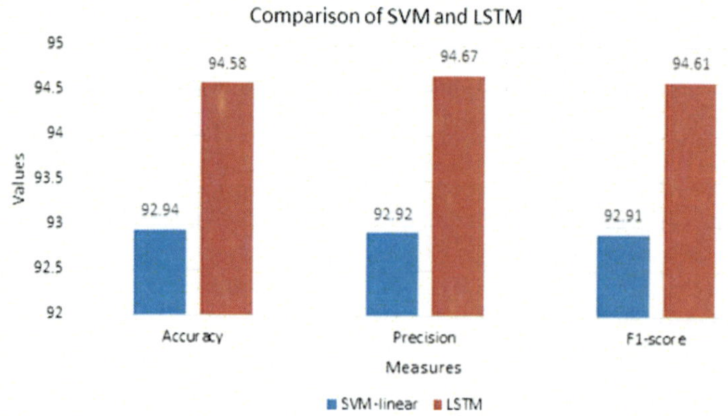

Fig. 2 This figure shows the accuracy, precision, and F1-score of SVM and LSTM models

Table 2 LSTM of different models

LSTM optimizers	Accuracy	Precision	F1-score
Nadam	92.1	92.29	92.12
Adam	92.51	92.7	92.51
Rmsprop	92.72	92.81	92.73
Adagrad	46.22	21.35	29.21
Adamax	89.78	89.43	89.53
Adadelta	45.75	20.93	28.72
Ftrl	45.7	20.89	28.67
SGD	86.28	85.75	85.45

optimizers is evaluated using accuracy, F1-score, and precision and compared. Thus, we got different accuracy measures for different optimizers.

Best among the LSTM optimizers is RMSprop which gives an accuracy of 92.72%, precision of 92.81%, and F1-score of 92.73%, and the least performing optimizer is Ftrl gives an accuracy of 45.7%, precision of 20.89%, and F1-score of 28.67% which is a least performing optimizer. It is shown in Table 2

4.2 *Hyperparameter Tuning*

A word embedding uses a dense vector representation in order to represent words and documents. These representations are sparse since the vocabularies are huge and a given word or a document should be defined in a large vector mostly composed of zero values. Inside the vector space, the position of the word is identified from text. The number of units in an LSTM is defined as the number of LSTM memory units for each input in the sequence in vertical manner attached to each other. Each unit passes the filtered information to next memory units. The embedding layer output dimension and the LSTM units are increased in the existing model (Model 1) and named as Model 2. Model 1 and Model 2 are the two models differing in their LSTM units and embedding layer dimensions. Table 3 shows the specifications of Model 1 and Model 2.

Figure 3 shows the precision comparisons between the Model 1 and Model 2 of LSTM for different optimizers. The Model 2 produced better precision than Model.

Table 3 Description of Model 1 and Model 2

LSTM model	Embedding layer output dimension	LSTM units
Model 1	20	150
Model 2	15	50

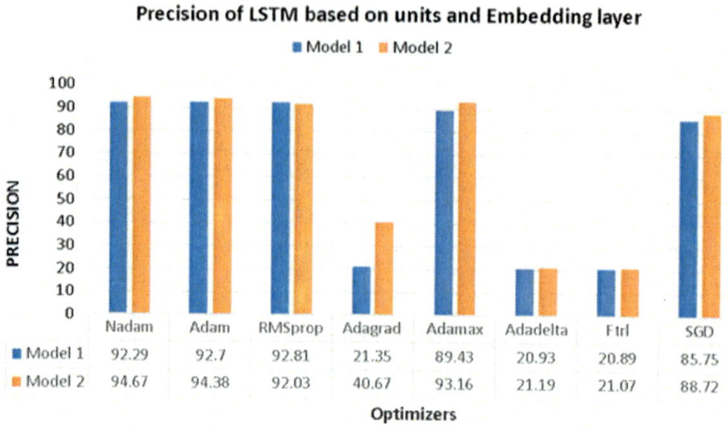

Fig. 3 This figure shows the precision comparisons of Model 1 and Model 2 for different optimizers

In Model 2, Nadam optimizer gives a precision of 94.67%, and Ftrl gives the least precision of 20.89% for Model 1.

Figure 4 shows the F1-score comparisons between the Model 1 and Model 2 of LSTM for different optimizers. Model 2 performed well than the Model 1. The optimizer which produced best precision is Nadam with F1-score of 94.61%, and RMSprop gives the least F1-score of 1.83% for Model 1.

Increasing the embedding layer output dimension and LSTM memory units shows a visible difference in the accuracy and other measures. The model with greater number of units and output dimension (Model 2) gave good results than the another model (Model 1). Nadam optimizer gave an accuracy of 94.58%, F1-score of 94.61,

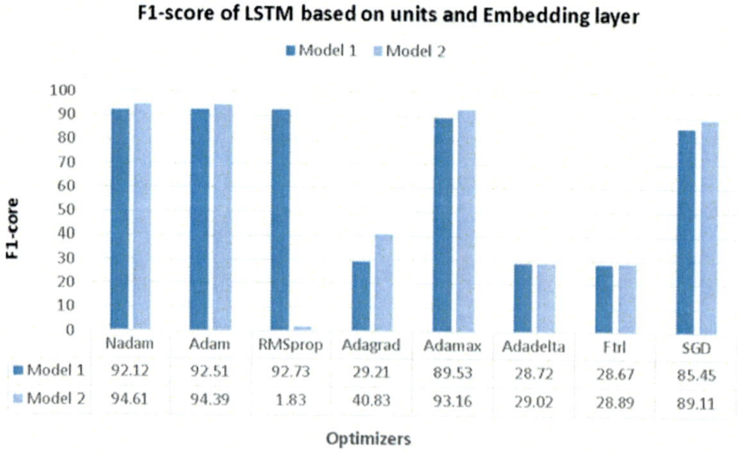

Fig. 4 This figure shows the F1-score comparisons of Model 1 and Model 2 for different optimizers

and precision of 94.61%. The accuracy comparisons between the optimizers of LSTM are shown in Fig. 5.

Loss and accuracy chart show us the measure of loss and accuracy during each epochs during the model fitting. The history of the model will contain these values in form of a dictionary which can give training loss, training accuracy, validation loss, and validation accuracy. These measures are plotted for the Model 2 for different optimizers separately. Figure 6 shows the loss and accuracy fluctuations for Nadam optimizer after hyperparameter tuning.

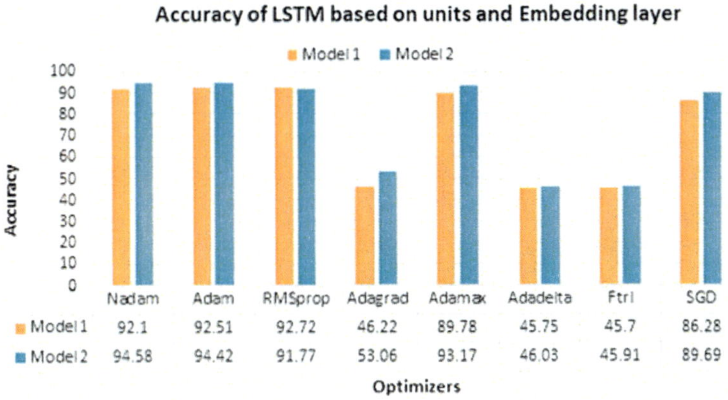

	Nadam	Adam	RMSprop	Adagrad	Adamax	Adadelta	Ftrl	SGD
Model 1	92.1	92.51	92.72	46.22	89.78	45.75	45.7	86.28
Model 2	94.58	94.42	91.77	53.06	93.17	46.03	45.91	89.69

Fig. 5 This figure shows the accuracy comparisons of Model 1 and Model 2 for different optimizers

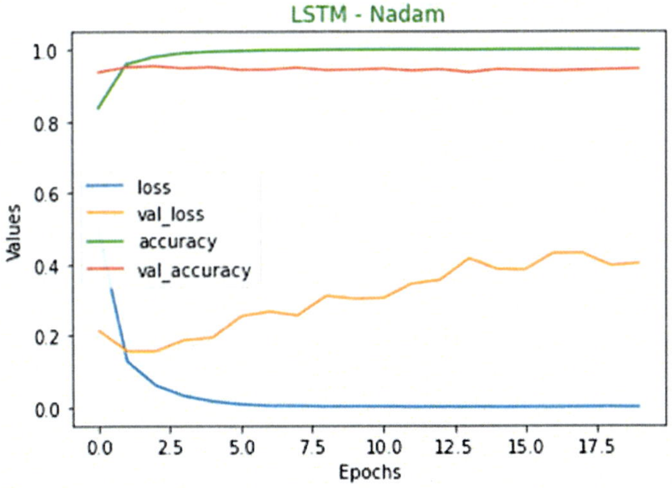

Fig. 6 This figure shows the loss and accuracy chart for Nadam optimizer

5 Conclusion

The aim of this project is finding the COVID's impact on public using sentiment analysis. The dataset is taken from IEEE dataport. SVM linear gave a maximum accuracy of maximum accuracy of 92.94%, F1-score of 92.92% and 92.9. While changing the optimizers of RNN, RMSprop optimizer was better than others. Accuracy, F1-score, and precision obtained are 92.72, 92.81, and 92.73%. Then increasing the embedding dimension in LSTM and increasing the LSTM units, the accuracy of the model gets increased in a good manner. Nadam optimizer gives best results after tuning the hyperparameters. Finally, the updated accuracy, F1-score, and precision of LSTM with Nadam optimizer are 94.58, 94.61, and 94.67%. The proposed system can be further enhanced for finding the impact of COVID-19 on stock markets and by using a bigger dataset.

References

1. Alhuri LA, Aljohani HR, Almutairi RM, Haron F (2020) Sentiment analysis of COVID-19 on Saudi trending hashtags using recurrent neural network. In: 2020 13th International conference on developments in eSystems engineering (DeSE). IEEE, pp 299–304
2. Chakraborty K, Bhatia S (2020) Sentiment analysis of COVID-19 tweets by deep learning classifiers—a study to show how popularity is affecting accuracy in social media. Appl Soft Comput 97:106754
3. Basiri ME, Nemati S, Abdar M, Asadi S, Acharrya UR (2021) A novel fusion-based deep learning model for sentiment analysis of COVID-19 tweets
4. Imran AS, Daudpota SM, Kastrati Z, Batra R (2020) Cross-cultural polarity and emotion detection using sentiment analysis and deep learning on COVID-19 related tweets. IEEE Access 8:181074–181090
5. Kaur H, Ahsaan SU, Alankar B, Chang V (2021) A proposed sentiment analysis deep learning algorithm for analyzing COVID-19 Tweets. Inf Syst Front, 1–13
6. Nirmaladevi K, Shanthi S, Agila T, Dharani RT, Dhivyapriya P (2020) Analysis and prediction of diabetes using machine learning. Test Eng Manage 83(9):14533–14538
7. Shanthi S, Nirmaladevi K, Pyingkodi M, Dharanesh K, Gowthaman T, Harsavardan B (2021) Machine learning approach for detection of keratoconus. IOP Conf Ser Mater Sci Eng 1055(1):012112. IOP Publishing
8. Mengistie TT, Kumar D (2021) Deep learning based sentiment analysis on COVID-19 public reviews. In: 2021 International conference on artificial intelligence in information and communication (ICAIIC). IEEE, pp 444–449
9. Abdelminaam DS, Neggaz N, Gomaa IAE, Ismail FH, Elsawy AA (2021) ArabicDialects: an efficient framework for Arabic dialects opinion mining on Twitter using optimized deep neural networks. IEEE Access 9:97079–97099
10. Roy A, Ojha M (2020) Twitter sentiment analysis using deep learning models. In: 2020 IEEE 17th India Council international conference (INDICON). IEEE, pp 1–6

Investigation of the Effect of Nanoclay on Composite Plates Under Low-Speed Impact Using Artificial Neural Networks

Ali Khoshnoudrad, Seyed Morteza Razavi, Ali Sadollah, and Fatemeh Taghiha

Abstract In this study, low-velocity impact on fiber metal laminate (FML) plate with nanoclay reinforcement particles has been investigated. Rectangular FML specimens have been made by hand lay-up method using two 2024 aluminum sheets and 4 layers of basalt fibers. The ultrasonic device is used for nanoclay in samples, and the specimens are constructed in different weight percentages of nanoclay particles from 0 to 5%. Low-velocity impact test has been conducted using drop weight device at different level of impact energies. Artificial neural networks are designed and used for the prediction of the response factor (i.e., absorbed impact energy), while input factors are weight percentage of nanoclay particles and applied level of impact energy. The obtained experimental and numerical results showed that nanoclay reinforcement particles in specimens containing different weight percentages of nanoclay possess more contact force and energy absorption with respect to the non-nanoclay specimens.

Keywords Low-velocity impact · Fiber metal laminate · Nanoclay · Basalt fiber · Artificial neural networks

1 Introduction

Composites are considered as a class of advanced materials in which a combination of simple materials is used to create new materials with superior mechanical and physical properties. Fiber metal laminate (FLM) is composed of layers of metal and polymer-based composite materials including reinforcing fibers. Many applications of deep stretching of multilayer sheets show the importance of research on these materials [1]. Today, with the spread of science and knowledge in the field of materials which lead to form new and intelligent materials, the desire to make materials with having high ratio of mechanical properties to their weights has increased, especially in the aerospace industry.

A. Khoshnoudrad · S. M. Razavi (✉) · A. Sadollah · F. Taghiha
Department of Mechanical Engineering, University of Science and Culture, Tehran, Iran
e-mail: m.razavi@usc.ac.ir

J. H. Kim et al. (eds.), *Proceedings of 7th International Conference on Harmony Search, Soft Computing and Applications*, Lecture Notes on Data Engineering and Communications Technologies 140, https://doi.org/10.1007/978-981-19-2948-9_32

By introducing of nanoparticles in material science, nano-reinforced polymers have been attracted attentions by the scientific and industrial communities [1]. In the case of nanocomposites, significant improvement of mechanical properties such as strength, rigidity, and thermal properties without negative effect on density and processability has been reported [2]. Reinforced polymers with nanoclay, in addition to having the aforementioned advantages, possess low permeability to water and gas passage [3, 4]. The main difference between composite reinforcing fibers and composite reinforcing nanomaterials is the low percentage of nanomaterials (i.e., usually around 2–5 wt%) [5].

The discussion of impact and resistance of a structure to impact loads are one of the most important challenges in mechanical engineering. Impacts, especially low-velocity impacts, are among the dynamic loads that play a very important role in the design and analysis of structures. Bagheri and Marouf [6] investigated mechanical behavior of epoxy-modified nanoclay composites. Their study showed the effect of the modified nanoclay presence in the resin epoxy increases both mechanical strength and fracture toughness compared with the pure epoxy resin.

Avila et al. [7] investigated the behavior of composite plates by impact at low velocities. They investigated the effect of changing the volume percentage of particles on the amount of contact force at different levels of impact energy. Fan et al. [8] considered the low-velocity impact response in the FML composites and simple composites. The results verified that the FML composites have higher perforation resistance than the simple composites.

Tsaris et al. [9] studied the impact resistance of FML composites under low-velocity impact to provide an optimal design for the impact resistance of aerospace applications. Sedighi et al. [10] investigated the impact resistance of FML composite with different types of metal and thicknesses under low-velocity impact and observed that increasing the thickness of aluminum has improved the performance of FML composite.

Lane [11] studied the impact behavior of silicate-reinforced polymer nanocomposites. In his research, in addition to determining the mechanical properties, the effect of the structure and arrangement of nanoparticles in the matrix have been considered. Zamani et al. [12] investigated the response of mechanical properties and high impact of polypropylene nanocomposites with multi-walled carbon nanotubes. In their study, the impact strength for the perforated nanocomposite sample was increased to prevent crack propagation.

Artificial intelligence (AI) plays an important role in real-engineering applications. An artificial neural networks (ANNs), a branch of AI, are a paradigm of mathematics or computation that is based on the computational and/or functional dimensions of biological neural networks. The use of ANNs gives the bonus of being able to represent dynamic relationships between input and output and is suitable when used for data classification, function approximation, and so forth. Hassoun [13] described the notion for the majority of ANNs. Besides, applications of ANNs in material science were extensively reviewed by Sha and Edwards [14].

This paper aims to conduct research on the effect of adding nanoclay particles in a polymer-based composite reinforced with basalt fibers by considering absorbed impact energy. Different weight percentages of nanoclay particles from zero to 5 percent have been examined with 18 independent experiments, and the effects of additive nanoparticles have been discussed. Besides, the ANNs are utilized to predict the response factor of the studied model and are compared with the actual experimental data.

The rest of this paper is organized as follows: Next section describes used materials and test method. Section 3 in brief explains processes of ANNs. Results and discussions along numerical and experimental results have been provided in Sect. 4. Finally, Sect. 5 concludes the finding of this research in accompany with future directions.

2 Materials and Test

2.1 Used Materials

In order to make samples, two layers of 2024 aluminum with a thickness of 0.75 mm were used with four layers of basalt fibers in between. Since basalt fibers are simple in texture, their properties are the same in length and width. The resin used was selected from the epoxy family and made by Chimax Company with EPR1080 brand and with EA1080 hardener. The chemical reaction and curing of this resin are performed at the temperature of 25–80 °C according to the manufacturer's suggestion. The ratio of hardener to resin was 15% by weight scale and after making the samples, sintering was performed at 25 °C. Also, in this research, a modified mineral nanoclay with the brand name of Bentonite made by Samchun Korea was used.

2.2 Resin Preparation

Nanoparticles must be added to the resin to make nanoclay-containing samples. First, the nanoparticles were dispersed in the solution using a mechanical stirrer, and then, the solution was sonicated using an ultrasonic device. The model of SONOPLUS HD 3200 ultrasonic device by BANDELIN company in Germany was utilized in this study. The maximum power of the device was set to 200 W, which produces the desired waves at a frequency of 20 kHz.

2.3 Low-Velocity Impact Test

The low-velocity impact test was performed with the drop impact device (see Fig. 1). The projectile is made of steel with a spherical tip and a diameter of 16 mm, and

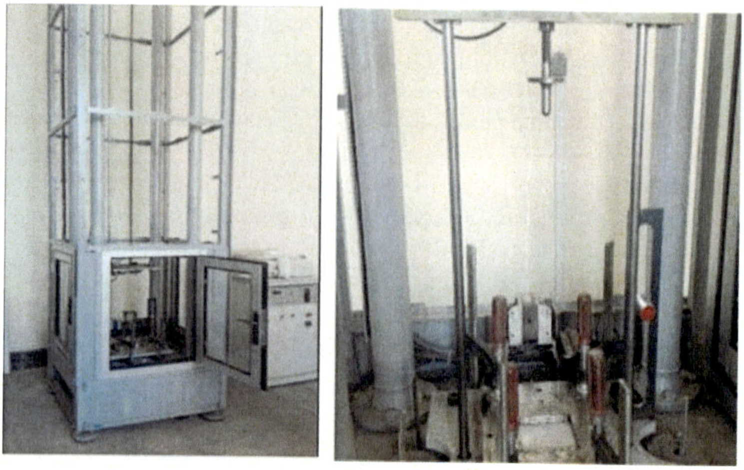

Fig. 1 Falling weight machine used in this paper

the mass of the projectile is ignored. An accelerometer sensor is built into the falling weight device. At the moment of collision, the data are transferred from the accelerometer to the data logger and then to the computer.

To obtain the effect of nanoclay addition, low-velocity pulses were performed at three energy levels of 14.6, 22.52, and 27.98 J. To obtain the maximum amount of energy that does not destroy the sample, a quasi-static test was performed, and the results were obtained by determining the height of the striker at 55 cm for all samples.

3 Artificial Neural Networks Implementation

The ANNs are a computational model which consist of layers of artificial neurons that mimic the behavior of biological neurons in the brain [15]. The ANNs architecture may consist of one or more hidden layers with artificial neurons in each layer. In the literature, many learning algorithms have been proposed for training the ANNs [16]. The training process involves finding the optimal values of the network parameters (i.e., weights and biases) such that the difference between the measured and predicted values of ANNs is minimized [17].

The Levenberg–Marquardt (LM) algorithm is one of the most efficient back-propagation algorithms for training the ANNs. It combines the gradient descent and Gauss–Newton training methods. The LM algorithm can train the small and medium-sized problems very effectively [18], (i.e., the reason of choosing LM training algorithm in this paper). During the training of an ANNs, the LM algorithm iteratively optimizes the network parameters in order to minimize the cost function. The training stops when the termination condition is satisfied. The sum of squared error and the mean squared error (MSE) are often used as cost functions. In this paper, the MSE function has been used.

The dataset was randomly divided into training and testing sets. 70% of the data records was used for training, and the remaining 30% was used for testing and validation (both 15%) of the model. The validation set was utilized to evaluate the generalization capability of the models and to stop training when overfitting occurs. In this paper, the optimal number of hidden neurons was found to be 5 for ANNs-LM, and one single hidden layer has been considered.

4 Results and Discussions

In this section, obtained experimental results from the experiments and the ANNs-LM are compared in terms of error percentage and linear regression among the network outputs and the corresponding experimental results. Table 1 shows specifications of some constructed samples. By observing Table 1, A01 code means sample with zero percent nanoclay particle and with first impact energy level (i.e., 14.63 J), while B12 means sample with 2 wt% nanoclay tested by second impact energy level (i.e., 22.52 J). Similarly, sample codes C11 and C15 mean samples with 1 wt% and 5 wt% nanoclay particles for third impact energy level (i.e., 27.99 J), respectively.

Looking at Table 1, differences in thickness in the samples are due to the manual fabrication method and laboratory conditions. In this study, the absorbed impact energy was considered as the response of the system, and the obtained experimental results are shown in Table 2.

Due to difficulties and complexities in the testing process, 18 experiments have been conducted. As representative figure for the rest samples, Fig. 2 demonstrates the force–displacement diagram for three different levels (i.e., *A*, *B*, and *C*) of impact energy for FML with (i.e., 1 wt% nanoclay particles) and without nanoclay particles. As can be seen in Fig. 2, sample with 1 wt% nanoclay shows larger amount of absorbed energy (area under force–displacement diagram) and a smaller amount of displacement with respect to the sample without nanoclay particle.

Table 1 Specification for some samples with and without nanoclay particles

Density (g/cm^3)	Thickness (mm)	Impact energy (J)	Impact mass (kg)	Nanoclay weight (%)	Sample code
0.52	3.10	14.63	2.71	0	A01
0.55	3.25	14.63	2.71	1	A11
0.53	3.23	22.52	4.17	0	B01
0.54	3.18	22.52	4.17	1	B11
0.55	3.35	27.99	5.18	0	C01
0.54	3.21	27.99	5.18	1	C11

Table 2 Experimental results obtained from 18 tests for the studied composite

Exp.	Factor 1: mat	Factor 2: applied impact energy (J)	Response: absorbed impact energy (J)	Exp.	Factor 1: mat	Factor 2: applied impact energy (J)	Response: absorbed impact energy (J)
1	A01	14.63	6.39	10	B12	22.52	10.18
2	B01	22.52	9.90	11	B13	22.52	11.03
3	C01	27.99	27.68	12	B14	22.52	11.89
4	A11	14.63	6.54	13	B15	22.52	12.36
5	A12	14.63	6.68	14	C11	27.99	28.22
6	A13	14.63	7.28	15	C12	27.99	29.85
7	A14	14.63	8.18	16	C13	27.99	30.15
8	A15	14.63	8.85	17	C14	27.99	31.18
9	B11	22.52	10.06	18	C15	27.99	32.01

Looking at Fig. 2, for smaller impact energies (i.e., 14.63 and 22.52 J), the effect of added nanoclay particles compared with zero nanoclay sample is not considerable, however, as the impact energy increases (i.e., impact energy of 27.99 J given in Fig. 2), the effect of adding nanoclay particles becomes apparent. Indeed, when the energy level increases, the samples show different behavior in response to a low-velocity impact. In fact, the impact resistance of the samples with 1 wt% nanoclay is higher than those without nanoclay particles. Furthermore, the dorsal surfaces of each specimen are tested and illustrated in Fig. 3 along with the extent of their damages.

It is worth mentioning that laminating is a common condition in the structure of low-velocity fiber metal composite that propagates between composite layers and metal-to-metal composite layers. The spread of delamination in composite layers of plain texture depends on several factors such as fiber strain, resin toughness, and overall sample thickness.

Concerning developed prediction model by ANNs, the MSE plot for all samples including training, validation, and test-sample errors is depicted in Fig. 4 for the absorbed impact energy response. The MSE of the network was started at a large value and decreased to a smaller value as shown in Fig. 4. In other words, Fig. 4 shows that the network is learning.

Besides, Fig. 5 displays scatter plots of the measured absorbed impact energy versus the predicted results of the ANNs-LM, respectively, for the training, validation, and testing sets (vertical and horizontal axes are the predicted output and corresponding targets, respectively). By judging Fig. 5, the ANNs-LM has a good generalization capability with having high R factor (correlation coefficient) which demonstrates that there is an excellent correlation between the measured and predicted values.

Good performance of the ANNs-LM model on the validation set indicates the good predictive capability of the generated model. The testing set was used to assess

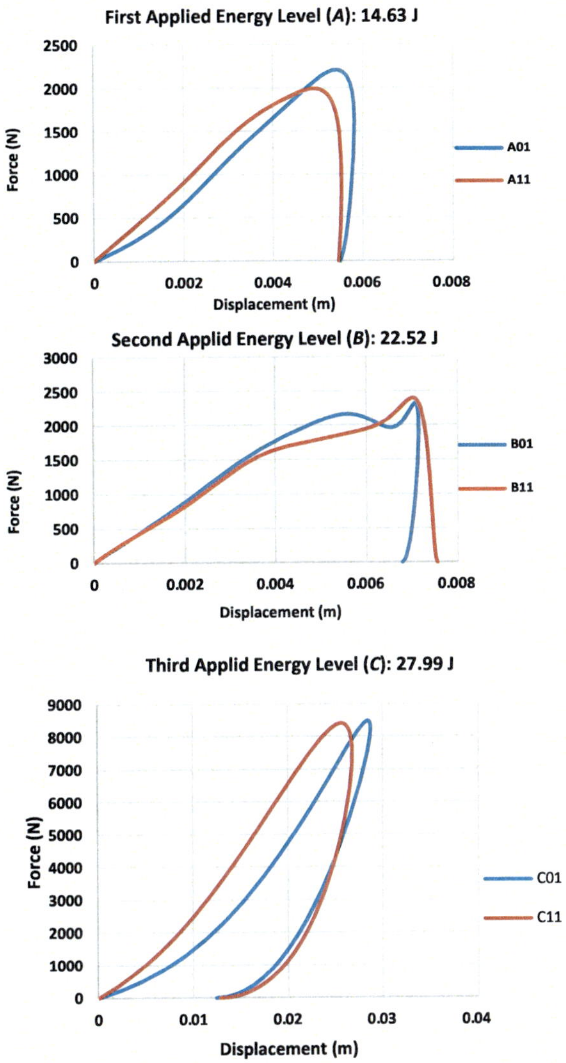

Fig. 2 Comparison diagram of force-sample displacement with (1 wt%) and without nanoclay particles under different impact energy levels

the predictive accuracy of the models on unseen data. Table 3 shows the comparison of experimental and predicted results for the absorbed impact energy response obtained using the ANNs-LM over unseen data. Three unseen test samples and their corresponding experimental data are compared as shown in Table 3. For all three samples given in Table 3, it is clear that the results obtained using the ANNs-LM model are consistent with the experimental values.

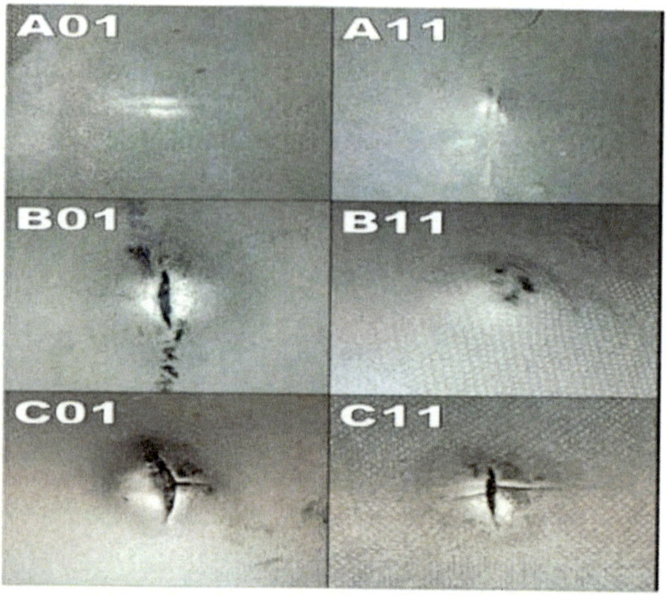

Fig. 3 Dorsal surfaces of each specimen

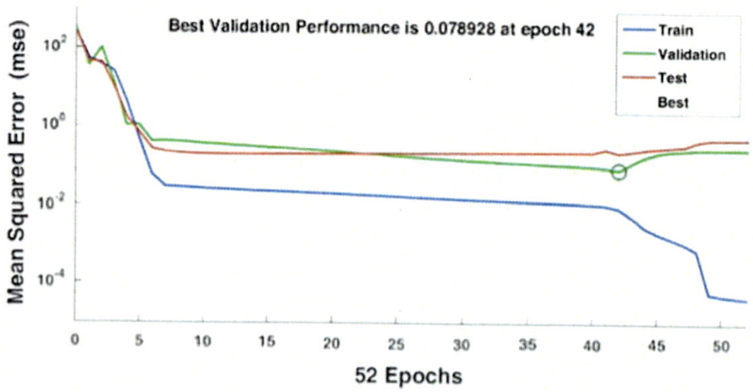

Fig. 4 MSE plot with respect to the number of epochs for training, validation, and test samples

5 Conclusions and Future Research

In this paper, the effect of adding nanoclay particles on fiber metal laminate (FML) basalt-aluminum under low-velocity impact has been investigated. Different impact energy levels (i.e., 14.63, 22.52, and 27.99 J) have been considered along with different weight percentages of nanoclay particles. Also, in this research, utilization of artificial neural networks (ANNs) trained by Levenberg–Marquardt (LM)

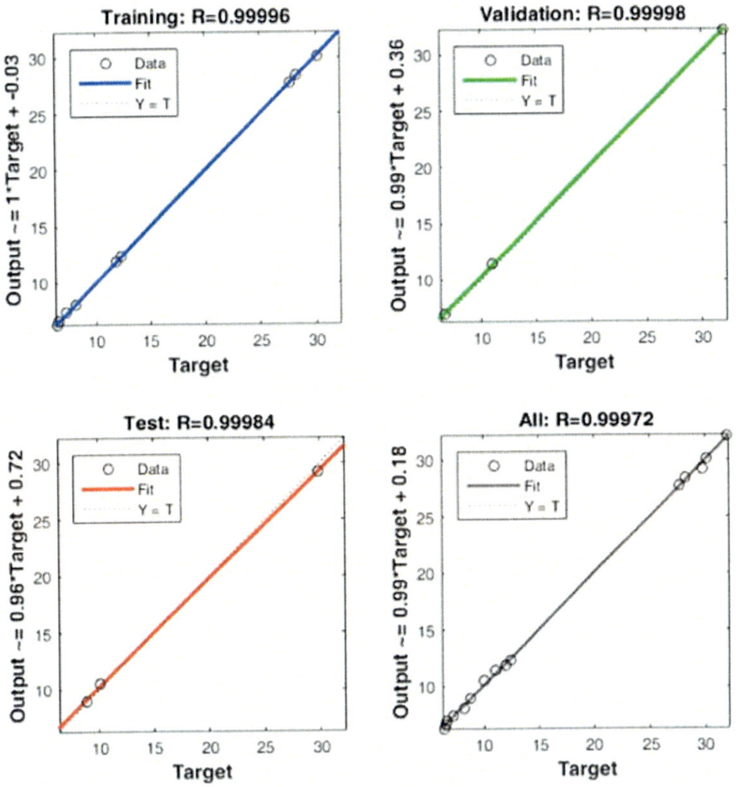

Fig. 5 Correlation coefficient plot for the absorbed impact energy response

Table 3 Comparison of experimental and predicted results for the absorbed impact energy

Data	Sample	Applied energy level (J)	ANNs-LM (Abs. error %)	Experimental (J)
1	B01	22.52	10.10 (2.02)	9.90
2	B12	22.52	10.96 (7.66)	10.18
3	C14	27.99	31.12 (0.19)	31.18

was investigated in order to predict the absorbed energy impact (J) of the given FML composite.

The input factors were the weight percentage of nanoclay particle (0–5 wt%) and applied energy level (J), while the response factor was the absorbed energy impact (J). Based on the obtained experimental and numerical results, the higher the impact energy, the greater the impact resistance of the particles with increasing the weight percentage of nanoclay particles. Indeed, the rate of energy absorption after impact in samples having higher nanoclay particles is higher than the samples with less and without nanoclay particles. Besides, better adhesion and strong bonding between layers are effective factors in the impact response in nanoclay-reinforced specimens. In terms of error percentage corresponding to the actual experimental results, the ANNs-LM verified its efficiency in response factor prediction.

As future direction, extreme learning machine considered as another machine learning approach can be introduced for validation and estimation purposes. Besides, other mechanical features of the studied FML composite-based nanoparticles will be investigated along with other types of nano-fillers as future direction of this paper.

References

1. Kommann X, Lindberg H, Berglund LA (2001) Synthesis of epoxy-clay nanocomposites: influence of the nature of the clay on structure. Polymer 42(10):4493–4499
2. Liu W, Hoa SV, Pugh M (2005) Organoclay-modified high performance epoxy nanocomposites. Compos Sci Technol 65:307–316
3. Isik I, Yilmazer U, Bayram G (2003) Impact modified epoxy/montmorillonite nanocomposites: synthesis and characterization. Ploymer 44:6371–6377
4. Zilg C, Mulhaupt R, Finter J (1999) Morphology and toughness/stiffness balance of nanocomposites based upon anhydride-cured epoxy resins and layered silicates. Chem Phys 200:661–670
5. Zhu R, Pan E, Roy AK (2007) Molecular dynamics study of the stress-strain behavior of carbon-nanotube reinforced Epon 862 composites. Mater Sci Eng A 447(1–2):51–57
6. Bagheri R, Takieh Marouf B (2007) Studies on the mechanical behavior of epoxy-clay nanocomposite. Iran J Polym Sci Technol 20:59–64 (in Persian)
7. Avila AF, Soares MI, Silva NA (2007) A study on nanostructured laminated plates behavior under low-velocity impact loadings. Int J Impact Eng 34(1):28–41
8. Fan J, Cantwell WJ, Guan Z (2011) The low-velocity impact response of fiber-metal laminates. J Reinf Plast Compos 30(1):26–35
9. Tsartsaris N, Meo M, Dolce F, Polimeno U, Guida M, Marula F (2011) Low-velocity impact behavior of fiber metal laminates. J Compos Mater 45(7):803–814
10. Sadighi M, Pamanen T, Alderliesten R, Sayeaftabi M, Benedictus R (2012) Experimental and numerical investigation of metal type and thickness effects on the impact resistance of fiber metal laminates. Appl Compos Mater 19:545–559
11. Lin JC (2008) Investigation of various silica-reinforced polymeric matrix nanocomposites. Compos Struct 84:125–131
12. Zamani MM, Fereidoon A, Sabet A (2012) Multi-walled carbon nanotube-filled polypropylene nanocomposites: high velocity impact response and mechanical properties. Iran Polym J 21:887–894
13. Hassoun MH (1995) Fundamentals of artificial neural networks. MIT Press, Cambridge
14. Sha W, Edwards KL (2007) The use of artificial neural networks in materials science based research. Mater Des 28(6):1747–1752
15. Rosenblatt F (1958) The perceptron: a probabilistic model for information storage and organization in the brain. Psychol Rev 65(6):386–408
16. Rumelhart D, Hinton G, Williams R (1986) Learning representations by back-propagating errors. Nature 323:533–536
17. Khanlou HM, Sadollah A, Ang BC, Kim JH, Talebianm S, Ghadimi A (2014) Prediction and optimization of electrospinning parameters for polymethyl methacrylate nanofiber fabrication using response surface methodology and artificial neural networks. Neural Comput Appl 25:767–777
18. Yu H, Wilamowski BM (2011) Levenberg-Marquardt training, 2nd edn. Industrial Electronics Handbook

Development of Deep Learning-based Self-adaptive Harmony Search

Taewook Kim, Hyeon Woo Jung, and Joong Hoon Kim

Abstract Nature-inspired optimization algorithms are widely used in various mathematical and engineering problems because of their usability and applicability. However, these optimization algorithms show different performances depending on the characteristics of the problem applied. There have been various effort to solve this problem by developing a new algorithm, applying other heuristics, changing parameters, etc. The deep learning-based self-adaptive harmony search (DLSaHS) developed in this study is another effort to tackle the problem by controlling the probability of heuristics by using recurrent neural network (RNN) and the parameter called checkpoint (CP). DLSaHS contains the heuristics obtained from harmony search (HS), genetic algorithm (GA), particle swarm optimization (PSO), and copycat harmony search (CcHS). DLSaHS was applied to the ten mathematical benchmark problems obtained from IEEE CEC 2021. The performance of DLSaHS is compared to the HS which showed better performance. Also, the optimal CP value obtained when applied to low-dimensional problems and the probability of heuristics according to the CP were derived to efficiently apply the DLSaHS, and it is confirmed that the error and standard deviation of the result, and computation time can be considerably reduced by applying them to high-dimensional problems.

Keywords DLSaHS · Probability of heuristics · RNN · Deep learning-based · CEC 2021 benchmark functions

T. Kim · H. W. Jung
Department of Civil, Environmental and Architectural Engineering, Korea University, Seoul, South Korea

J. H. Kim (✉)
School of Civil, Environmental and Architectural Engineering, Korea University, Seoul, South Korea
e-mail: jaykim@korea.ac.kr

J. H. Kim et al. (eds.), *Proceedings of 7th International Conference on Harmony Search, Soft Computing and Applications*, Lecture Notes on Data Engineering and Communications Technologies 140, https://doi.org/10.1007/978-981-19-2948-9_33

1 Introduction

Optimization is an essential process for the best efficiency in a variety of fields, such as minimizing construction costs or maximizing performance. To solve the optimization problem, various methods have been applied, but in the modern era when computer technology has developed, meta-heuristic algorithms are being widely used. Among them, a meta-heuristic algorithm is an algorithm that finds an optimal solution by imitating natural phenomena, such as Genetic Algorithm (GA; Holland [5]), Harmony Search (HS; Geem et al. [4]), Particle Swarm Optimization (PSO; Eberhart and Kennedy [3]), and Differential Evolution (DE; Storn and Price [7]). Meta-heuristic algorithms were successfully solved various optimization problems in various fields such as mathematics, economics, and engineering. However, the No Free Lunch (NFL) theory can be applied to all algorithms in that one algorithm cannot show the best performance for a wide range of problems characteristics. Wolpert and Macready [8] showed that there was no outstanding algorithm for all problems through geometric interpretation.

Rice [6] has conducted a study on the methodology of selecting the appropriate optimization algorithm according to the problem to be solved. The research to select an optimization algorithm that shows the best performance according to the problem characteristics is proceeding actively. The reason why the performance of one algorithm varies according to the characteristics of the problem can be seen as the difference by the structure or heuristic of each algorithm. Cowling et al. [2] introduced the concept of a hyper-heuristic algorithm with the expression "heuristics to choose heuristics". Hyper-heuristic can be divided into high-level and low-level. A high-level search strategy is to select/generate heuristics that exist at a low-level at every iteration, and lower-level creates new decision variables of the problem. Chaurasia and Kim [1] applied the hyper-heuristic algorithm to the set packing problem (SPP) and the minimum weight dominating set (MWDS) problem to verify the performance. The proposed algorithm improves the performance of the algorithm by updating the probability to choose a low-level heuristic.

The performance of the optimization algorithm can be improved by selecting heuristic properly. In this study, deep learning-based self-adaptive harmony search (DLSaHS) selects heuristic of other optimization algorithms that perform well for various fields using recurrent neural network (RNN), one of the deep learning techniques. The reason for using RNN is that RNN has a cyclic structure and is effective in processing data that changes according to time series (which can be considered as a number of function evaluations). Therefore, the checkpoint (CP) is set during the process of searching for the solution, and RNN is used every time the CP is reached, and the probability to choose the heuristics in DLSaHS is newly calculated through the learned RNN so that effective operators can be used. The developed optimization algorithm was applied to the mathematical benchmark function provided in the Congress on Evolutionary Computation (CEC) 2021 and showed superior performance than the existing HS-based algorithms.

2 Deep Learning-Based Self-adaptive Harmony Search

Deep learning-based self-adaptive harmony search (DLSaHS) was developed by combining heuristics of various meta-heuristic algorithms that are used in the world. There is 6 heuristics in DLSaHS that improvise a new solution, harmony memory consideration (HMC), crossover (CO), equation of particle swarm optimization (PSO), pitch adjusting (PA), copycat (Cc), and random search (RS). It has 3 major processes, such as initialization, improvising, and deep learning. Figure 1 shows the flowchart of DLSaHS.

2.1 *Initialization*

At the initialization phase, the parameter is set (harmony memory size, velocity vector, coefficient for PSO, and the number of heuristics, checkpoint). The probability set to choose the heuristic is defined as P_H, where H is a heuristic (the maximum value of H is 6). It is also part of the parameter, but at the initial phase every P_H has uniform value ($P_1 = 0.1667$, $P_2 = 0.1667$, ..., $P_6 = 0.1667$). Using these P_H, DLSaHS improvise a new memory until the first checkpoint (CP).

2.2 *Improvising*

Every iteration, DLSaHS chooses a heuristic to generate new memory using roulette wheel selection (RWS). After one memory vector is generated, it compares with the

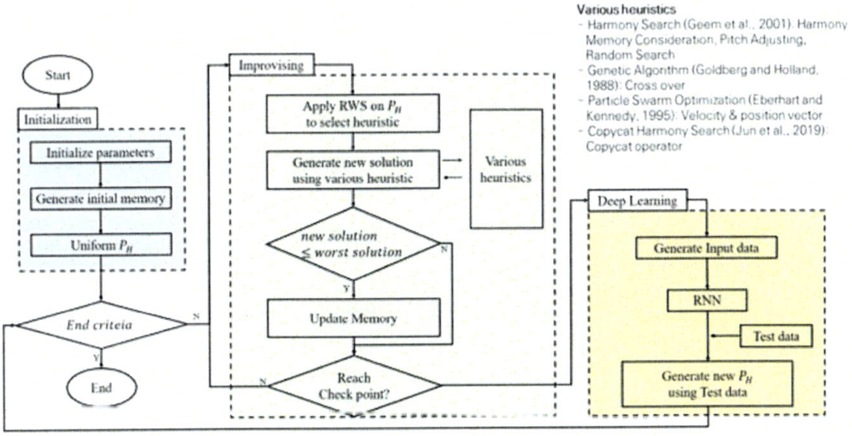

Fig. 1 Flowchart of deep learning-based self-adaptive harmony search

worst solution in the existing memory. If the new memory is better than the worst one, the memory will be updated. If the new memory is updated, it will be labeled as 1 (if not, it will be labeled as 0), which will be the target data to train the RNN.

The improvising phase will be continued until the next CP. CP is one of the parameters, which leads to the deep learning phase when iteration reaches the CP.

2.3 Deep Learning

In the deep learning phase, it generates the new P_H. In DLSaHS, RNN is used. The input data consist of 2 parts. One is the vector of heuristics used in 1 iteration, and the other is whether the solution has developed when the vector is used. The latter is used as target data. By using these input data, RNN is trained. After RNN is trained, random test data are generated and tested on RNN. If one of the test data vectors gets 1 as target data, it will be stored, and if not, it will be discarded. When a certain level of test data is gathered, a new P_H is calculated through this test data set.

3 Mathematical Benchmark Problem

All benchmark functions are aimed at finding the minimum value and consist of one unimodal, three basics, three hybrid, and three composition functions (Table 1). The calculated error of the benchmark functions with the given four kinds of characteristics is close to zero, and it is judged that the performance of the algorithm is excellent. The error means the difference between the solution obtained through an algorithm and the optimal solution of the function, and if the calculated error is smaller than 10^{-8}, it is considered to be 0. The search range is $[-100, 100]^D$, and the dimension (D) is 10 and 30. The results are mean, best, worst, median, and standard deviation (Std.) of error for 30 runs.

4 Application Results

DLSaHS was applied to the CEC 2021 problem and presented the results of the basic part; the results of the calculated error for 30 runs were presented.

Table 2 shows the result of solving the 10-D function. It found the optimal solution for all functions. The developed algorithm worked properly. Table 3 was derived by applying it to 30-D function. Even though the D of the problem has increased, the algorithm derived a close result to the optimal solution. As the D of the problem increases, the complexity of the function increases, and the Std. increases accordingly. The developed algorithm derives a new P_H for each CP, which is one parameter and confirms that the performance of the algorithm varies according to the function.

Table 1 Summary of the benchmark function of IEEE CEC 2021

	No.	Functions
Unimodal function	1	Shifted and rotated Bent Cigar function
Basic functions	2	Shifted and rotated Schwefel's function
	3	Shifted and rotated Lunacek bi-Rastrigin function
	4	Expanded Rosenbrock's plus Griewangk's function
Hybrid functions	5	Hybrid function 1 (modified Schwefel, Rastrigin, high-conditioned elliptic function)
	6	Hybrid function 2 (expanded Schaffer, HGBat, Rosenbrock, modified Schwefel's function)
	7	Hybrid function 3 (expanded Schaffer, HGBat, Rosenbrock, modified Schwefel, high-conditioned elliptic function)
Composition functions	8	Composition functions 1 (Rastrigin, Griewank, modified Schwefel's function)
	9	Composition functions 2 (Ackley, high-conditioned elliptic, Griewank, Rastrigin's function)
	10	Composition functions 3 (Rastrigin, Happycat, Ackley, Discus, Rosenbrock's function)

Table 2 Results for 10-D functions (basic)

Func.	Best	Worst	Median	Mean	Std.	CP
1	0	0	0	0	0	5
2	3.55.E−01	7.14.E+01	1.13.E+02	4.15.E+02	1.08.E+02	5
3	8.21.E−02	1.19.E+01	1.13.E+01	2.50.E+01	6.48.E+00	5
4	1.18.E−01	7.31.E−01	9.20.E−01	1.85.E+00	5.32.E−01	5
5	0	0	4.07.E−02	1.48.E−01	5.54.E−02	10
6	4.24.E−05	1.64.E−02	5.03.E−02	2.77.E−01	7.27.E−02	2
7	2.34.E−06	6.28.E−03	1.02.E−01	6.86.E−01	1.52.E−01	2
8	0	0	0	0	0	5
9	0	0	0	0	0	10
10	2.57.E−01	2.55.E+01	2.39.E+01	4.78.E+01	1.46.E+01	2

The results presented in Tables 2 and 3 summarize the results of the best performance by changing the value of CP to 2, 5, and 10. The best performance for two different dimension problems showed that optimal CP is the same. It indicates that to derive the best performance by applying DLSaHS, it can derive optimal CP for low-dimensional problems, and efficient results can be obtained by increasing the dimension of problems.

The advantage of the developed algorithm is that it can collect and utilize P_H that showed the best performance. By using the known P_H, the efficiency of computation time was secured by reducing the time used to learn RNN in the deep learning phase,

Table 3 Results for 30-D functions (basic)

Func.	Best	Worst	Median	Mean	Std.	CP
1	4.00.E+02	1.83.E+10	9.65.E+03	3.13.E+07	4.68.E+07	5
2	3.60.E+01	1.85.E+04	6.61.E+03	3.47.E+03	5.83.E+03	5
3	7.84.E+00	2.01.E+03	1.89.E+03	1.83.E+03	9.62.E+02	5
4	4.49.E−05	1.41.E+01	6.24.E+00	5.98.E+00	7.30.E+00	5
5	1.86.E+02	5.65.E+05	3.62.E+04	5.41.E+04	8.03.E+04	10
6	6.70.E+00	2.01.E+04	3.02.E+03	2.35.E+03	4.16.E+03	2
7	1.09.E+01	3.24.E+05	1.54.E+04	2.17.E+04	1.04.E+05	2
8	1.10.E+03	4.02.E+03	1.60.E+03	1.20.E+03	4.85.E+01	5
9	3.99.E+02	9.27.E+03	2.37.E+03	3.22.E+03	3.82.E+02	10
10	7.38.E+02	3.60.E+03	2.38.E+03	1.34.E+03	1.89.E+02	2

and the performance has also been improved. Table 4 is the result of collecting the case which had obtained the good result among the P_H obtained before, and deriving the average value and applying the newly obtained P_H to the 30-D problem. As seen, the best value has decreased even though it is the same 30-D problem, and Std. has also been significantly reduced.

The final results of the developed algorithm (Table 4) and the results obtained from HS (Table 5) were compared. The results showed that the DLSaHS was more efficient than the HS. The best and worst error was small for all functions, and Std showed a smaller value which meant DLSaHS showed stable performance.

Table 4 Results for 30-D functions (basic)—using known heuristic probability

Func.	Best	Worst	Median	Mean	Std.
1	6.66.E+01	5.39.E+03	6.43.E+02	1.16.E+03	1.56.E+03
2	3.60.E+00	7.71.E+02	2.36.E+02	2.48.E+02	1.88.E+02
3	1.12.E+00	8.73.E+01	7.57.E+01	7.04.E+01	2.29.E+01
4	4.49.E−06	4.14.E−01	2.23.E−01	2.39.E−01	1.49.E−01
5	2.07.E+01	1.13.E+04	2.01.E+03	3.38.E+03	3.21.E+03
6	3.35.E+00	4.56.E+02	1.04.E+02	1.96.E+02	1.54.E+02
7	5.43.E+00	1.35.E+04	5.50.E+02	1.67.E+03	2.73.E+03
8	1.00.E+02	1.03.E+02	1.00.E+02	1.00.E+02	1.01.E+00
9	1.33.E+02	2.06.E+02	1.58.E+02	1.61.E+02	1.66.E+01
10	8.20.E+01	1.24.E+02	8.81.E+01	8.90.E+01	6.74.E+00

Table 5 Results for 30-D functions (basic)—harmony search

Func.	Best	Worst	Median	Mean	Std.
1	2.37.E+05	6.14.E+05	4.70.E+05	4.51.E+05	1.02.E+05
2	6.03.E+02	2.88.E+03	1.52.E+03	1.52.E+03	5.96.E+02
3	7.26.E+01	1.41.E+02	9.78.E+01	9.72.E+01	1.46.E+01
4	9.51.E−01	1.96.E+00	1.36.E+00	1.40.E+00	2.20.E−01
5	6.78.E+04	7.39.E+05	2.88.E+05	3.06.E+05	1.84.E+05
6	3.36.E+01	1.29.E+03	6.11.E+02	6.00.E+02	2.66.E+02
7	3.52.E+04	8.34.E+05	2.44.E+05	2.97.E+05	1.93.E+05
8	1.06.E+02	6.19.E+03	1.51.E+03	1.59.E+03	1.94.E+03
9	4.44.E+02	6.20.E+02	5.96.E+02	5.77.E+02	5.33.E+01
10	3.84.E+02	3.89.E+02	3.87.E+02	3.87.E+02	7.71.E−01

5 Conclusions

In this study, DLSaHS was developed to solve the optimization problem by calculating the probability of heuristic according to the characteristics of the optimization problem applied by using deep learning. The developed algorithm improvises the new memory through the equal probability at the beginning but redefines P_H every time it reaches the newly introduced parameter CP. It stores all data where memory is updated until it reaches the CP. The data stored are what heuristic had created a new solution. Increasing the probability of heuristics that helps to create solutions close to the optimal solution will ultimately help to find the optimal solution at the end of optimization. Redefining P_H will be repeated at each CP, and the probability of a heuristic that helps to derive a good solution at that point will be raised. Through this, DLSaHS shows excellent performance by controlling the probability of heuristics that are possessed according to the characteristics of the optimization problem.

DLSaHS was applied to the 10 functions provided by IEEE CEC 2021. The dimension of the problem was increased to 10 and 30. The value of CP, which was the best result of DLSaHS, was not affected by the dimension of the problems. Therefore, it was confirmed that the efficient use of DLSaHS was to solve the high-dimensional problem after finding the optimal CP value by applying it to the low-dimensional problem first. In addition, the algorithm was applied to find the optimal solution using known P_H. The result showed that (1) computation time was greatly reduced, (2) error value was reduced, and (3) Std. was reduced by 30% on average.

DLSaHS is expected to apply to a wide range of problems by analyzing and adding heuristics of various algorithms in future. Therefore, for the future study, various algorithms that have shown good performance in the past or newly developed will be analyzed and benchmarked to improve DLSaHS. In addition, DLSaHS will be applied to engineering problems to verify performance.

References

1. Chaurasia SN, Kim JH (2019) An evolutionary algorithm based hyper-heuristic framework for the set packing problem. Inf Sci 505:1–31
2. Cowling P, Kendall G, Soubeiga E (2001) A hyperheuristic approach for scheduling a sales summit. In: Selected papers of the third international conference on the practice and theory of automated timetabling, PATAT 2000. Lecture notes in computer science
3. Eberhart R, Kennedy J (1995) A new optimizer using particle swarm theory. In: MHS'95. Proceedings of the sixth international symposium on micro machine and human science. IEEE, pp 39–43
4. Geem ZW, Kim JH, Loganathan GV (2001) A new heuristic optimization algorithm: harmony search. Simulation 76(2):60–68
5. Holland J (1975) Adaptation in natural and artificial systems: an introductory analysis with application to biology. In: Control and artificial intelligence
6. Rice JR (1976) The algorithm selection problem. In: Advances in computers, vol 15. Elsevier, pp 65–118
7. Storn R, Price K (1997) Differential evolution—a simple and efficient heuristic for global optimization over continuous spaces. J Global Optim 11(4):341–359
8. Wolpert DH, Macready WG (1997) No free lunch theorems for optimization. IEEE Transactions on Evolutionary Computation 1(1):67–82 https://doi.org/10.1109/4235.585893

Optimal Meter Placements Based on Multiple Data-Driven Statistical Methods for Effective Pipe Burst Detection in Water Distribution System

Sehyeong Kim and Donghwi Jung

Abstract Pipe burst events in the water distribution system (WDS) should be rapidly detected to secure the system functionality and reliability. Effective and efficient pipe burst detection (PBD) techniques can reduce water losses and prevent potential secondary failure (e.g., sinkhole) caused by ground liquefaction. This paper proposed an optimal meter placement method for PBD based on six data-driven statistical process control (SPC) methods. Total five meter placements are deduced by the optimization process, harmony search (HS) algorithm. They consist of two heterogeneous meters measuring different hydraulic quality characteristic values: pressure and flow rate. The optimal meter sets sought were compared with respect to two detection effectiveness indicators: true detection probability and false detection probability. The final optimal meter set should be generally effective for all utilized methodologies, therefore, the objective function is generated with those indicators and weights of them to reflect their effectiveness fairly. Three sets of synthetic time series data were produced from a network hydraulic EPANET model of Austin in Texas to demonstrate the proposed method: historical, normal, and abnormal data.

Keywords Harmony search algorithm · Water distribution system · Optimal meter placements · Pipe burst detection · Statistical process control method

1 Introduction

Effectively detecting and responding to pipe burst events in water distribution systems (WDS) is important because they reduce water pressure, which could directly affect consumers' water use, and cause secondary damage, such as sinkholes. Supervisory

S. Kim
Department of Civil, Environmental and Architectural Engineering, Korea University, Anam-ro 145, Seongbuk-gu, Seoul 02841, South Korea

D. Jung (✉)
School of Civil, Environmental and Architectural Engineering, Korea University, Anam-ro 145, Seongbuk-gu, Seoul 02841, South Korea
e-mail: sunnyjung625@korea.ac.kr

J. H. Kim et al. (eds.), *Proceedings of 7th International Conference on Harmony Search, Soft Computing and Applications*, Lecture Notes on Data Engineering and Communications Technologies 140, https://doi.org/10.1007/978-981-19-2948-9_34

control and data acquisition (SCADA) system can observe the real-time change of pressure and flow rates caused by pipe bursts. SCADA-based WDS anomaly detection techniques include artificial neural networks [9–11], state estimation [1, 17, 18], Bayesian approach [6, 13], time series modeling [3, 14], and statistical process control (SPC) methods [7, 8, 12, 15, 16] (Jung et al. 2013) have been studied in some works of literature.

Although some apparently large bursts are commonly detected by most methods, various detection techniques are specialized to detect different magnitude or durations of burst events, respectively, based on their different detection mechanisms. For example, one method can easily capture a sudden abrupt signal while the other method is specialized in detecting a long-lasting minor shift in pressure and pipe flow. Therefore, simultaneously applying multiple detection techniques with various detection characteristics would be beneficial in significantly improving overall detection effectiveness. In addition, PBD methods are affected by not only the data-analyzing mechanisms but also the structure of the distribution network. Definitely, the core of the data-driven methodologies is information from the obtained data. In other words, the number of meters is generally limited due to various (especially economic) reasons, and how to place them has a crucial impact on the detection effectiveness. However, little effort has been devoted to the determination of optimal meter placements using multiple detection methodologies simultaneously.

Jung et al. [7] compared the detectability of six different SPC methods for PBD concerning their detectability indicators and detection time. Three univariate methods including the Western Electric Company (WEC) rules, the cumulative sum (CUSUM), and the exponentially weighted moving average (EWMA), and three multivariate methods including the Hotelling T^2 control chart, the multivariate CUSUM, and the multivariate EWMA methods were utilized and compared. Synthetic hydraulic quality characteristic data of the modified Austin network [2] were generated with EPANET. The EWMA and CUSUM methods were the best in the aspect of detection efficiency and effectiveness. However, up to five flow rate and pressure meters were assumed to be placed at predefined locations, and the impact of the meter placements was not studied or compared.

Hagos et al. [5] determined the optimal meter placement for PBD in WDS using WEC rules and binary integer programming model. Shewhart control charts were developed for the hydraulic network model, and WEC rules were applied to produce false alarm and detection matrices. Then, the binary integer optimization algorithm optimized the meter placements which maximizes the detection probability or minimize the rate of false alarm. With the objective function which maximizing the detection probability, pressure meters were located at the nodes that far from the source while flow meters were placed near to the source. On the contrary, the result from the objective function minimizing the false alarm rate was the opposite. This study also had the limitation that the optimization process was applied for only one SPC method. Also, the objective function was separately focused on the detection probability and the rate of false alarm, not considering them simultaneously as a single detection effectiveness indicator.

Therefore, this paper proposes a method to determine optimal meter locations for multiple data-driven SPC methods for effective burst detection. The synthetic data of Austin network, Texas will be generated for historical and two types of test data (normal and abnormal), considering the three different types of user-types: residential, commercial, and industrial. Then, SPC methods establish their control limits using historical data and determine whether the test data are out of the limit. True detection probability (TDP) and false detection probability (FDP) are calculated with the abnormal (burst) and normal (no burst) test data, respectively. Harmony search (HS) algorithm is used as the optimization algorithm, and the decision variables consist of five placements and the hyperparameters of some SPC methods. Its objective function is calculated with the combination of the positive TDP, negative FDP, and weight values. Weights reflect the individual proportion of each indicator in total summated value, to consider the actual reliability. Six SPC methods that were used in Jung et al. [7] are utilized for this optimization process, and the finally deduced meter set will reflect their performances and are generally effective for all of them.

2 Methodology

2.1 Statistical Process Control (SPC) Methods

Statistical process control method is one of the most important management techniques, especially in manufacturing industries. It directly means that the methodology to maintain the production processes by monitoring and controlling the quality characteristics, with the assistance of enough data and statistical preprocessing. In the case of the WDS, its production process and quality characteristics are the same as the system functionality (i.e., water supply) and the hydraulic measures such as pressure or flow rate. Quantitatively, the common purpose of various SPC methods is to decide whether the quality of the system is within the control limit or fluctuates. Generally, poor system functionality causes mean-shifted characteristic values, and the control limits of the SPC methods established based on the data average can detect anomalies. In this study, a total of six SPC methods that consisted of three univariate and the three multivariate methods are materials of this optimal meter placement problem. All data should be normalized before they are applied to SPC methods.

As mentioned above, three univariate methods that use a single sort of measurement are utilized to detect the pipe burst events: WEC rules, CUSUM, and EWMA. The first univariate method is WEC rules, which was developed in 1958 by the Western Electric Company. Based on the Shewhart control chart, four rules examine the different number of single or consecutive scores with different criteria, whether they are out of their control limits. There are two control limits in the upper and lower side relative to the centerline (i.e., mean), and all rules are one-side rules. It means that if the rule requires two or more deviated consecutive scores to account for the data as a non-random anomaly, those scores should exist outside the control limit-band in the same direction. Next, the CUSUM control chart method directly includes the past data. Two user-set hyperparameters are used to preprocess the data and set the control limit. Preprocessed and normalized data are conditionally accumulated over time. If the accumulated score exceeds the user-set control limit, it is considered as the quality characteristic data from the abnormal circumstance. Lastly, the EWMA control chart method also has two user-set hyperparameters similar to CUSUM, however, it utilizes past and present scores. EWMA control limits change with the scores' moving average over time. Thus, unlike the WEC rules, the detection effectiveness of CUSUM and EWMA can be very different according to the appropriateness of those parameters. Therefore, they must be estimated and considered as decision variables to obtain fine detection results.

Then, the three multivariate SPC methods that analyze two or more sorts of measurement are also applied to the WDS hydraulic measurements: Hotelling T^2 control chart, multi-CUSUM, and multi-EWMA. Those multivariate methods including the Hotelling T^2 control chart are based on the concept of the Mahalanobis distance. It is the standardized distance between the multiple variables that have interrelationships to each other. Unlike the three univariate methods mentioned above, multivariate methods' control limits only exist as positive values because the Mahalanobis distance is non-negative. The Hotelling T^2 method issues the burst alarm when the Mahalanobis distance that transformed from the normalized quality characteristic measurement exceeds the control limit. In this case, the control limit is estimated by the chi-square distribution, which is related to the confidence level and the length of the historical record for each measurement. Multi-CUSUM and multi-EWMA methods are the multivariate versions of CUSUM and EWMA, using hyperparameters similar to their univariate version.

2.2 Detection Effectiveness Indicators

The effectiveness of the burst detection methodologies is related to how accurate the method separates the true burst events and random fluctuation patterns. It can be evaluated with two aspects: true detection probability (TDP) and false detection probability (FDP). TDP means the proportion of correctly detected burst events to total burst events. On the other hand, FDP means the proportion of incorrectly detected normal events to total normal events. Considering that the above two effectiveness indicators increase only when the burst alarm is issued, each indicator is the same as the true positive probability and false-positive probability, which are related to Type II error and Type I error. As a result, the high effectiveness of the detection method means it has high TDP and low FDP.

2.3 *Optimization: Harmony Search Algorithm (HSA)*

The HS is an optimization algorithm inspired by the musical improvisation process where the musicians try to find a better harmony. HS generally stores chords (i.e., the combination of decision variables) found in the training sequence in a storage called harmony memory (HM). If the new chord generated for each iteration is better than the worst chord stored in the existing HM, the new chord is added, and the worst chord is removed. A new solution that occurs at each iteration can be generated by an HM consideration operator or a random search operator, with the probability of harmony memory considering rate (HMCR) generating the former and the latter with an extra probability. In the former case, when creating a new chord, it is based on the chord in HM, and in the latter case, it is randomly generated in the new area. Therefore, the random search operator maintains the globality of the search by preventing the algorithm from falling into the local solutions that may occur as the iterative calculation continues, and the harmony memory operator maintains the locality of the search by increasing the probability of finding nearby optimal solutions through a combination of excellent solutions. Meanwhile, the pitch adjusting rate (PAR) is another local explorer considering values adjacent to the existing solution to find fine solutions such as fine adjustment of the pitch of each instrument to make a harmonic chord, whenever the HM consideration operator is applied. In addition, securing an appropriate balance between global and local exploration through the stochastic application of the operators can be said to be the largest advantage of the HS.

3 Study Network and Data Generation

Here, all SPC methods are applied to the hydraulic model of the study network, in order to obtain the detection effectiveness indicators. Austin network (Fig. 1) is used for the application of the optimal meter placements method. Austin network is a small-scale WDS consisted of a total of 127 nodes, 156 pipes connected with a single reservoir, with several looped internal subareas and branched distal ends. The total average demand of the system is a total of 3041 LPM, and it can be assumed that there are three types of customers using the water with each different their own diurnal pattern. Therefore, three types (i.e., residential, commercial, and industrial) of demand patterns are considered to generate the synthetic hydraulic data from EPANET.

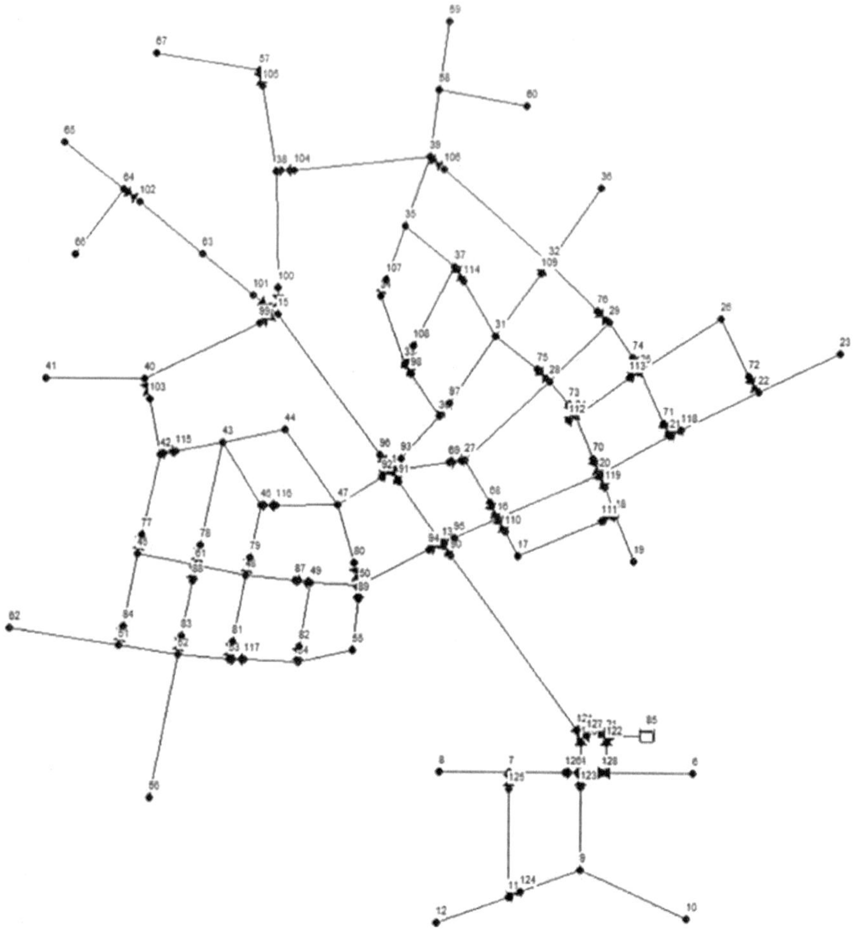

Fig. 1 Layout of the Austin network obtained by EPANET

4 Application Results

To determine the five locations of optimal meters for high effectiveness for all six utilized SPC methods, 1000 sets of the hydraulic data were synthesized for normal and abnormal test data and also normal historical data, respectively. Each data set consists of a time series with 576-time steps (i.e., 5-min interval measurements for 2 days). Normal historical data are used to set the control limits of the SPC methods, and two types of test data are generated for the calculation of detection effectiveness indicators. The burst circumstances are simulated with the emitter function of EPANET. A positive integer will be assigned as an emitter coefficient to a single random node, in the range of 1–50, and we assume that each SPC method applies its

Table 1 Parameter conditions of the harmony search

Parameter	Value	Interval
HM size	100	–
# of iteration	100,000	–
HMCR	0.7–0.9	0.1
PAR	0.05–0.20	0.05
BW	1–3	1

detection rules from the occurrence time of burst. That is, there is no leading time of normal condition for methods to detect.

Then, HM in the HS basically includes the five meter placements where the hydraulic data are obtained from, and SPC methods will be applied to them. Additionally, four of the total six SPC methods have hyperparameters that are concerned with the data preprocessing and control limit setting. Therefore, the seven user-set hyperparameters are also included in the decision variables. The HM size is 100, to consider the not small number of the harmony size. Other parameters are represented in Table 1, including HMCR and PAR. Those two parameters are applied by three values, respectively, to estimate the finest HS conditions and optimal solutions. Lastly, the objective function (Eq. 1) is generated with the combination of TDP and FDP. With the basic form of summation of the positive TDP and negative FDP, additional weight is applied. It is because the functionality of the method from the final optimal meter sets should not only be evenly good for all SPC methods but also be effective itself. The weight value of each SPC method is calculated by considering its indicator's proportion to the total summation of all indicators. It amplifies the positive impact of larger indicators than others, and vice versa.

$$\text{Maximize} \quad \sum_i \left(\frac{\text{TDP}_i}{\sum \text{TDP}_i} \times \text{TDP}_i - \frac{\text{FDP}_i}{\sum \text{FDP}_i} \times \text{FDP}_i \right) \tag{1}$$

However, considering the number of notes in the chord, the total calculation load for this optimization process is significantly large. In other words, a total of six SPC methods have to be applied for every 1000 sets of normal and abnormal test data, which are consisted of 576-time steps, in a single iteration of HS. Hence, the critical components are deduced by the sensitivity matrix of the study network, and only they become the candidates of the decision variables. The sensitivity matrix is generated with the response of the components (i.e., nodes and pipes) according to a single pipe's openness status, for the nodal pressure [4]. The critical pressure resulted from the sensitivity matrix is represented in Table 2, and there are a total of 22 critical nodes and 16 critical pipes.

Finally, the optimal meter placements are deduced within the conditions mentioned above. The best result is shown from the HS condition with the HMCR of 0.9, PAR of 0.15, and the bandwidth of 2. The objective function value is around 82.37, where the mean total TDP is 82.56% and that of FDP is 1.32%. The optimal

Table 2 Critical components deduced by sensitivity analysis based on single pipe failure events

Component	ID number
Node	10, 12, 21, 23, 26, 28, 29, 32, 34, 36, 37, 39, 43, 45, 48, 51, 56, 59, 60, 61, 65, 67
Pipe	1, 6, 8, 11, 13, 14, 15, 16, 32, 51, 73, 74, 75, 82, 85, 86

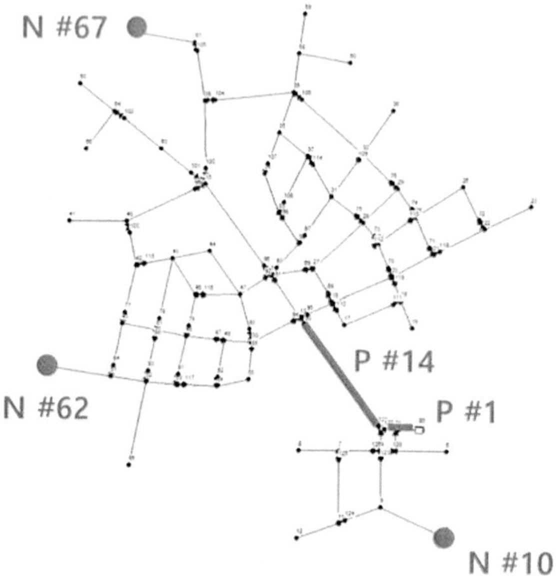

Fig. 2 Five locations optimal meter placements resulted from the HS algorithm

solutions including the meter placements and the SPC hyperparameters are represented in Fig. 2. The data from the three nodes (10, 62, and 67) for pressure meters and two pipes (1 and 14) for flow rate meters provide the most appropriate information for effective PBD. There are some characteristics with the optimal components. The three nodes are located at the end of the network's branches, while they are not concentrated in a specific location. Reversely, two pipes are the mainline located near the reservoir. Pipe 1 is directly connected to the reservoir, and pipe 14 is the only upward mainline that is responsible for the most proportion of the water supply.

5 Conclusions

PBD effectiveness is related to the size of the burst event and the disposition of the hydraulic measuring meters. Thus, if multiple meters can be installed in the system, it is important that identify their placements for effectual detection. This study suggested the model computing optimal meter placements for multiple SPC methods, which can obtain the high functionality of all methodologies. This determination model was applied to the synthetic data of the real network, considering three different demand patterns in it, and solved the optimal placements for five hydraulic meters. Total six SPC methods were utilized including three univariate and three multivariate methods. HS algorithm is used to find the best meter placements and the SPC hyperparameters, which can maximize the TDP and minimize the FDP for all methods. The objective function of this optimization process was consisted of the combination of positive TDP and negative FDP, with the weight value reflecting the proportion of their respective performance. Before the determination process, the sensitivity matrix generated by the manipulation of the single pipe failure circumstances extracted the critical components. They are utilized for the entire domain of decision variables, and the total calculation efficiency was significantly improved by this selection. As a result of the optimization, three nodes for pressure meters and two pipes for pipe flow rate meters and the SPC parameters for them were deduced as an optimal solution.

The optimal nodes were located at the endpoint of the network, which was spread far from each other. Differently, the optimal pipes existed at the initial part with respect to the reservoir, and they were responsible for most of the water supply. It means that they sensitively reflect the response of the burst event. If the pipe is ruptured in the WDS, the nodal pressure is decreased due to the crack, and the pipe's flow rate is increased to replenish the leaked water volume. In other words, the terminal nodes and the initial main pipes can reflect the non-random fluctuation of the quality characteristics better than any other components. Hence, assuming that the WDS is divided into upper and lower parts based on pipe 14, node 62, 67, and pipe 14 provide the burst information of the upper part, while node 10 gives that of the lower part. Pipe 1 is assumed to provide the information of both parts, due to its bidirectional connectivity.

The proposed optimization process also has some limitations. Firstly, the detection methods' performance can be more effective, in some cases, if they used a smaller number of meters. Thus, more cases with various cases of the number of meters should be studied and compared. Also, this paper did not investigate the probability of all WDS components due to the efficiency issue. The result of this method should be analyzed in the aspect of the ability to find the global optimum.

Acknowledgements This work was supported by the National Research Foundation of Korea (NRF) grant funded by the Korea government (MSIT) (No. 2020R1C1C1006481).

References

1. Andersen JH, Powell RS (2000) Implicit state-estimation technique for water network monitoring. Urban Water 2(2):123–130
2. Brion LM, Mays LW (1991) Methodology for optimal operation of pumping stations in water distribution systems. J Hydraul Eng 117(11):1551–1569

3. Christodoulou SE, Kourti E, Agathokleous A (2017) Waterloss detection in water distribution networks using wavelet change-point detection. Water Resour Manage 31(3):979–994
4. Diao K, Jung D, Farmani R, Fu G, Butler D, Lansey K (2021) Modular interdependency analysis for water distribution systems. Water Res, 117320
5. Hagos M, Jung D, Lansey KE (2016) Optimal meter placement for pipe burst detection in water distribution systems. J Hydroinf 18(4):741–756
6. Hutton C, Kapelan Z (2015) Real-time burst detection in water distribution systems using a Bayesian demand forecasting methodology. Proc Eng 119:13–18
7. Jung D, Kang D, Liu J, Lansey K (2015) Improving the rapidity of responses to pipe burst in water distribution systems: a comparison of statistical process control methods. J Hydroinf 17(2):307–328
8. Misiunas D, Vítkovský J, Olsson G, Lambert M, Simpson A (2006) Failure monitoring in water distribution networks. Water Sci Technol 53(4–5):503–511
9. Mounce SR, Day AJ, Wood AS, Khan A, Widdop PD, Machell J (2002) A neural network approach to burst detection. Water Sci Technol 45(4–5):237–246
10. Mounce SR, Khan A, Wood AS, Day AJ, Widdop PD, Machell J (2003) Sensor-fusion of hydraulic data for burst detection and location in a treated water distribution system. Inform Fusion 4(3):217–229
11. Mounce SR, Boxall JB, Machell J (2010) Development and verification of an online artificial intelligence system for detection of bursts and other abnormal flows. J Water Resour Plan Manag 136(3):309–318
12. Palau CV, Arregui FJ, Carlos M (2012) Burst detection in water networks using principal component analysis. J Water Resour Plan Manag 138(1):47–54
13. Poulakis Z, Valougeorgis D, Papadimitriou C (2003) Leakage detection in water pipe networks using a Bayesian probabilistic framework. Probab Eng Mech 18(4):315–327
14. Quevedo J, Puig V, Cembrano G, Blanch J, Aguilar J, Saporta D, Benito G, Hedo M, Molina A (2010) Validation and reconstruction of flow meter data in the Barcelona water distribution network. Control Eng Pract 18(6):640–651
15. Romano M, Kapelan Z, Savić DA (2010) Real-time leak detection in water distribution systems. In: Water distribution systems analysis, pp 1074–1082
16. Romano M, Kapelan Z, Savić DA (2014) Automated detection of pipe bursts and other events in water distribution systems. J Water Resour Plan Manag 140(4):457–467
17. Ye G, Fenner RA (2011) Kalman filtering of hydraulic measurements for burst detection in water distribution systems. J Pipeline Syst Eng Pract 2(1):14–22
18. Ye G, Fenner RA (2014) Weighted least squares with expectation-maximization algorithm for burst detection in UK water distribution systems. J Water Resour Plan Manag 140(4):417–424

Comparison of Classical and Adaptive Parameter Setting for Harmony Search on a Structural Optimization Problem

Ayla Ocak, Gebrail Bekdaş, and Sinan Melih Nigdeli

Abstract In this study, a truss system is optimized with the Harmony Search Algorithm (HS) inspired by the process of searching for the best harmony. It is compared with the Adaptive Harmony Search Algorithm whose initial values of parameters are defined and gradually decrease. The optimization process is aimed to optimize the volume by calculating the optimum volume of the elements of the truss system. In design, the effect of design factors specific to the harmony search algorithm on optimization was investigated. For this purpose, optimization was performed with different harmony memory consideration ratio (HMCR) values. By adapting the harmony search algorithm, optimum results are obtained for the values that gradually decrease according to the HMCR and fret width (FW) values. Optimum volume function with the lowest standard deviation values was obtained with adaptive harmony search optimization in the mean of the optimum value found for both optimizations.

Keywords Truss system · Harmony search · Adaptive harmony search · Structural engineering · Optimization

1 Introduction

One of the most important parameters in the construction of structures is the cost. Both the harmful environmental effects arising from the use of materials and the financial losses to be brought by the cost necessitate the optimization process. There are many optimization methods. Metaheuristic algorithms, which have been frequently used for this process in recent years, play an active role in solving various engineering

A. Ocak (✉) · G. Bekdaş · S. M. Nigdeli
Department of Civil Engineering, Istanbul University-Cerrahpaşa, 34320 Avcılar, İstanbul, Turkey
e-mail: aylaocak@outlook.com

G. Bekdaş
e-mail: bekdas@istanbul.edu.tr

S. M. Nigdeli
e-mail: melihnig@istanbul.edu.tr

J. H. Kim et al. (eds.), *Proceedings of 7th International Conference on Harmony Search, Soft Computing and Applications*, Lecture Notes on Data Engineering and Communications Technologies 140, https://doi.org/10.1007/978-981-19-2948-9_35

problems. In particular, various optimization studies have been carried out as a solution to the carbon footprint which has become a problem in today's world, which symbolizes cost optimization and environmental damage [1–6]. Metaheuristic algorithms, the effectiveness of which has been proven by many studies, have also been used in the design of lattice structures [7–10].

The harmony search algorithm (HS), one of the metaheuristic algorithms developed by Geem et al., is a preferred algorithm in the design of engineering structures [11]. It is an efficient algorithm that can be adapted because it contains a design factor. The effect of the HS algorithm on design optimization has been supported by many studies [12–16]. Optimization studies have been carried out in lattice systems using harmony search algorithm as well as adaptive harmony search and hybrids of this algorithm [17–22].

In this study, a truss system is optimized on the Matlab program for the most cost-effective with harmony search and adaptive harmony search algorithms by choosing different memory consideration rates [23]. The volume of the system is minimized and the volume is directly related with the cost. As a result of the research, the effects of two methods and assigning different HMCR values to the system on the optimum volume were investigated.

2 Methodology

2.1 Optimum Design of Truss Systems via a Modified Harmony Search Algorithm

The harmony search algorithm is a metaheuristic algorithm developed by Geem et al. in 2001, based on a sound-sounding harmony search process [11]. In the optimization process, any harmony is first memorized. For the design, the lower and upper limit values and the design factors of the algorithm such as harmony memory consideration ratio (HMCR) and fret width (FW) is defined in the system. As in Eq. (1), randomly assigned harmony vectors are created and stored in a matrix. The X_{new}, new solutions X_{max} and X_{min} used in Eq. (1) represent the maximum and minimum limit values, respectively.

$$X_{\text{new}} = X_{\min} + \text{rand}(1)(X_{\max} - X_{\min}) \quad \text{if HMCR} > r1 \tag{1}$$

Fret width (FW) is also used to create the harmony vector. In the harmony vector equation shown in Eq. (2), *n*. For harmony, a new harmony vector is created with random assignments between 0 and 1 values.

$$X_{\text{new}} = X_n + \text{rand}(1)\text{FW}(X_{\max} - X_{\min}) \quad \text{if HMCR} \le r1 \tag{2}$$

The HMCR value is effective in the selection of Eqs. (1) and (2) to create the harmony vector. The choice of the equation is decided according to the magnitude relationship between the new solutions and the HMCR. The $r1$ value used to compare is a randomly chosen value between 0 and 1. By repeating these operations for the number of iterations, the new harmony vector is written in the function to be minimized, compared with the old solutions, and the new solutions are updated. Thus, the solution matrix takes its final form.

The adaptive harmony search algorithm is obtained by gradually decreasing this value by giving the initial value of the HMCR and FW values ($\mathrm{HMCR_{in}}/\mathrm{FW_{in}}$). The equations defined to the system regarding this reduction are shown in Eqs. (3) and (4).

$$\mathrm{HMCR} = \mathrm{HMCR_{in}}\left(1 - \frac{t}{mt}\right) \tag{3}$$

$$\mathrm{FW} = \mathrm{FW_{in}}\left(1 - \frac{t}{mt}\right) \tag{4}$$

Here t indicates the number of iterations and mt indicates the maximum number of iterations.

3 The Numerical Example

In this research, optimization was made by taking a truss system as an example and optimizing the volume within the design constraints. An example of the truss system used in the optimization application is shown in Fig. 1.

The algorithm parameters defined for the optimization problem are shown in Table 1. Optimization design constraints are given in Eqs. (5), (6), and (7), and the lower and upper limits of the objective function and design variables are given in Eq. (8).

$$g_1 = \frac{\sqrt{2}A_1 + A_2}{\sqrt{2}A_1^2 + 2A_1A_2}P - \sigma \leq 0 \tag{5}$$

$$g_2 = \frac{A_2}{\sqrt{2}A_1^2 + 2A_1A_2}P - \sigma \leq 0 \tag{6}$$

$$g_3 = \frac{1}{A_1 + \sqrt{2}A_2}P - \sigma \leq 0 \tag{7}$$

$$\text{Minimize } f(A_1, A_2) = \left(2\sqrt{2}A_1 + A_2\right)L \quad 0 \leq A_1 \leq 1 \quad \text{and} \quad 0 \leq A_2 \leq 1 \tag{8}$$

P in the equations represents the load in the system, L represents the bar length, σ refers to the stress and A_1, A_2 refers to the bar cross-sectional areas.

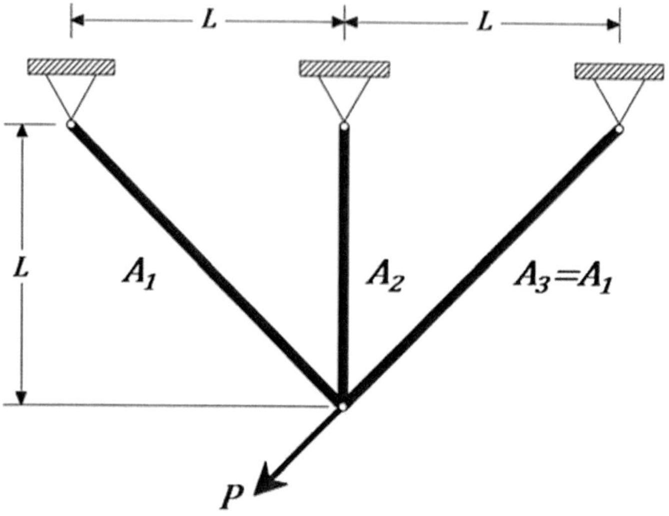

Fig. 1 Truss system model

Table 1 The algorithm parameters

Symbol	Definition	Value
HMCR	Harmony memory considering rate	0–1
$HMCR_{in}$	Initial harmony memory considering rate	0–1
FW	Fret width	0.05
FW_{in}	Initial fret width	0.05
mt	Maximum iteration number	10,000
t	Iteration number	1–10,000
HMS	Harmony memory size	15

For the optimization process, it was run 20 times with 10,000 iterations and optimum results were obtained. The mean values and standard deviations of the objective function optimized for different HMCR values with the classical harmony search and adaptive harmony search algorithms are shown in Table 2.

4 Conclusions

Metaheuristic algorithms are an optimization method that has a simple and understandable layout and has remarkable effects in reaching the optimum result. Among the types used in recent years, the harmony search algorithm has made it stand out among metaheuristic algorithms by giving even better results by modifying its design

Table 2 The optimum results

HMCR	HS		HS-adaptive	
	Mean (m^3)	Standard deviation	Mean (m^3)	Standard deviation
0.00	275.1361753	0.1140	274.4608522	0.0016
0.25	264.3549837	0.1377	264.0366638	0.0034
0.50	264.2417244	0.1403	263.9872296	0.0014
0.75	264.1635793	0.1717	263.9582158	0.0012
1.00	264.2087072	0.1824	263.9100925	0.0010

factors. The correct selection of the design parameters and adaptively lowering the FW and HMCR values from the initial values to 0, and the controls made at intermediate values, constituted a step toward improving the harmony search algorithm. In this study, for the cost (volume) minimization of an engineering problem selected as a truss system, the classical harmony search algorithm was optimized for 5 different HMCR values, the same values were assigned to the system as the initial value of the HMCR, and with the adaptive harmony algorithm was also optimized.

The results from the optimization are summarized as follows.

- The optimum volume value was obtained in the adaptive harmony search algorithm where the HMCR value was chosen as 1. It has been determined that the adaptive harmony search algorithm gives better results than the classical harmony search algorithm.
- The worst results were obtained when the HMCR value was 0 for both optimizations. From this situation, it was understood that disabling global optimization by taking 0 makes it difficult to obtain the optimum value.
- While an increasing graph is observed between the standard deviation values of the population means in classical harmony searching, there is a regular decreasing process in adaptive harmony searching. Adaptive harmony searching appears to generate a population with less standard deviation mean than the classical harmony search process.

When all the results are examined, it can be said that the adaptive harmony search algorithm gives better results than the classical harmony search in the HMCR values defined for the problem. It has been understood that the correct definition of design factors is effective in obtaining optimum results.

References

1. Mahjoubi S, Barhemat R, Guo P, Meng W, Bao Y (2021) Prediction and multi-objective optimization of mechanical, economical, and environmental properties for strain-hardening cementitious composites (SHCC) based on automated machine learning and metaheuristic algorithms. J Clean Product 329:129665

2. Ghasemof A, Mirtaheri M, Mohammadi RK, Mashayekhi MR (2021) Multi-objective optimal design of steel MRF buildings based on life-cycle cost using a swift algorithm. Structures 34:4041–4059
3. Yeates C, Schmidt-Hattenberger C, Weinzierl W, Bruhn D (2021) Heuristic methods for minimum-cost pipeline network design—a node valency transfer metaheuristic. Netw Spat Econ, 1–33
4. Temur R, Bekdaş G (2016) Teaching learning-based optimization for design of cantilever retaining walls. Struct Eng Mech 57(4):763–783
5. Serpik IN, Mironenko IV, Averchenkov VI (2016) Algorithm for evolutionary optimization of reinforced concrete frames subject to nonlinear material deformation. Proc Eng 150:1311–1316
6. Zhang J, Huang Y, Aslani F, Ma G, Nener B (2020) A hybrid intelligent system for designing optimal proportions of recycled aggregate concrete. J Clean Prod 273:122922
7. Bekdaş G, Yucel M, Nigdeli SM (2021) Evaluation of metaheuristic-based methods for optimization of truss structures via various algorithms and Lèvy flight modification. Buildings 11(2):49
8. Miguel LFF, Lopez RH, Miguel LFF (2013) Multimodal size, shape, and topology optimization of truss structures using the Firefly algorithm. Adv Eng Softw 56:23–37
9. Serpik IN, Tarasova NV (2020) Optimization of steel trusses with a choice of multi-stage prestressing conditions. Mag Civ Eng 96(5)
10. Roy S, Kundu CK (2020) State-of-the-art review on the use of optimization algorithms in steel truss. Int J Sci Technol Res 9(3):160–165
11. Geem ZW, Kim JH, Loganathan GV (2001) A new heuristic optimization algorithm: harmony search. Simulation 76(2):60–68
12. Siddique N, Adeli H (2015) Applications of harmony search algorithms in engineering. Int J Artif Intell Tools 24(06):1530002
13. Nigdeli SM, Bekdas G, Kim S, Geem ZW (2015) A novel harmony search based optimization of reinforced concrete biaxially loaded columns. Struct Eng Mech Int J 54(6):1097–1109
14. Cakiroglu C, Islam K, Bekdaş G, Kim S, Geem ZW (2021) Metaheuristic optimization of laminated composite plates with cut-outs. Coatings 11(10):1235
15. Arama ZA, Kayabekir AE, Bekdaş G, Kim S, Geem ZW (2021) The usage of the harmony search algorithm for the optimal design problem of reinforced concrete retaining walls. Appl Sci 11(3):1343
16. de Almeida FS (2019) Optimization of laminated composite structures using a harmony search algorithm. Compos Struct 221:110852
17. Degertekin SO, Minooei M, Santoro L, Trentadue B, Lamberti L (2021) Large-scale truss-sizing optimization with enhanced hybrid HS algorithm. Appl Sci 11(7):3270
18. Toklu YC, Bekdaş G, Yücel M, Nigdeli SM, Kayabekir AE, Kim S, Geem ZW (2021) Total potential optimization using metaheuristic algorithms for solving nonlinear plane strain systems. Appl Sci 11(7):3220
19. Talatahari S, Goodarzimehr V, Taghizadieh N (2020) Hybrid teaching-learning-based optimization and harmony search for optimum design of space trusses. J Optim Ind Eng 13(1):177–194
20. Cao H, Qian X, Zhou YL, Yang H (2018) Applicability of subspace harmony search hybrid with improved deb rule in optimizing trusses. J Comput Civ Eng 32(4):04018021
21. Artar M (2016) A comparative study on optimum design of multi-element truss structures. Steel Compos Struct 22(3):521–535
22. Kaveh A, Naeimi M (2015) Sizing optimization of skeletal structures with a multi-adaptive harmony search algorithm
23. The MathWorks, Matlab R2018a (2018) Natick, MA

Optimum Discrete Design of Steel Planar Trusses Comprising Earthquake Load Impact

Serdar Carbas and Musa Artar

Abstract In this study, the success of teaching–learning-based optimization (TLBO) and biogeography-based optimization (BBO) metaheuristic methods in optimum discrete sizing design of a steel planar truss comprising earthquake load impact has been investigated. To do this, a 46-element steel planar truss has been handled as a design example. Like many other stochastic optimization methods, the TLBO and BBO techniques imitate specific natural events. In TLBO, the processes are carried out by mimicking a class consisting of teachers and students; on the other hand, the BBO simulates the distribution of species in nature based on biodiversity. The stress and displacement constraints in American Institute of Steel Construction-Allowable Stress Design (AISC-ASD) provisions are considered as structural behavior constraints. Both algorithms select design profiles from a discrete list containing steel W-shaped sections. For obtaining the minimum weighted optimum structural design, the algorithms encoded in MATLAB are supplied with open application programming interface (OAPI) functions that enable mutual data transfer with a structural analysis software (SAP2000) to practically get the structural responses under the effect of load combinations containing earthquake load. The optimal truss designs yielded with TLBO and BBO algorithms are compared with those already existed in the literature. Accordingly, it has been concluded that the TLBO and BBO algorithms give successful solutions.

Keywords Structural design optimization · Steel planar trusses · Metaheuristic algorithms · OAPI · Earthquake load

S. Carbas (✉)
Karamanoglu Mehmetbey University, 70200 Karaman, Turkey
e-mail: scarbas@kmu.edu.tr

M. Artar
Bayburt University, 69000 Bayburt, Turkey

J. H. Kim et al. (eds.), *Proceedings of 7th International Conference on Harmony Search, Soft Computing and Applications*, Lecture Notes on Data Engineering and Communications Technologies 140, https://doi.org/10.1007/978-981-19-2948-9_36

1 Introduction

Recently, the optimum sizing design of steel structures has been a highly trending research subject in structural engineering. The increasing importance of steel materials used at construction sites in developing world is the main actor in this. For this reason, in recent studies, it is aimed to ensure that all required structural behavior constraints such as stress, displacement, and geometric in optimally sized structural systems comply with the practice code specifications, so that a structure acquires the minimum design weight [1]. Advances in computer technology have allowed the practical applicability of various algorithmic techniques to reach optimum designs of the structures. These so-called nature-inspired metaheuristic algorithms are generally non-deterministic methods that mimic natural events and can be applied to discrete variable problems quite easily [2, 3]. Especially in last two decades, more novel stochastic solution algorithms have been applied by researchers in achieving optimum designs of different types of steel structures [4–9]. Among those, the teaching–learning-based optimization (TLBO) and biogeography-based optimization (BBO) algorithms have widely been used to obtain solutions of engineering design problems.

Moreover, the design optimization of steel structures under earthquake load impact has been a popular topic in structural engineering field. Considering the earthquake load impact in optimum design process leads more realistic structural designs [10, 11]. However, it is obvious that incorporating the earthquake load impact into the optimization process at the structural analyzes causes the computational intensity.

The principal goal of this study is to investigate the optimum sizing design of a steel planar truss under impact of earthquake load through TLBO and BBO algorithms. For this purpose, it is ensured that the algorithms coded in MATLAB [12] can run simultaneously through open application programming interface (OAPI) [13] functions in order to mutually receive the structural analysis responses from a structural analysis program SAP2000 [14]. The stress and displacement constraints are implemented from AISC-ASD practice code of specifications [15]. The algorithms select the W-shaped steel profiles from AISC for structural elements. Eventually, the optimal designs achieved via TLBO and BBO algorithms without affecting the earthquake load are also compared with those attained previously via different algorithms.

2 Mathematical Definition of Design Optimization Problem to AISC-ASD

The optimum discrete sizing design problem of a steel planar truss for weight minimization is specified as follows;

$$\min W = \sum_{k=1}^{ng} A_k \sum_{i=1}^{nk} \rho_i L_i \quad (1)$$

where W denotes the truss weight, A_k stands for cross-sectional area of group k, $\rho_{i,}$ and L_i identify the density and length of member i, ng is total group number, nk is the total member number in group k.

The stress and displacement constraints taken from specifications of AISC-ASD are, respectively, presented as below;

The stress constraints can be described as

$$g_m = \frac{\sigma_m}{\sigma_{m,\text{all}}} - 1.0 \leq 0.0, \quad m = 1, \ldots, ne \quad (2)$$

here, σ_m and $\sigma_{m,\text{all}}$ are, respectively, the calculated and allowable axial stresses for mth truss member.

The stress constraints for tension and compression members;

(i) The allowable stress can be described for tension members,

$$\sigma_{t,\text{all}} = 0.6F_y \quad (3)$$

here, F_y is yield stress.

(ii) The allowable stress can be described for compression members,

$$\sigma_{c,\text{all}} = \frac{\left[1 - \frac{\lambda_m^2}{2C_c^2}\right]F_y}{\left[\frac{5}{3} + \frac{3\lambda_m}{8C_c} + \frac{\lambda_m^3}{8C_c^3}\right]} \quad \text{for inelastic buckling } (\lambda_m \geq C_c) \quad (4)$$

$$\sigma_{c,\text{all}} = \frac{12\pi^2 E}{\lambda_m^2} \quad \text{for elastic buckling } (\lambda_m < C_c) \quad (5)$$

here, $\lambda_m = (K_m L_m)/r_m$ for $m = 1, \ldots, ne$ and $C_c = [(2\pi 2E)/F_y]^{1/2}$ in which λ_m represents the slenderness ratio, K_m is the effective length factor and taken as 1.0 for truss members, r_m is the minimum radii of gyration, and C_c is the critical slenderness ratio parameter.

Also, the displacement constraints are defined as follows,

$$g_j = \frac{\delta_{jl}}{\delta_{ju}} - 1.0 \leq 0.0, \quad j = 1, \ldots, n \quad (6)$$

here, δ_{jl} is displacement of jth degree of freedom, δ_{ju} is upper displacement limit, n is restricted displacements number.

3 Nature-Inspired Metaheuristic Optimization Algorithms

Here, the standard versions of TLBO [16] and BBO [17] algorithms that are adscititious to nature-inspired metaheuristic optimization algorithms are introduced.

3.1 Teaching–Learning-Based Optimization (TLBO) Algorithm

The TLBO algorithm is a population-based procedure mimicking the influence of a teacher over learners (students) [16, 18]. The population acts for class of learners. In a design optimization problem, various design variables are taken into account as various themes proffered to the learners (students). The objective function fitness value is analog with the achievement of learners. The best solution can be reckoned teacher in the whole population. The TLBO algorithm is composed of two stages; (i) in teacher stage class learns from a teacher and (ii) learner (student) stage, the learning event happens via the interactivity among the learners (students).

In teacher stage, a beneficial teacher strives to raise the level of learners with respect to knowledge based on a random operation. Suggesting that n number of learners (size of population, $k = 1, 2, \ldots, n$) are provided with m number of themes (design variables). If T_i is accepted as teacher and M_i is accepted as mean of achievements of learners at any consequent teaching–learning cycle i, T_i attempts to progress to mean M_i to a higher grade. M_{new} as a new mean comes evident after the teaching of T_i. According to the difference between M_{new} and $M_{i,}$ the solution is updated as given in Eq. (7).

$$\text{Mean}_{\text{difference}} = r_i(M_{\text{new}} - T_F M_i) \quad (7)$$

where T_F is factor of teaching that determines the mean value to be changed, r_i is a uniform random number in the range of [0, 1]. The T_F can be assigned as 1.0 or 2.0. It is not a parameter in the algorithm which is computed randomly as $T_F = \text{round}[1 + \text{rand}(0, 1)\{2 - 1\}]$. Through $\text{Mean}_{\text{difference}}$ computed as in Eq. (7), the existing solution is modified as shown in Eq. (8).

$$x_{\text{new},i} = x_{\text{old},i} + \text{Mean}_{\text{difference}} \quad (8)$$

In stage of learners, the learning process among the learners via interactivity to each other is simulated. The more experience the learner has, the more it learns. Two learners are selected randomly as x_i and x_j on condition that $i \neq j$. Then, the modification of learners can be defined as in below.

$$x_{\text{new},i} = x_{\text{old},i} + r_i(x_i - x_j) \quad \text{if } f(x_i) < f(x_j) \quad (9)$$

$$x_{\text{new},i} = x_{\text{old},i} + r_i\left(x_j - x_i\right) \quad \text{if } f(x_i) > f(x_j) \tag{10}$$

here, if x_{new} generates better function value, it is accepted. This procedure is repeated for all learners in the population.

3.2 Biogeography-Based Optimization (BBO) Algorithm

The BBO algorithm imitates the island biogeography theory [17, 19]. The migration and extinction of species among islands are depicted with a biogeographical. When the islands friendly to the earth and life, their habitat suitability index (HSI) is assumed as high. The suitability index variable (SIV) is expressed as variables characterizing habitability. The SIV and HSI can be taken into account as habitat's free and dependent variables, respectively. Obviously, the higher the HSI, the larger the species amount otherwise the lower the species amount. Since the habitats filled up with species, their immigration rate is low and HSI is high. In Fig. 1, the relevance between migration and immigration rates, and number of species is presented. The I is the maximum immigration rate, E is the maximum emigration rate, S_0 is the equilibrium species number, and $S_{\max}$ is the maximum number of species.

Each solution is decided to be modified or not according to the immigration rate. The λ is the immigration probability of an independent variable x. The roulette wheel selection is performed to select a candidate solution in N number of candidate solutions to be emigrated with an emigration probability μ.

$$P(x_j) = \frac{\mu_j}{\sum_{i=1}^{N} \mu_i} \quad \text{for } j = 1, \ldots, N \tag{11}$$

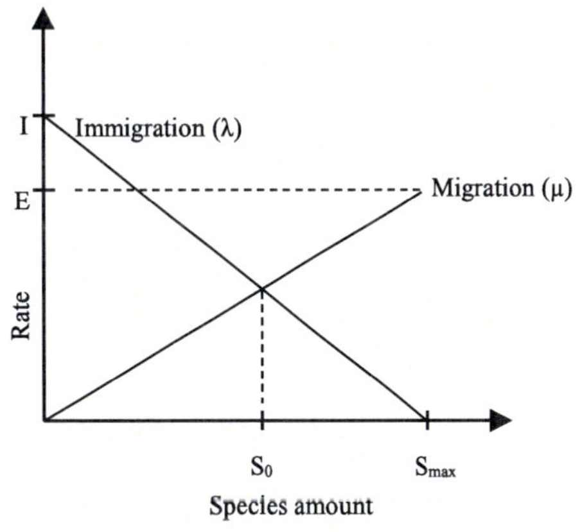

Fig. 1 Model of species in a single habitat

To raise the abundance of species in islands, the mutation factor is employed. This factor enhances the diversity in population as shown in Eq. (12).

$$m(s) = m_{\max}\left(\frac{1 - P_s}{P_{\max}}\right) \tag{12}$$

here, $m_{\max}$ is a user defined parameter. P_s is the species amount of the habitat, and $P_{\max}$ is the maximum number of species.

4 Open Application Programming Interface (OAPI)

In seismic design optimization of a steel planar truss structure, the structural analysis responses are required for the selection of steel sections from the design pool. A structural analysis program SAP2000 [14] is used to comprise the response spectrum for earthquake load impact. For this purpose, the open application programming interface (OAPI) [13] functions generate an interface so that any modeling and analyzing value acquired from one software can be used by another encoding languages like MATLAB [12]. In this study, the OAPI functions pertaining to the SAP2000 are used to produce an interface to the TLBO and BBO algorithms, encoded with MATLAB, in order for making ready the structural responses of the steel planar truss structure to be utilized in optimization process. So, thanks to the OAPI, it is ensured that data is transferred between SAP2000 and MATLAB in a safe and accurate manner until the predefined termination criteria is reached by an optimization algorithm.

5 Design Example

In order to execute a comparative study on TLBO and BBO algorithms for obtaining optimum discrete sizing designs, a steel planar truss structure is taken as design example where the earthquake load impact is added into the existing load combinations. The optimal designs attained via both algorithms are, also, compared with those obtained previously reported ones. The population size, N, is set to 50 for both algorithms [18, 19]. Besides, in the BBO algorithm, the mutation probability per solution per independent variable is selected as 0.01 and the number of elites that specify how many of the best solutions to keep from one generation to the next is set to $0.1 \times N$ for the design example [19]. The material properties and available discrete profile lists are arranged identical as in the previously reported studies.

The steel planar truss structure includes 46 members and 25 joints [20]. Structural elements have been grouped by considering structural symmetry so as to create a total of 23 design variables. Due to the structural symmetry, only the left side joints coordinates are illustrated in Table 1. In Fig. 2, the structural geometry and member grouping belonging to 23 design variables are depicted. The upper bound of the

Table 1 Left side joints coordinates of 46-member steel planar truss structure

Joint	x (m)	y (m)	Joint	x (m)	y (m)
1	0.00	0.00	8	3.92	0.24
2	0.28	0.13	9	4.62	0.61
3	0.25	0.18	10	5.38	0.38
4	1.01	0.05	11	6.07	0.76
5	1.70	0.32	12	6.10	0.46
6	2.46	0.09	13	6.35	0.635
7	3.16	0.47			

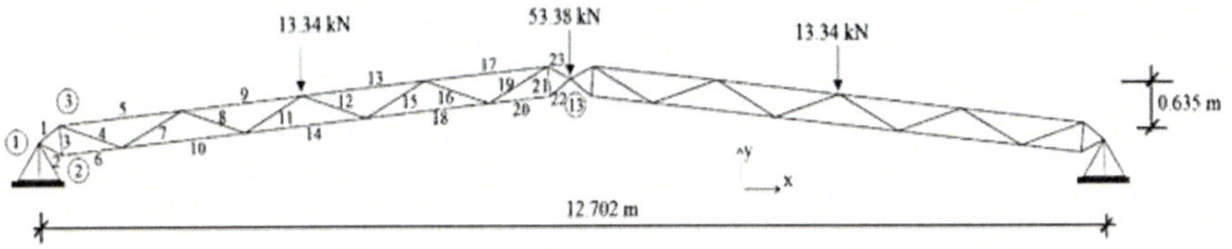

Fig. 2 Geometry and member grouping of 46-member steel planar truss structure

displacement is enforced as 0.254 cm on all joints along *x*- and *y*-directions of the truss structure. In order to randomly select the steel *W*-profiles for design variables, an economical discrete profile list containing 137 steel sections whose lower and upper areas are, respectively, 39.74 cm^2 (6.16 $in.^2$) and 1387.09 cm^2 (215 $in.^2$). As stated before, AISC-ASD structural specification is utilized to impose the stress and displacement constraints on the structural members. The loading strategy of the structure as also shown in Fig. 2. The loads are applied on joints 7 and 19 as 13.34 kN and on joint 19 as 53.38 kN in negative *y*-direction.

This steel planar truss structure is previously designed for minimum structural weight without considering the earthquake load impact via different metaheuristic algorithms, such as particle swarm optimizer (PSO), harmony search optimization (HSO), and genetic algorithms (GAs) [21]. The modulus of elasticity (E) is taken as 203,893.6 MPa (29,000 ksi), and yield stress (F_y) is taken as 253.1 MPa (36 ksi) for utilized steel material. The attained optimum discrete sizing designs with TLBO and BBO as well as the previously announced PSO, HSO, and GAs algorithms without earthquake load impact are shown in Table 2. From Table 2, it is seen that the minimum truss weight accomplished by the TLBO algorithm with 47.06 kN which is followed by BBO algorithm with very slight heavy truss design having weight of 47.08 kN. Whereas, same truss was more heavily designed by PSO, HSO, and GAs algorithms. From Table 2, it is worth mentioning that the displacement constraint is very active on optimization process for both TLBO and BBO algorithms.

In this study, the 46-member steel planar is once again designed by proposed TLBO and BBO algorithms with adding +0.7E as earthquake load impact to related

Table 2 Obtained optimum discrete sizing designs of 46-member steel planar truss structure

Group no.	Previously reported optimum designs [21]			Present study			
				Without earthquake load impact		With +0.7E additional earthquake load	
	PSO	HSO	GAs	TLBO	BBO	TLBO	BBO
1	W33 × 130	W33 × 130	W30 × 124	W14 × 132	W21 × 147	W12 × 106	W10 × 100
2	W14 × 145	W36 × 150	W24 × 112	W21 × 132	W18 × 119	W24 × 117	W30 × 108
3	W14 × 68	W21 × 68	W18 × 55	W18 × 71	W14 × 74	W10 × 60	W10 × 68
4	W6 × 9	W6 × 9	W6 × 9	W10 × 12	W8 × 10	W10 × 12	W10 × 12
5	W27 × 114	W10 × 112	W30 × 108	W24 × 103	W24 × 117	W30 × 108	W24 × 117
6	W30 × 124	W30 × 124	W27 × 117	W24 × 103	W18 × 119	W14 × 109	W24 × 117
7	W6 × 9	W6 × 9	W6 × 9	W10 × 12	W8 × 10	W10 × 12	W8 × 10
8	W6 × 9	W6 × 9	W6 × 9	W8 × 10	W8 × 10	W8 × 10	W8 × 13
9	W24 × 104	W27 × 102	W18 × 106	W21 × 147	W12 × 106	W30 × 148	W14 × 132
10	W30 × 132	W30 × 132	W12 × 136	W24 × 94	W18 × 97	W12 × 106	W14 × 109
11	W6 × 9	W6 × 9	W6 × 9	W8 × 10	W10 × 12	W8 × 10	W10 × 12
12	W6 × 9	W6 × 9	W6 × 9	W8 × 10	W8 × 10	W8 × 10	W10 × 12
13	W14 × 90	W12 × 87	W12 × 106	W40 × 149	W21 × 132	W18X130	W14 × 145
14	W36 × 135	W12 × 136	W30 × 124	W14 × 99	W18 × 97	W10 × 88	W24 × 94
15	W6 × 9	W6 × 9	W6 × 9	W8 × 10	W8 × 10	W8 × 10	W8 × 10
16	W6 × 9	W6 × 9	W6 × 9	W8 × 10	W8 × 10	W8 × 10	W8 × 10
17	W21 × 101	W27 × 102	W27 × 102	W14 × 120	W18 × 130	W14 × 145	W14 × 109
18	W21 × 122	W30 × 124	W24 × 131	W12 × 96	W10 × 112	W10 × 100	W30 × 108
19	W6 × 9	W6 × 9	W6 × 9	W8 × 10	W10 × 12	W8 × 10	W6 × 12
20	W30 × 116	W30 × 116	W33 × 118	W24 × 117	W12 × 120	W12 × 106	W14 × 145

(continued)

Table 2 (continued)

Group no.	Previously reported optimum designs [21]			Present study			
				Without earthquake load impact		With +0.7E additional earthquake load	
	PSO	HSO	GAs	TLBO	BBO	TLBO	BBO
21	W8 × 67	W18 × 65	W18 × 60	W12 × 58	W21 × 62	W21 × 83	W14 × 68
22	W36 × 135	W14 × 145	W14 × 132	W14 × 132	W14 × 145	W24 × 103	W12 × 106
23	W36 × 135	W12 × 136	W36 × 135	W14 × 132	W40 × 149	W18 × 130	W24 × 103
Weight (kN)	47.17	47.17	47.52	47.06	47.08	47.17	47.44
Max. disp. (cm)	–	–	–	0.254	0.254	0.254	0.254

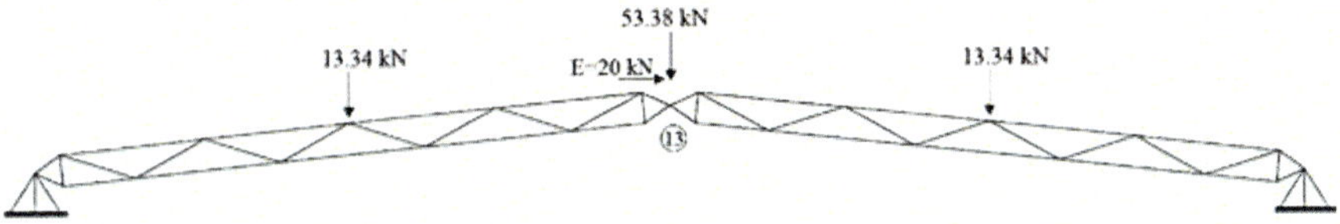

Fig. 3 46-member steel planar truss structure loading with earthquake load impact

load combinations. Here, E is taken as 20 kN, and it is additionally applied to the top joint 13 in positive x-direction as shown in Fig. 3.

In last two columns of Table 2, the optimum discrete sizing designs acquired via TLBO and BBO algorithms with earthquake load impact are given. While the TLBO algorithm reaches the optimum structural weight of 47.17 kN, the BBO algorithm yields a little heavier structural design weight as 47.44 kN. By addition of earthquake loading, the proposed TLBO and BBO algorithms for the discrete sizing of structural elements have tried to satisfy all design constraints, and naturally, the size of some structural elements collected under member groups as design variables has increased. But this structural weight increasing is not too much since the difference of optimal design weights through TLBO and BBO algorithms with and without earthquake load impact are only 0.24% and 0.76%, respectively. From Table 2, it can also be figured out that the achieved displacement values using both algorithms are also still dominant under activation of earthquake loading. The design history graph that displays the changes of the structural weight during design optimization process through TLBO and BBO algorithms is given in Fig. 4. The convergence rates of the optimization algorithms to the optimum solution are clearly observed in this figure.

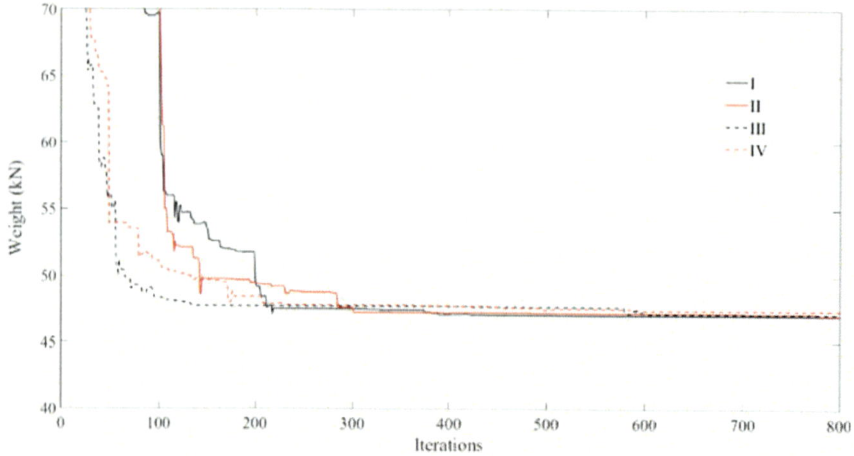

I: TLBO (without earthquake load impact), II: BBO (without earthquake load impact)

III: TLBO (+0.7E additional earthquake load impact), IV: BBO (+0.7E additional earthquake load impact)

Fig. 4 Design histories of 46-member steel planar truss structure

6 Concluding Remarks

In this study, nature-inspired teaching–learning-based optimization (TLBO) and biogeography-based optimization (BBO) metaheuristic algorithms are examined on obtaining optimal designs of a steel planar truss structure under external loading combinations included earthquake load impact. In the design, the stress and displacement constraints are implemented from AISC-ASD practice code provisions to control the structural behavior. The sections assigned to the structural element groups, which are the design variables, by solutions algorithms are randomly selected among 137 W-shaped steel profiles. To execute more accurate structural analyzes, both algorithms encoded in MATLAB are enabled to work together with the SAP2000 structural analysis program in a way to exchange data mutually through the OAPI functions. The optimal design results obtained with TLBO and BBO algorithms for design example are in parallel with those previously reported in the literature for without earthquake load impact case, but they are lighter in terms of total structural weight. When the acquired optimum design results of the design example are examined for the load case in which earthquake load impact is considered, it is observed that there is a certain amount of increase in the weight of the tested steel planar truss structure. In light of these results, it has been deduced that the algorithms and solution methodology proposed in this study are suitable for accomplishing the design optimization of steel planar truss structures in the cases of with and without earthquake load impact.

References

1. Saka MP, Carbas S, Aydogdu I, Akin A, Geem ZW (2015) Comparative study on recent meta-heuristic algorithms in design optimization of cold-formed steel structures. In: Computational methods in applied sciences, pp 145–173. https://doi.org/10.1007/978-3-319-18320-6_9
2. Carbas S, Toktas A, Ustun D (eds) (2021) Nature-inspired metaheuristic algorithms for engineering optimization applications. Springer, Singapore. https://doi.org/10.1007/978-981-33-6773-9
3. Saka MP, Carbas S, Aydogdu I, Akin A (2016) Use of swarm intelligence in structural steel design optimization. In: Modeling and optimization in science and technologies, pp. 43–73. https://doi.org/10.1007/978-3-319-26245-1_3
4. Aydogdu I, Ormecioglu TO, Carbas S (2021) Electrostatic discharge algorithm for optimum design of real-size truss structures. In: Carbas S, Toktas A, Ustun D (eds) Nature-inspired metaheuristic algorithms for engineering optimization applications. Springer, Singapore, pp 93–109. https://doi.org/10.1007/978-981-33-6773-9_5
5. Artar M, Carbas S (2021) Discrete sizing design of steel truss bridges through teaching-learning-based and biogeography-based optimization algorithms involving dynamic constraints. Structures 34:3533–3547. https://doi.org/10.1016/J.ISTRUC.2021.09.101
6. Kaveh A, Ghazaan MI (2018) Meta-heuristic algorithms for optimal design of real-size structures. Springer International Publishing. https://doi.org/10.1007/978-3-319-78780-0
7. Kaveh A, Dadras Eslamlou A (2020) Metaheuristic optimization algorithms in civil engineering: new applications. Springer International Publishing, Cham. https://doi.org/10.1007/978-3-030-45473-9
8. Yang XS, Bekdas G, Nigdeli SM (eds) (2016) Metaheuristics and optimization in civil engineering. Springer. https://doi.org/10.1007/978-3-319-26245-1
9. Nigdeli SM, Bekdaş G, Kayabekir AE, Yucel M (eds) (2021) Advances in structural engineering-optimization; emerging trends in structural optimization. Springer International Publishing, Cham. https://doi.org/10.1007/978-3-030-61848-3
10. Gholizadeh S, Salajegheh E (2010) Optimal seismic design of steel structures by an efficient soft computing based algorithm. J Constr Steel Res 66:85–95. https://doi.org/10.1016/J.JCSR.2009.07.006
11. Gholizadeh S, Fattahi F (2015) Optimum design of steel structures for earthquake loading by grey wolf algorithm. Asian J Civ Eng (Build Hous) 16:663–679
12. MATLAB (2009) The language of technical computing
13. OAPI—technical knowledge base. Computers and Structures, Inc. https://wiki.csiamerica.com/display/kb/OAPI. Last accessed 06 Mar 2021
14. SAP2000 (2008) Integrated finite element analysis and design of structures
15. AISC-ASD (1989) Manual of steel construction: allowable stress design. Chicago, Illinois
16. Rao RV, Savsani VJ, Vakharia DP (2011) Teaching-learning-based optimization: a novel method for constrained mechanical design optimization problems. CAD Comput Aided Des 43:303–315. https://doi.org/10.1016/j.cad.2010.12.015
17. Simon D (2008) Biogeography-based optimization. IEEE Trans Evol Comput 12:702–713. https://doi.org/10.1109/TEVC.2008.919004
18. Rao RV (2016) Teaching-learning-based optimization algorithm and its engineering applications. Springer International Publishing. https://doi.org/10.1007/978-3-319-22732-0_2
19. Zheng Y, Lu X, Zhang M, Chen S (2018) Biogeography-based optimization: algorithms and applications. Springer, Singapore. https://doi.org/10.1007/978-981-13-2586-1
20. Suleman A, Sedaghati R (2005) Benchmark case studies in optimization of geometrically nonlinear structures. Struct Multidiscip Optim 30:273–296. https://doi.org/10.1007/s00158-005-0524-2
21. Carbas S, Hasancebi O (2013) Optimal design of steel trusses using stochastic search techniques. In: The fourth international conference on mathematical and computational applications. Manisa, Turkey

Online Newton Step Based on Pseudo-Inverse and Elementwise Multiplication

Charanjeet Singh and Anuj Sharma

Abstract The recent advances in online machine learning have open scope to experiment with different types of multimedia data. This paper has been presented in this direction using a popular second-order method as Newton step that guarantees a regret bound. We have used first- and second-order matrices of dimensions m (classes) by d (features) with pseudo-inverse technique and elementwise multiplication. This first- and second-order matrices size results in less storage and faster computations. The extensive experimentation in this paper has been performed on benchmarked datasets, where mistake rate, number of updates, and computation time are evaluated. The result indicates that our method outcome is at-par with state-of-the-art algorithms.

Keywords Online learning · Elementwise multiplication · Pseudo inverse · Hessian matrix · Online convex optimization

1 Introduction

Online learning (OL) includes sequential decisions by choosing point in a convex set S. This results in revealing the concave payoff function after each decision. The idea is to minimize the regret where regret is the difference between average payoff and best-fixed point. Advances in the field of online learning witnessed promising techniques. One of the popular techniques is online gradient descent (OGD) with optimal regret rate as the square root of the sequence length [1]. Interestingly regret

C. Singh (✉) · A. Sharma
Department of Computer Science and Application, Panjab University, Chandigarh, India
e-mail: charan@pu.ac.in
URL: https://sites.google.com/view/anujsharma/

A. Sharma
e-mail: anujs@pu.ac.in

C. Singh
Department of Mathematics, Panjab University, Chandigarh, India

J. H. Kim et al. (eds.), *Proceedings of 7th International Conference on Harmony Search, Soft Computing and Applications*, Lecture Notes on Data Engineering and Communications Technologies 140, https://doi.org/10.1007/978-981-19-2948-9_37

rate was further improved using the exp-concave function to logarithmic value of sequence length [2]. This study did resulted in an understanding of two important observations:

- First- and second-order methods could have logarithmic regret bound.
- Second-order methods are expensive as it requires computations of Hessian.

Also, other promising work in online learning using first-order and second-order did help to understand that second-order advantages cannot be ignored against memory usage. Many variants of first- and second-order online learning techniques were introduced. Online learning has been extensively studied in the last decades. Perceptron [3] is the first algorithm in this direction proposed by cognitive psychologist Frank Rosenblatt in early 1958. It includes the euclidean instance space $X \in \mathbb{R}^d$ to maintain the linear classifier in such space. A New Approximate Maximal Margin Classification Algorithm (ALMA) [4] is an incremental approach proposed by Claudio Gentile in 2001. It works on linearly separable data by approximating the maximal margin hyperplane w.r.t. norm $p \geq 2$. The Relaxed Online Maximum Margin [5] (ROMMA) is an incremental approach for classification using linear threshold function. ROMMA iteratively chooses a hyperplane that correctly classifies the previous instance with the maximum margin. ROMMA formulated the maximum margin by minimizing the length of the weight vector subject to the number of linear constraints. Online gradient descent [1] (OGD) algorithm formulated as an online convex optimization task. In online convex optimization task, there is a convex set $S \subset \mathbb{R}^n$ and a convex function $f : S \rightarrow \mathbb{R}$. The goal of convex optimization is to minimize the f in a convex set S. Second-order perceptron (SOP) [6] is an extension of the standard perceptron algorithm, which is expressed as the interaction between the target vector and eigenvalue of the correlation matrix of the data. It utilizes the second-order geometrical properties of the data which are missed in the first-order algorithm. Passive–aggressive (PA) [7] follows the principle of margin-based algorithms using the set of linear predictors. Online Newton Step [2] (ONS) is the second-order algorithm, which is formulated in the online convex optimization. This utilized the second-order information of the loss function and is based on the Newton method for optimization. It attains a logarithmic regret bound for the α exp-concave loss function. In 2008, Crammer *et.al.* proposed a new online learning approach that measures the confidence of the algorithm in its current parameter estimation, called confidence weighted (CW) [8]. It maintains a Gaussian distribution over parameter vector with mean $\mu \in \mathbb{R}^d$ and standard deviation $\sigma \in \mathbb{R}^d$. It uses the covariance matrix $\Sigma \in \mathbb{R}^{d \times d}$ for the distribution, with diagonal σ and zero for non-diagonal entries. The another variant of CW is SCW [9], which is deal with non-separable data. Different from AROW which directly adds loss and confidence regularization and thus loses the adaptive margin property; SCW exploits adaptive margin by assigning different margins for different instances via a probability formulation. This raises a vital observation as the improvement in regret bound is the only criteria in the selection of an algorithm or real-time issues as memory requirements and execution time are also important. Naturally, the optimum answer is the inclusion of both as minimized regret bound and less memory-time of computations. One of the recent works in this direction includes the elementwise multiplication approach [10], where a non-square matrix has been used as a replacement to Hessian matrix. The logistic loss function used in this approach results in less computation time as it requires less memory to store the matrix. The regret bound as the square root of sequence length is found in this approach. This work includes an extra regularization term that escapes the second-order matrix to be rank-deficient. Also, no rank-deficient cases are discussed in this approach. This motivates us to work on the following two questions:

1. Can Newton step using logistic function with pseudo inverse and elementwise multiplication help in OL?
2. Is it possible to use online Newton step without extra regularization terms and escape from rank-deficient cases?

This paper is an effort in this direction to solve these two questions. Our approach guarantees both square root and logarithmic sequence length as regret bounds. In addition, the proposed work uses a logistic function using the online Newton step without extra regularization terms. We have used elementwise multiplication as done by [10] and use replacement of inverse function as pseudo-inverse function [11]. The experimentation further adds clarity to the understanding of the present work. Therefore, our work clearly indicates the use of logistic function in the online newton step theoretically and its implementation. The novelty of this paper is as follows:

- Logistic function as $f(W)$ use with no regularization terms.
- The weight updating decision depends on 0–1 loss function.
- Elementwise multiplication and pseudo inverse used to handle non-square matrices.
- Experimentation includes two types of step size to result in square root and logarithmic regret bounds.

This paper includes five sections including this section as an introduction. Section 2 presents needful preliminaries to understand the notations and foundations of the present work. This section also includes online learning algorithm working. The proposed algorithm has been discussed in Sect. 3. Section 4 includes experimentation, and last Sect. 5 concludes the paper with findings.

2 Preliminaries and Notions

In this section, we present the notions that are used throughout this paper. We also describe the basic framework for multi-class online learning.

2.1 Multi-class Online Learning

The multi-class learning techniques have been extensively studied in the last few years [8, 9]. In multi-class learning, instances from feature space S take the label from finite set Y, $|Y| = K$, where K is the number of classes. We use the standard approach [12] for generalizing binary class classification and assume a feature function $f(x, y) \in R^d$ mapping instance $x \in X$ and label $y \in Y$ into a common space.

Algorithm 1 A framework for Online multi-class framework

Initialization: $W_1 = 0$
for $t = 1, 2, \ldots, T$ **do**
  Pick an input instance: $x_t \in X$
  Predict class label: $\hat{y}_t = argmax_{i=1}^{k}\{W_{t,i}x_t\}$
  revealed true class label: $y_t \in Y$
  calculate the Current instance Loss: $y_t \neq \hat{y}_t$
  if $l_t > 0$ **then**
    Update Classifier: $W_{t,i}$ + Selected Loss function
  end if
end for

Algorithm 1 shows an online learning framework for multi-class classification, where the learning is executed in rounds. On each round the learner pick an instance x_t from input set X such that $x_t \in X$, after that model make a prediction on x_t and compute $\hat{y}_t$ which is always largest prediction value. After the prediction of true class, y_t is revealed from a finite set of y_t, $|Y| \in K$, and the learner then computes the suffered loss based on some loss function. Based on the loss value, the learner decides how and when to update the classifier at the end of each iteration. The main goal is to minimize the cumulative mistake rate $\sum_{t=1}^{T}\{t : y_t \neq \hat{y}_t\}$.

2.2 *Notations*

In Table 1, we present the main symbols used in this paper. We have used bold upper case characters for matrix (e.g., ($\mathbf{A}$)) representation, and $\mathbf{A^{-1}}$ and $\mathbf{A^T}$ denote the inverse and transpose of matrix, respectively; small bold case character represents the vector ($\mathbf{x}$). $||x||$ is used for Euclidean norm of a vector that is dual to itself. ∇ denotes the first-order derivative and ∇^2 denotes the second-order derivatives of the loss function.

In our case, the dimension of the second-order derivative is $K \times d$, where K is the number of classes in the dataset and d is the number of features in feature vector $\mathbf{x_t}$.

2.3 *Pseudo Inverse*

The Moore–Penrose pseudo inverse of the matrix when the matrix may not be invertible. If A is invertible, then the pseudo inverse is equal to the matrix inverse. However, the pseudo inverse is defined even when A is not invertible.

Table 1 Symbol definitions

Notations	Definitions
S	Convex feasible set
W	Weight vector
X	Feature vector:$x_1, x_2, \ldots, x_t$
l_t	0–1 loss function: $l_t = (y_t \neq \hat{y}_t)$
F	$f(w_t) = \log(1 + \exp(-y_t \langle x_t W^T \rangle))$
$\nabla(f_t)$	Gradient of the objective function
$\nabla^2(f_t)$	Second-order derivative of the objective function
Y	Class label $y_1, y_2, \ldots, y_t$
$R_G(T)$	Regret bound
K	Classes in dataset
d	Feature in each vector x_t
A	Second-order derivative of $f(w_t)$
A^+	H^+ is the pseudo-inverse
g	First order derivative of $f(w_t)$
A^T	Transpose of matrix H
$\odot$	Elementwise multiplication

The pseudo-inverse, $A+$, of an $m \times n$ matrix, is defined by the unique $n \times m$ matrix satisfying the following four criteria.

$$AA^+A = A \quad A^+AA^+ = A^+ \quad (AA^+)^T = AA^+ \quad (A^+A)^T = A^+A \tag{1}$$

If A is an $m \times n$ matrix where $m > n$ and A is of full rank, then

$$A^+ = A^T(AA^T)^{-1} \tag{2}$$

and the solution of $Ax = b$ is $x = A + b$. In this case, the solution is not exact. It finds the solution that is closest in the least-squares sense.

3 Online Newton Step Based on Pseudo Inverse and Elementwise Multiplication (ONS-PI)

The Newton method and its variants are widely used in the classification problem. The Newton method is a second-order approach, which iteratively computes the value of an objective function and its inverse which is time-consuming. The proposed work is inspired from the [6, 11] which incrementally learns the pseudo inverse solution to the weight vector, which is exactly the same computed using singular value decomposition (SVD) and also uses the elementwise multiplication for minimizing the computation time for second-order derivative.

A generic framework for proposed ONS-PI algorithm is summarized in Algorithm 1. The approach of ONS-PI is based on elementwise multiplication and pseudo-inverse mathematical properties. Newton method is a second-order optimization problem that iteratively updates the weight vector $W \in \mathbb{R}^{K \times d}$. In the proposed method, a problem is to minimize the objective function f to find the descent direction d_t and update the weight vector $W_t \in \mathbb{R}^{K \times d}$ as:

$$W_{t+1} = W_t + \eta_t d_t, \tag{3}$$

where d_t is the Newton search direction and η_t is the step size. The advantage of using the Newton approach is that the objective function f locally approximated around each iteration of weight vector W_t as

$$f(W_t + d_t) \approx q_{(W_t)}(d_t) = f(W_t) + g_t d_t + \frac{1}{2} d_t^T H d_t, \tag{4}$$

where g_t is the gradient and H is the Hessian at each iteration t. To move in the descent direction, we minimize (4), i.e., d_t. The computation complexity of the Hessian is $\mathcal{O}(d)^2$. It is not possible to work with this complexity in real-world datasets. A solution to this problem is addressed through different directions mainly [2].

A recent work in this direction [10] where a Hessian matrix is replaced with matrix element of dimension $K \times d$ and used elementwise multiplication [10]. This results in storing less volume of data and encouraging the work in this direction. The computation of the inverse of a matrix is direct [10] and does not guarantee rank-deficient condition.

Algorithm 2 Online Newton step based on pseudo inverse and elementwise multiplication (ONS-PI)

Input: learning rate η_t;
Initialize: $W_1 = 0, tol = 10^8$
for t=1 to T **do**
  Pick instance $x_t \in X$;
  Predict: $\hat{y}_t = argmax_{i=1}^{k}\{W_{t,i}x_t\}$
  Revealed true class: $y_t \in Y$
  Compute Loss: $l_t = (y_t \neq \hat{y}_t)$
  if $l_t > 0$ **then**
    Compute: $f(W_t) = log(1 + exp(y_t < x_t W_t^T >)$
    Compute: $g = \nabla f(W_t)$
    Compute: $A = \nabla^2 f(W_t)$
    if $||A|| < tol$ **then**
      $tol = ||A||$
    end if
    Compute: $\eta_t = \frac{0.05}{tol \cdot \sqrt{t}}$
    Compute Pseudo Inverse: $A^+ = A^T(AA^T)^{-1}$
    Compute Euclidean norm: $||A^+|$
    Update the Prediction Model: $W_{t+1} = W_t + eta_t(||A||^+ x_t)$
  end if
end for

This work includes elementwise multiplication and implements pseudo-inverse of the second-order matrix. This helps to escape from the rank-deficient condition and better in learning time. In addition, our objective function $f(w)$ is used without any regularization term. To find an inverse of A, a non-square matrix in nature, the pseudo inverse has been used [6]. The output of pseudo-inverse is not rank-deficient using this technique. The pseudo-inverse works as:

$$A^+ = VD^+U^T \tag{5}$$

where U, V, and D, respectively, the left singular vector, the values, and the right singular vectors of matrix A. Here, D^+ is the pseudo inverse of D and D is a diagonal matrix, and thus, D^+ can be calculated by taking the reciprocal of the nonzero value of D. The pseudo-inverse has been used in the second-order perceptron in online learning [6]. The work in [6] does not include logistic loss function implementation. The loss function 0–1 has been used to check the necessity of weight updating. The dimension of g and A is $K \times d$ where g and A refer to first- and second-order matrices as mentioned in Table 1. The dimension of matrix A^+ is $K \times d$ which is the same as matrix A. Further computation of matrix $A^+ \odot g$ results in same dimension as $K \times d$. As nature step size could vary subject to algorithms, we have possibly use $\frac{0.05}{\text{tol}\cdot\sqrt{t}}$. The online learning algorithm by Zinkewich [1] and Hazan [2] results in regret bound in order as $\sqrt{T}$ and $\log(T)$ with respective step size $\frac{2}{2\sqrt{T}}$ and $\frac{1}{t}$. The proof of the regret bound is already available in the literature with chosen step sizes; therefore, we have focused our work to prove the efficiency of logistic function using Newton step with experimentation. The present study experimentation performed with $\eta_t = \frac{0.05}{\text{tol}\cdot\sqrt{t}}$ step size. The proposed work uses logistic function using online Newton step without extra regularization term. We have used elementwise multiplication as done by [10] and use replacement of inverse function as pseudo-inverse function [6, 11].

4 Experiments

In this section, we have presented the empirical results of the proposed technique which were performed on multi-class classification datasets, i.e., satimage [13], MNIST [14], acoustic [15], and protein [16]. Table 2 represents the statistical properties of the datasets used in experimentation.

We have compared our proposed algorithm with the two following baselines:

- The ONS-PI computes the inverse of a matrix using Moore–Penrose pseudo-inverse and the dimension of second-order as well as first-order matrices which are A and b from $d \times d$ to $K \times d$.
- In ONS-PI, we have used the elementwise multiplication denoted by $\odot$ in Table 1 for reduction of matrices multiplication complexity.

To make a fair comparison, our algorithm adopted the same experimental setting compared to other second-order online learning algorithms. We randomly select each

Table 2 Summary of datasets used in the experimentation

Dataset	n = #instances	d = #features	K = #classes	Type
Satimage	3104	36	6	Classification
MNIST	60,000	780	10	Classification
Acoustic	78,823	50	3	Classification
Protein	17,766	357	3	Classification

feature vector x_t at a time t and the learning rate $\eta_t = \frac{0.05}{\text{tol}\cdot\sqrt{t}}$. After that, all experimental results have been drafted by averaging the 10 runs with the help of LIBOL [12]. We add our proposed algorithm to the LIBOL library and compared it with all the state-of-the-art second-order online learning algorithms. We compared the proposed second-order modified Newton step algorithm with the existing second-order online learning algorithms [8–10, 17]. In addition, the online Newton step is the baseline technique with full information feedback. To accelerate the computational efficiency, the modified Newton step employs a pure Newton formula with the pseudo-inverse.

$$W_{t+1} = W_t + \eta_t \cdot A_t^+ x_t \tag{6}$$

To compute A_t^+ in time $O(Kd)$ where K is the number of classes in multiclass dataset and x_t is feature vector in each iteration t.

We have compared our algorithm against state-of-the-art algorithms using library [12]. We evaluated different second-order online learning methods in the terms of classification task. We selected the loss function $l(w_t) = (y_t \neq \hat{y}_t)$ as a classifier. In this experimentation, the step size of ONS-PI is $\eta_t = \frac{0.05}{\text{tol}\cdot\sqrt{t}}$. We found that the modified Newton step using pseudo inverse is more efficient in terms of running time and memory consumption. Hence, the present approach provides a principle way to deal with large-scale online learning problems. Experimental results have been presented in tabulated form, comparing the mistake rate, number of updates, and time. Table 2 shows the summarized detail of datasets.

Table 3 shows the empirical results of the satimage dataset, which contains 3104 instances and each instance contains 34 features and 6 classes. Table 3 represents the algorithm name, mistake rate, number of updates, and time in seconds for respective algorithms. Our approach performed better in learning time compared to other methods. For mistake rate, we are close to AROW, SCW-I SCW-II, and CW algorithm.

Table 4 presents the results of the MNIST dataset, which contains 60,000 instances, and each instance contains 780 features with 10 classes. In the MNIST dataset results, our approach performed better than the other algorithms. Also, our method achieved the lowest mistake rate for MNIST [14] dataset and was close to the ONS-EM algorithm.

Table 5 shows the empirical results of the acoustic dataset, which contains the 78823 instances, 50 features of each instance, and 3 classes. We observed that our algorithm again outperformed CW, AROW, SCW-I, and SCW-II in training time.

Table 3 Results of satimage dataset

Dataset	Satimage (#instances = 3104, #features = 36, #classes = 6)		
Algorithm	Mistake	#updates	Time (s)
CW	0.190 ± 0.004	910.5 ± 21.5	0.178 ± 0.022
AROW	0.169 ± 0.008	2076.5 ± 84.7	0.225 ± 0.022
SCW1	0.152 ± 0.005	786.7 ± 31.6	0.182 ± 0.046
SCW2	0.157 ± 0.003	992.1 ± 35.6	0.182 ± 0.023
ONS_EM	0.176 ± 0.008	1272.7 ± 26.8	0.140 ± 0.039
ONS_PI	0.197 ± 0.010	612.6 ± 31.5	0.158 ± 0.007

Table 4 Results of MNIST dataset

Dataset	MNIST (#instances = 60,000, #features = 780, #classes = 10)		
Algorithm	Mistake	#updates	Time (s)
CW	0.133 ± 0.002	15301.8 ± 65.9	378.248 ± 5.665
AROW	0.416 ± 0.063	28542.3 ± 2910.3	695.664 ± 65.755
SCW1	0.186 ± 0.005	13696.0 ± 252.4	338.694 ± 6.220
SCW2	0.130 ± 0.001	15214.1 ± 60.9	376.580 ± 7.429
ONS_EM	0.114 ± 0.004	20334.7 ± 396.7	5.356 ± 0.101
ONS_PI	0.139 ± 0.003	8337.0 ± 194.0	10.440 ± 0.753

Table 5 Results of acoustic dataset

Dataset	Acoustic (#instances = 78,823, #features = 50, #classes = 3)		
Algorithm	Mistake	Updates	Time (s)
CW	0.413 ± 0.001	48414.5 ± 132.3	6.358 ± 0.145
AROW	0.321 ± 0.001	77379.3 ± 64.1	7.919 ± 0.186
SCW1	0.347 ± 0.007	34982.5 ± 1457.3	5.579 ± 0.342
SCW2	0.322 ± 0.001	69226.8 ± 107.7	7.833 ± 0.266
ONS_EM	0.341 ± 0.006	40832.0 ± 461.2	3.498 ± 0.068
ONS_PI	0.444 ± 0.002	34992.3 ± 150.1	5.593 ± 0.075

Table 6 shows the results of the protein dataset, which contains 17,766 instances, and each instance contains 357 features and 3 classes. We also surplus in terms of time comparison and close to ONS-EM and better than CW with respect to mistake rate.

Figures 1, 2, 3, and 4 show the comparison of CW, AROW, SCW and SCW2, and ONS-EM algorithms. Our proposed ONS-PI shows in solid cyan line in Figs. 1, 2, 3, and 4. All figures include three parts: first part of the figure shows cumulative mistake rate, second part of the figure shows number of updates, and third part of the figure shows time in seconds.

Table 6 Results of protein dataset

Dataset	Protein (#instances = 17,766, #features = 357, #classes = 3)		
Algorithm	Mistake	Updates	Time (s)
CW	0.431 ± 0.002	11736.8 ± 48.1	20.146 ± 0.339
AROW	0.342 ± 0.002	17007.0 ± 18.4	28.182 ± 0.456
SCW1	0.374 ± 0.002	9337.7 ± 47.3	16.273 ± 0.236
SCW2	0.348 ± 0.002	11843.3 ± 76.6	20.279 ± 0.300
ONS_EM	0.402 ± 0.005	15508.6 ± 65.5	1.390 ± 0.009
ONS_PI	0.4143 ± 0.0029	7360.80 ± 51.12	2.4039 ± 0.1179

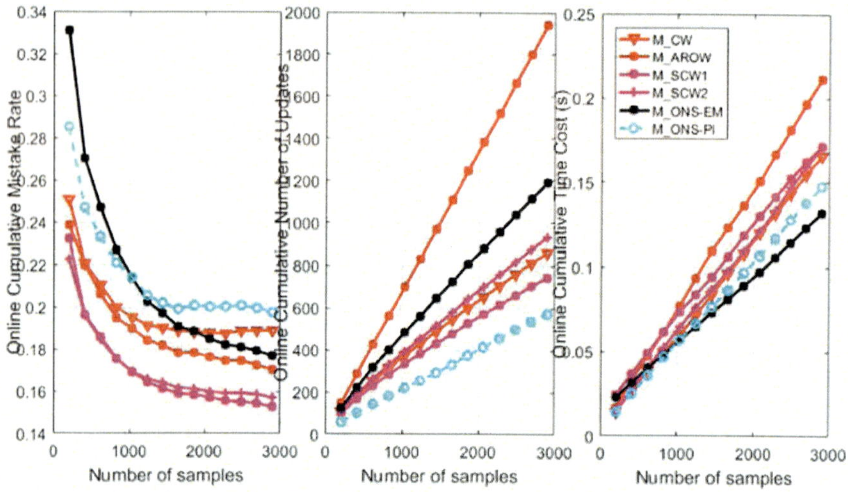

Fig. 1 SATIMAGE dataset results

Figure 1 shows the comparison between second-order online learning algorithms discussed in the literature with the multiclass classification dataset satimage. We outperformed in terms of time taken when compared to other methods but close to the ONS_EM.

Figure 2 shows the results of the MNIST dataset, and our results are better in mistake rate and time comparison compared to CW, AROW, SCW-I, and SCW-II. We are close to ONS-EM in mistake rate as well as learning time.

Figure 3 shows the results of acoustic data, in which we performed better in time and close to ONS-PI

Figure 4 shows the result of protein dataset, in which our approach performed better than the other algorithms and close to ONS-PI.

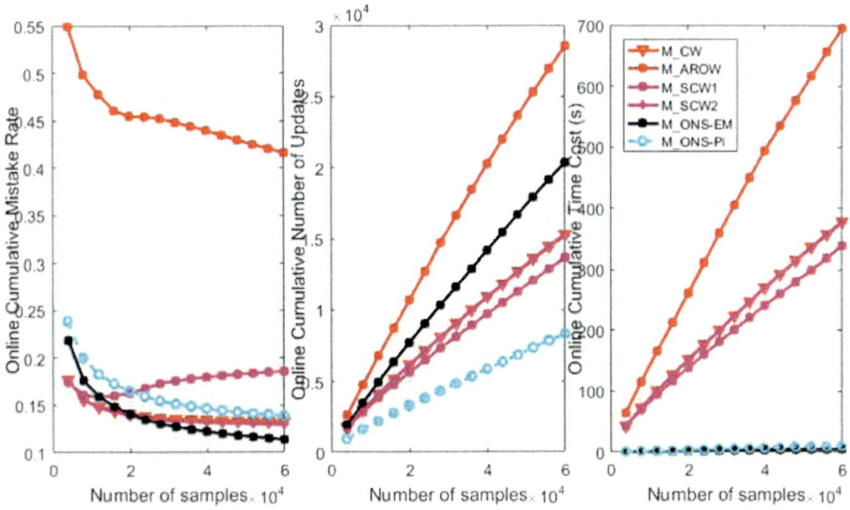

Fig. 2 MNIST dataset results

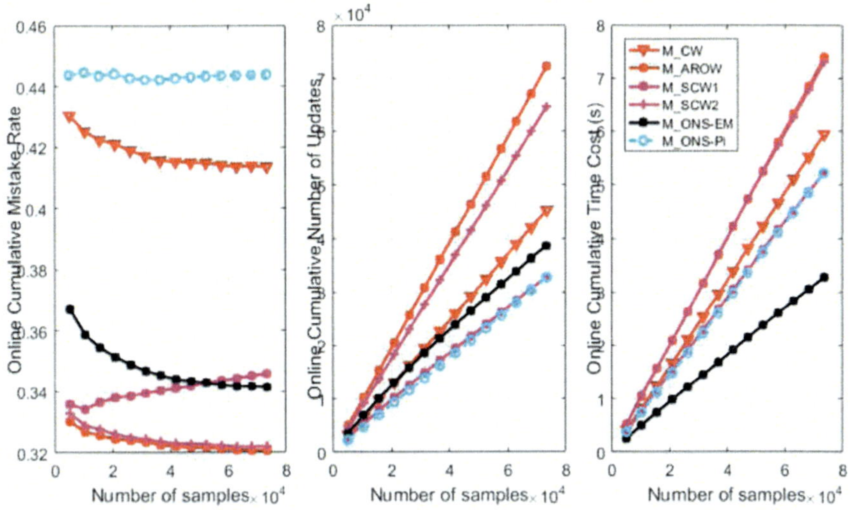

Fig. 3 Acoustic dataset results

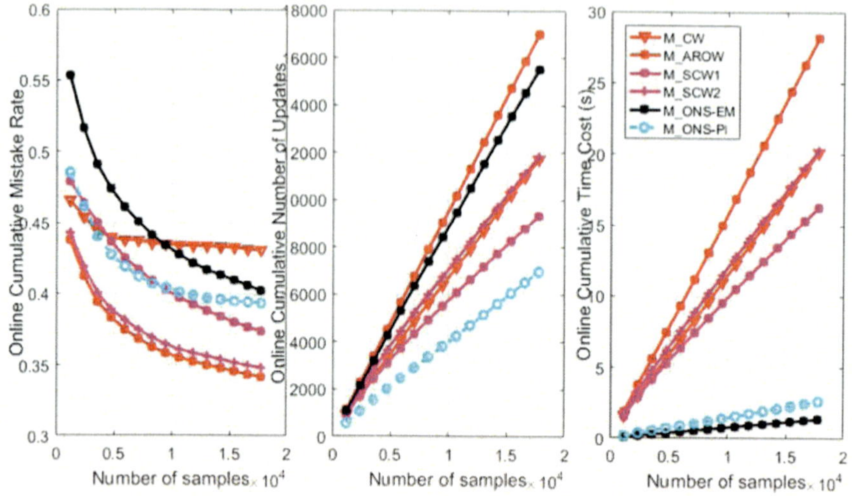

Fig. 4 Protein dataset results

5 Conclusion

The proposed method uses elementwise multiplication and implements pseudo-inverse of a second-order matrix. This helps to escape from the rank-deficient condition and better in learning time. We reduce the size of second-order matrices as $K \times d$, using elementwise multiplication, and compute the inverse of a matrix using the pseudo inverse to handle the rank-deficient condition. We analyze its empirical evaluation by conducting a set of experimentation for learning time and performance. We also experimented on large-scale dataset, and all results illustrate that our proposed method runs efficiently on a common machine. The results depict a better learning time, and performance is at-par than other second-order methods. This further strengthens our approach as second-order learning, which can be used in real-time applications.

References

1. Zinkevich M (2003) Online convex programming and generalized infinitesimal gradient ascent. In: ICML'03 proceedings of the twentieth international conference on international conference on machine learning, pp 928–935
2. Kale S, Hazan E, Agarwal A (2007) Logarithmic regret algorithms for online convex optimization. Mach Learn 69(2–3):169–192
3. Rosenblatt F (1958) The perceptron: a probabilistic model for information storage and organization in the brain. Psychol Rev 65(6):386
4. Gentile C (2001) A new approximate maximal margin classification algorithm. J Mach Learn Res 2:213–242

5. Li Y, Long PM (2002) The relaxed online maximum margin algorithm. Mach Learn 46(1–3):361–387
6. Gentile C, Cesa-Bianchi N, Conconi A (2005) A second-order perceptron algorithm. SIAM J Comput 34(3):640–668
7. Keshet J, Shalev-Shwartz S, Singer Y, Crammer K, Dekel O (2006) Online passive-aggressive algorithms. J Mach Learn Res 7:551–585
8. Pereira F, Dredze M, Crammer K (2008) Confidence-weighted linear classification. In: Proceedings of the 25th international conference on machine learning. ACM, pp 264–271
9. Hoi SCH, Wang J, Zhao P (2016) Soft confidence-weighted learning. ACM Trans Intell Syst Technol (TIST) 8(1):15
10. Singh C, Sharma A (2020) Modified online Newton step based on elementwise multiplication. Comput Intell 36:1010–1025
11. Greville TNE (1960) Some applications of the pseudoinverse of a matrix. SIAM Rev 12(1):15–22
12. Zha P, Hoi SC, Wang J (2014) Libol: a library for online learning algorithms. J Mach Learn Res 15(1):495–499
13. Hsu C-W, Lin C-J (2002) A comparison of methods for multi-class support vector machines. IEEE Trans Neural Networks 13(2):415–425
14. LeCun Y, Bottou L, Bengio Y, Haffner P (1998) Gradient-based learning applied to document recognition. Proc IEEE 86(11):2278–2324
15. Duarte M, Hu YH (2004) Vehicle classification in distributed sensor networks. J Parall Distrib Comput 64(7):826–838
16. Wang J-Y (2002) Application of support vector machines in bioinformatics. Master's thesis. Department of Computer Science and Information Engineering, National Taiwan University
17. Dredze M, Crammer K, Kulesza A (2009) Adaptive regularization of weight vectors. Mach Learn, pp 1–33
18. Davis C (1962) The norm of the Schur product operation. Numer Math 4:343. https://doi.org/10.1007/BF01386329

Evaluating the Performance of LSTM and GRU in Detection of Distributed Denial of Service Attacks Using CICDDoS2019 Dataset

Malliga Subrmanian, Kogilavani Shanmugavadivel, P. S. Nandhini, and R. Sowmya

Abstract A Distributed Denial of Service (DDoS) attack occurs when an intruder or a group of attackers attempts to block legitimate users from accessing a service. A DoS attack is carried out by a single system, while a DDoS attack is carried out by numerous systems. DDoS attacks can be directed at several OSI layers. Deep learning has played a crucial role in the advancement of intrusion detection technologies in recent years. The main purpose of this work is to detect and identify DDoS attacks in the OSI model's application, network, and transport layers using deep learning models. The proposed models have been evaluated against the CICDDoS2019 dataset which consists of application, network and transport layer DDoS attacks. For the CICIDDOS2019 dataset, Long-Short-Term memory and Gated Recurrent Unit attained an average accuracy of 99.4% and 92.5%, respectively. We also compared the suggested models' performance to that of a few other higher accuracy models and found that the proposed models have higher accuracy with fewer epochs. In addition, the performance of the proposed system is also evaluated for various types of DDoS attacks in the CICDDoD2019 dataset and LSTM is found to produce good accuracy.

Keywords Distributed Denial of Service (DDoS) · Deep neural network · Long short-term memory · Gated recurrent unit · CICDDoS2019 · Accuracy

1 Introduction

An Artificial Neural Network (ANN) is a computer system that is capable of learning and making intelligent and prudent decisions through the use of algorithms. Deep learning is a subtype of neural network. A neural network with multiple hidden layers of nodes is used in a deep learning model. The term "deep" refers to the number of layers in the neural network. Deep learning was developed primarily for the purpose of managing massive amounts of data and running complex algorithms

M. Subrmanian (✉) · K. Shanmugavadivel · P. S. Nandhini · R. Sowmya
Department of CSE, Kongu Engineering College, Perundurai, Erode, India
e-mail: mallinishanth72@gmail.com

J. H. Kim et al. (eds.), *Proceedings of 7th International Conference on Harmony Search, Soft Computing and Applications*, Lecture Notes on Data Engineering and Communications Technologies 140, https://doi.org/10.1007/978-981-19-2948-9_38

to optimize performance. Deep learning models are used for both feature extraction and classification. Data flows from the input layer to the output layer of a Deep Neural Network (DNN), and no loops or cycles are produced. A Recurrent Neural Network (RNN) is a type of ANN in which the connections between the nodes form a directed graph and can learn sequential data over time steps. When the back-propagation algorithm iterates across all the neurons in the neural network to update their weights, the vanishing gradient problem occurs. Long Short-Term Memory (LSTM) has been designed to address the problem of vanishing gradients in RNNs. The GRU is a more recent type of RNN that resembles an LSTM. The GRUs obliterated the cell state in favor of using the hidden state to transfer data.

Distributed Denial of Service (DDoS) is a malicious attempt to degrade or interrupt a server's legitimate traffic. DDoS attacks are classified into three types: volume-based attacks, transport/network-layer attacks, and application layer attacks. By flooding a resource, such as a server, with fake traffic, volume-based attacks overload it. ICMP, UDP, and spoofed-packet flood attacks are such attacks. DDoS attacks at the protocol levels such as SYN floods and Smurf DDoS, flood a targeted server with packets. Application layer attacks are initiated by overflowing apps with bogusly created requests. DDoS attacks may also appear to be benign traffic. Even a big volume of legitimate requests from legitimate users will bring the system to a grinding halt. As a result, research into the detection of DDoS attacks is becoming increasingly critical. Machine learning is a widely used technology for detecting DDoS attacks through the use of statistical features. On the other side, machine learning methods employ shallow representation models, which do not result in increased accuracy. However, deep learning models have recently proved their ability to discriminate between DDoS and legitimate traffic by discriminating between high-level and low-level properties.

Additionally, we noticed that the bulk of DDoS attack detection techniques evaluate performance using the NSL-KDD and KDDCUP 99 datasets. Since these datasets have been created long back and do not reflect the recent trends in the attack pattern, we use CICDDoS2019. We create two models using LSTM, and GRU to train and recognize the types of attacks in this dataset. DDoS attacks can be classified as network/transport layer or application layer attacks, depending on which tier the DDoS attacker targets. The CICDDoS2019 dataset comprises traces of attacks on the application, transport, and network layers. Additionally, we employ the proposed approaches to detect other sorts of attacks in the dataset, including LDAP, UDP, Portmap, NETBIOS, and UDP-LAG. Classification of harmful packets and neural network training serve as the foundation for deep learning-based attack diagnostics. The remaining part of the paper is set out as follows: Sect. 2 examines state-of-the-art deep learning models developed to detect DDoS attacks using contemporary and CICDDoS2019 datasets. We provide a brief summary of the dataset in Sect. 3. Section 4 describes how LSTM and GRU are used to classify the attacks in the dataset. In Sect. 5, we analyze the performance of the proposed models and present our results. Finally, we provide our conclusions in Sect. 6.

2 Related Works

Using big data and neural networks, Hsieh and Chan et al. [1] suggested a DDoS detection technique. The detection system was written in R and is based on Apache Spark. Ferrag et al. investigated a variety of deep learning models for detecting cyber security intrusions and reviewed the two new real-time traffic data sets CSECI-CIDS2018 and the Bot-IoT [2]. Corin et al. proposed a system called "LUCID" that classified traffic flows as malicious or benign based on CNN features [3]. Asad et al. developed a framework for detecting DDoS attacks using a deep learning method. For CICIDS2017 dataset, this system employed a feed-forward back-propagation network to identify application layer DDoS attacks [4]. Anthi et al. developed an IoT IDS using machine learning algorithms. The system can differentiate network scanning, probing, and simple DDoS attacks [5]. Doshi et al. demonstrated a packet-level DoS system capable of consistently distinguishing between normal and DoS attack data received from IoT devices [6]. They demonstrated that by leveraging IoT-specific network behaviors and feature selection, it is possible to accurately detect DDoS attacks in IoT network data using a variety of machine learning algorithms. Elsayed et al. developed a DDoSNET, an IDS for detecting DDoS assaults in a Software Defined Network, using the CICDDoS2019 data set. This approach made use of RNN and Auto Encoder [7]. Jiang et al. created a system for anomaly identification using back-propagation neural networks. This framework incorporated traffic and user activity data extracted from web server logs. The authors assessed the performance of the proposed model using the CICIDS2017 data set [8].

Shone et al. (2018) proposed the Non-symmetric Deep Auto Encoder as an unsupervised feature learning approach. Classification of this model, which has been generated using stacked NDAEs, was performed using the Random Forest algorithm [9]. Hsieh and Chan [10] suggested two techniques for detecting Distributed Reflection DoS (DrDoS) attacks on the Internet of Things. In the first way, they employed a hybrid Intrusion Detection System (IDS) to detect IoT-DoS attacks, and in the second method, they trained a deep learning model based on LSTM on the most current data set to detect various forms of DrDoS attacks. Muraleedharan and Janet [11] describes a technique for identifying HTTP-related DoS threats using a deep classification technique based on traffic statistics. The classifier is evaluated using CICIDS2017 data. An attempt by Cil et al. [12] proposed employing a DNN as a deep learning model for detecting DDoS attacks on a batch of packets obtained from network traffic. The attack types were classified in this study with an accuracy rate of 94.57%. Bolodurina et al. [13] examined the effect of data balancing approaches in the network traffic categorization issue on various types of DDoS using the CICDoS2019 dataset, which contains data on reflection- and exploitation-based attacks and achieved a maximum accuracy of 98.62%. Sindian and Sindian [14] suggested an improved Deep Sparse Autoencoder-based approach for identifying DDoS attacks that utilizes two hidden layers. The primary objective of this work is to employ an autoencoder to extract representative features from the CICDDoS2019 dataset, to minimize classification error, and to correctly detect DDoS attacks. This investigation achieved a detection accuracy of 98%. Kolias et al. [18] presented a DDoS attack on the IoT using Mirai and botnets. This attack was carried out via a network of interconnected devices connected to the Internet and a specific malware called "Mirai botnet," which pummeling the targeted servers with traffic until they succumbed to the strain. Hou et al. [19] proposed a method for identifying DDoS traffic using NetFlow feature selection and machine learning. Additionally, they evaluated the CICIDS2017 dataset in conjunction with real-world NetFlow logs given by a large ISP in China, Unicorn. The experiment is conducted utilizing a random forest as the detector, and the findings obtained using the CICIDS2017 demonstrate 99% accuracy. Ferrag et al. [20] suggested deep learning model for cyber security intrusion detection and used two new real-world traffic datasets: CSECICIDS2018 and Bot-IoT. From the reviews, we believe that deep learning has potential to detect recent types of DDoS attacks and can be well explored to improve the detection accuracy. The different researches are used for detecting the application, network and transport layer-based DDoS attacks. Nevertheless, we could find only a few research works on CICDDoS2019 dataset. Moreover, these research attempts have found that whether a particular instance is an attack or not, but not the type of the attacks. But we find the different types of attacks also.

3 Dataset Description

Given the nature of DoS/DDoS attacks, this section provides a brief overview of benchmark datasets for designing systems for detecting DoS/DDoS attacks. DARPA, KDD, and NSL-KDD are just a few publicly available datasets that have been utilized as benchmarks. In this study, the dataset CICDDoS2019 [16] has been used to develop and assess deep learning models. This dataset includes the most recent and benign prevalent DDoS attacks on the application and transport/network layers. The CICDDoS2019 data collection includes 12,794,627 traces of DDoS attacks. Among the 86 features are flow length, total forward packets, total reverse packets, and so on [16]. The dataset is classified as benign, NETBIOS, UDP, UDP-LAG, LDAP, NTP, SSDP, TFTP, and MSSQL.

4 Deep Learning Models for DDoS Attack Detection

The primary objective of this study is to analyze the CICDDoS2019 data sets using deep learning. We employ three deep learning models in this study to detect DDoS attacks at the application, network, and transport layers. These models include a deep neural network generated using the feed-forward back-propagation approach and a RNN constructed using LSTM and GRU.

4.1 Preprocessing

Preprocessing is a technique for preparing data for analysis. When data is in its "raw" state, it is of limited utility for analysis. As a result, we perform preprocessing on the data before passing it to a classification system. To avoid overfitting, categorical features such as the source IP address, the destination IP address, the port number, and the source port are excluded. We have removed instances with attributes containing NAN and infinite values because we have a sufficient number of records. Due to the fact that all features should be within the same range, we can scale them using a variety of scaling strategies, including standard scalar and Minmax Scalar. Minmax Scalar has been utilized in our study to convert all of the features to a same set. Then, using label encoding, we substitute a numeric value between 0 and the number of class labels for the category values of the class labels. CICDDOS2019 has a class designation of 0 (for benign traffic) and 1 (for attack traffic). The term "attack traffic" refers to all forms of attacks. The next sections describe the deep learning models used to categorize the chosen dataset in this work.

4.2 Recurrent Neural Network Using LSTM and GRU

In a typical neural network, each neuron's output is determined by the current input, with no relationship between the input and the neuron's previous output. However, if we want to predict the next word in a sentence, we must recall the previous words to do so correctly. Thus, RNN was introduced. Unlike feed-forward neural networks, RNN have cyclic connections thus making them more appropriate for modeling sequence of data. As a result, in recent years, RNN has played an important role in machine translation, robot control, time series prediction, and other areas. Long Short-Term Memory (LSTM) and Gated Recurrent Unit (GRU) are two types of RNN.

LSTM: When training, conventional RNN faces the issue of vanishing gradients. The LSTM algorithm is used to solve this problem. RNN cannot carry information from earlier time steps to later ones for long-term dependencies if a series is long enough. Hence, LSTM is a good choice for sequences with long-term dependencies.

GRU: An LSTM without an output gate is known as a GRU. At each time step, GRU is able to write the contents of memory cells to the larger net. The vanishing gradient problem in RNN is also addressed by GRU. Since both are built similarly and, in some cases, produce equally good results. RNN and its variants are shown in Figs. 1 and 2.

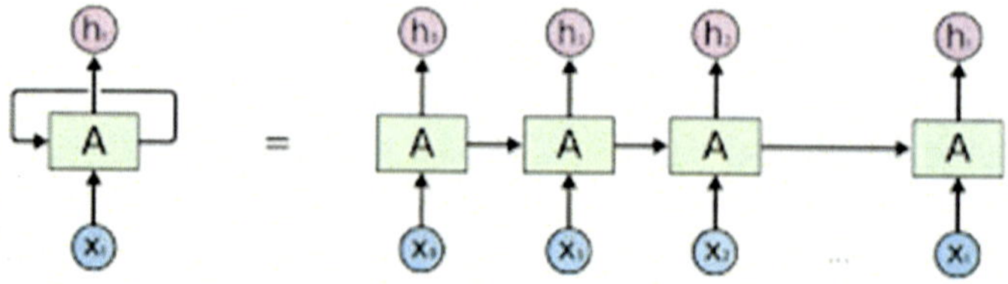

Fig. 1 A simple RNN

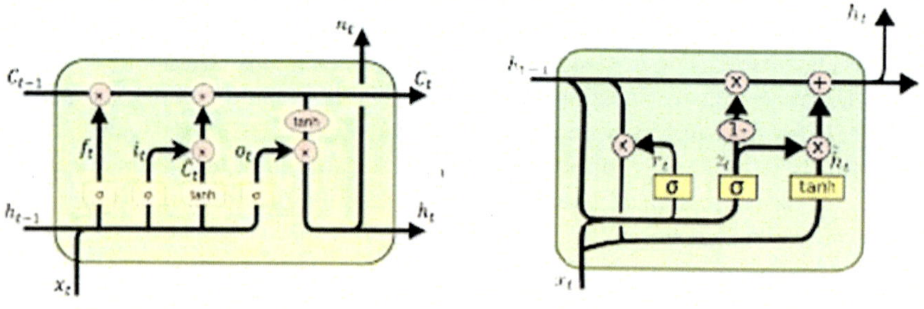

(a). A LSTM Cell (b). A GRU Cell

Fig. 2 RNN and its variants

5 Results and Discussion

This section presents and discusses the outcomes of the experiments conducted using the neural network models that have been developed. DELL EMC 740 with 128 GB RAM and 32 Gb GPU RAM is utilized to train models in this article. Hyper-parameters are critical in deep learning algorithms because they describe the details of training and have a direct impact on model output. We used Bayesian optimization to select optimal values for hyper-parameters that would yield good accuracy. It is a search method that employs Bayes' theorem to find the minimum or maximum of an objective function. This study uses accuracy as the objective function, which we would like to maximize. The Bayesian approach may increase the time spent on selecting the hyper-parameters but reduce the time spent evaluating the objective function, resulting in a lower computational cost. The set of hyper-parameters tuned in this study is optimizer, learning rate, activation function, and number of neurons. Other parameters like batch size, epochs, etc. are fixed. Table 1 summarizes the hyper-parameters tuned in our work and their search spaces.

A summary of the LSTM and GRU training descriptions can be seen in Table 2.

For the first set of experiments, we have translated the eight labels of the attack class into "attacks" in the CICDDoS2019 dataset. So, in the output layer, we used binary cross-entropy as a loss function. The data set CICDDoS2019 is divided into training and test sets in an 80/20 ratio. Training data is given for the proposed neural

Table 1 Hyper-parameters and their search space

Parameter	Search space	Description
Optimizer	Adam, RMSProp, SGD	To optimize the input weights by comparing the prediction and the loss function
Learning rate	1e−3, 1e−4, 1e−5, 1e−6	To determine the step size at each iteration while minimizing the loss function
Activation function	Relu, Elu and Tanh	To introduce non-linearity into the output of neurons
Number of neurons	64,128, 56, 512, 1024	To compute the weighted average of input

Table 2 Tuned hyper-parameters in LSTM and GRU

Parameters	Values
Optimizer	Adam
Loss	Binary cross-entropy/categorical cross-entropy
Activation	Relu
Learning rate	0.001
The number of neurons	1024

network models. Their performance against DDoS attacks for test data sets were assessed after training of LSTM and GRU. We used the hyper-parameter settings in Table 2 to run the built models. The models were run with a varied number of layers, and the results are shown in Table 3. From Table 3, it is understood that LSTM provides better classification accuracy for CICDDoS2019 data set. Due to the long-term dependent nature of LSTM, it produces good accuracy. In second set of experiments, we have evaluated the detection performance of the proposed models against each type of the attacks. Various metrics such as Accuracy, Precision, Recall, and F1-score were evaluated and presented in Tables 4, 5, 6, 7, 8, 9, 10 and 11.

As can be seen from Tables 4, 5, 6, 7, 8, 9, 10 and 11, LSTM outclassed GRU for all the types of attacks. Because of the long-term dependencies of the dataset, LSTM performed well. Another reason for better accuracy by LSTM is that the size of the

Table 3 Performance evaluation of LSTM and GRU on CICDDoS2019 dataset

Metrics	Epochs = 20					
	LSTM			GRU		
Number of layers	1	2	3	1	2	3
Training accuracy (%)	99.1	99.32	99.41	92.25	92.51	92.73
Testing accuracy (%)	99.02	99.43	99.38	92.87	93.07	93.55
Precision (%)	98.8	99.04	98.83	93.8	93.9	94.1
Recall (%)	98.56	98.3	99.12	92	92.67	92.9
F1-score (%)	99.01	99.3	99.23	92.9	93.29	93.56

Table 4 Performance evaluation of LSTM and GRU against LDAP

No. of layers	No. of epochs	Training accuracy (%)		Testing accuracy (%)		Precision (%)		Recall (%)		F1-score (%)	
		LSTM	GRU	LSTM	GRU	LSTM	GRU	LSTM	GRU	LSTM	GRU
1	5	90.07	90.84	90.67	91.20	84.41	84.4	99.21	99	91.56	91.5
1	10	90.80	90.96	91.05	91.35	84.46	84.4	99.43	99	91.55	91.5
2	5	90.55	90.14	90.71	90.78	84.42	84.4	99.12	99	91.50	91.5
2	10	92.62	90.50	93.13	91.34	84.44	84.4	99.37	99	91.52	91.5

Table 5 Performance evaluation of LSTM and GRU against NETBIOS

No. of layers	No. of epochs	Training accuracy (%)		Testing accuracy (%)		Precision (%)		Recall (%)		F1-score (%)	
		LSTM	GRU	LSTM	GRU	LSTM	GRU	LSTM	GRU	LSTM	GRU
1	5	95.87	94.19	96.09	94.94	92.9	92.9	99.9	99.9	96.3	96.35
1	10	96.29	94.75	99.15	95.0	92.9	92.9	99.9	99.9	96.3	96.3
2	5	95.13	94.45	94.85	99.30	92.9	92.9	99.9	99.9	96.3	96.3
2	10	94.02	95.03	94.99	94.74	92.9	92.9	99.9	99.95	96.3	96.3

Table 6 Performance evaluation of LSTM and GRU against NTP

No. of layers	No. of epochs	Training accuracy (%)		Testing accuracy (%)		Precision (%)		Recall (%)		F1-score (%)	
		LSTM	GRU	LSTM	GRU	LSTM	GRU	LSTM	GRU	LSTM	GRU
1	5	96.15	95.79	96.50	96.58	99.8	99.8	93.2	93.2	96.4	96.4
1	10	96.50	95.34	95.63	96.18	99.8	99.8	93.2	93.2	96.4	96.4
2	5	95.89	96.18	96.52	96.66	99.8	99.8	93.2	93.2	96.4	96.4
2	10	96.24	96.11	96.62	96.31	99.8	99.8	93.2	93.2	96.4	96.4

Table 7 Performance evaluation of LSTM and GRU against SSDP

No. of layers	No. of epochs	Training accuracy (%)		Testing accuracy (%)		Precision (%)		Recall (%)		F1-score (%)	
		LSTM	GRU	LSTM	GRU	LSTM	GRU	LSTM	GRU	LSTM	GRU
1	5	92.71	92.32	94.16	93.71	89	89	99	99	94.1	94.1
1	10	94.42	90.35	95.32	92.86	89	89	99	99	94.1	94.1
2	5	91.24	94.35	95.44	95.37	89	89	99	99	94.1	94.1
2	10	93.84	93.04	96.87	95.80	89	89	99	99	94.1	94.1

Table 8 Performance Evaluation of LSTM and GRU against MSSQ L

No. of layers	No. of epochs	Training accuracy (%)		Testing accuracy (%)		Precision (%)		Recall (%)		F1-score (%)	
		LSTM	GRU	LSTM	GRU	LSTM	GRU	LSTM	GRU	LSTM	GRU
1	5	93.18	92.81	94.11	93.01	88.1	88.1	99.9	99.9	93.6	93.6
1	10	92.87	92.99	93.04	94.31	88.1	88.1	99.9	99.9	93.6	93.6
2	5	92.06	92.42	92.93	93.04	88.1	88.1	99.9	99.9	93.6	93.6
2	10	92.92	92.94	93.05	93.05	88.1	88.1	99.9	99.9	93.6	93.6

Table 9 Performance evaluation of LSTM and GRU against UDP-LAG

No. of layers	No. of epochs	Training accuracy (%)		Testing accuracy (%)		Precision (%)		Recall (%)		F1-score (%)	
		LSTM	GRU	LSTM	GRU	LSTM	GRU	LSTM	GRU	LSTM	GRU
1	5	93.25	88.78	93.77	88.07	90.7	90.7	98	98	94.2	94.2
1	10	93.25	91.46	94	92.75	90.7	90.7	98	98	94.2	94.2
2	5	90.32	88.75	90.74	93.31	90.7	90.7	98	98	94.2	94.2
2	10	91.80	91.49	94.01	92.98	90.7	90.7	98	98	94.2	94.2

Table 10 Performance evaluation of LSTM and GRU against UDP

No. of layers	No. of epochs	Training accuracy (%)		Testing accuracy (%)		Precision (%)		Recall (%)		F1-score (%)	
		LSTM	GRU	LSTM	GRU	LSTM	GRU	LSTM	GRU	LSTM	GRU
1	5	92.9	91.57	93.72	93.33	94.6	94.6	99.9	99.9	97.2	97.2
1	10	95.67	93.91	95.73	94.78	94.6	94.6	99.9	99.9	97.2	97.2
2	5	92.56	91.39	93.97	95.45	94.6	94.6	99.9	99.9	97.2	97.2
2	10	94.69	94.56	96.19	95.52	94.6	94.6	99.9	99.9	97.2	97.2

Table 11 Performance evaluation of LSTM and GRU against TFTP

No. of layers	No. of epochs	Training accuracy (%)		Testing accuracy (%)		Precision (%)		Recall (%)		F1-score (%)	
		LSTM	GRU	LSTM	GRU	LSTM	GRU	LSTM	GRU	LSTM	GRU
1	5	87.75	87.66	89.83	90.41	83.4	83.4	99.4	99.4	90.7	90.7
1	10	88.85	90.27	91.72	91.07	83.4	83.4	99.4	99.4	90.7	90.7
2	5	89.28	90.40	91.16	91.46	83.4	83.4	99.4	99.4	90.7	90.7
2	10	89	89.81	91.02	91.085	83.4	83.4	99.4	99.4	90.7	90.7

Table 12 Comparison of proposed models

Models	Overall accuracy (%)
Proposed models:LSTM **GRU**	**99.40** **92.5**
Shurman and Khrais [10]	99.19
Cil et al. [12]	94.57
Bolodurina et al. [13]	98.62
Sindian et al. [14]	98

To show the results of the proposed models, the accuracy has been highlighted

dataset is huge. After evaluating the performance of the developed models for each type of attacks, we have also compared their performance against the systems that used the same data set and the results of comparison are presented in Table 12.

Only the models with high accuracy were used for comparison. However, these models are found to have more hidden layers or epochs. The proposed models were able to provide comparable performance with fewer layers or epochs. We assume that fine-tuning hyper-parameters has resulted in improved accuracy. In addition, since we used GPU, the proposed models took less time to practice. Even though, GRU uses time sequence to predict the class, in our attempt, it has given less accuracy compared with other models. Since, the proportion of benign and attack traffic is not balanced in the dataset, the GRU training might not have taken place appropriately. Using sampling techniques like SMOTE, random sampling, etc., the dataset may be balanced and then training can be carried out. This we would like to take up as a future work. The positive side of GRU is that it uses less training parameter and therefore uses less memory and executes faster than LSTM.

6 Conclusion and Future Work

In this work, we investigated the use of deep learning models to create a DDoS attack detection system in this work since deep learning has shown its promise in many domains. We looked at recent research on the topic and discovered that deep learning models can be used to boost the performance of existing systems. To boost accuracy even further, we created RNN-based LSTM and GRU. We evaluated the models in a number of network configurations parameters. These models LSTM and GRU gave an accuracy of 99.4% and 92.5%, respectively, for CICDDoS2019 data set. In the future, we want to expand the models to include very recent data sets with a large number of DDoS attack traces. Other learning models, such as auto encoder and stacked auto encoder, will be tried in the future to increase accuracy and discover new types of DDoS attacks. Also, sampling techniques like SMOTE can be applied on the dataset to address the data imbalance issues. Since the CICDDoS2019 dataset contains more than 80 features and all the features will not contribute for classification, feature selection algorithms may be employed to reduce the feature set and improve the training accuracy.

References

1. Hsieh C-J, Chan T-Y, Detection of DDoS attacks based on neural network using apache spark. In: International conference on applied system innovation (ICASI), 26–30 May 2016. https://doi.org/10.1109/ICASI.2016.7539833
2. Ferrag MA, Maglaras L, Moschoyiannis S, Janicke H (2020) Deep learning for cyber security intrusion detection: approaches, data sets, and comparative study. J Inf Secur Appl 50(4)
3. Doriguzzi-Corin R, Millar S, Scott-Hayward S, Martinez-del-Rinconand J, Siracusa D, LUCID: a practical, lightweight deep learning solution for DDoS attack detection IEEE Trans Netw Serv Manage. https://doi.org/10.1109/TNSM.2020.2971776
4. Asad M, Asim M, Javed T, Beg MO, Mujtaba H, Abbas S (2020) Deep detect: detection of distributed denial of service attacks using deep learning. Comput J 63(7):983–994
5. Anthi E, Williams L, Burnap P, Pulse: an adaptive intrusion detection for the internet of things. In: Living in the internet of things: cybersecurity of the IoT—2018 conference, 28–29 Mar 2018. https://doi.org/10.1049/cp.2018.0035
6. Doshi R, Apthrope N, Feamster N (2018) Machine learning DDoS detection for consumer Internet of Things devices. In: 2018 IEEE security and privacy workshops (SPW). https://doi.org/10.1109/SPW.2018.00013
7. Elsayed MS, Le-Khac N-A, Dev S, Jurcut AD (2020) DDoSNET—a deep learning model for detecting network attacks. In: 2020 IEEE 21st international symposium on "a world of wireless, mobile and multimedia networks" (WoWMoM), 31 Aug–3 Sept 2020. https://doi.org/10.1109/WoWMoM49955.2020.00072
8. Jiang J, Yu Q, Yu M, Li G, Chen J: ALDD: A hybrid traffic-user behavior detection method for application layer DDoS. In: 2018 17th IEEE international conference on trust, security and privacy in communications. https://doi.org/10.1109/TrustCom/BigDataSE.2018.00225
9. Shone N, Ngoc TN, Phai VD, Shi Q (2018) A deep learning approach to network intrusion detection. IEEE Trans Emerg Topics Comput Intell 2(1)
10. Shurman M, Khrais R, Yateem A (2020) DoS and DDoS attack detection using deep learning and IDS. Int Arab J Inf Technol 17(4A)
11. Muraleedharan N, Janet B (2020) A deep learning based HTTP slow DoS classification approach using flow data. https://doi.org/10.1016/j.icte.2020.08.005
12. Cil AE, Yildiz K, Buldu A, Detection of DDoS attacks with feed forward based deep neural network model. Expert Syst Appl 169:114520
13. Bolodurina I, Shukhman A, Parfenov D, Zhigalov A, Zabrodina L (2020) Investigation of the problem of classifying unbalanced datasets in identifying distributed denial of service attacks. J Phys Conf Ser 1679:042020
14. Sindian S, Sindian S (2020) An enhanced deep Autoencoder-based approach for DDoS attack detection. WSEAS Trans Syst Control 15
15. Kim J, Kim J, Kim H, Shim M, Choi E (2020) CNN-based network intrusion detection against denial of service attacks. Electronics 9(6):916
16. Dong S, Abbas K, Jain R (2019) A survey on Distributed Denial of Service (DDoS) attacks in SDN and cloud computing environments. IEEE Access 7:80813–80828. https://doi.org/10.1109/ACCESS.2019.2922196
17. Bhardwaj A, Goundar S (2017) Comparing single tier and three tier infrastructure designs against DDoS attacks. Int J Cloud Appl Comput 7:59
18. Kolias C, Kambourakis G, Stavrou A, Voas J (2017) DDoS in the IoT: Mirai and other botnets. Computer 7(7):80–84. https://doi.org/10.1109/MC.2017.201

19. Hou J, Fu P, Cao Z, Xu A (2018) Machine learning based DDos detection through NetFlow analysis. In: MILCOM 2018—2018 IEEE military communications conference (MILCOM), pp 1–6. https://doi.org/10.1109/MILCOM.2018.8599738
20. Ferrag MA, Maglaras L, Moschoyiannis S, Janicke H (2020) Deep learning for cyber security intrusion detection: approaches, datasets, and comparative study. J Inf Secur Appl 50:102419

Investigation of Required Numbers of Runs for Verification of Jaya Algorithm for a Structural Engineering Problem

Muhammed Çoşut, Gebrail Bekdaş, and Sinan Melih Nigdeli

Abstract In this study, a construction element using carbon fiber reinforced polymer (CFRP) material was optimized to increase the shear capacity of T-section reinforced concrete beams using the Jaya algorithm. For the objective function, it is aimed that the values of the spacing of the most suitable material strips, the width of the strips, and the angle of the strips are optimized to create a minimum area increasing the shear capacity with the desired amount. The algorithm has different values for 100 iterations and 1000 iterations when several runs (5, 10, 15, 20, 25, 30, 40, 50) of running the program are evaluated. In low iterations, it is observed that the standard deviations as a result of different runs are very big values, but it approaches the minimum area as the number of runs is increased. It has been observed that when the number of iterations is big, as the number of runs increases, it approaches the minimum area very close and the standard deviation takes very small values.

Keywords T-beam · CFRP · Optimization · Minimum area design · Jaya algorithm

1 Introduction

Many studies have been conducted to better sustain life and to better protect against factors such as disasters, explosions, natural disasters, and these studies are progressing day by day. As a result of increasing urbanization, the use of environmentally friendly and sustainable materials in building elements in general has led to an emphasis on different studies [1]. All works performed are designed to comply

M. Çoşut (✉) · G. Bekdaş · S. M. Nigdeli
Department of Civil Engineering, Istanbul University-Cerrahpaşa, 34320 Avcılar, Istanbul, Turkey
e-mail: muhammed.cosut@ogr.iuc.edu.tr

G. Bekdaş
e-mail: bekdas@iuc.edu.tr

S. M. Nigdeli
e-mail: melihnig@iuc.edu.tr

J. H. Kim et al. (eds.), *Proceedings of 7th International Conference on Harmony Search, Soft Computing and Applications*, Lecture Notes on Data Engineering and Communications Technologies 140, https://doi.org/10.1007/978-981-19-2948-9_39

with the regulations of the country in which they are performed. In this way, the application stage is started. But sometimes, even if the structure is designed in accordance with all the formulations, the structure may be damaged for some reason. This may be due to some application errors at the application stage or because the elements used for the structure do not meet the specifications. Under the earthquake loads on reinforced concrete elements, shear damage occurs suddenly, and the structure is damaged [2]. It has been seen that it is necessary to increase the cross-sectional dimensions to increase the capacity, but as a result of different studies, this will not be much needed. To prevent these situations, many studies have been conducted on the need to strengthen the structural elements in case of deterioration. One of them is the application of carbon fiber reinforced polymer (CFRP) used in building elements. In addition, it extends the service life of metallic bridges due to their very good performance [3–5]. CFRP is formed as a result of combining suitable fibers to contain at least 9% carbon. It has been observed that the strength and ductility of structural elements are increased by using CFRP [6]. This application contributes to the undesirable deterioration, corrosion, and strengthening of structures and structural elements. The use of this element is to optimize the space in such a way that it is minimal with the help of algorithms. The use of CFRP with a minimum area can be found using metaheuristic algorithms. Metaheuristic algorithms arise as a result of the creation of simple rules inspired by natural phenomena [7, 8]. In this study, the best objective function will be achieved by using the Jaya algorithm, which is a metaheuristic algorithm. The Jaya algorithm is named after a word that means victory [9]. This algorithm is evaluated for the requirements of multiple runs tı trust the correctiveness of the optimum results.

2 Optimum Carbon Fiber Reinforced Polymer Design for T-Beam

Beams that transfer loads from the floor to the columns can withstand the shear force and bending moments. Increasing the capacity of the building element can be done in different ways. The capacity of the T-beam can be increased by adding CFRP strips to the beam. This capacity increase for shear force is carried out using a specific formulation. The objective function will be calculated in such a way that the CFRP material with the minimum area is used. By running algorithms in the form of 5, 10, 15, 20, 25, 30, 40, 50 runs, checks were made and compared how close the objective function came to the most minimized state. These runs were performed for 100 iterations and 1000 iterations. For this purpose, our design variables are the distance of the CFRP variables depending on the width of the CFRP and the angle of the CFRP. Algorithms will be realized by designing according to variable boundaries. The constants and variables are shown in Table 1. This application detail is shown in Fig. 1.

Table 1 Constants and variables for CFRP

Definition	Symbol	Unit	Value
Breadth	b_w	mm	200
Height	h	mm	500
Effective depth	d	mm	450
Thickness of CFRP	t_f	mm	0.165
Reduction factor	R	–	0.5
Thickness of slab	h_f	mm	100
Comp. strength of concrete	f'_c	MPa	20
Effective tensile strength of CFRP	f_{fe}	MPa	3790
Width of CFRP	w_f	mm	10–100
Spacing of CFRP	s_f	mm	0–d/4
Angle of CFRP	β	°	0–90
Additional shear force	$V_{additional}$	kN	100
Shear force capacity of rebar	V_s	kN	50

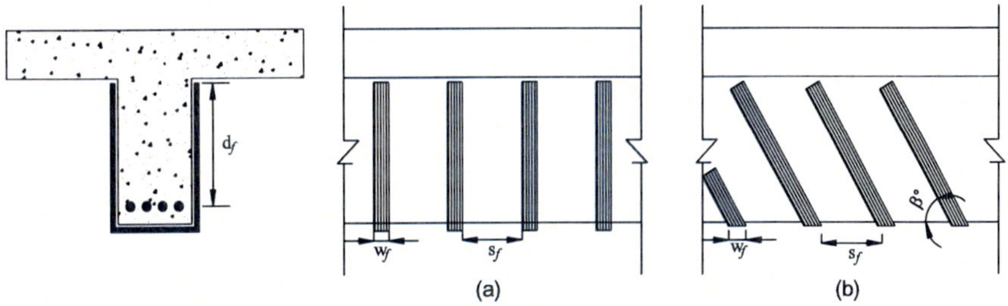

Fig. 1 T-beam with CFRP

The objective function (Eq. 1) is an account for the use of CFRP material with a minimum area that provides constraint values.

$$F_x = \frac{w_f\left(\frac{2d_f}{\sin\beta} + b\right)}{s_f} \times 1000 \tag{1}$$

There are three design constraints (Eqs. 2, 3, and 4) in this application.

$$g_1(x): s_f \leq \frac{d}{4} \tag{2}$$

$$g_2(x) : V_{\text{additional}} < 0.7R\frac{(2t_f w_f f_{fe})(\sin\beta + \cos\beta)d_f}{s_f + w_f} \tag{3}$$

$$g_3(x): \frac{(2t_f w_f f_{fe})(\sin\beta + \cos\beta)}{s_f + w_f} \leq \frac{2\sqrt{f_c'}b_w d}{3} - V_s \quad (4)$$

3 Jaya Algorithm

The Jaya algorithm is one of the metaheuristic algorithms. This algorithm always tries to approach the best solution to achieve success and to move away from bad solutions to get away from failure. The Jaya algorithm does not have any control parameters and is easy to implement [10, 11]. For these reasons, it is one of the most common algorithms to use. The formula of the Jaya algorithm shows in Eq. (5).

$$X_{i,\text{new}}' = X_{i,j} + r()\left(X_{i,g_{\text{best}}} - |X_{i,j}|\right) - r()\left(X_{i,g_{\text{worst}}} - |X_{i,j}|\right) \quad (5)$$

$X_{i,g_{\text{best}}}$	Best value for existing design variables
$X_{i,g_{\text{worst}}}$	Worst value for existing design variables
$X_{i,j}$	Candidate vector that is randomly selected
$X_{i,\text{new}}'$	The value of the new design variables
$r()$	The random number.

4 Numerical Results and Conclusion

In order to increase the sliding capacity of the T-section reinforced concrete beam, CFRP material will be used. By writing an algorithm with a minimum area to provide limit values for this material, the minimum area required to run the optimization process in the algorithm will be reached no matter how many times. Approach ratios were checked for different units and different iteration values.

For 100 iterations and 1000 iterations, processes were repeatedly run. Design variables such as CFRP width, spacing, and angle are presented for the best solution. Best F_{x}, worst F_{x}, and standard deviation were recorded for comparison.

It is observed that as a result of different runs in 100 iterations, the standard deviation becomes too large, and as the number of runs increases, there is some improvement in the objective function. It is shown in Table 2.

It is observed that for all the different decommissioning values in 1000 iterations, the objective function is very close to the objective function we want to achieve, and the differences between the standard deviation are reduced. It is shown in Table 3.

Table 2 For 100 iteration

Runs	w_f	s_f	β	Best F_x	Worst F_x	St. dev.
5	99.26	101.8	65.8	477,606	511,156	13,718
10	73.6	68	63.4	477,601	486,310	2827
15	100	103.8	64.8	477,605	503,760	7713
20	18.5	17.6	70.6	477,580	667,814	44,431
25	95.3	97.4	65.9	477,596	503,205	6905
30	82.9	63.4	80.2	477,572	517,550	10,932.2
40	96.4	99.1	65.5	477,578	525,456	9929.5
50	62.6	65.4	63.7	477,569	528,597	9327.2

Table 3 For 1000 iteration

Runs	w_f	s_f	β	Best F_x	Worst F_x	St. dev.
5	61.9	63.7	65.5	477,567.7	477,594	11.5
10	101.3	104.6	65.2	477,567.62	477,575	2.89
15	108.8	112.5	65.2	477,567.62	477,596	8.9
20	35.3	36.5	65.2	477,567.62	478,809	276
25	108.8	112.5	65.1	477,567.8	477,590	6
30	108.4	112.1	65.2	477,567.62	477,581	3.5
40	108.8	112.5	65.2	477,567.61	477,603	7.25
50	55.8	57.7	65.2	477,567.61	477,647	12.3

5 Conclusion

CFRP element was used to strengthen the T-section beam, and a construction system was created to ensure the use of CFRP material with a minimum area and an application that provides constraint values was carried out. The Jaya algorithm is suitable for this restricted problem. The purpose of the study is to investigate that how many times it is necessary to run the optimization process to reach the lowest area correctly. For 100 iterations and 1000 iterations, the comparison was performed by running 5, 10, 15, 20, 25, 30, 40, 50 times separately. It is observed that if the number of iterations and the number of runs are increased, it approaches the minimum area with small standard deviations. However, it has been observed that the standard deviations remain at a very big value in low iterations. In the case of an increase in the number of iterations, decreases in the standard deviations were observed and came very close to the objective function. In this case, it is concluded that if the iteration numbers are large in the algorithm studies, the objective function will reach its value better, and the standard deviations will be reset in the final algorithm. For 100 iterations, the standard deviation values show decreasing and increasing until 40 runs, and 40 and 50 runs have similar standard deviation. According to the findings of this study, the number of runs that are needed for the correct evaluation is 40 according to 100 iteration results.

References

1. Junaid MT, Karzad AS, Elbana A, Altoubat S, Experimental study on shear response of GFRP reinforced concrete beams strengthened with externally bonded CFRP sheets. ScienceDirect
2. Ssad
3. Deng J, Li J, Zhu M (2021) Fatigue behavior of notched steel beams strengthened by a prestressed CFRP plate subjected to wetting/drying cycles. Compos Part B: Eng 230:109491
4. Teng J, Yu T, Fernando D (2012) Strengthening of steel structures with fiber-reinforced polymer composites. J Constr Steel Res 78:131–143
5. Hosseini A, Ghafoori E, Al-Mahaidi R, Zhao X-L, Motavalli M (2019) Strengthening of a 19th-century roadway metallic bridge using nonprestressed bonded and prestressed unbonded CFRP plates. Constr Build Mater 209:240–259
6. Al-Ridha ASD, Mahmoud KS, Atshan AF (2020) Effect of carbon fiber reinforced polymer (CFRP) laminates on behaviour of flexural strength of steel beams with and without end anchorage plates. Sciencedirect
7. Zhang Y, Chi A, Mirjalili S (2021) Enhanced Jaya algorithm: a simple but efficient optimization method for constrained engineering design problems. Sciencedirect 233:107555
8. Eskandar H, Sadollah A, Bahreininejad A, Hamdi M (2012) Water cycle algorithm—a novel metaheuristic optimization method for solving constrained engineering optimization problems. Comput Struct 110–111:151–166
9. Ding Z, Li J, Hao H (2019) Structural damage identification using improved Jaya algorithm based on sparse regularization and Bayesian inference. Sciendirect 132:211–231
10. Kang F, Wu Y, Li J, Li H (2021) Dynamic parameter inverse analysis of concrete dams based on Jaya algorithm with Gaussian processes surrogate model. Adv Eng Inf 49:101348
11. Rao RV, Waghmare GG (2016) A new optimization algorithm for solving complex constrained design optimization problems. Eng Optim 49:60–83

Performance Comparison of Different Convolutional Neural Network Models for the Detection of COVID-19

S. V. Kogilavani, R. Sandhiya, and S. Malliga

Abstract Coronavirus, a new virus, has emerged as a pandemic in recent years. Humans are becoming infected with the virus. In the year 2019, the city of Wuhan reported the first ever incidence of coronavirus. Coronavirus-infected people have symptoms that are related to pneumonia, and the virus affects the body's respiratory organs, making breathing difficult. A real-time reverse transcriptase polymerase chain reaction (RT-PCR) kit is used for diagnosis of disease. Due to a shortage of kits, suspected patients are unable to be treated in a timely manner, which results in spreading of disease. To come up with an alternative, radiologists looked at the changes in radiological imaging like CT scans. The suspected patient's computed tomography (CT) scan is used to distinguish between a healthy individual and a coronavirus patient using deep learning algorithms. For COVID-19, a lot of deep learning methods have been proposed. The proposed work utilizes CNN architectures like ResNet101, InceptionV3, VGG19, NASNet, and VGG16. Dataset contains 3873 total CT scan images with the class labels as COVID-19 and non-COVID-19. The dataset is divided into train, test, and validation. Accuracy obtained for ResNet101 is 77.42%, InceptionV3 is 78%, VGG19 is 82%, NASNet is 89.51%, and VGG16 is 97.68%, respectively. From the obtained analysis, the results show that the VGG16 architecture gives better accuracy compared to other architectures.

Keywords Deep learning · CNN—convolutional neural network · Artificial intelligence · COVID-19 · Lung CT scan

---

S. V. Kogilavani (✉) · R. Sandhiya · S. Malliga
Department of Computer Science and Engineering, Kongu Engineering College, Perundurai, Erode, India
e-mail: kogilavani.sv@gmail.com

R. Sandhiya
e-mail: sandhiyar.20mcse@kongu.edu

J. H. Kim et al. (eds.), *Proceedings of 7th International Conference on Harmony Search, Soft Computing and Applications*, Lecture Notes on Data Engineering and Communications Technologies 140, https://doi.org/10.1007/978-981-19-2948-9_40

1 Introduction

Coronaviruses are microorganisms that can infect the intestines or lungs and cause illnesses. The infections in the lungs can range from a simple cold to a life-threatening condition. Corona infections are often accompanied by symptoms of the respiratory system. The symptoms of individuals may have a minor, self-limiting illness with negative effects like influenza on rare occasions. Fever, cough, and difficulty breathing are among the symptoms as are respiratory issues, weariness, and a sore throat [1].

The use of X-rays and computed tomography scans is one of the most essential approaches to diagnose COVID-19. Chest imaging is a quick and efficient method that is suggested by medical health regulations, and it has been highlighted in a number of literature as the first instrument in epidemic screening. Different computer vision approaches are used such as segmentation and classification. An automated technique that can provide fragmentation and measurement of the infection region of patients every three to five days as well as monitoring the evolution of infected patients through CT scan imaging and clinical detection is required. COVID-19 is a difficult virus to diagnose even for expert doctors [2].

Many studies have been undertaken on the use of deep learning in the interpretation of radiological images that have been undertaken to solve the constraints of COVID-19 medical techniques based on radiological images. Among the most significant deep learning algorithms, the most effective approach for detecting cancer and illnesses in chest radiographs CNN architecture is used. Data processing in relation to deep learning algorithms notably CNN has received much interest. The contribution of this research work is to

- Collect the COVID-19 sample dataset from Kaggle which contains 3873 CT scan images.
- Preprocess the dataset to make all images in same size.
- The preprocessed dataset is split into training, validation, and test data.
- The training dataset is fed into different CNN architectures like ResNet101, InceptionV3, VGG19, NASNet, and VGG16.
- The trained models are validated using validation dataset with 50 epochs.
- Now, the models are tested by supplying test data.

The rest of this paper is formulated in the following way: The results of recent COVID-19 identification study are presented in the literature review. The proposed system section discusses about the detection of COVID-19. This section also gives full description of the classification models utilized in the proposed system. The steps involved in the architecture are described in the section on system architecture. The results and discussion section contains a complete analysis and comparison of CNN models' performance. Finally, in the conclusion section, a brief summary of the proposed work and future work is specified.

2 Literature Review

In early 2020, the coronavirus outbreak has become a worldwide epidemic. The sickness was declared as a significance of international public health emergency by the World Health Organization, and the condition was considered as health emergency. Automatic detection of lung infections through CT scans provides an excellent opportunity to extend traditional healthcare methods to address COVID-19. But, CT has many problems [1]. CNN is used to detect tumors in lungs, pneumonia, tuberculosis, emphysema, or other pleural diseases. The disadvantages of the CT system are as follows: The contrast of the soft tissues is lower than that of the MRI because it is an X-ray exposure to radiation [2].

The suspected patient's X-ray and CT scan can be used to distinguish between a healthy person and a coronavirus patient using deep learning algorithms. Deep learning models are employed in creating diagnosis systems of coronavirus. DenseNet121, VGG16, Xception, EfficientNet, and NASNet are the architectures employed, and multiclass classification is used. Positive individuals of COVID-19, normal patients, and other patients are also considered. Chest X-ray images indicating pneumonia, flu, and other chest-related disorders belong into another category. Accuracies obtained for the architectures are VGG16 achieves 79.01%, EfficientNet achieves 93.48, Xception achieves 88.03, NASNet achieves 85.03, and DenseNet121 achieves 89.96%, respectively [3].

For disease diagnosis, the algorithms presented include the DNN based on imaging features of fractal and the CNN that directly uses lung imaging. The suggested architecture of CNN with higher accuracy of 93.2% and sensitivity of 96.1% outperforms the DNN technique which has 83.4% precision and 86% sensitivity. CNN architecture is offered during the segmentation phase to detect contaminated tissue in the lung picture. The results show that this method can detect almost 4444 infected areas with an accuracy rate of 83.84%. And the finding is used to monitor and control the growth of the patient's protected area [4].

Preprocessing, dictionary building, and picture classification are the three key stages of the classification approach based on a set of features. Features are manually retrieved and passed to a classifier neural network in the suggested method, and it is compared to modern methods in an experimental environment. On three datasets, the technique has an accuracy of 96.1, 99.84, and 98%. These results are superior than those obtained using modern approaches. In a visual word bag, the SURF method is utilized to extract objects. Because the SURF technique is dependent on gradients, the obtained solution is noise sensitive. The offered approaches may misclassify the image if the image quality is inadequate. In this case, picking the right pretreatment procedure can help you get better outcomes [5].

A new multi-core deep neural network method is proposed to detect the coronavirus disease which is also known as COVID-19 using chest CT images. In this paper, the characteristics were extracted from lung CT images using a CNN. A predefined DenseNet201 CNN architecture based on transfer learning is employed for this purpose. The ELM method classifier depends on various activation algorithms which calculate the performance of architecture. When applying the MKsELMDNN model, the accuracy score reached is 98.36% according to the data [6].

Machine learning techniques based on X-ray imaging are utilized as a decision support mechanism to assist radiologists in speeding up the diagnosis process. A critical review is of 12 conventional CNN designs which were first proposed for natural image processing with the goal of assisting radiologists in distinguishing COVID-19 disease from radiographic pictures of the chest. COVID-19 X-ray pictures were used as well as a huge dataset of non-COVID-19 viral illnesses, bacterial infections, and routine radiographs was also used. When trained on a tiny image dataset, a simple CNN design can outperform architectures like Xception and DenseNet. Finally, despite their excellent classification accuracy, therapists should not examine CNN conclusions until they can visually analyze the region of the input image acquired by the CNN [7].

For automatic COVID-19 categorization, different deep learning methods by extracting their features were compared. MobileNet, ResNetV2, VGGNet, ResNet, InceptionV3, DenseNet, Xception, Inception, and NASNet had been selected from a huge list of convolutional neural networks in order to produce the most accurate feature, which is an important part of learning. The features that are collected are fed into a series of machine learning classifiers to determine whether the subjects were COVID-19 cases or controls. To promote a stronger generalization capacity for unknown data, this strategy achieved task-specific data preprocessing approaches. The accuracy of the method was tested using the publicly available COVID-19 dataset of chest X-ray and CT images. With a bagging tree classifier, DenseNet121 obtained 99%, while ResNet50 achieved % accuracy [8].

Chest X-ray images have recently emerged as a promising option for COVID-19 screening when combined with current AI techniques particularly DL algorithms. The classification of COVID-19 from normal cases was evaluated using eight architectures like AlexNet, GoogleNet, SqueezeNet, VGG16, ResNet50, MobileNetV2, ResNet34, and InceptionV3, respectively. The models were evaluated on publicly accessible chest X-ray images with ResNet34 getting the best results with an accuracy of 98.33% [9].

A CT scan involves slides of hundred scans, and using such scans to diagnose COVID-19 can cause delays in hospitals. The artificial intelligence tools could help radiologists to diagnose COVID-19 infection in these images more quickly and correctly. Using artificial intelligence, this study offers a technique for identifying COVID-19 and non-COVID-19 classes. The suggested AI method predicts COVID-19 in each 3D CT scan image using the ResNet50 deep learning model. Using image-level predictions, an AI method detects COVID-19 in 3D CT volume. With an accuracy of 96%, the suggested deep learning model detects the disease on CT images [10].

The ten possible best convolutional neural networks used to distinguish the infection are AlexNet, GoogleNet, VGG19, ResNet101, VGG16, ResNet50, MobileNetV2, ResNet18, SqueezeNet, and Xception. All of the networks performed best except ResNet101 and Xception. For discriminating the disease, ResNet101 had an accuracy of 99.4%, while Xception had an accuracy of 99%. ResNet101 is a moderate model for identifying and detecting COVID-19 infections in radiology departments which may be used as a replacement [11].

DenseNet, InceptionV3, and Inception-ResNetV4 were recommended as three different models. In the investigation, the chest X-ray radiographs were used to diagnose individuals with coronavirus and pneumonia. Using 5-fold cross-validation, these three models create and evaluate ROC curve analyses and uncertainty matrices. In simulations, the pre-trained DenseNet architecture had the optimum classification efficiency of 92%, while the other two models InceptionV3 and Inception-ResNetV4 achieved accuracy of 83.47% and 85.57%, respectively [12].

Radiological imaging in advanced artificial intelligence techniques can aid in the precise disease detection as well as by overcoming the problem of shortage in expert physicians on rural areas. Based on raw chest X-ray images, this paper offers a new method for automated COVID-19 identification. The suggested technique aims to offer correct diagnostics for the binary and multiclass classification in binary and multiclass environments. The model gives accuracy of 98.08% for binary classes and 87.02% for the instance of multiclass [13].

The binary classification method was trained using 3877 CT and X-ray images including 1917 images of COVID-19 patients. The binary classification had a 99.64% overall accuracy, 99.58% recall, 99.56% precision, 99.59% F1-score, and 100% ROC. COVID-19 contains instances of 1917, normal healthy people class contains instances of 1960, and pneumonia contains 2200 instances. By using this instances, the classifier was tested on a total of 6077 images for different classifications. The multiclass has a 99.87% ROC, 98.28% accuracy, 98.25% recall, 98.22% F1-score, and 98.22% precision [14].

3 Proposed System

3.1 Deep Learning Techniques

Artificial intelligence technique that resembles how humans acquire knowledge and deep learning is similar to machine learning techniques. Data science, which covers statistics and predictive modeling, includes deep learning as a key component. In deep learning, a convolutional neural network is a kind of deep neural network that is used to analyze visual imagery. A deep learning method CNN takes input image and assigns weight to various objects in the image, allowing it to differentiate between them. Because of its great accuracy, CNNs are used to classify and identify images [15].

3.2 Classification

Deep learning architectures, namely ResNet101, InceptionV3, VGG19, NASNet, and VGG16, are used to classify the data. Transfer learning is used to train these models. Each model has been trained for a total of 50 epochs.

ResNet101. ResNet is a well-known neural network that serves as the basis for many computer vision applications. There are 101 layers in the neural network. The network can be loaded with a pre-trained version that has been trained on over a thousand images from the database [11].

InceptionV3. The InceptionV3 is intended to maximize the use of computational services within the network by expanding the network's width and depth while maintaining the same computing processes. The term "inception modules" was developed by the network's architects to represent an optimized network topology with skipped connection that may be utilized as a construction block. To reduce dimensions in workable level for computing, this inception module is replicated physically by layering with frequent max-pooling layers [8].

NASNet. The Google ML team created the NAS network. Reinforcement learning is used to build the network architecture. The network adjustment is made based on the changes in the effectiveness of child block. The parental block evaluates the effectiveness of the children block. RNN and CNN are the network's components. Various changes to the architecture were made to gain the optimum performance from the network including weights, regularization methods, layers, optimizer functions, and so on. Reinforced evolutionary methods are used to select the best candidates and to choose the best cells utilizing various NASNet variants like A, B, and C algorithms [16].

VGG16 and VGG19. VGG16 and VGG19 are CNN models, and the model was created by the VGG at Oxford University. The network's replacement, AlexNet, was founded in 2012. The model may also be used for transfer learning because certain frameworks, like as Keras, provide pre-trained weights that can be utilized to construct custom models with little changes. VGG16 consists of 16 layers, whereas VGG19 consists of 19 layers [3].

3.3 Proposed System Architecture

Chest tomography CT scan images is given as the input (see Fig. 1). CT images are preprocessed. The image must fit the network's input size in order to train it and generate predictions on the data. The data is rescaled to fit the network's input size. In proposed system, the size of the input image is 224 × 224. So, the data is rescaled according to the input size. CNN architectures like ResNet101, InceptionV3, VGG19, NASNet, and VGG16 are performed for detection of COVID-19 [17–20].

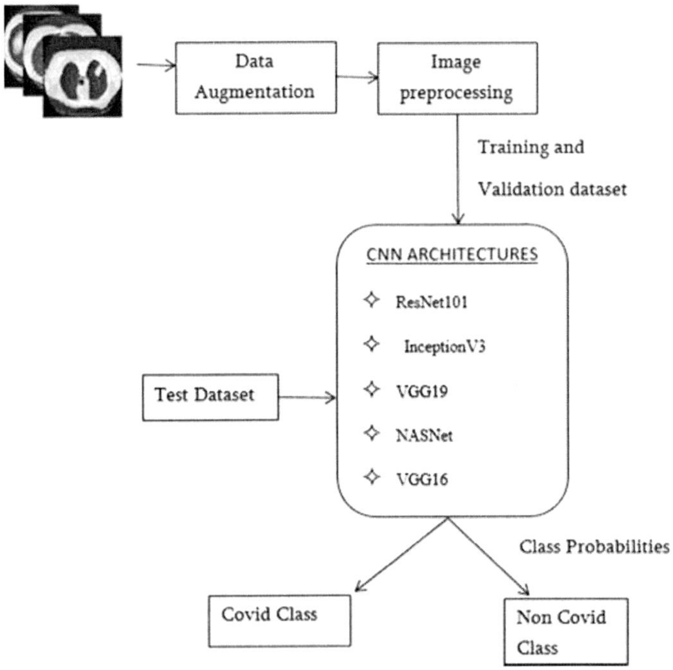

Fig. 1 Proposed system architecture

Data augmentation such as cropping, horizontal flipping generally produces new images by zooming in and out based on the kind of input parameter given in Table 1. The amount of training examples used in one iteration is referred to as batch size. For all models, a batch size of 16 is chosen. Optimizers are algorithms used to change attributes of neural networks such as weight and learning rate to reduce the losses. Adam optimizer is used for optimization.

To reduce the nonlinearity into the output of a neuron, certain activations functions are used. The output layer's activation function determines the kind of predictions the model can make. In the proposed system, softmax function is used as the activation function for all the models. Softmax is used in the last layer which is the output layer to predict a multinomial probability distribution. Loss is the prediction error of network,

Table 1 Parameters for training the model

Performance	CNN models
Batch size	16
Image dimension	224 × 224
Optimizer	Adam
Activation function	Softmax
Loss function	Binary cross-entropy

Table 2 Kaggle dataset description

Dataset	COVID-19	Non-COVID-19
Training	930	915
Validation	156	164
Test	166	150
Total	1252	1229

and loss function is the method used to calculate the error. And the loss function also calculates gradients. Gradients are used to update the weights of neural network. In binary classification tasks, binary cross-entropy compares predicted probability to actual class output, which might be either 0 or 1. Image augmentation is the process of expanding the available dataset for training the model. The dataset is divided into training, validation, and testing. The collection of samples used to learn how to suit the parameters is referred to as training. Validation is a collection of examples used to fine-tune a classifier's parameters. The data is trained and validated for 50 epochs, and the class probability of the images is generated. By evaluating the performance measures like accuracy, precision, recall, and F1-score, the results were calculated.

4 Performance Evaluation Measures

4.1 Dataset

The dataset collection includes lung CT scan images. CT scan utilizes an advanced X-ray technology to carefully diagnose sensitive internal organs. Dataset was taken from Kaggle [17] and consists of 3873 images. COVID-19 and non-COVID-19 are the two categories in which the data is divided. The COVID-19 class includes CT scan images from COVID-19 patients, while the non-COVID-19 class includes CT scan images from healthy individuals. There are 1958 CT scan images in the COVID-19 class and 1915 CT scan images in the non-COVID-19 class. The model is trained on 70% of lung CT scans, validated on 15% of lung CT scans, and tested on 15% of lung CT scans as illustrated in Table 2.

4.2 Preprocessing

Squared image or the image with some predetermined aspect ratio is scaled with approximate height and width. Image filtering preprocessing technique is used to filter the size of all input samples. In the proposed system, the images are rescaled to 224 × 224.

Table 3 Augmented dataset description

Dataset	COVID-19	Non-COVID-19	Total
Training	1257	1234	2491
Validation	345	346	691
Test	356	335	691
Total	1958	1915	3873

Table 4 Confusion matrix representation

	Predicted	
Actual	Yes	No
Yes	TP	TN
No	FN	FP

4.3 Image Augmentation

It is the process of expanding the existing dataset for training the model. To generate new samples of images, existing data is altered using generative adversarial networks (GAN) augmentation techniques. GAN consists of two neural models, and the goal of the technique is to learn from the training data and generation of new data with same characteristics of training data. The description of augmented dataset is illustrated in Table 3.

4.4 Performance Measures

There are several methods to evaluate a model's performance. Accuracy, precision, recall, and F1-score are the measures considered for evaluation of chest CT scan images. In general, confusion matrix is represented as in Table 4.

Where TP denotes true positive, TN denotes true negative, FN represents false negative, and FP represents false positive.

Precision—The ratio of correctly predicted positive cases is known as precision as given in Eq. (1).

$$\text{Precision} = \frac{\text{TP}}{\text{TP} + \text{FP}} \tag{1}$$

Recall—The ratio of accurately detected positive cases is known as recall as given in Eq. (2).

$$\text{Recall} = \frac{\text{TP}}{\text{TP} + \text{FN}} \tag{2}$$

F1-score—F1-score is the harmonic mean of precision and recall and is given in Eq. (3).

$$\text{F1score} = 2 \times \frac{\text{Precision} \times \text{Recall}}{\text{Precision} + \text{Recall}} \tag{3}$$

Accuracy—The percentage of correct predictions in the total number of predictions is called accuracy, and it is specified in Eq. 4.

$$\text{Accuracy} = \frac{\text{TP} + \text{TN}}{\text{TP} + \text{FP} + \text{TN} + \text{FN}} \tag{4}$$

5 Results and Discussion

The confusion matrix that lists the number of correct and wrong classification model predictions was also calculated. The confusion matrix obtained for all CNN models is illustrated (see Fig. 2).

By the analysis, VGG16 gives better results as the number of parameters trained in VGG16 is less, and it takes less time to train the samples. So, it is better than other CNN models.

The accuracies obtained for the proposed CNN models are illustrated (see Fig. 3). VGG16 model has trained with 138 million parameters, so the model performs better than other CNN models. With huge 138 billion parameters available in VGG16 makes this model a slower one, but this model is best one to produce better accuracy. ResNet101 model gives accuracy of 77.42%, InceptionV3 model gives accuracy of 78%, VGG19 model gives accuracy of 82%, NASNet model gives accuracy of 89.51%, and VGG16 model gives accuracy of 97.68%.

6 Conclusion

In the proposed research work, high-performance deep learning architectures like ResNet101, InceptionV3, VGG19, NASNet, and VGG16 are employed to detect COVID-19. Preprocessing is applied to make all images into same size. Then, training dataset is applied to all the CNN models, and the models are validated using validation dataset. The highest accuracy is obtained from VGG16, i.e., 97.68%. Hence, the proposed system identifies VGG16 model as the best model to classify the given CT scan images into COVID-19 and non-COVID-19. But, the limitation of the proposed work is that it does not identify COVID-19-affected area in Lungs. So, further enhancement is needed to detect affected areas in Lungs by considering large dataset.

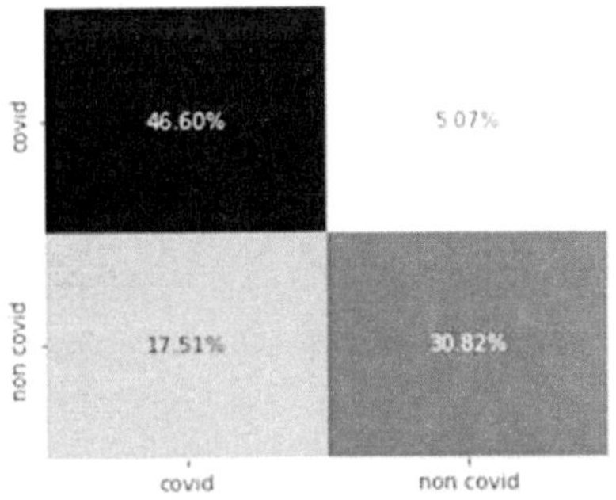

a) ResNet101 architecture predicts 322 correctly identified covid images and 213 correctly identified Non-covid images.

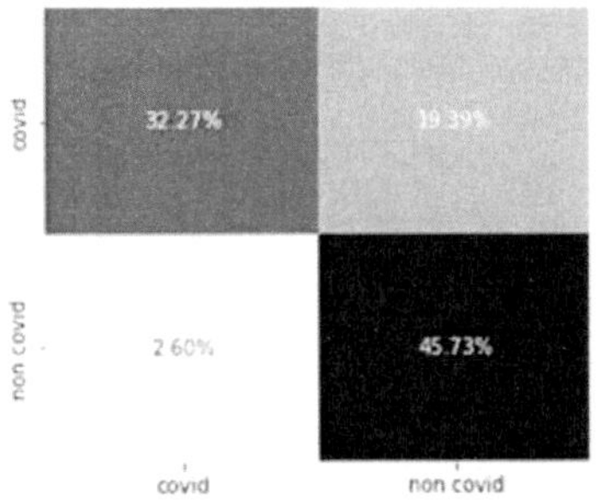

b) InceptionV3 architecture predicts 322 correctly identified covid images and 213 correctly identified Non-covid images

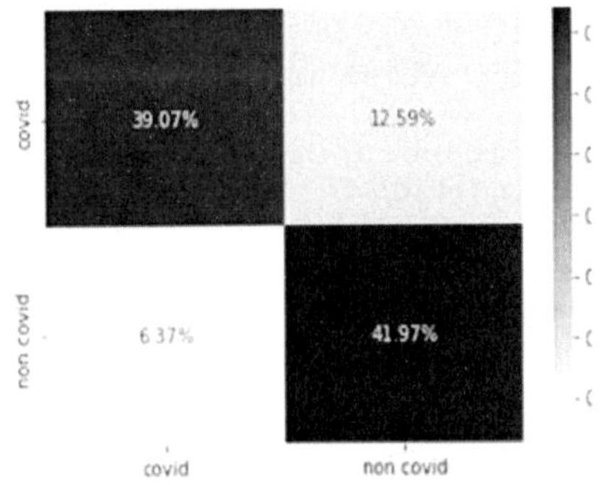

c) VGG19 architecture predicts 270 correctly identified covid images and 290 correctly identified Non-covid images.

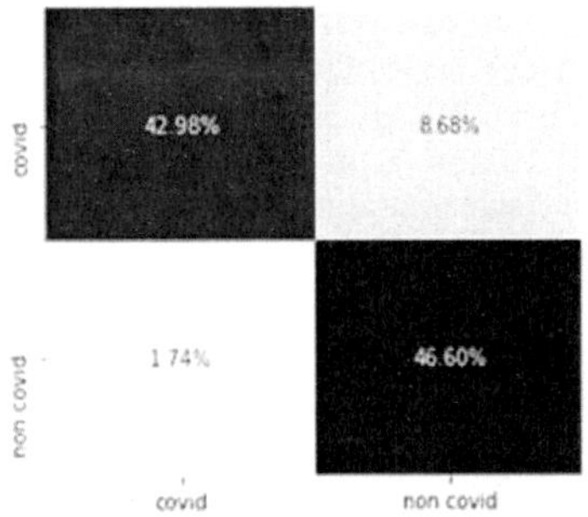

d) NASNet architecture predicts 297 correctly identified covid images and 322 correctly identified Non-covid images.

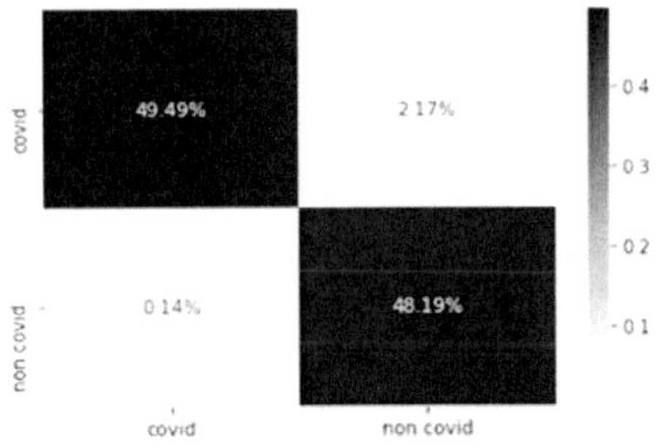

e) VGG16 is more desirable than other models and the model predicts 342 correctly identified Covid images and 333 correctly identified Non-covid images out of 691 samples.

Fig. 2 Confusion matrices of CNN models

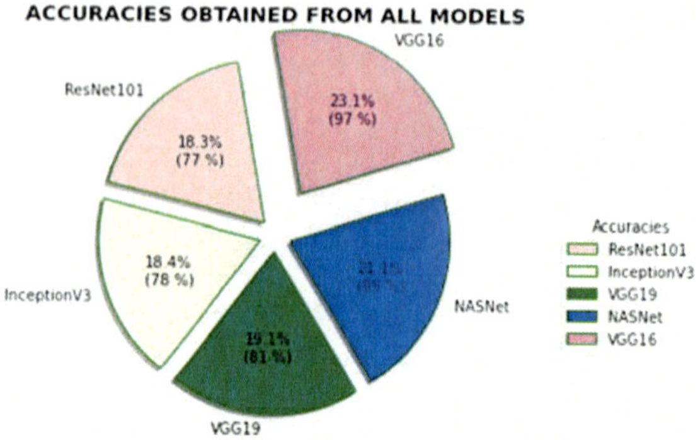

Fig. 3 Accuracies obtained for all CNN models

References

1. Priya C et al (2021) Automatic optimized CNN based COVID-19 lung infection segmentation from CT image. Mater Today Proc
2. Oulefki A et al (2021) Automatic COVID-19 lung infected region segmentation and measurement using CT-scans images. Pattern Recogn 114:107747
3. Nigam B et al (2021) COVID-19: automatic detection from X-ray images by utilizing deep learning methods. Expert Syst Appl 176:114883
4. Hassantabar S, Ahmadi M, Sharifi A (2020) Diagnosis and detection of infected tissue of COVID-19 patients based on lung X-ray image using convolutional neural network approaches. Chaos Solitons Fractals 140:110170
5. Nabizadeh-Shahre-Babak Z et al (2021) Detection of COVID-19 in X-ray images by classification of bag of visual words using neural networks. Biomed Signal Process Control 68:102750
6. Turkoglu M (2021) COVID-19 detection system using chest CT images and multiple kernels-extreme learning machine based on deep neural network. IRBM
7. Majeed T et al (2020) Covid-19 detection using CNN transfer learning from X-ray images. medRxiv
8. Kassania SH et al (2021) Automatic detection of coronavirus disease (COVID-19) in X-ray and CT images: a machine learning based approach. Biocybernetics Biomed Eng 41(3):867–879
9. Nayak SR et al (2021) Application of deep learning techniques for detection of COVID-19 cases using chest X-ray images: a comprehensive study. Biomed Signal Process Control 64:102365
10. Serte S, Demirel H (2021) Deep learning for diagnosis of COVID-19 using 3D CT scans. Comput Biol Med 132:104306
11. Ardakani AA et al (2020) Application of deep learning technique to manage COVID-19 in routine clinical practice using CT images: results of 10 convolutional neural networks. Comput Biol Med 121:103795
12. Albahli S, Ayub N, Shiraz M (2021) Coronavirus disease (COVID-19) detection using X-ray images and enhanced DenseNet. Appl Soft Comput 110:107645
13. Ozturk T et al (2020) Automated detection of COVID-19 cases using deep neural networks with X-ray images. Comput Biol Med 121:103792
14. Thakur S, Kumar A (2021) X-ray and CT-scan-based automated detection and classification of COVID-19 using convolutional neural networks (CNN). Biomed Signal Process Control 69:102920
15. Fouladi S et al (2021) Efficient deep neural networks for classification of COVID-19 based on CT images: virtualization via software defined radio. Comput Commun
16. Polsinelli M, Cinque L, Placidi G (2020) A light CNN for detecting COVID-19 from CT scans of the chest. Pattern Recogn Lett 140:95–100

17. Soares E, Angelov P (2020) SARS-COV-2 Ct-scan dataset. Kaggle. https://doi.org/10.34740/KAGGLE/DSV/1199870
18. Mishra NK, Singh P, Joshi SD (2021) Automated detection of COVID-19 from CT scan using convolutional neural network. Biocybernetics Biomed Eng 41(2):572–588
19. Shahid O et al (2021) Machine learning research towards combating COVID-19: virus detection, spread prevention, and medical assistance. J Biomed Inform 117:103751
20. Abraham B, Nair MS (2020) Computer-aided detection of COVID-19 from X-ray images using multi-CNN and Bayesnet classifier. Biocybernetics Biomed Eng 40(4):1436–1445

A Novel Cosine Swarm Algorithm for Solving Optimization Problems

Priteesha Sarangi and Prabhujit Mohapatra

Abstract In this paper, a robust swarm-inspired algorithm has been proposed known as Cosine algorithm (CA) to solve the optimisation problem. The CA generates several initial random agents' solution and requires all of them to change towards or outwards the ideal solution by means of mathematical model on Cosine function. A number of adaptive and random variables are also added into this method to promote exploitation and exploration of the search space at certain optimization milestones. The results of performance metrics and test functions demonstrate that the developed algorithm is capable of successfully exploring diverse areas of a search space, avoiding local optima, converging towards the worldwide optimum and exploiting potential parts of a search space through optimisation.

Keywords Swarm intelligence · Evolutionary algorithms · Sine–Cosine algorithm

1 Introduction

The procedure of determining ideal values for the given system's parameters from all feasible values in order to decrease or increase system's output is referred to as optimisation. Optimisation issues may be observed in many fields of study, making the progress of optimization approaches both crucial and fascinating for academics [1]. Because of the disadvantages of traditional optimization models, such as stagnation of local optima and the requirement for deriving the search space, there has been an increased interest in stochastic optimisation methodologies during the last two decades. Theoretical research in the literature may be classified into three broad categories: hybridising diverse algorithms, enhancing existing approaches and introducing new algorithms [2]. A newer algorithm has been inspired by evolutionary

P. Sarangi · P. Mohapatra (✉)
Department of Mathematics, Vellore Institute of Technology, Vellore, India
e-mail: prabhujit.mohapatra@gmail.com; prabhujit.mohapatra@vit.ac.in

P. Sarangi
e-mail: pritishasarangi@gmail.com

J. H. Kim et al. (eds.), *Proceedings of 7th International Conference on Harmony Search, Soft Computing and Applications*, Lecture Notes on Data Engineering and Communications Technologies 140, https://doi.org/10.1007/978-981-19-2948-9_41

occurrences, collective behaviour of creatures, human-related notions and physical principles. Regardless of the requirement for more function assessments, the research reveals that population-based technique are superior to individual-based algorithms to solve the real-world issues because they have the ability to examine the search space (exploration), exploit the global best and avoid the local best. Single objective optimisation is concerned with optimising only single objective. This word refers previously multi-objective optimization, which involves optimising more than single objective [3, 4]. Parameters and restrictions are also included in the single objective optimisation process. The variables (unknowns) of optimisation problems (systems) that must be optimised are referred to as parameters. The limitations specify the possibility of the derived objective value. Limitations include stress constraints while developing aerodynamic systems and the spectrum of variables [5].

A single objective optimisation can be expressed as a minimization problem as given below:

$$\text{Minimize: } H(z_1, z_2, z_3, \ldots z_{n-1}, z_n) \tag{1}$$

$$\text{Subject to: } g_j(z_1, z_2, z_3, \ldots z_{n-1}, z_n) = 0 \quad j = 1, 2, \ldots r \tag{2}$$

$$f_j(z_1, z_2, z_3, \ldots z_{n-1}, z_n)0 \quad j = 1, 2, \ldots n \tag{3}$$

$$Lb_j \leq z_j \leq Ub_j \quad j = 1, 2, \ldots, m \tag{4}$$

where the number of variables is m, n is the number of inequality constraints, r signifies the number of equality constraints, the lower bound of the jth variable is Lb_j and the upper bound of the jth variable is Ub_j. In Eqs. (2) and (3), there are two types of constraints: equality and inequality. A search space for a particular problem is constructed by a collection of variables, objectives and constraints. Unfortunately, because of the large complexity of the variables, drawing the search space is frequently unfeasible [6, 7].

The search area of a real-world issues has been quite difficult. A large number of local optima, discontinuity and global best positioned on the borders of constraints, a big number of constraints, isolation of the global optimum and misleading valleys towards local optima are certain difficulties of real search areas. To find the global best, an optimization technique should be equipped with appropriate operators for dealing with all of these complications [8, 9].

2 Cosine Algorithm

Swarm-based optimisation approaches begin the optimisation process with a collection of arbitrary agents. This arbitrary set is assessed periodically by an objective function and enhanced by a set of rules that serves as the foundation of an optimization process. There is no assurance of finding a solution in a single run because swarm-based optimisation approaches search for the optima of optimisation issues stochastically. However, with the sufficient number of optimisation steps (iterations) and random solutions increases, the possibility of determining the global optimum rises [10].

In this algorithm, the optimisation problem works with two phases: exploitation and exploration. In the phase of exploitation, random variations are significantly fewer than in the exploration phase and, there are incremental modifications in the random solutions. On the other hand, to find the interesting areas of the search area, an optimization method mixes the random results in the set of solutions suddenly with a high rate of uncertainty in the exploration phase [11].

For both the phases, the following equations are proposed:

$$Z_i^{t+1} = Z_i^t + r_1 * \cos(r_2) * r_3 Q_i^t - Z_i^t \quad (5)$$

where Z_i^t indicate the current solution's position in ith dimension and tth iteration, Z_i^{t+1} is the updated position, Q_i^t signifies the point of destination in ith dimension and tth iteration, $r_1/r_2/r_3$ are the random variables and || is the absolute value.

In the above equation, this algorithm uses three parameters:r_1,r_2 and r_3. Every parameter has its own significance. r_1 is the random number that controls the exploitation and exploration during the search process using the expression given below:

$$r_1 = b - t\frac{b}{t_{\max}} \quad (6)$$

where b is constant, t is current iteration and $t_{\max}$ is the maximum iteration.

r_2 defines the random number that lies in $[0,2\pi]$ and decides the direction of moment either towards the current solution (exploitation) or away from the current solution (exploration). The parameter r_3 provides weight to the destination which emphasises the exploration ($r_3 > 1$) and exploitation ($r_3 < 1$). Figure 1 represents the effect of period of Sine and Cosine [12]. Although a two-dimensional example is shown in the Figure, it should be emphasised that this equation may be expanded to greater dimensions. Figure 2 represents the range of Cosine function. When it is from $[-1, 1]$, this technique exploit the search space.

3 Experimental Setup and Result Discussion

Several test cases have been used to confirm the performance of a proposed method in the field of optimization employing evolutionary algorithms and meta-heuristics [13, 14]. This is related to the stochastic character of this technique, which imposes

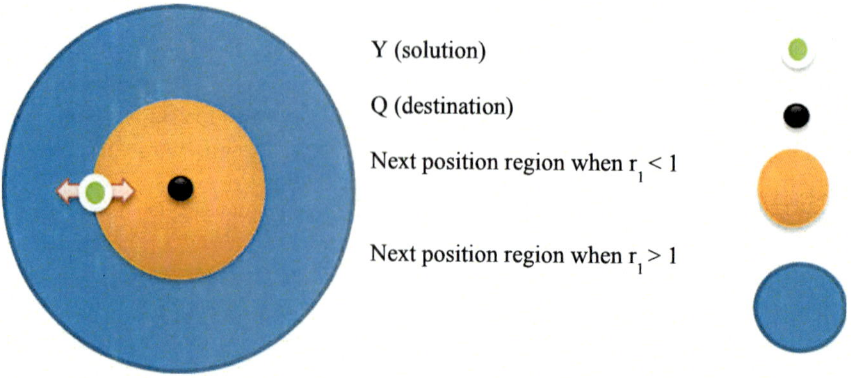

Fig. 1 Effect of period of Sine and Cosine

Fig. 2 Cosine with range [−1, 1]

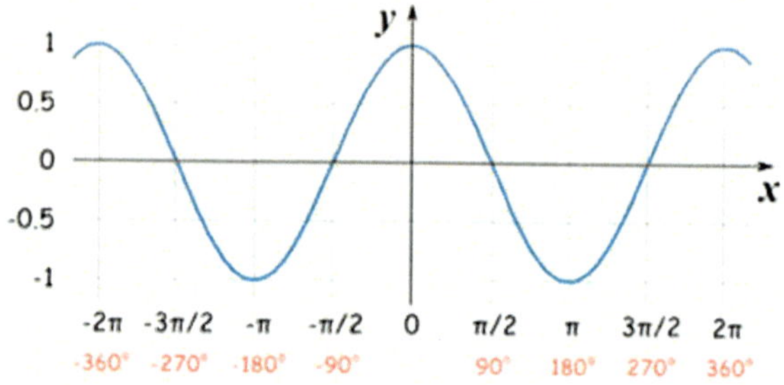

the use of a suitable and adequate number of case studies and test functions to ensure that the superior outcomes are not the product of chance (Tables 1 and 2). In this experiment, several benchmark functions have been tested such as unimodal (H1-H7)

Table 1 Unimodal functions

Equation name	Benchmark	dim	Interval	$H_{\min}$
Sphere	$H_1(z) = \sum_{i=1}^{n} z_i^2$	30	[−100,100]	0
Schwefel's problem 2.22	$H_2(z) = \sum_{i=1}^{n} \lvert z \rvert + \prod_i^n \lvert z_i \rvert$	30	[−10,10]	0
Schwefel's problem 1.2	$H_3(z) = \sum_{i=1}^{n} \left(\sum_{j-1}^{i} z_j\right)^2$	30	[−100,100]	0
Schwefel's problem 2.21	$H_4(z) = \max_i\{\lvert\lvert z_i \rvert, 1 \le i \le n\}$	30	[−100,100]	0
Rosen brock	$H_5(z) = \sum_{i=1}^{n-1} [100(z_{i+1} - z_i^2)^2 + (z_i - 1)^2$	30	[−30,30]	0
Step	$H_6(z) = \sum_{i=1}^{n} ([z_i + 0.5])^2$	30	[−100,100]	0
Noise	$H_7(z) = \sum_{i=1}^{n} i z_i^4 + \text{random}[0, 1)$	30	[−1.28,1.28]	0

Table 2 Multi-modal functions

Equation name	Benchmark	Dim	Interval	H_{min}
Generalised Schwefel's problem	$H_8(z) = \sum_{i=1}^{n} -z_i \sin(\sqrt{\lvert z_i \rvert})$	30	[−500,500]	−2094.9145
Rastrigin	$H_9(z) = \sum_{i=1}^{n} [z_i^2 - 10\cos(2\pi z_i) + 10]$	30	[−5.12,5.12]	0
Ackley	$H_{10}(z) = -20\exp\left(-0.2\sqrt{\frac{1}{n}\sum_{i=1}^{n} z_i^2}\right) - \exp\left(\frac{1}{n}\sum_{i=1}^{n}\cos(2\pi z_i)) + 20 + e\right)$	30	[−32,32]	0
Griewank	$H_{11}(z) = \frac{1}{4000}\sum_{i=1}^{n} z_i^2 - \prod_{i=1}^{n}\cos\left(\frac{z_i}{\sqrt{i}}\right) + 1$	30	[−600,600]	0
Generalized Penalised function 1	$H_{12}(z) = \frac{\pi}{n}\{10\sin\pi y_1) + \sum_{i=1}^{n-1}(y_i - 1)^2$ $[1 + 10\sin^2(\pi y_{i+1})] + (y_n - 1)^2\} +$ $\sum_{i=1}^{n} u(z_i, 10, 100, 4)$ $y_i = 1 + \frac{z_i + 1}{4}$ $u(z_i, a, k, m) = \begin{cases} k(z_i - a)^m & z_i > a \\ 0 & -a < z_i < a \\ k(-z_i - a)^m & z_i < -a \end{cases}$	30	[−50,50]	0
Generalized Penalised function 2	$H_{13}(z) = 0.1\{\sin^2(3\pi z_1) + \sum_{i=1}^{n}(z_i - 1)^2$ $[1 + \sin^2(3\pi z_i + 1)] + (z_n - 1)^2$ $[1 + \sin^2(2\pi z_n)]\} + \sum_{i=1}^{n} u(z_i, 5, 100, 4)$	30	[−50,50]	0

and multi-modal functions (H8, H9, H10 and H11) [15, 16]. A total of thirty search candidates have been allowed to control the global optimum across 1000 iterations while solving the test functions. For the verification of results, the CA algorithm has been compared to Sine Cosine algorithm (SCA), genetic algorithm (GA) [17] and PSO [18]. All these algorithms are run over thirty times and the statistical results have been collected and reported in Table 3. In Table 3, the CA algorithm shows the superior results on 4 out of 7 in case of unimodal and in multi-modal, the results show 3 out of 4 functions. In Figs. 3 and 4, the graph represents the comparison of sphere function and Rastrigin function of CA and SCA, respectively, in which there is more convergence rate in CA as compared to other algorithms such as GA, PSO and SCA. These outcomes strongly indicate that the CA technique has high convergence and exploitation due to the peculiarities of the unimodal test functions.

Table 3 Benchmark function's results

Functions	CA		SCA		PSO		GA	
	AVG	STDV	AVG	STDV	AVG	STDV	AVG	STDV
H1	0.0000	0.0000	0.0000	0.0000	0.0003	0.0 011	0.8078	0.4393
H2	0.0000	0.0000	0.0000	0.0001	0.0693	0.2164	0.5406	0.2363
H3	0.0000	0.0000	0.0371	0.1372	0.0157	0.0158	0.5323	0.2423
H4	0.0000	0.0000	0.0965	0.5823	0.0936	0.4282	0.8837	0.7528
H5	7.14598	0.119219	0.0005	0.0017	0.0000	0.0000	0.6677	0.4334
H6	0.261552	0.080669	0.0002	0.0001	0.0004	0.0 033	0.7618	0.7443
H7	0.000301	0.000324	0.0000	0.0014	0.0398	0.0634	0.5080	0.1125
H8	−2189.53	62.39521	1.0000	0.0036	1.0000	0.0036	1.0000	0.0055
H9	0.0000	0.0000	0.0000	0.7303	0.3582	0.8795	1.0000	0.6881
H10	0.0000	0.0000	0.3804	1.0000	0.1045	0.0541	0.8323	0.0686
H11	0.0000	0.0000	0.0000	0.0051	0.0521	0.0448	0.7679	0.2776

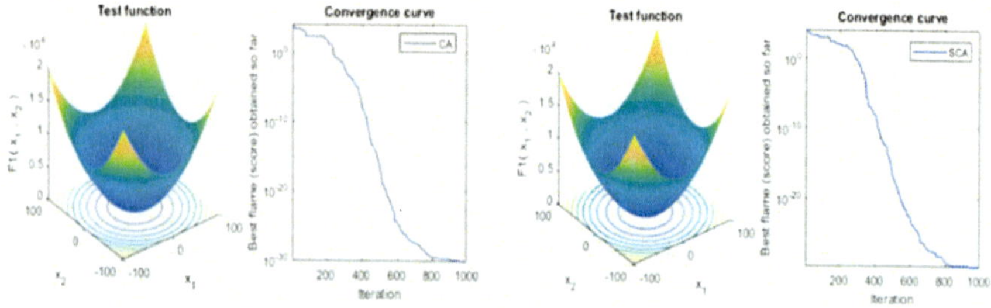

Fig. 3 Comparison of sphere functions of CA and SCA

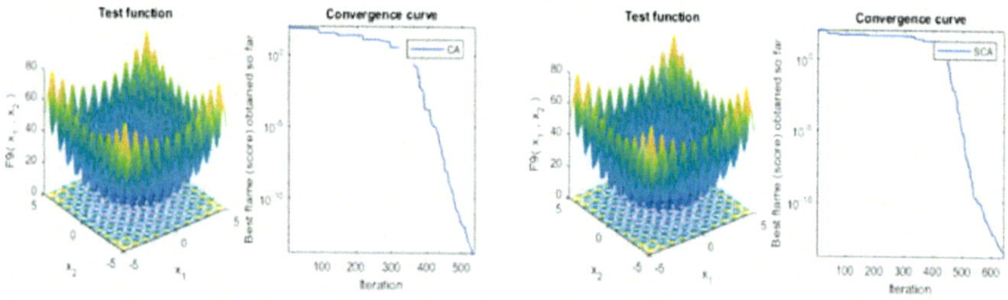

Fig. 4 Comparison of Rastrigin functions of CA and SCA

4 Conclusion

This paper represents a robust algorithm to solve optimization problems. The search agents in the proposed CA algorithm were needed to update their locations in relation to the best solution achieved so far as the target point. To guarantee that the search space is exploited and explored, the mathematical model of position update fluctuated the solutions towards or outwards the goal point. The convergence and divergence of the search candidates in the CA algorithm are also facilitated by a number of adaptive factors. According to the unimodal test functions, this technique has been converged significantly quicker than GA, PSO and SCA algorithms. Similar behaviour has been observed in the multi-modal test functions, demonstrating the algorithm's strong exploration and avoidance of local optima. Future research can be directed in numerous ways as a result of this work. On a certain set of issues, this approach may not be able to outperform other algorithms, but it is worth testing and applying to a variety of issues. The novelty of this paper is that this approach can be proposed in binary and multi-objective. The CA algorithm can be improved the usage of levy flight, mutation and evolutionary operators. This algorithm gives the better results as compared to other algorithms due its simple implementation, time consumption and convergence speediness.

Acknowledgments The authors acknowledge VIT for providing VIT SEED GRANT for carrying out this research work.

References

1. Mirjalili S (2016) SCA: a sine cosine algorithm for solving optimization problems. Knowl-Based Syst 96:120–133
2. Boussaïd I, Lepagnot J, Siarry P (2013) A survey on optimization metaheuristics. Inf Sci 237:82–117
3. Parpinelli RS, Lopes HS (2011) New inspirations in swarm intelligence: a survey. Int J Bio-Inspired Computation 3(1):1–16
4. Yang X-S et al (eds) (2013) Swarm intelligence and bio-inspired computation: theory and applications. Newnes
5. Li H-R, Gao Y-L (2009) Particle swarm optimization algorithm with exponent decreasing inertia weight and stochastic mutation. In: 2009 second international conference on information and computing science, vol 1. IEEE
6. Chen S (2012) Particle swarm optimization with pbest crossover. In: 2012 IEEE congress on evolutionary computation. IEEE
7. Shi XH et al (2005) An improved GA and a novel PSO-GA-based hybrid algorithm. Inf Process Lett 93(5):255–261
8. Blum C, Roli A (2008) Hybrid metaheuristics: an introduction. Hybrid metaheuristics. Springer, Berlin, pp 1–30
9. Abd Elaziz M, Oliva D, Xiong S (2017) An improved opposition-based sine cosine algorithm for global optimization. Expert Syst Appl 90:484–500
10. Sindhu R et al (2017) Sine–cosine algorithm for feature selection with elitism strategy and new updating mechanism. Neural Comput Appl 28(10):2947–2958

11. Attia A-F, El Sehiemy RA, Hasanien HM (2018) Optimal power flow solution in power systems using a novel Sine-Cosine algorithm. Int J Electr Power Energy Syst 99:331–343
12. Li S, Fang H, Liu X (2018) Parameter optimization of support vector regression based on sine cosine algorithm. Expert Syst Appl 91:63–77
13. Nenavath H, Jatoth RK (2018) Hybridizing sine cosine algorithm with differential evolution for global optimization and object tracking. Appl Soft Comput 62:1019–1043
14. Rizk-Allah RM (2018) Hybridizing sine cosine algorithm with multi-orthogonal search strategy for engineering design problems. J Comput Design Eng 5(2):249–273
15. Gupta S, Deep K (2019) A hybrid self-adaptive sine cosine algorithm with opposition based learning. Expert Syst Appl 119:210–230
16. Gupta S, Deep K, Engelbrecht AP (2020) A memory guided sine cosine algorithm for global optimization. Eng Appl Artif Intell 93:103718
17. Gupta S et al (2020) A modified sine cosine algorithm with novel transition parameter and mutation operator for global optimization. Expert Syst Appl 154:113395
18. Abualigah L, Diabat A (2021) Advances in sine cosine algorithm: a comprehensive survey. Artif Intell Rev 1–42

Investigation of Parametric Effect in Optimum Retaining Wall Design Using Harmony Search Algorithm

Esra Uray, Serdar Carbas, and Murat Olgun

Abstract This study focuses on the investigation of the parametric effect of the harmony search algorithm on design variables for optimum cantilever retaining wall design. Variance analyses have been performed for the detection of design parameters which are taken as discrete variables in the optimization process and percentages of the parameter effect have been calculated, as well. Optimum results obtained by the harmony search algorithm have been examined according to the parameter effect of design variables statistically determined via the Taguchi method. Eventually, the attained results have shown that changing the value of the effective design parameter in the design positively contributes to the optimization process. In addition, it has been observed that developing a new solution for the design parameter which has little effect on the design is not significant in converging to the optimum solution.

Keywords Optimum design · Retaining wall · Harmony search algorithm · Taguchi method · Parameter effect

1 Introduction

In geotechnical engineering, which is one of the most important branches of civil engineering, retaining wall designs have very common use in terms of resisting horizontal soil loads between two different soil levels. In the presence of parameters such as earthquake load, surcharge load, terrain condition, slope characteristics, soil stratification, construction time, building usage purpose, obtaining a safe and optimum design which is satisfied wall stability criteria like sliding, overturning, and slope

E. Uray (✉)
KTO Karatay University, Konya 42020, Turkey
e-mail: esra.uray@karatay.edu.tr

S. Carbas
Karamanoglu Mehmetbey University, Karaman 70100, Turkey

M. Olgun
Konya Technical University, Konya 42250, Turkey

J. H. Kim et al. (eds.), *Proceedings of 7th International Conference on Harmony Search, Soft Computing and Applications*, Lecture Notes on Data Engineering and Communications Technologies 140, https://doi.org/10.1007/978-981-19-2948-9_42

stability safety factors [1] is a complex engineering problem. The Harmony search algorithm (HSA), which is presented by Geem et al. [2] as a metaheuristic optimization algorithm, has recently been widely utilized in an effective and powerful manner, in solving complex geotechnical engineering problems encountered in real life [3]. In literature, it is possible to find several studies which investigate optimum cantilever retaining wall design utilizing metaheuristic optimization algorithms [4–10]. The performance of the optimum searching process for the optimum value is as important as obtaining optimum results utilizing optimization algorithms. In this study, parametric effect of design variables has been investigated using optimum results presented by Uray et al. [11] and the statistical results obtained by the Taguchi method.

2 Experimental Procedure

2.1 Cantilever Retaining Wall Design and Taguchi Method

Taguchi method [12] based on statistical is a robust and alternative method that increases the quality in production, reduces research costs, and provides obtaining of the effect of the parameters taken into account in production on the result. The orthogonal arrays presented by Taguchi are utilized in acquiring these benefits, which are selected according to the number of parameters and their levels.

$L_d(a)^k$ or L_d impend general representation of orthogonal array which includes a, k, and d correspond to the number of levels, parameters, and trials (experiments), respectively. Since parameters of the cantilever retaining wall (CRW) design demonstrated in Fig. 1 are determined as five parameters with four-parameter levels (Table 1), the L_{16} orthogonal array (OA) is selected for this study [11]. All combinations of five parameters with four levels correspond to 1024 (4^5) designs and it is possible that parameter effect end optimum values of design parameters are obtained with just sixteen designs utilizing the Taguchi method and OA.

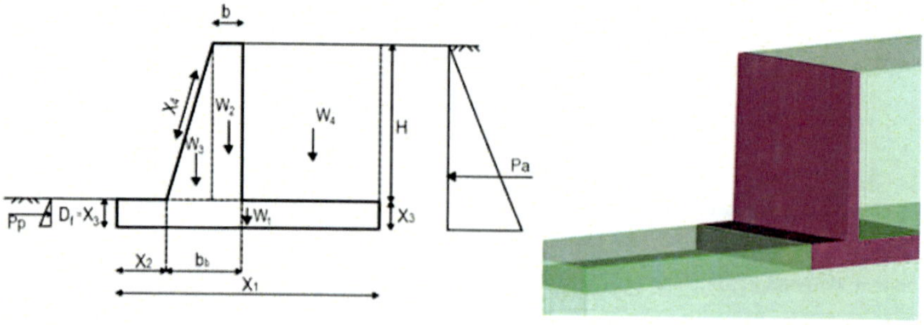

Fig. 1 Design variables of cantilever retaining wall design with action load on the wall

Table 1 Design parameters with their levels

Design parameter	Level 1	Level 2	Level 3	Level 4
X_1: base length	0.25 H	0.50 H	0.75 H	1.0 H
X_2: toe extension	0.15 X_1	0.30 X_1	0.45 X_1	0.60 X_1
X_3: base thickness	0.06 H	0.09 H	0.12 H	0.15 H
X_4: front face angle (%)	0	1	2	4
Ø: angle of internal friction (°)	20	27	34	41

CRW design should be satisfied stability criteria in order to safely complete its service life. This stability criterion is basically defined as withstanding resistance to arising effect from lateral soil loads, which are occurred with acting loads on the wall.

Sliding (F_s), overturning (F_o), and slope stability (F_{ss}) safety factors have been regarded in the CRW design. Sliding and overturning safety factors are calculated using Eqs. 1 and 2, consecutively. While F_s is defined as the ratio of resisting loads ($F_{\text{resistant}}$) to slide to shear force (F_{sliding}), F_o is calculated as the ratio of the resistant moment ($M_{\text{resistant}}$) against to overturning of the wall to overturning moment ($M_{\text{overturning}}$).

$$F_s = F_{\text{resistant}}/F_{\text{sliding}},\ F_{\text{resistant}} = [(W_1 + W_2 + W_3 + W_4)\tan\delta + P_p],\ F_{\text{sliding}} = P_a \quad (1)$$

$$\begin{aligned} F_o = M_{\text{resistant}}/M_{\text{overturning}},\ M_{\text{resistant}} &= [0.5X_1W_1 + (b_b - 0.5b + X_2)W_2 \\ &+ (0.667(b_b - b) + X_2)W_3 + 0.5(X_1 + X_2 + b_b)W_4 \\ &+ 0.333P_pX_3],\ M_{\text{overturning}} = [P_aH/3] \end{aligned} \quad (2)$$

In numerical analyses, b, δ, γ_{soil}, and γ_{concrete} parameters are taken as 0.25 m, 2/3 Ø, 18 kN/m^3, and 25 kN/m^3, respectively. Slope stability safety factors (F_{ss}) have been obtained using GEO5 [13] analysis software according to the Bishop method.

In Table 2, the revised design matrix of L16 OA according to CRW design parameters given Table 1, safety factors of CRW, and calculated *S/N* ratios using values of safety factors are tabulated [11].

In Taguchi design, after determining a suitable orthogonal array for the problem, the signal/noise (*S/N*) ratio is divided into three groups as smaller is better, nominal is best, larger is better is selected considering the desirable target of the design. *S/N* ratio suggested by Taguchi is the control criteria used in evaluation parameter effect on the design. In this study, the *S/N* ratio has been calculated for the case of larger is better because the cantilever retaining wall safety factors have been considered as response value (Eq. 3). In Eq. 3, while Y corresponds to the response value, n is the number of repetitions of trials.

Table 2 Revised L16 OA according to CRW design parameters

No.	X_1(H)	X_2(X_1)	X_3(H)	X_4 (%)	Ø (°)	F_s	F_o	F_{ss}	*S/N* (F_s)	*S/N* (F_o)	*S/N* (F_{ss})
1	0.25	0.15	0.06	0	20	0.22	0.35	0.75	−13.152	−9.119	−2.499
2	0.25	0.30	0.09	1	27	0.34	0.42	1.09	−9.370	−7.535	0.749
3	0.25	0.45	0.12	2	34	0.52	0.48	1.48	−5.680	−6.375	3.405
4	0.25	0.60	0.15	4	41	0.97	0.53	1.96	−0.265	−5.514	5.845
5	0.50	0.15	0.09	2	41	2.48	3.11	2.18	7.889	9.855	6.769
6	0.50	0.30	0.06	4	34	1.08	2.24	1.54	0.669	7.005	3.750
7	0.50	0.45	0.15	0	27	0.59	1.36	1.27	−4.583	2.671	2.076
8	0.50	0.60	0.12	1	20	0.24	0.92	0.84	−12.396	−0.724	−1.514
9	0.75	0.15	0.12	4	27	1.15	3.67	1.51	1.214	11.293	3.580
10	0.75	0.30	0.15	2	20	0.54	2.55	1.06	−5.352	8.131	0.506
11	0.75	0.45	0.06	1	41	2.34	6.14	2.10	7.384	15.763	6.444
12	0.75	0.60	0.09	0	34	1.11	3.65	1.58	0.907	11.246	3.973
13	1.00	0.15	0.15	1	34	3.04	8.31	2.26	9.658	18.392	7.082
14	1.00	0.30	0.12	0	41	4.77	11.18	2.67	13.570	20.969	8.530
15	1.00	0.45	0.09	4	20	0.57	4.38	1.00	−4.883	12.829	0.000
16	1.00	0.60	0.06	2	27	0.78	4.94	1.23	−2.158	13.875	1.798

$$S/N = -10\log\left[\Sigma\left(1/Y^2\right)/n\right] \quad (3)$$

In Taguchi design, variance analysis (ANOVA) is conducted, and optimum design parameters are obtained using calculated average *S/N* ratios.

2.2 *Harmony Search Algorithm Implementation for Optimum Cantilever Retaining Wall Design*

The flowchart of harmony search algorithm (HSA) [2] considered in optimization analyses is shown in Fig. 2 [11]. Considered the HSA parameters in optimization analyses are harmony memory matrix size (HMS = 30), harmony memory considering rate (HMCR = 0.95), and pitch adjusting rate (PAR = 0.30) [14]. In HSA, 5000 iterations were operated until they reach the maximum iteration number, and all iterations were performed for each independent run of 500. The wall stem height (*H*) and the angle of internal friction of retained soil (Ø) values are taken as 6 m and 20°, 22°, 24°, 26°, 28°, 30°, 32°, 34°, 36°, 38°, 40°, 42°, respectively.

After HSA parameters are defined in the algorithm, the harmony memory matrix (HM) whose dimensions are HMS × NVAR is generated. The HM is filled values selected randomly from the formed design space which is formed regarding discrete values given in design variables (NVAR) tabulated in Table 3 [11]. The objective

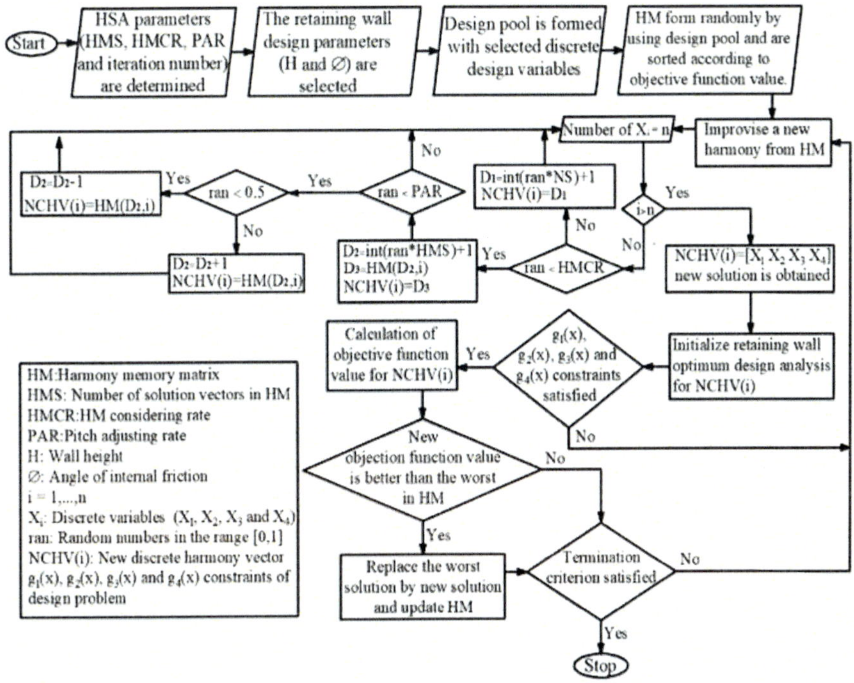

Fig. 2 Flowchart of the HSA for the optimum design of a CRW

Table 3 Design space for discrete design variables of optimum CRW design

Design variables	Discrete values	Index
$X_1(H)$	0.25: 0.05:1.00	1:1:16
$X_2(X_1)$	0.15: 0.05:0.60	1:1:10
$X_3(H)$	0.06: 0.015: 0.15	1:1:7
$X_4(\%)$	0:1:4	1:1:5

function and design constraints (Table 4) are calculated for each row of HM which corresponds to a solution vector with different values of NVAR [11]. Calculated objective function values for the HM have evaluated their penalty values which are determined according to design constraints values in case of not satisfying. Based on this check, the best solution and the worst solution are assigned considering all solution vectors in HM. In the process of finding the best solution, Deb's rules [15] are utilized as constraint handling methods.

Mathematical expressions tabulated in Table 4 have been utilized for design constraints and the objective function instead of safety factors values obtained numerically and analytically. These mathematical models have been improved by Uray et al. [11] considering the average *S*/*N* ratios. Result of conducted verification analyses,

Table 4 Objective function and design constraints of optimum CRW design

Objective function	Design constraints
$f_{\min} = 0.33F_s + 0.33F_o + 0.33F_{ss}$	$g_x(1) = 1 - F_s / 1.3$
	$g_x(2) = 1 - F_o / 1.3$
	$g_x(3) = 1 - F_{ss} / 1.3$
	$g_x(4) = F_s / 3.0 - 1$
	$g_x(5) = F_o / 3.0 - 1$
	$g_x(6) = F_{ss} / 3.0 - 1$
	$g_x(7) = (X_2 + 0.25 + X_4 H)/X_1 - 1$

mathematical models have been suggested in obtaining stability control of CRW which involved sliding, overturning, and slope stability safety factors with approximately 10% error. The general representation of the mathematical models for safety factors is given by Eq. 4, for λ which calculates differently according to the safety factors sliding, overturning, and slope stability.

$$F_{(s,o,ss)} = 10^{(\lambda/20)} \tag{4}$$

After the best and the worst solution have been determined, the improvisation process of a new solution vector for the first iteration is initialized regarding possibilities of HMCR and PAR algorithm parameters. For the current design variable (X_i, $i = 1,\ldots,$ NVAR), if the randomly generated number (ranges between 0 and 1) is smaller than HMCR, a new value from HM is selected, taking into account the probability of HMCR. Else a new value is randomly selected from a design pool with the possibilities of (1-HMCR). For the new value selected from HM, a new value is specified considering of neighboring of the current selection with the probability of PAR. The objective function and penalty values are determined for the new solution vector which includes the new value of the design variables (X_1, X_2, X_3, X_4) and it is checked whether the new solution vector is better than the best solution assigned before. If it is better than the previous solution, the new solution vector is stored in HM as the best one instead of the worst solution, and the best and the worst solution are updated in the algorithm. Otherwise, it is checked in terms of providing termination criteria, and the searching process of optimum solutions is repeated until the current iteration number reaches the maximum iteration number.

3 Results

3.1 Taguchi Analyses

In the investigation of parameter effect on the CRW design, the average *S/N* ratios have been determined utilizing *S/N* ratios given in Table 3 for F_s, F_o, and F_{ss} (Fig. 3).

In Table 5, variance analyses results are given with variance (ν), rank (R) and parameter effect (PE). It is obvious from variance analyses results given in Table 5 that the most effective parameter is the angle of internal friction (Ø) for F_s and F_{ss} and the base length (X_1) has the most impact on F_o. Besides, it has been seen that the front face angle (X_4) has not any impact on all safety factors. For maximum safety

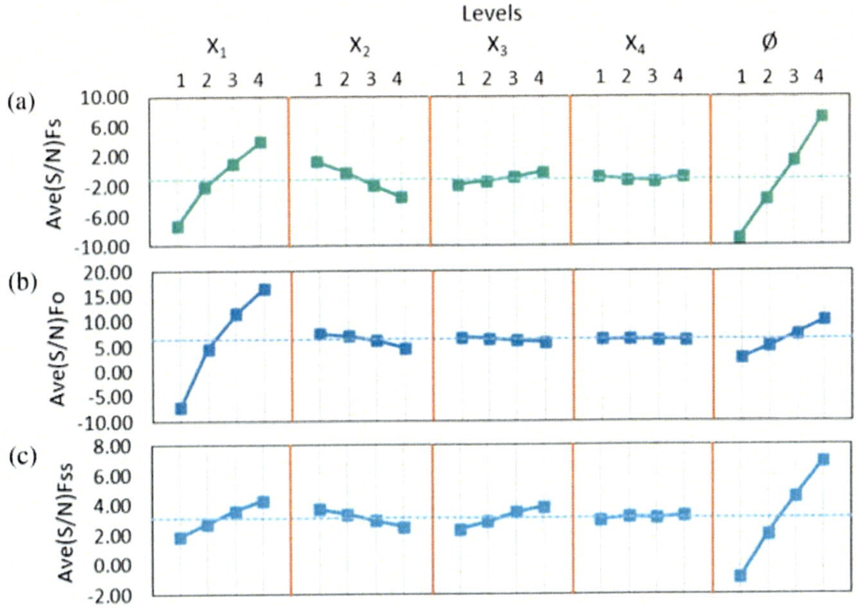

Fig. 3 Average *S/N* ratios of design parameter levels; **a** for F_s, **b** for F_o, **c** for F_{ss}

Table 5 Results of variance analyses CRW design

	F_s			F_o			F_{ss}		
P	*R*	ν	PE (%)	*R*	ν	PE (%)	*R*	ν	PE (%)
X_1	2	91.0	30	1	420.8	90	2	4.6	9
X_2	3	18.1	6	2	6.5	1	4	1.1	2
X_3	4	2.1	1	3	0.7	0	3	1.8	3
X_4	5	0.3	0	4	0.0	0	5	0.1	0
Ø	1	190.1	63	5	41.6	9	1	44.6	86

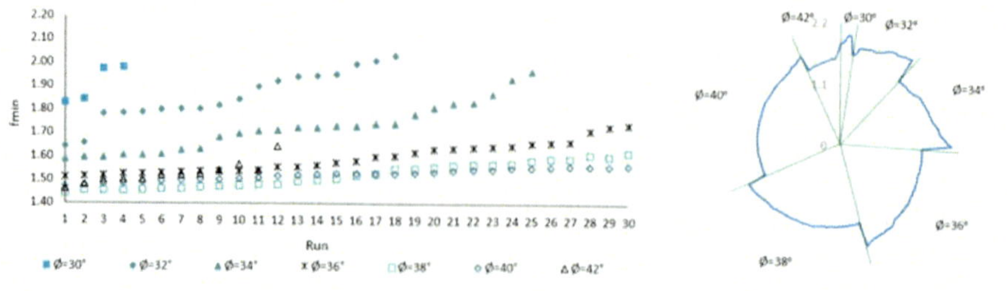

Fig. 4 Objective function values for 30 runs

factors of CRW design, value of Ø has been obtained as 41° with approach of the Taguchi optimum designs. It is a reasonable conclusion since the retained soil which has 41° is good quality for the CRW designs. However, Ø design parameter is the dependent factor in real life because of the possibility of the soil which will have any quality. For this reason, the optimum wall designs have been investigated for the Ø value considered keeping it constant in optimization analyses.

3.2 Optimization Analyses

The results of optimization analyses presented by Uray et al. [11] have been evaluated in terms of parametric effect based on statistical values obtained by the Taguchi method. Optimum CRW designs obtained for independent runs are given in Fig. 4 for objective functions and in Fig. 5 for the design variables.

When the objective function values given in Fig. 4 are examined, it is seen that there is an essential change in the value of $f_{\min}$ at different Ø values and the most minimum $f_{\min}$ is obtained for Ø = 42°. This is interpreted as the result of the Ø effect, which is obtained the most effective parameter in Taguchi designs.

When obtained optimum wall dimensions have been examined in terms of having the different values, it has been seen from Fig. 5 that there is a ranking as X_4, X_3, X_2, and X_1 in descending order according to case being the more different values each other. This situation has been interpreted that it is due to the development solution effort of the algorithm for searching more new values of design parameter (variable) which has less effect on the objective function (Table 5). That is, while the scattering of the X_1 values, which is one of the parameters effectively obtained in the Taguchi design, found for different runs, is more stable, it is seen that the values of X_4, which has the least effect on the design, have a more complex and irregular distribution.

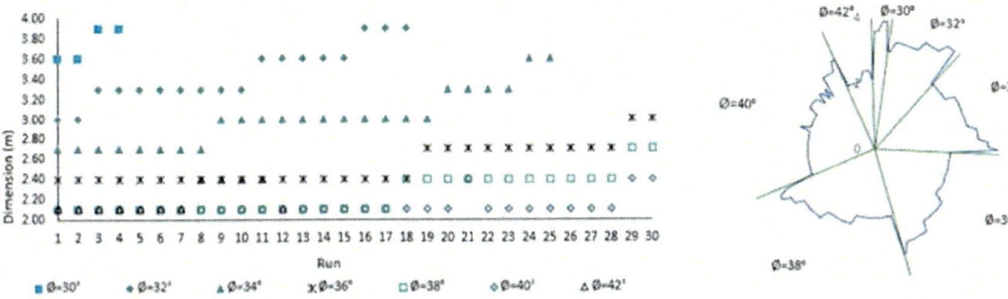

(a) X_1

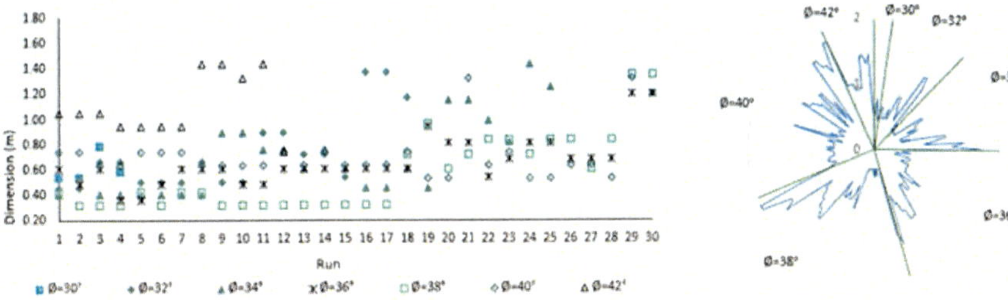

(b) X_2

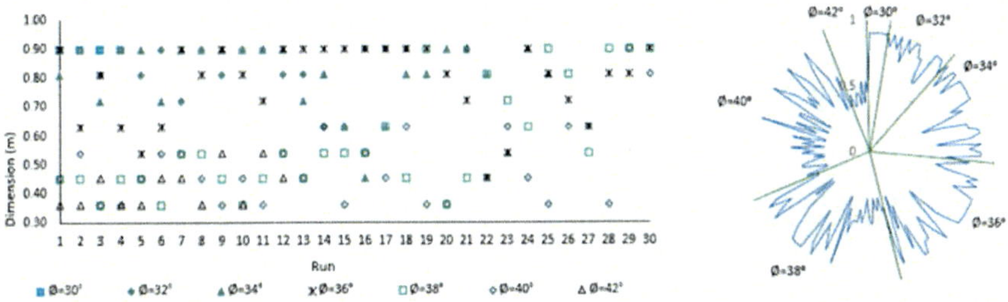

(c) X_3

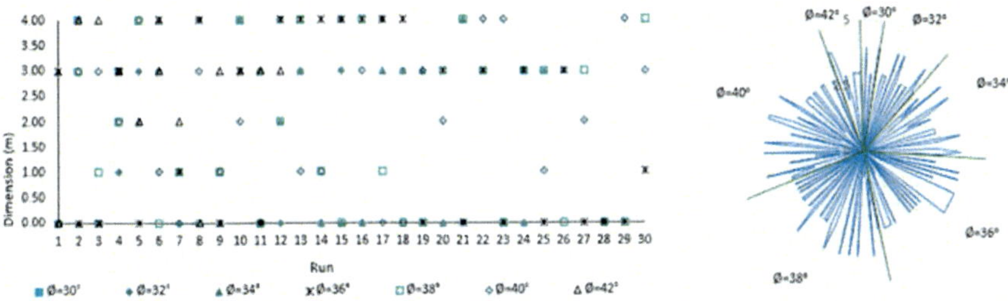

(d) X_4.

Fig. 5 Parametric investigation of discrete design variables for 30 runs

Table 6 Comparison between optimum and other design variables values in terms of objective function and parameter effect

Index				Values					f	Percentage of change (%)
X_1	X_2	X_3	X_4 (%)	X_1 (m)	X_2 (m)	X_3 (m)	X_4 (%)	Ø (°)		
8	1	7	**1**	3.60	0.54	0.90	0.0	30	1.8285	0.8
8	1	7	**5**	3.60	0.54	0.90	4.0	30	1.8419	
6	1	7	5	3.00	0.45	0.90	4.0	32	1.6573	10.1
7	1	7	5	3.30	0.50	0.90	4.0	32	1.8441	
5	1	**6**	4	2.70	0.41	0.81	3.0	34	1.5896	0.9
5	1	**7**	4	2.70	0.41	0.90	3.0	34	1.6046	
5	1	6	4	2.70	0.41	0.81	3.0	34	1.5896	10.7
6	1	6	4	3.00	0.45	0.81	3.0	34	1.7804	
4	2	**4**	1	2.40	0.48	0.63	0.0	36	1.5142	4.3
4	2	**7**	1	2.40	0.60	0.90	0.0	36	1.5824	
3	**2**	2	1	2.10	0.42	0.45	0.0	38	1.4423	2.4
3	**1**	2	1	2.10	0.32	0.45	0.00	38	1.4779	
4	9	7	**1**	2.40	1.32	0.90	0.0	40	1.5408	1.2
4	9	7	**5**	2.40	1.32	0.90	4.0	40	1.5596	

4 Discussion

It has been seen from Table 6 listed for the comparison of some of the CRW optimum designs given in Fig. 5 that 10% change occurred in the f_{min} value due to the different value of only X_1, which is an effective parameter in the design. On the other hand, approximately 1% change is observed with the change of only X_4 value, which is the least effective parameter. The design variable whose value changes while the others stay same for comparison between two designs have been shown in bold in the Table 6.

5 Conclusions

In the study that the safety factors were based on statistical mathematical models, the optimum cantilever retaining wall designs have been evaluated in terms of optimization process by the harmony search algorithm. Statistically improved mathematical models presented by Uray et al. [11] have been taken as design constraints and objective function of the CRW optimization problem.

The parametric effect of the design variables of the wall found using the Taguchi method has been studied in terms of optimum designs obtained for different operating values. As a result, a relationship has been observed between the effect of the design variables of the optimization problem on the design and the optimum search process. Considering the parametric effect in the heuristic optimization algorithm is thought that the optimum designs may be achieved with fewer iterations.

References

1. Das BM, Sivakugan N (2018) Principles of foundation engineering (9th edn). Cengage Learning
2. Geem Z, Kim J, Loganathan G (2001) A new heuristic optimization algorithm: harmony search. SIMULATION 76:60–68. https://doi.org/10.1177/003754970107600201
3. Afzal M, Liu Y, Cheng JCP, Gan VJL (2020) Reinforced concrete structural design optimization: a critical review. J Clean Prod 260:120623. https://doi.org/10.1016/J.JCLEPRO.2020.120623
4. Akin A, Saka MP (2010) Optimum design of concrete cantilever retaining walls using the harmony search algorithm. Civil-Comp Proc 93:1–21. https://doi.org/10.4203/ccp.93.130
5. Temur R, Bekdas G (2016) Teaching learning-based optimization for design of cantilever retaining walls design and analysis of nonlinear structural systems view project teaching learning-based optimization for design of cantilever retaining walls. Struct Eng Mech 57:763–783. https://doi.org/10.12989/sem.2016.57.4.763
6. Aydogdu I (2017) Cost optimization of reinforced concrete cantilever retaining walls under seismic loading using a biogeography-based optimization algorithm with Levy flights. Taylor Fr 49:381–400. https://doi.org/10.1080/0305215X.2016.1191837
7. Uray E, Çarbaş S, Erkan İH, Tan Ö (2019) Parametric investigation for discrete optimal design of a cantilever retaining wall. Chall J Struct Mech 5:108. https://doi.org/10.20528/cjsmec.2019.03.004
8. Uray E, Hakli H, Carbas S (2021) Statistical investigation of the robustness for the optimization algorithms. In: Carbas S, Toktas A, Ustun D (eds) Nature-inspired metaheuristic algorithms for engineering optimization applications. Springer, Singapore, pp 201–224. https://doi.org/10.1007/978-981-33-6773-9_10
9. Uray E, Çarbaş S, Erkan İH, Olgun M (2020) Investigation of optimal designs for concrete cantilever retaining walls in different soils. Chall J Concr Res Lett 11. https://doi.org/10.20528/cjcrl.2020.02.003
10. Uray E, Carbas S, Geem ZW, Kim S (2022) Parameters optimization of Taguchi method integrated hybrid harmony search algorithm for engineering design problems. Math 10:327. https://doi.org/10.3390/MATH10030327
11. Uray E, Tan Ö, Çarbaş S, Erkan H (2021) Metaheuristics-based pre-design guide for cantilever retaining walls. Tek Dergi 32. https://doi.org/10.18400/tekderg.561956
12. Taguchi G (1986) Introduction to quality engineering: designing quality into products and processes. The Organization
13. GEO5, Geotechnical Software. https://www.finesoftware.eu/geotechnical-software/, https://www.finesoftware.eu/geotechnical-software/
14. Lee K, Geem Z (2005) A new meta-heuristic algorithm for continuous engineering optimization: harmony search theory and practice. Comput Methods Appl Mech Eng 194:3902–3933. https://doi.org/10.1016/j.cma.2004.09.007
15. Deb K (2000) An efficient constraint handling method for genetic algorithms. Comput Methods Appl Mech Eng 186:311–338. https://doi.org/10.1016/S0045-7825(99)00389-8

Forecast-Based Reservoir Operation with Dynamic Forecast Horizon

Narjes Ghaderi and S. Jamshid Mousavi

Abstract Forecast-based reservoir operation has been recognized as a way to deal with the inflow-to-reservoir uncertainty. This paper investigates how dynamically optimized forecast horizons (FHs) can influence the performance of a forecast-based real-time reservoir operation optimization model. Two types of models, i.e., the fixed FH model and the time-dependent variable FH model, are developed and compared in Rudbar Dam operations in Iran. The vector of optimal FH, each element of which is the FH for a specific month of a year, is determined using harmony search (HS) algorithm within a real-time reservoir operation optimization model. K-nearest neighbor (KNN) method is applied as a forecasting approach to predict monthly inflows to the reservoir. The objective function of the optimization model is minimizing the deviation of controlled releases from water demands. Results show that the real-time model benefiting from a dynamically optimized FH vector outperforms the one having a fixed FH by 27% in terms of the value of objective function in a yearly operation model.

Keywords Forecast horizon · Real-time operation · Harmony search · KNN

1 Introduction

Real-time reservoir operation is widely used for purposes such as flood control, water supply, and hydropower generation [11, 12]. Forecast-based operations have been used to deal with hydrologic variability and uncertainty. Operators would be able to manage the system's water storage more flexibly by including inflow forecast information in reservoir operations [3]. However, inflow forecasts are not perfect, and the forecast error increases when longer forecast horizons (FHs) are applied. There are several studies regarding how to incorporate inflow forecasts into real-time reservoir operation and to determine an appropriate forecast horizon toward improving

N. Ghaderi · S. Jamshid Mousavi (✉)
Department of Civil and Environmental Engineering, Amirkabir University of Technology, Tehran, Iran
e-mail: jmosavi@aut.ac.ir

J. H. Kim et al. (eds.), *Proceedings of 7th International Conference on Harmony Search, Soft Computing and Applications*, Lecture Notes on Data Engineering and Communications Technologies 140, https://doi.org/10.1007/978-981-19-2948-9_43

the system performance [3, 4, 13, 14, 16]. In order to determine the optimal FH in forecast-based reservoir operations, You and Cai [13] presented both theoretical and numerical frameworks considering various factors including inflow, water demand, and reservoir capacity affecting the best FH. They evaluated the relationship between FH and the reservoir capacity and concluded that a larger reservoir capacity requires a longer FH, although factors such as streamflow uncertainty can complicate the situation. Even though the forecast lead time should provide sufficient information to make a particular decision, there is the potential for growth of forecast uncertainty and error over time that may reduce the utility of forecast information [14, 15]. Therefore, we must account for the trade-off between the requirement for reliable information and the limitations of the forecast capabilities beyond a specific time horizon. Zhao et al. [16] established an effective forecast horizon (EFH) concept, i.e., a forecast horizon beyond which the system performance does not improve due to increasing uncertainty. In addition, they found that there is a complicated relationship between forecast uncertainty and forecast horizon. The longer the forecast horizon, more information on future is provided for decision-making on one hand, but the larger forecast uncertainty is imposed on the other hand. Anghileri et al. [1] used a forecast-based adaptive control framework and evaluated how long-term streamflow forecasts could be used to design reservoir operations for water supply in snow-dominated rivers basins. They found that the forecast value is affected by the operation objectives, hydrological characteristics, and other structural features of the water system. Gavahi et al. [4] examined the effect of forecast time horizon on the performance of an adaptive forecast-based monthly real-time reservoir operation model. Their inflow forecasting model was K-nearest neighbor (KNN), and they used the genetic algorithm to determine the optimal forecast time horizon. Jafari et al. [8] investigated the role of precipitation forecast system characteristics—forecast time horizon and forecasting type—in real-time operation of urban drainage systems. They used a simulation–optimization model that incorporated a rainfall prediction module, a rainfall-runoff simulator, and the harmony search (HS) optimization algorithm. They concluded that due to the system's limited capacity, increasing the forecast time horizon only improves operation to a certain point beyond which it has no positive effect. In a multipurpose reservoir system, Doering et al. [3] suggested a theory of time-varying sensitivity analysis of control policies that could be used to determine the value of alternative FHs. They concluded that with respect to the goals being pursued, the value of inflow information may differ widely. It can decline or disappear entirely when system constraints (e.g., lower limits on reservoir levels and spillways) limit the use of forecasts to drive decision-making, particularly during extreme events.

This paper presents a forecast-based real-time reservoir operation optimization model that addresses the influence of season-dependent dynamically optimized FHs. The objective function is water supply to monthly water demands during a one-year operation planning horizon. The paper is organized as follows: first, we introduce the framework and formulation of the presented forecast-based real-time reservoir operation model. Second, the KNN algorithm as an inflow forecasting model is explained. Subsequently, the HS optimization algorithm is presented to optimize the forecast horizon in each month. Finally, discussion and conclusion are provided.

2 Forecast-Based Real-Time Reservoir Operation Optimization Model

The real-time reservoir operation approach aims to account for uncertainty of inflows to the reservoir through benefiting the latest available information for deciding in the near future [4]. Because the inflow forecast quality decreases generally as the forecast lead time increases, it is essential to update the forecasts as new data and information on future inflows and the current system state are acquired [1, 11, 12]. We propose an adaptive reservoir operation model in which the inflow forecasts and the release decisions are updated in each time step based on the latest information being obtained. The proposed model comprises three main phases. In the first phase and at the beginning of each time period (month) t, the latest historical data series is collected. Then, a prediction model forecasts the monthly inflows ($\vec{I}$) to the reservoir from the current time step up to the end of the operation horizon (OH), i.e., $\vec{I} = \{I_t, I_{t+1}, \ldots I_{t+\mathrm{OH}}\}$. It is worth mentioning that the OH in our study is 12 months. In the second phase, the predicted inflow vector ($\vec{I}$) is given as an input to a reservoir operation optimization model that determines the optimal reservoir releases ($\vec{Q}$) from the current time step up to the end of the OH, i.e., $\vec{Q} = \{Q_t, Q_{t+1}, \ldots Q_{t+\mathrm{OH}}\}$. In the last phase, only the first element of the vector of optimal releases (Q_t) is applied, and the system is transferred to a state at the beginning of the next time step (month), and the future inflow forecasts are updated. This procedure is repeated for every time step ahead until the last step when $t = \mathrm{OH}$. Figure 1 shows the framework and steps of the presented forecast-based real-time operation optimization model.

2.1 *Problem Formulation*

In this study, a single reservoir system functioning for water supply purposes is considered. Using a nonlinear optimization model, the problem is designed to minimize the sum of squared water release deviation from a predefined water demand vector over a 12-month operation horizon. The optimization model formulation is as follows:

$$\mathrm{Min} \sum_{t=1}^{T} (Q_t - D_t)^2 \tag{1}$$

S.t.:

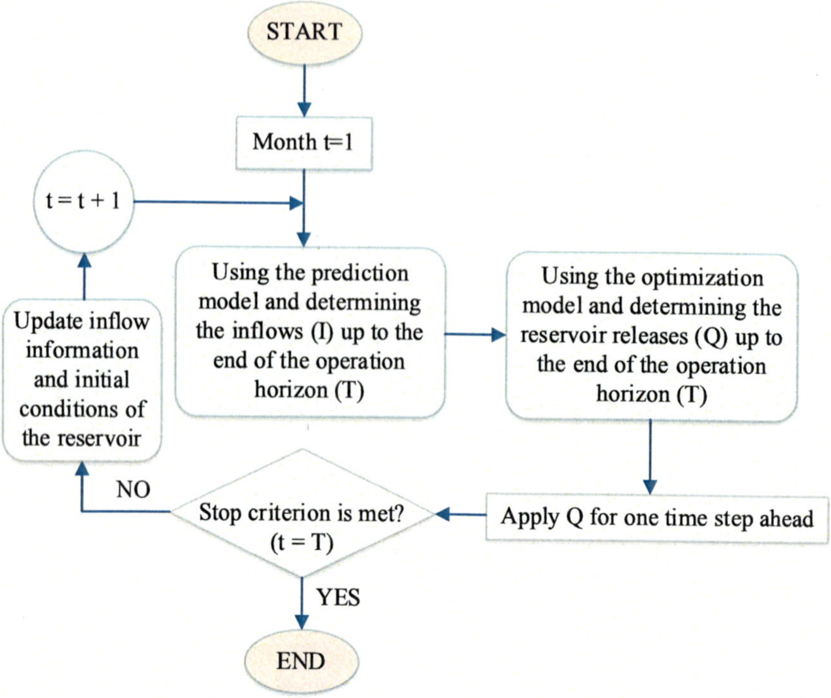

Fig. 1 Flowchart of the forecast-based real-time reservoir operation optimization model

$$S_{t+1} = S_t + I_t - Q_t \quad (2)$$

$$S_{\min} \leq S_t \leq S_{\max} \quad (3)$$

$$Q_t \leq D_t \quad (4)$$

where T is the number of operation horizons, Q_t is the reservoir release, D_t is the monthly demand, S_t is the beginning-of-month reservoir storage, I_t is the inflow to the reservoir, and $S_{\min}$ and $S_{\max}$ are the minimum and maximum storage volumes of the reservoir, respectively.

3 KNN Regression Algorithm

A forecast-based real-time reservoir operation optimization scheme can perform well only if it benefits from an appropriate streamflow forecasting approach over short time horizons. In such a scheme, artificial intelligence and machine learning

(ML) algorithms should focus on improving runoff forecasts [7]. Over the last few decades, KNN as one of the simplest ML approaches to time series forecasting has been widely used [2, 4, 5, 10]. KNN is a data-driven, non-parametric approach that determines the best estimate for a dependent variable by comparing independent observations (predictors) with historical data. The basic premise of using KNN in time series prediction is that each time series contains iterative patterns, so we can find the previous k patterns similar to the current structure of the series based on a distance measure and use them to predict future behavior [10]. The present study uses a recursive approach for multistep-ahead forecasting, i.e., predicting the next N values over the operation horizon. This strategy (also called iterative or multi-stage approach) is the simplest and oldest way of forecasting. In this approach, for forecasting H steps ahead, we first apply the KNN model to forecast inflow only for one-step ahead. Afterward, the previously forecasted value is used as part of the input variables for forecasting the next step's value using the same one-step-ahead model. The process is repeated until the entire horizon [2]. A disadvantage of the recursive method is the cumulative impact of the forecast errors on future predictions. In this study, the KNN approach is evaluated over a 60-year historical monthly inflow record to the Rudbar Dam in Iran from 1955 to 2015, and ten years of data are used for validation. Due to the seasonal nature of the data, for each dependent variable, the last 12 months' inflows are considered as predictors (independent variables). Euclidean distance is used as a measure of proximity, and the best value of parameter K is four, based on a trial and error approach.

4 Harmony Search Algorithm

There is a strong correlation between forecast time horizon, reservoir characteristics, and operating objectives [1, 3, 8, 13]. As a result, selecting a forecast horizon that provides adequate information for reliable releases is essential to make the best use of forecast information [14]. In this study, we use the HS optimization algorithm to determine monthly forecast horizons. The HS algorithm was first proposed by Geem et al. [6]. This algorithm was conceptualized based on the musical process of searching for a perfect state of harmony. Similarly, the goal of an optimization process is finding a global solution based on an objective function [9]. In order to determine the optimal FH vector, first, the HS generates a random initial solution for monthly forecast horizons based on the size of harmony memory (HMS). Each monthly FH value as a decision variable can take any value from 1 to 12. Next, the forecast-based real-time reservoir operation optimization model is run for each solution to determine the variables required to evaluate the HS objective function. The HS objective function is to minimize the sum of squared water release deviation from predefined water demand over the 12-month operation horizons after applying the adaptation in each time step. Then, to create new harmony, each decision variable of the solution is required to choose a value from one of three categories: (1) any value from the HM (harmony memory consideration rate (HMCR)), (2) an adjacent value of one value from the HM [pitch adjustment rate (PAR)], or (3) a totally random value out of the possible range of values. Whenever a new harmony is introduced, the HS objective function must be evaluated again. The process is repeated until the convergence or stopping criteria are met. We determined suitable HS parameter values by performing several trial runs, and we set HMS, HMCR, and PAR to 100, 0.8, and 0.2, respectively.

5 Results and Discussion

In this study, 59 years of historical inflow data to the Rudbar Dam in west of Iran are used to develop two types of forecast-based real-time reservoir operation optimization models for 60th year (2015) with: (A) a fixed forecast time horizon irrespective of the time step of operation and (B) dynamic, variable forecast time horizon for each month of operation. For the fixed FH model, the forecast horizon is equal to the number of operation horizon, i.e., 12 months for every month. However, the forecast horizon can be different for each time step of the second model, so the best horizon should be determined for each month. As a decision variable in the HS algorithm, each monthly FH value can take any value from 1 to 12. The HS objective function was defined in Eq. (1) after applying the adaptation scheme in every time step.

The optimal forecast horizon vector obtained by HS algorithm is FH = [4, 7, 8, 2, 2, 2, 2, 1, 1, 1, 9, 7]. In this case, considering that the reservoir has a limited storage capacity, for those months when water demand is significantly different from the mean monthly inflow to the reservoir, a longer forecast horizon has been obtained as the optimal forecast horizon to make decisions based on the knowledge of system behavior in the future. On the other hand, for the months when water demand is close to the mean monthly inflow, shorter optimal FHs have been resulted. According to previous studies, the forecast horizon depends on various factors, including reservoir capacity, objective function, water demands, and inflow patterns. Figure 2 shows the convergence curve of the HS algorithm over 300 iterations where the algorithm finds the optimal solution after 171 iterations. Table 1 reports the results of the objective function value for the two mentioned models. It is seen that Model B having a variable FH outperforms the fixed FH model (Model A) as it has improved the objective function value by 27%. This outcome demonstrates the value of utilizing dynamically-tuned FHs in forecast-based reservoir operation optimization. In this strategy, the optimal value of FH in each month is determined depending on the specific system characteristics in that month such as the available regulating capacity of the reservoir and the pattern of water demands over future months of operation remained.

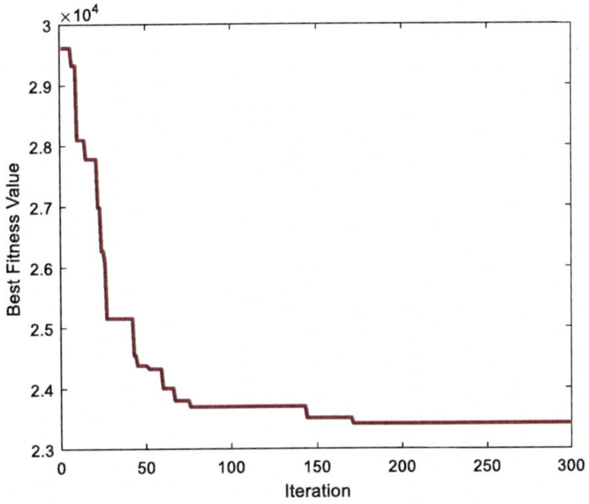

Fig. 2 Convergence curve of the harmony search algorithm to the optimal solution

Table 1 The results of real-time reservoir operation models

Model type	Objective function $\times 10^3$
A	32.17
B	23.42

6 Conclusion

This paper addressed how dynamically optimized variable forecast horizons influenced the performance of a forecast-based real-time reservoir operation optimization model. Two types of forecast-based operation models with (1) fixed FH and (2) variable FH were developed and tested in which HS optimization algorithm was used to determine the optimal forecast horizons for each month of the operation. For the case studied, longer FHs were found to be optimal for those months when water demand was significantly different from the mean monthly inflow compared to other months. Results revealed that the model benefiting from the variable, season-dependent optimized FH outperformed the fixed, constant FH model (27% improvement of the objective function). Such a model makes use of the fact that each month's optimal forecast horizon depends on the system state and characteristics belonging to that specific month such as available regulating capacity and the pattern of current and future water demands, and extending the FH beyond the optimal horizon does not lead to an additional positive impact.

References

1. Anghileri D, Voisin N, Castelletti A, Pianosi F, Nijssen B, Lettenmaier DP (2016) Value of long-term streamflow forecasts to reservoir operations for water supply in snow-dominated river catchments. Water Resour Res 52(6):4209–4225. https://doi.org/10.1002/2015WR017864
2. Ben Taieb S, Bontempi G, Atiya AF, Sorjamaa A (2012) A review and comparison of strategies for multi-step ahead time series forecasting based on the NN5 forecasting competition. Expert Syst Appl 39(8):7067–7083. https://doi.org/10.1016/j.eswa.2012.01.039
3. Doering K, Quinn J, Reed PM, Steinschneider S (2021) Diagnosing the time-varying value of forecasts in multiobjective reservoir control. J Water Res Plann Manag 147(7). https://doi.org/10.1061/(asce)wr.1943-5452.0001386
4. Gavahi K, Mousavi SJ, Ponnambalam K (2019) The role of sreamflow forecast horizon in real-time reservoir operation. Sustain Safe Dams Around World, 1603–1610https://doi.org/10.1201/9780429319778-143
5. Gavahi K, Mousavi SJ, Ponnambalam K (2018) Comparison of two data-driven streamflow forecast approaches in an adaptive optimal reservoir operation model, vol 3, pp 755–745. https://doi.org/10.29007/wrn8
6. Geem ZW, Kim JH, Loganathan GV (2001) A new heuristic optimization algorithm: Harmony search. Simulation 76(2):60–68. https://doi.org/10.1177/003754970107600201
7. Ilich N, Basistha A (2021) Importance of multiple time step optimization in river basin planning and management: a case study of Damodar River basin in India. Hydrol Sci J 66(5):809–825. https://doi.org/10.1080/02626667.2021.1895438
8. Jafari F, Mousavi SJ, Kim JH (2020) Investigation of rainfall forecast system characteristics in real-time optimal operation of urban drainage systems. Water Resour Manage 34(5):1773–1787. https://doi.org/10.1007/s11269-020-02528-1
9. Lee KS, Geem ZW (2005) A new meta-heuristic algorithm for continuous engineering optimization: Harmony search theory and practice. Comput Methods Appl Mech Eng 194(36–38):3902–3933. https://doi.org/10.1016/j.cma.2004.09.007
10. Martínez F, Frías MP, Pérez MD, Rivera AJ (2019) A methodology for applying k-nearest neighbor to time series forecasting. Artif Intell Rev 52(3):2019–2037. https://doi.org/10.1007/s10462-017-9593-z
11. Vedula S, Mohan S (1990) Real-time multipurpose reservoir operation: a case study. Hydrol Sci J 35(4):447–462. https://doi.org/10.1080/02626669009492445
12. Wei CC, Hsu NS (2008) Multireservoir real-time operations for flood control using balanced water level index method. J Environ Manage 88(4):1624–1639. https://doi.org/10.1016/j.jenvman.2007.08.004
13. You JY, Cai X (2008) Determining forecast and decision horizons for reservoir operations under hedging policies. Water Res Res 44(11). https://doi.org/10.1029/2008WR006978
14. Zhao Q, Cai X, Li Y (2019) Determining Inflow forecast horizon for reservoir operation. Water Resour Res 55(5):4066–4081. https://doi.org/10.1029/2019WR025226
15. Zhao T, Cai X, Yang D (2011) Effect of streamflow forecast uncertainty on real-time reservoir operation. Adv Water Resour 34(4):495–504. https://doi.org/10.1016/j.advwatres.2011.01.004
16. Zhao T, Yang D, Cai X, Zhao J, Wang H (2012) Identifying effective forecast horizon for real-time reservoir operation under a limited inflow forecast. Water Res Resour 48(1). https://doi.org/10.1029/2011WR010623

Recommending Best Quality Image Using Tracking and Re-identification Approaches

Alim Manjiyani, Abhishek Naik, Swathi Jamjala Narayanan, and Boominathan Perumal

Abstract The society we live in has several issues as apart as catching thieves and capturing the best moments of a marathon or any special occasion. In this paper, we aim to tackle the problems, which involve person re-identification, tracking, detection, and image quality evaluation. These domains find their efficacy in a variety of situations such as crime investigations, wildlife photography, sports events, and many more. Our paper endeavors to provide a solution, which will be able to track, detect, re-identify the person under perception along with suggesting the best quality frame possible in the given input, which encompasses the target. The proposed architecture is composed of four modules. The video input received goes through detection module frame by frame. The target selected by the user goes tracked throughout, until it stays in the frame using the tracking module. The person re-id module kicks in when the target is out of the frame and starts scanning every person detected for a precise match. Finally, the image quality module is run over the frames, which holds the target and returns with the suggestion for the best quality images. We have used YOLOv5 for detection, IoU for tracking, deep learning algorithm for person re-id module and BRISQUE algorithm for image quality module.

Keywords Person re-identification · You only look once (YOLO) v5 · Intersection over Union (IoU) · Deep learning · Blind/referenceless image spatial quality evaluator (BRISQUE)

1 Introduction

In this age of advanced technologies such as Artificial Intelligence and Machine Learning, issues, which require human effort, are reducing day by day. The need to detect and recognize a person in a given video footage with a good accuracy lies under the domain of person re-identification, which is one such issue that has room for improvement. The process of identifying and tracking the thief/suspect in a video

A. Manjiyani · A. Naik · S. J. Narayanan (✉) · B. Perumal
School of Computer Science and Engineering, VIT, Tamil Nadu, Vellore 632014, India
e-mail: swathi.jns@gmail.com

J. H. Kim et al. (eds.), *Proceedings of 7th International Conference on Harmony Search, Soft Computing and Applications*, Lecture Notes on Data Engineering and Communications Technologies 140, https://doi.org/10.1007/978-981-19-2948-9_44

footage can be very toiling. Therefore, we are motivated to come up with a solution to this problem using the modern day techniques of Machine Learning and Artificial Intelligence. This research work aims on providing a solution to the end-users by providing a software capable of tracking, detecting, re-identifying the person under observation along with recommending the best possible quality image present in the given input.

The existing work on the topic of developing a framework which can efficiently track, detect, re-identify and provide best quality image of the target, lies only on its respective components.

Ahmed et al. [1] proposed a profound learning engineering for Person Re-Identification, which comprises two novel layers for catching connections between two perspectives: a cross-input neighborhood contrasts layer, and an ensuing layer that sums up these distinctions. It showed comparatively moderate accuracy when compared to modern day person re-id algorithms. Also, [2] presented a novel three-branch structure named 'Profound Person' to adapt exceptionally discriminative profound highlights for individual Re-ID. End up being cutting edge 2019 by outflanking different strategies on MARKET-1501, CUHK03, and DUKEMTMC-re-id datasets. Moreover, [3] proposed a new model for generating deep person image by creating pose-normalized individual pictures for person identification. No additional data collection and tuning required when testing on a new dataset and was the 2018 state-of-the-art.

Similarly, [4–7] proposed various methods for person re-identification such as semantically aligned person re-id, re-identifying in a crowd, self-critical attention learning, respectively, and were competitive during their state-of-the-art. Ren et al. proposed Regional Convolutional Neural Network for multiple object real-time detection. Redmon and Farhadi [8] introduced an incremental improvement in the then existing yolov2 and presented the yolov3. Recently, detection algorithms such as YOLOv5 and PP-YOLO have made their names as the state of art for the detection.

In the world of image quality metric determination, first [16] paper presents a model which takes 3 credits to decide nature of picture in particular, loss of connection, luminance contortion, and difference mutilation. All inclusive, free technique is utilized to survey the nature of picture. It very well may be utilized in mix with other picture preparing calculations too. Also, it provides less mathematical complexity. In 2012, Mittal et al. proposed the BRISQUE algorithm for assigning a relative image quality value which does not require cross-referencing. Since, the BRISQUE is an unsupervised algorithm it found its way in our framework.

The most common form of tracking, which has a little to no possibilities of ID switching and a good eye for occlusion handling is the Intersection over Union (IoU) technique. But [9] demonstrates the flaws of the widely used Intersection over Union (IoU) metric for bounding box overlapping and tracking. Introduces a new Generalized Intersection over Union method (GIoU) to overcome the flaws of IoU.

As we moved forward with taking the best out of the existing models and works from the survey we found few gaps which can be filled to improve these models/works. Firstly, [1]'s architecture for person re-identification has a comparatively moderate accuracy (84–86%) which can be improved as better model alternatives are available. Also, recently YOLOv5 and PP-YOLO techniques are available which outperforms the YOLOv3 [8] in terms of speed and accuracy. Secondly, better blind image de-noising algorithms available, which can improve its efficiency [10], which is surprising as it is not addressed since 2012.

The tracking technique used (IoU) [9] can only be used for 2-D object detection frameworks. The article [11] lacks an attention mechanism to naturally choose more segregating body parts instead of only relying on the features along the vertical direction, and improvements can be done for occlusion and inaccurate detections [3]; Though models with better accuracy and time complexity are available, excluding the discriminative features caused id switching and occlusion problems. Finally, returning to person re-identification [5], due to the complex network being used in this paper, it is difficult to keep such a complex code up to date in the real world. Also, this increased complexity makes the code base less dynamic [6]. This technique only concentrates its attention on features present on the target body and not on the size and shape of the target itself. Therefore, this may lead to false positives when two individuals have the same outer features such as children in a classroom wearing the same uniform [7]. Since an extra spine network gives a solid administrative sign dependent on its perception is being utilized in this model, time intricacy and space intricacy of the codebase increments.

2 Proposed Architecture

The deep neural network architecture [1] used, codifies person re-id problem as binary classification. The research work is to decide if the given two pictures address a similar individual. The framework first passes the given pictures through two layers of attached convolution with max pooling. The outcome is then passed through cross-input neighborhood contrasts followed by the calculation of fix synopsis highlights, across-fix highlights and higher-request connections. At last, the SoftMax layer gives a last gauge of whether the given pictures address a similar person. Each of these is clarified henceforth and the architecture is given in Fig. 1.

2.1 Person Re-identification

Tied Convolution: The convolution layers process higher level information of the given pictures. This computation is done using tied convolution so that these features are comparable across layers henceforth. This technique uses the same weights for both images to ensure that both the inputs pass through a similar filter. As a result, we obtain 25 features maps having 12 rows and 37 columns each.

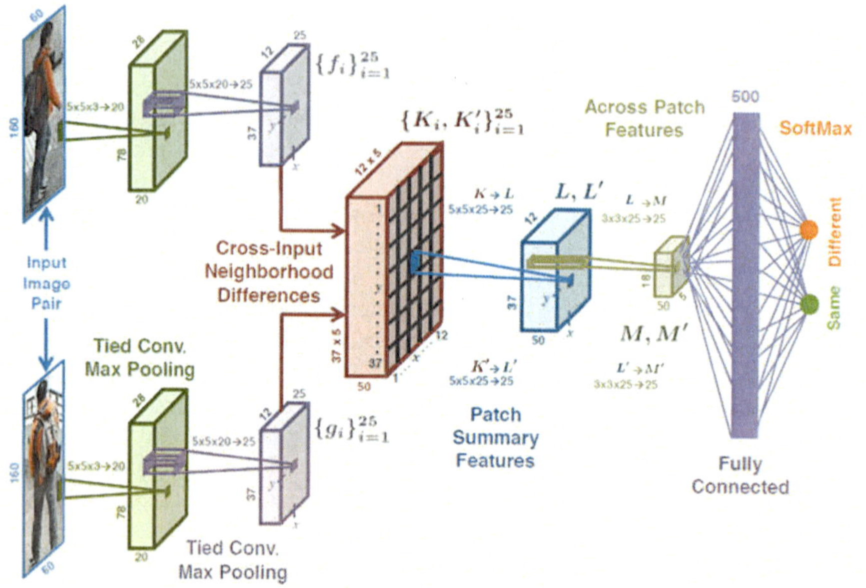

Fig. 1 Person re-identification architecture

Cross-Input Neighborhood difference: The feature maps obtained from the previous layer helps in learning the relationship between the inputs. Differences in feature values across the two views around a neighborhood of each feature location is computed by a cross-input neighborhood difference, resulting a set of 25 neighborhood variation maps (Ki). Similarly, we compute Ki' in which the feature maps are in reverse order. Now, each of these 50 neighborhood difference maps of size $12 \times 37 \times 5 \times 5$ are transited along a Rectified Linear unit (ReLu).

Patch Summary Features: To summarize the neighborhood difference maps obtained from the preceding layer and to construct a comprehensive representation of the contrast in each 5×5 block, a patch summary layer is used. This layer plays out the planning from $K \in R12 \times 37 \times 5 \times 5 \times 25 \rightarrow L \in R12 \times 37 \times 25$ by convolving K with 25 channels of size $5 \times 5 \times 25$, with a step of 5. We evaluate L' and K' by reversing the order of neighborhood difference maps. Finally, we pass L and L' through a ReLu.

Across-Patch Features: Across-patch features enable us to learn spatial relationships among neighborhood differences. In this layer, we convolve L with 25 channels of size $3 \times 3 \times 25$ utilizing a unit step. A decrease in stature and width by a factor of 2 is achieved by passing the above result through a max pooling kernel. This also results in $M \in R5 \times 18 \times 25$ which is nothing but 25 feature maps of size 5×18. M' and L' are obtained however, channels for the planning $L \rightarrow M$ and $L' \rightarrow M'$ are not tied.

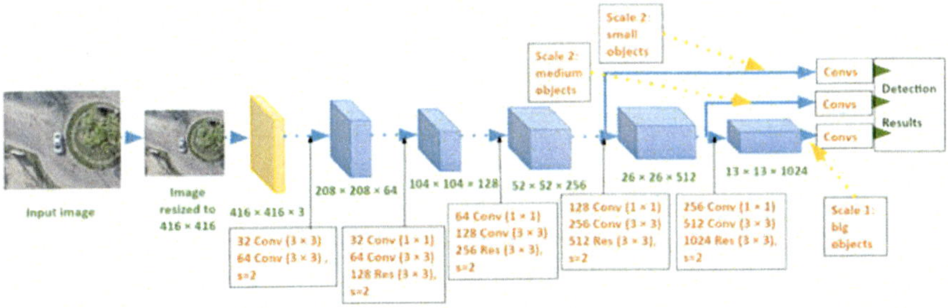

Fig. 2 YOLOv3 architecture

Higher Order Relationship: Here, we apply a completely associated layer after M and M'. This catches higher-request connections by joining data from patches that are a long way from one another and by consolidating data from M with data from M'. The resultant component vector of size 500 is passed through a ReLu nonlinearity. The outcome of ReLu is again passed through two SoftMax units which aggregates into a new layer and returns the probability of the two views being the same.

2.2 *Module 2: Detection Using YOLO v5*

You Only Look Once (YOLO) v5 is the state-of-the-art person and object detection technique. YOLO v5 is an upgrade to the YOLO v3 shown in Fig. 2. Yolo V5 uses a variant of darknet and has 53-layer network trained on ImageNet. An addition of 53 layers is performed for the detection part resulting in 106 fully convolutional layers which serve as the backbone for this architecture.

1×1 detection kernels are applied on 3 different sizes of features maps at three different junctions in the network. Binary cross-entropy is used for the calculation of classification loss and logistic regression for predicting object confidence and class predictions. The input is transformed into batch of images of shape ($m \times 416 \times 416 \times 3$) and passed to a Convolutional Neural Network (CNN). The last two dimensions of the obtained output are flattened to get an output volume of size ($19 \times 19 \times 425$). As a result, a list of bounding boxes is obtained as an output. To avoid any selection of overlapping bounding boxes, non-max suppression and IoU is performed.

2.3 *Module 3: Tracking*

Figure 3 intersection over Union (IoU) is an assessment grade used to compute the correctness of an object detector on a particular dataset. IoU is the proportion of space of crossing point of ground truth bounding and anticipated bounding box to

Fig. 3 Intersection over union (IoU)

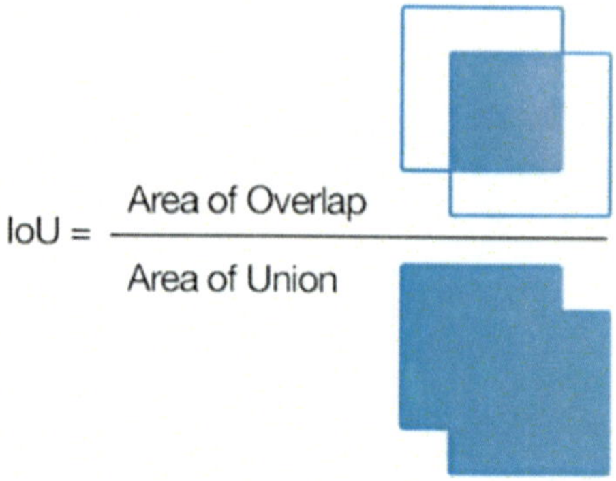

their space of association. It can also be used to measure the Gaussian distance between two bounding boxes corresponding to different frames at different times.

2.4 Module 4: BRISQUE Architecture

The BRISQUE (Blind/Referenceless Image Spatial Quality Evaluator) algorithm is a distortion-generic blind/no-reference (NR) quality assessment algorithm which works in spatial domain. This algorithm does not include any distortion features such as ringing, blur or blocking, and only gives a mathematical value for the naturalness in the image which can be accounted by the distortions present in it.

Initially, local mean subtraction and divisive normalization is used to calculate (1) locally normalized luminance $\left(\hat{I}(i, j)\right)$.

$$\hat{I}(i, j) = \frac{I(i, j) - \mu(i, j)}{\sigma(i, j) + C} \tag{1}$$

where $i \in 1, 2 \ldots Y$, $j \in 1, 2 \ldots X$, are spatial files, Y, X are the picture stature and width separately, $C = 1$ is a consistent and

$$\mu(i, j) = \sum_{k=-K}^{K} \sum_{l=-L}^{L} \omega_{k,l} I_{k,l}(i, j) \tag{2}$$

$$\sigma(i, j) = \sqrt{\sum_{k=-K}^{K} \sum_{l=-L}^{L} \omega_{k,l}\left(I_{k,l}(i, j) - \mu(i, j)\right)^2} \tag{3}$$

where $w = \{w_{k,l} | k = -K, .., K, l = -L, .., L\}$ is a 2-D circularly-symmetric Gaussian weighting capacity tested out to 3 standard deviations and rescaled to unit volume and $K = L = 3$.

A generalized Gaussian distribution (GGD) can be used to effectively capture a broader spectrum of distorted image statistics, which often exhibit changes in the tail behavior (i.e., kurtosis) of the empirical coefficient distributions where the GGD with zero mean is given by:

$$f\left(x;\alpha,\sigma^2\right)=\frac{\alpha}{2\beta\gamma(1/\alpha)}\exp\left(-\left(\frac{|x|}{\beta}\right)^{\alpha}\right) \tag{4}$$

where

$$\beta=\sigma\sqrt{\frac{\gamma(1/\alpha)}{\gamma(3/\alpha)}} \tag{5}$$

and $\gamma(.)$ is the function:

$$\gamma(a)=\int_0^{\infty} t^{a-1}e^{-t}\mathrm{d}t \quad a>0 \tag{6}$$

This algorithm does not require any database and can be used for identification of distortion in the input image as well.

2.5 Integrated Framework

Finally, the integrations of the above components result in the architecture as given in Fig. 4. Initially, the video is fetched, and a selection GUI is presented to the user to select the target. After the target selection, a program tracks the target until it stays in the frame. Once the tracking fails, re-identifier works in collaboration with the

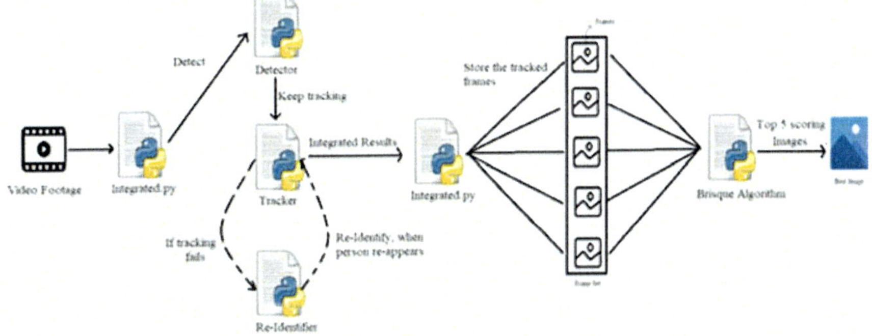

Fig. 4 Proposed architecture

detector to detect the target as soon as it re-enters the frame or the occlusion is no more. Following up with this re-identification, target is tracked again throughout, and this cycle continues until the final frame of the input is processed. All the frames stored, are passed to BRISQUE algorithm to obtain a score for each of the frames. These scores are then sorted, and the final recommendation of best quality image is presented to the user.

3 Experimental Results and Discussion

Figure 5 shows the selection GUI for the first video input. When a particular ID is chosen for re-identification namely ID 1, the following activities take place: The program tracks the person until it can (i.e., until it gets occluded or is out of the frame). When the tracking fails, the person re-identification module detects the person as soon as it re-enters the frame and again starts tracking. These frames are stored in folder '0' and '1', details of which is discussed later. After this, an Image quality score is assigned to all the frames stored.

After all the stored frames are assigned an image quality score, which stood as top 5 are stored in folder '2' and are the recommendations of the recommender module.

Fig. 5 Target selection screen for test case 1

Fig. 6 Output folders

Fig. 7 Cropped images of target in folder '0'

Fig. 8 Full frame images in folder '1'

Fig. 9 Top 5 recommendations are stored in folder '2'

Figure 6 shows the 3 output folders created. The folder '0' stores only the snippets of the person tracked (Fig. 7). Folder '1' stores the frame corresponding to the snippets in folder '0' (Fig. 8). Finally, the folder '2' contains the top 5 recommended images (Fig. 9).

4 Conclusion and Future Works

The proposed architecture can help in achieving the solution to the problem mentioned in this research work. Also, it provides an efficient way to track, detect, re-identify and calculate image quality scores. To overcome the limitations of light-weight trackers such as Boosting Tracker, MIL Tracker, KCF Tracker, GOTURN Tracker. which provides good fps but at the cost of decreased accuracy, in this research work we found that, using Intersection over Union along with template matching proved to be very efficient for tracking and the YOLOv5 detection module can be used to add as the extra support to the trackers. The methodologies proposed are used for detection and Image quality measurement. In this work, we achieved result of combination of solutions to different sub-problems like object detection, tracking, etc., where the architecture used to merge those sub-problems plays a crucial role in tackling the problem as a whole. Moreover, as a prospect of future work we can extend this work to operate on multiple camera feeds with efficiency.

Acknowledgements The authors thank VIT for providing 'VIT SEED GRANT' for carrying out this research work. We gratefully acknowledge the support of NVIDIA Corporation with the donation of the Titan Xp GPU used for this research on Recommending Best Quality Images using Tracking and Re-Identification Approaches.

References

1. Ahmed E, Jones M, Marks TK (2015) An improved deep learning architecture for person re-identification. In: Proceedings of the IEEE conference on computer vision and pattern recognition, pp 3908–3916
2. Li J, Wang J, Tian Q, Gao W, Zhang S (2019) Global-local temporal representations for video person re-identification. In: Proceedings of the IEEE/CVF international conference on computer vision, pp 3958–3967
3. Qian X, Fu Y, Xiang T, Wang W, Qiu J, Wu Y, Jiang Y-G, Xue X (2018) Pose-normalized image generation for person re-identification. In: Proceedings of the European conference on computer vision (ECCV), pp 650–667
4. Zhao R, Oyang W, Wang X (2016) Person re-identification by saliency learning. IEEE Trans Pattern Anal Mach Intell 39(2):356–370
5. Zhang Z, Lan C, Zeng W, Chen Z (2019) Densely semantically aligned person re-identification. In: Proceedings of the IEEE/CVF conference on computer vision and pattern recognition, pp 667–676
6. Mazzon R, Tahir SF, Cavallaro A (2012) Person re-identification in crowd. Pattern Recogn Lett 33(14):1828–1837
7. Wang C-Y, Mark Liao H-Y, Wu Y-H, Chen P-Y, Hsieh J-W, Yeh I-H (2020) CSPNet: a new backbone that can enhance learning capability of CNN. In: Proceedings of the IEEE/CVF conference on computer vision and pattern recognition workshops, pp 390–391
8. Redmon J, Farhadi A (2018) Yolov3: an incremental improvement. arXiv preprint arXiv:1804.02767
9. Rezatofighi H, Tsoi N, Gwak JY, Sadeghian A, Reid I, Savarese S (2019) Generalized intersection over union: A metric and a loss for bounding box regression. In: Proceedings of the IEEE/CVF conference on computer vision and pattern recognition, pp 658–666
10. Mittal A, Moorthy AK, Bovik AC (2012) No-reference image quality assessment in the spatial domain. In: IEEE transactions on image processing, vol 21, no 12:4695–4708
11. Bai X, Yang M, Huang T, Dou Z, Yu R, Xu Y (2020)Deep-person: learning discriminative deep features for person re-identification. Pattern Recogn 98:107036

A Modified Whale Optimisation Algorithm to Solve Global Optimisation Problems

S. Gopi and Prabhujit Mohapatra

Abstract Whale optimization algorithm (WOA) is a novel and competitive swarm-based optimisation method that exceeds several previous metaheuristic algorithms in terms of simplicity and efficiency. Whale optimisation algorithm, a revolutionary nature-inspired algorithm, which mimics the behaviour patterns of humpback whales. WOA will interference with local optimization and greatly reduce accuracy for global optimization issue. To solve this type of problem, in this work, a new update equation has been developed named as modified whale optimisation algorithm (MWOA). Also, MWOA has been tested some CEC 2005 benchmark functions with dimension ranging from 2 to 30. The experimental outcomes show that the MWOA produce improved outcomes in terms of optimum value, convergence speed, and stability.

Keywords Mathematical optimization · Exploration space · Global exploration · Optimization · Tested functions

1 Introduction

Metaheuristic algorithms inspired by nature to solve optimization issues by simulating genetic or physical events. They are classified into three types: evolution-based approaches, physics-based methods, and group-based methods. The rules of natural evolution motivate evolution-based approaches. The search begins with a randomly created population that evolves over several groups. Then, the strength of these systems is that the greatest people are constantly joined to generate the next group of people. This enables the population to be optimised across generations. Genetic algorithms [1] are the most widely used evolution-inspired approach. Further, well-known methods are genetic programming algorithm [2], evolution strategy algorithm [3],

S. Gopi · P. Mohapatra (✉)
Department of Mathematics, School of Advanced Sciences (SAS), Vellore Institute of Technology, Vellore, Tamil Nadu, India
e-mail: prabhujit.mohapatra@gmail.com; prabhujit.mohapatra@vit.ac.in

S. Gopi
e-mail: gopi.s2020@vitstudent.ac.in

J. H. Kim et al. (eds.), *Proceedings of 7th International Conference on Harmony Search, Soft Computing and Applications*, Lecture Notes on Data Engineering and Communications Technologies 140, https://doi.org/10.1007/978-981-19-2948-9_45

and biogeography-based optimizer algorithm [4]. The physics-based algorithms are approaches mimic the physical laws of the world. The most popular methods are gravitational local search algorithm [5, 6], charged system search algorithm [7], black hole algorithm [8], ray optimization algorithm [9], curved space optimization algorithm [10]. Swarm-based algorithms are approaches that replicate the public behaviour of animals in the groups are included in the third category of nature-inspired technologies. Then, the most widely used algorithm is particle swarm optimization algorithm [11] which was inspired by the public behaviour of grouping birds. To discover the optimum answer, it employs a number of elements (applicant results) that fly about the exploration space such as the optimal position. Temporarily, they are all tracing the optimal position (best result) along their own pathways. Further, elements evaluate both their individual best outcome and the best outcome create by the group thus far. Ant colony optimization algorithm (ACO) [12], initially suggested by Dorigo et al., is another prominent swarm-based approach. The public behaviour of ants in an ant colony inspired by ACO. It is also worth noting that there are additional meta-heuristic approaches stimulated by human activities that have been published in the literature. Such type of algorithms are Teaching learning-based optimization algorithm [13, 14], harmony search algorithm [15], Tabu search [16–18], group search optimizer [19, 20], firework algorithm [21].

2 Whale Optimisation Algorithm (WOA)

In this part firstly, we addressed the motivation of the proposed method. After that mathematical model frame work is presented.

2.1 Inspiration of WOA

Whales are magnificent animals. Whales are regarded as the world's biggest animals. An adult whale may reach a length of 30 m and a weight of 180 tonnes. Whales are primarily thought of as predators. Because they must breathe from the ocean's surface, they not ever sleep. Partial of the mind, in reality, only sleeps. Whales are fascinating animals because they are thought to be very intelligent and emotional. Additional interesting aspect is whale social behaviour. They might live alone or in sets. They are, however, typically encountered in sets. Some of them, such as killer whales, can live in a group for the rest of their lives. Humpback whales are the biggest baleen whales. An adult humpback whale is about the size of a school bus. Krill and tiny fish herds are their preferred prey. This animal is seen in Fig. 1.

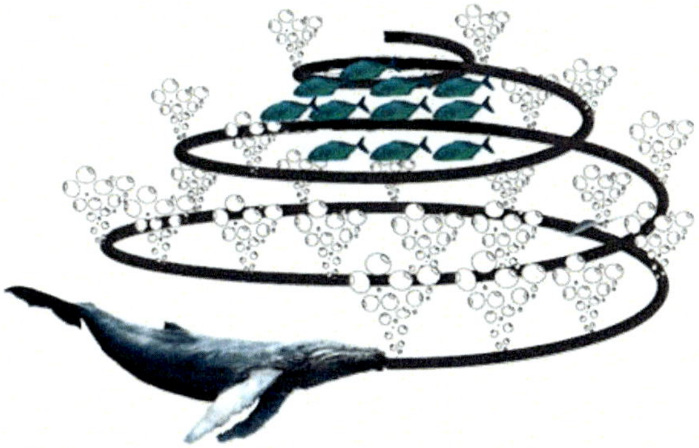

Fig. 1 Bubble-net feeding behaviour

2.2 *Mathematically Model for WOA Algorithm*

The mathematical model of searching for target, surrounding a target, and spiral bubble-net feeding approach is provided initially in this following section.

Surrounding prey:

Humpback whales can recognise the existence of a target and encircle it. Because the optimal project's location in the search space is unknown at the start, the WOA approach assumes the best search agent is found, and the other search agents will seek to update their locations in respect to the best search agent. The following equations demonstrate this behaviour.

$$\mathcal{D} = \left|\mathcal{C} \cdot \mathcal{T}^{*}(l) - \mathcal{T}(l)\right| \tag{1}$$

$$\mathcal{T}(l + 1) = \mathcal{T}^{*}(l) - \mathcal{B} \cdot \mathcal{D} \tag{2}$$

In the above equation, l denotes the present iteration, $\mathcal{B}$ and $\mathcal{C}$ are vectors, $\mathcal{T}^{*}$ is the location vector of the best result found so far, $\mathcal{T}$ is the position vector. It is important to note that $\mathcal{T}$ must be change in each iteration if there is improved outcome exists. Then, $\mathcal{B}$ and $\mathcal{C}$ are designed as follows:

$$\mathcal{B} = 2p \cdot random - p \tag{3}$$

$$\mathcal{C} = 2 \cdot \text{random} \tag{4}$$

where p approaches from 2 to 0 over the sequence of iterations and random is a random vector in [0,1].

The same concept may be applied to searching a space with n dimensions, with search agents moving in principal component analysis around the best answer found that so far. Humpback whales employ the bubble-net approach to attack the target, as explained in the preceding section. This strategy can be mathematically expressed as follows:

Bubble-net attacking procedure:

Two strategies have been developed to mathematically replicate the bubble-net behaviour of humpback whales.

1. Shrinking surrounding technique:
   This behaviour is obtained by lowering the value of p in Eq. 3. It is worth noting that $\mathcal{B}$'s fluctuation range is a random number in the interval $[-p, p]$ where p reduces from 2 to 0 during in the sequence of iterations. Assuming random values for $\mathcal{B}$ in $[-1,1]$, exploration agent's new position can be resolute anywhere amongst the agent's initial position and the position of the present best agent. Figure 2 displays the potential locations from $(\mathcal{T}, \mathcal{U})$ to $(\mathcal{T}^*, \mathcal{U}^*)$ that $0 \le \mathcal{B} \le 1$ may achieve in a 2D space.
2. Spiral updating location:

As shown in Fig. 3, this technique first computes the distance amongst the whale at $(\mathcal{T}, \mathcal{U})$ and the target at $(\mathcal{T}^*, \mathcal{U}^*)$. Then, a spiral calculation is constructed amongst the location of the whales and the target, as shown below:

$$\mathcal{T}(l+1) = \mathcal{D}' \cdot e^{bm} \cdot \cos(2\pi m) + \mathcal{T}^*(l) \tag{5}$$

where $\mathcal{D}' = |\mathcal{T}^*(l) - \mathcal{T}(l)|$ and $\mathcal{D}'$ are the distance amongst the ith whale and the target, b is a constant, m is a random variable in the interval -1 to 1.

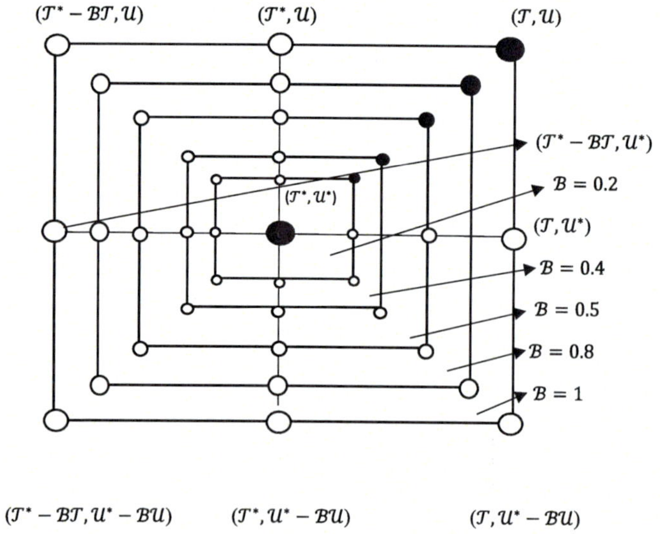

Fig. 2 Shrinking surrounding technique

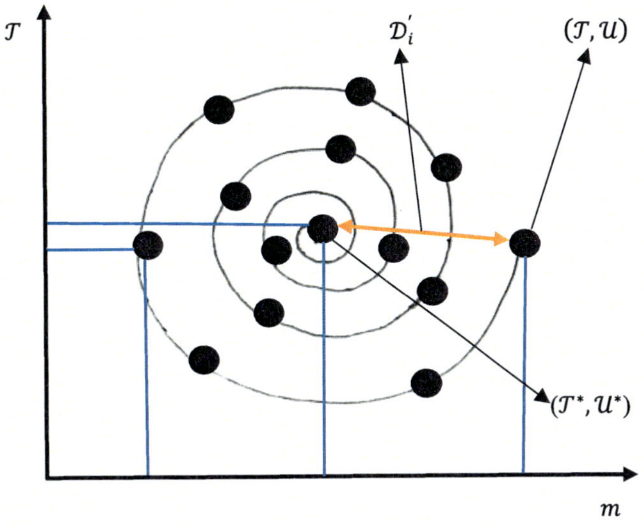

Fig. 3 Spiral changing location

To describe this concurrent behaviour, we assume that there is a 50% chance of selecting either the spiral modal or the shrinking surrounding technique to change the location of whales throughout optimisation. The following is the mathematically model for spiral updating mechanism:

$$\mathcal{T}(l+1) = \begin{cases} \mathcal{T}^*(l) - \mathrm{B} \cdot \mathrm{D} & \text{if } P < 0.5 \\ \mathcal{D}' \cdot \mathrm{e}^{\mathrm{bm}} \cdot \cos(2\pi \mathrm{m}) + \mathcal{T}^*(l) & \text{if } P \geq 0.5 \end{cases} \tag{6}$$

where $\mathcal{P}$ is random value in [0,1].

Search for target:

This technique, as well as $|\mathcal{B}| > 1$, encourages exploration and agree the WOA procedure to do a global exploration. The following is the mathematically model:

$$\mathcal{D} = |\mathcal{C} \cdot \mathcal{T}_{\text{rand}} - \mathcal{T}| \tag{7}$$

$$\mathcal{T}(l+1) = \mathcal{T}_{\text{rand}} - \mathcal{B} \cdot \mathcal{D} \tag{8}$$

where $\mathcal{T}_{\text{rand}}$ is a random location vector.

3 Modified Whale Optimization Algorithm (MWOA)

In this section, whale optimization algorithm is modified by the change the location of the present search agent equation. Then proposed by mathematically model for MWOA algorithm.

3.1 Mathematical Model for MWOA

The mathematical model of searching for target, encircling target, and spiral bubble-net feeding behaviour is provided as follows.

Searching for target:

Whales search for the target at random based on their current position. This humpback whale trait is employed to modified the algorithm exploring capacity. This behaviour can be expressed mathematically as follows:

$$\mathcal{D} = |\mathcal{C} \cdot \mathcal{S}_{\text{random}} - \mathcal{S}| \tag{9}$$

$$\mathcal{S}(\text{k} + 1) = \mathcal{S}_{\text{random}} - \mathcal{A} \cdot \mathcal{D} \tag{10}$$

where $\mathcal{S}$ shows a location vector of the people, $\mathcal{S}_{\text{random}}$ is a vector which is selected randomly from the present people, k denoted by present iteration, $\mathcal{D}$ is the distance amongst the random and present individual of the people.

The two parameters, $\mathcal{A}$ and $\mathcal{C}$, known as coefficient vectors, are calculated as follows:

$$\mathcal{A} = 2\text{a}_1 \cdot \text{random} - \text{a}_1 \tag{11}$$

$$\mathcal{C} = 2 \cdot \text{random} \tag{12}$$

where a_1 value is a number that decreases linearly from 2 to 0 per iteration and random indicates a random number in $[-1,1]$.

Surrounding the target

Throughout this segment, the best result discovered thus far is presumed to be the result nearest to the ideal value. The remaining members of the people update their places in relation to the current best solution. This behaviour is formally characterised as follows:

$$\mathcal{D} = \left|\mathcal{C} \cdot \mathcal{S}^{*}(\text{k}) - \mathcal{S}(\text{k})\right| \tag{13}$$

$$\mathcal{S}(\mathrm{k}+1) = \mathcal{S}^*(\mathrm{k}) - \mathcal{A} \cdot \mathcal{D} \tag{14}$$

where $\mathcal{S}^*$ is the present best result.

Bubble-net attacking technique

Humpback whales migrate in a helix-shaped pattern in response to a bubble-net attack. When searching in MWOA algorithm, the bubble-net method is used as follows:

$$\mathcal{D}^* = \mathcal{S}^*(\mathrm{k}) - \mathcal{S}(\mathrm{k}) \tag{15}$$

$$\mathcal{S}(\mathrm{k}+1) = \mathcal{D}^* \mathrm{e}^{\mathrm{bl}} \sin(2\pi \mathrm{l}) + \mathcal{S}^*(\mathrm{k}) \tag{16}$$

where the value of b is a constant. l is a random number in [−1,1].

Depending on the value of $|\mathcal{A}|$, switching amongst the search and manipulation (exploitation) processes is selected during the search process. If $|\mathcal{A}| \geq 1$ exists, the search process begins, allowing for global search using Eqs. 9 and 10. When $|\mathcal{A}| < 1$, the positions of people are updated using Eqs. 14 or 16. MWOA method changes between surrounding prey and bubble-net attacking strategies based on a probability value of 0.5 for each technique. This feature expressed as follows:

$$\mathcal{S}(\mathrm{k}+1) = \begin{cases} \mathcal{S}^*(\mathrm{k}) - \mathrm{A} & \text{if}\, \alpha \geq 0.5 \\ \mathcal{D}^* \mathrm{e}^{\mathrm{bl}} \sin(2\pi \mathrm{l}) + \mathcal{S}^*(\mathrm{k}) & \text{if}\, \alpha \geq 0.5 \end{cases} \tag{17}$$

where α is a random number in [0,1].

4 Results and Discussion

In general, Table 1 shows, the first two functions are unimodal objective function, next three functions are multimodal objective function, and the last five functions are fixed-dimension multimodal objective functions. In this paper, discussed at the CEC 2005 benchmark functions. The CEC 2005 objective functions are classified into 4 categories: unimodal functions, multimodal functions, fixed-dimension multimodal functions, and composite functions. In this paper, by solving 10 mathematical optimization problems (i.e. ten tested benchmark functions), the mathematical efficiency of the MWOA procedure established in this paper was verified. MWOA was compared to cutting-edge swarm-based optimization techniques. Table 1 summarises the test issues by presenting the range of variation, $f_{\min}$ denotes optimum value, and V_no denotes the number of design variables (population size). For all methods, the population size is different from randomly generated see Table 1, and maximum number of function evaluations are 1000 have been utilised. Each tested functions

Table 1 Ten tested functions (CEC 2005 tested benchmark function)

Function type	Function	V_no	Range	$f_{\min}$
UBF	$T_1(\mathcal{K}) = \sum_{l=1}^{n} \mathcal{K}_1^2$	30	[−100,100]	0
UBF	$T_2(\mathcal{K}) = \sum_{l=1}^{n} \lvert\mathcal{K}_1\rvert + \prod_{l=1}^{n} \lvert\mathcal{K}_1\rvert$	30	[−10,10]	0
MBF	$T_3(\mathcal{K}) = \sum_{l=1}^{n} [\mathcal{K}_1^2 - 10\cos(2\pi\mathcal{K}_1) + 10]$	30	[−10,10]	0
MBF	$T_4(\mathcal{K}) = -20\exp\left(-0.2\sqrt{\frac{1}{n}\sum_{l=1}^{n}\mathcal{K}_1^2}\right) - \exp\left(\frac{1}{n}\sum_{l=1}^{n}\cos(2\pi\mathcal{K}_1)\right) + 20 + e$	30	[−5.12,−5.12]	0
MBF	$T_5(\mathcal{K}) = \frac{1}{4000}\sum_{l=1}^{n}\mathcal{K}_1^2 - \prod_{l=1}^{n}\cos\left(\frac{\mathcal{K}_1}{\sqrt{1}}\right) + 1$	30	[−32,32]	0
FMBF	$T_6(\mathcal{K}) = \left(\frac{1}{500} + \sum_{m=1}^{25}\frac{1}{m+\sum_{m=1}^{2}(\mathcal{K}_1 - \mathrm{a}_{\mathrm{lm}})^6}\right)^{-1}$	2	[−65,65]	1
FMBF	$T_7(\mathcal{K}) = 4\mathcal{K}_1^2 - 2.1\mathcal{K}_1^4 + \frac{1}{3}\mathcal{K}_1^6 + \mathcal{K}_1\mathcal{K}_2 - 4\mathcal{K}_2^2 + 4\mathcal{K}_2^4$	2	[−5,5]	−1.0316
FMBF	$T_8(\mathcal{K}) = \left(\mathcal{K}_2 - \frac{5.1}{4\pi^2}\mathcal{K}_1^2 + \frac{5}{\pi}\mathcal{K}_1 - 6\right)^2 + 10\left(1 - \frac{1}{8\pi}\right)\cos\mathcal{K}_1 + 10$	2	[−5,5]	0.398
FMBF	$T_9(\mathcal{K}) = \{1 + [\mathcal{K}_1 + \mathcal{K}_2 + 1]^2[19 - 14\mathcal{K}_1 + 3\mathcal{K}_1^2 - 14\mathcal{K}_2 + 6\mathcal{K}_1\mathcal{K}_2 + 3\mathcal{K}_2^2]\}$ $\times \{30 + [2\mathcal{K}_1 - 3\mathcal{K}_2]^2 \times [18 - 32\mathcal{K}_1 + 12\mathcal{K}_1^2 + 48\mathcal{K}_2 - 36\mathcal{K}_1\mathcal{K}_2 + 27\mathcal{K}_2^2]\}$	2	[−2,2]	1
FMBF	$T_{10}(\mathcal{K}) = -\sum_{l=1}^{4} a_l \exp\left(-\sum_{m=1}^{3} c_{lm}(\mathcal{K}_{\mathrm{m}} - \mathrm{p}_{\mathrm{lm}})\right)^2$	3	[1,3]	−3.86

are run 10 times then to calculate average and standard deviation listed in Table 2. Figure 4 shows 2D version of parameter space in each test functions. The MWOA results compared with WOA algorithm [22], PSO algorithm [11], GSA algorithm [23], DE algorithm [24], and FEP algorithm [25] mentioned in Table 2. Then, finally ten functions are tested, eight functions results are good, and two functions results are better than others. The best outcome mentioned in bold letter in the Table 2.

The objective space of the MWOA algorithm was shown in Fig. 5 to show the algorithm convergence rate. Please keep in mind that the average of the best result produced in each iteration across 10 runs. When optimising the test functions, the MWOA algorithm exhibits three different convergence characteristics, as seen in this diagram. To begin with, as the number of iterations grows, the MWOA algorithm convergence tends to quicken. This is due to the adaptive of the exploration space in the early stages of iteration and converges to the optimum more quickly later passing almost half of the iterations.

5 Conclusion

This work introduced a robust swarm-based optimisation technique inspired by humpback whale chasing behaviour. The proposed approach has been updated. The WOA contained 3 operators to imitate the humpback whales hunt for the target, encircling the target, and bubble-net hunting behaviour. Comprehensive research was done on ten mathematical benchmark functions to assess the proposed algorithm's exploitation, exploration, local optimization avoidance, and behaviour of convergence. MWOA has been proved to be sufficiently inexpensive with other cutting-edge metaheuristic approaches.

Table 2 Comparison of best results obtained in ten tested benchmark functions

F	MWOA		WOA		DE		GSA		PSO		FEP	
	ava	std	ava	std	ava	std	ava	std	ava	std	ava	std
T_1	**8.41E−80**	2.19E−79	1.42E−30	4.90E−30	8.1E−14	5.8E−14	2.54E−16	9.68E−17	0.000137	0.000202	0.00058	0.00014
T_2	**1.58E−54**	2.20E−54	1.06E−21	2.39E−21	1.5E−09	9.9E−10	0.055655	0.194074	0.042144	0.045421	0.0081	0.00077
T_3	**0.00E−00**	0.00E+00	0.00E+00	0.00E+00	69.2	38.8	25.96841	7.470068	46.70423	11.62938	0.046	0.012
T_4	**5.51E−15**	2.39E−15	7.4043	9.897572	9.7E−15	4.2E−14	0.062087	0.23628	0.276015	0.50901	0.018	0.0021
T_5	**0.00E+00**	0.00E+00	0.000289	0.001586	0.00E+00	0.00E+00	27.70154	5.040343	0.009215	0.007724	0.016	0.022
T_6	**0.998**	9.48E−06	2.111973	2.498594	0.998004	3.3E−16	5.859838	3.831299	3.627168	2.560828	1.22	0.56
T_7	**−1.0316**	2.34E−16	−1.03163	4.2E−07	−1.03163	3.2E−13	−1.03163	4.88E−16	−1.03163	6.25E−16	−1.03	4.9E−07
T_8	0.397895	1.58E−05	0.397914	2.7E−05	0.397887	9.8E−09	0.397887	0.00E+00	0.397887	0.00E+00	0.398	1.5E−07
T_9	**3**	0.00E+00	3	4.22E−15	3	2.0E−15	3	4.17E−15	3	1.33E−15	3.02	0.11
T_{10}	−3.85874	3.44E−03	−3.85616	0.002706	N/A	N/A	−3.86278	2.29E−15	−3.86278	2.58E−15	−3.86	0.000014

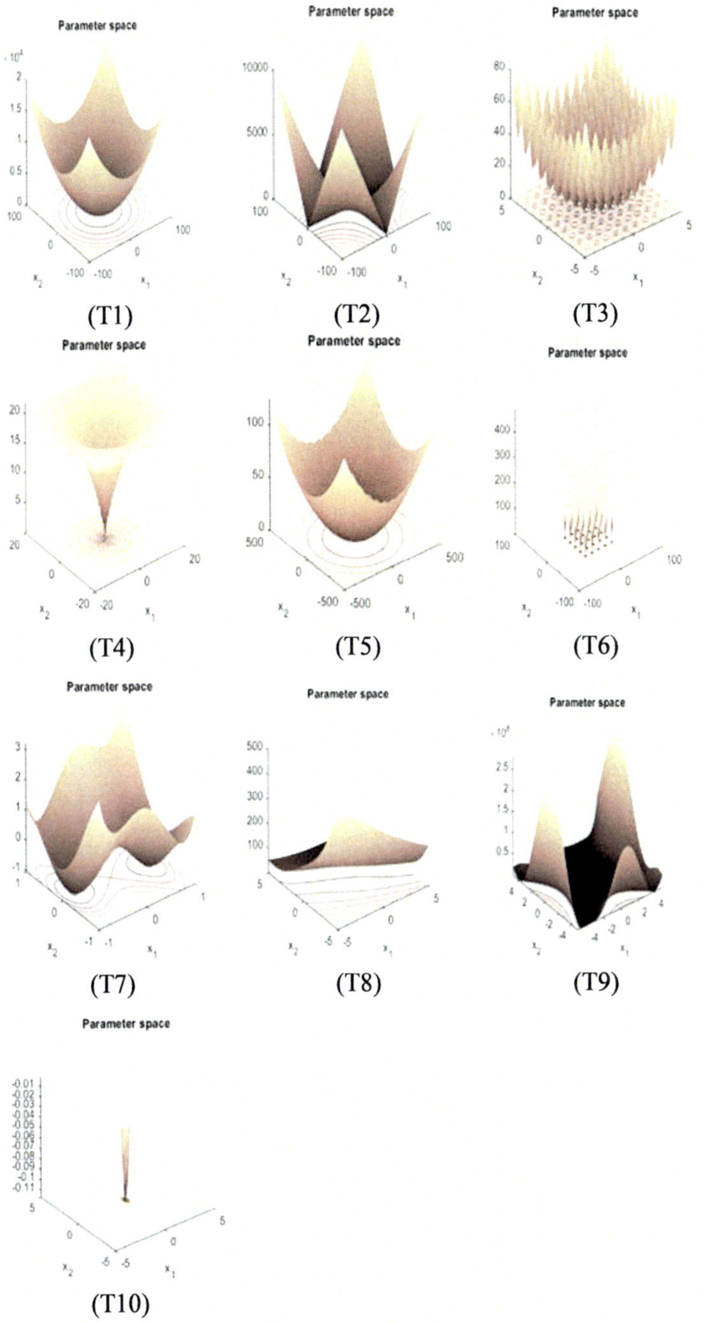

Fig. 4 Typical 2D version of ten tested functions

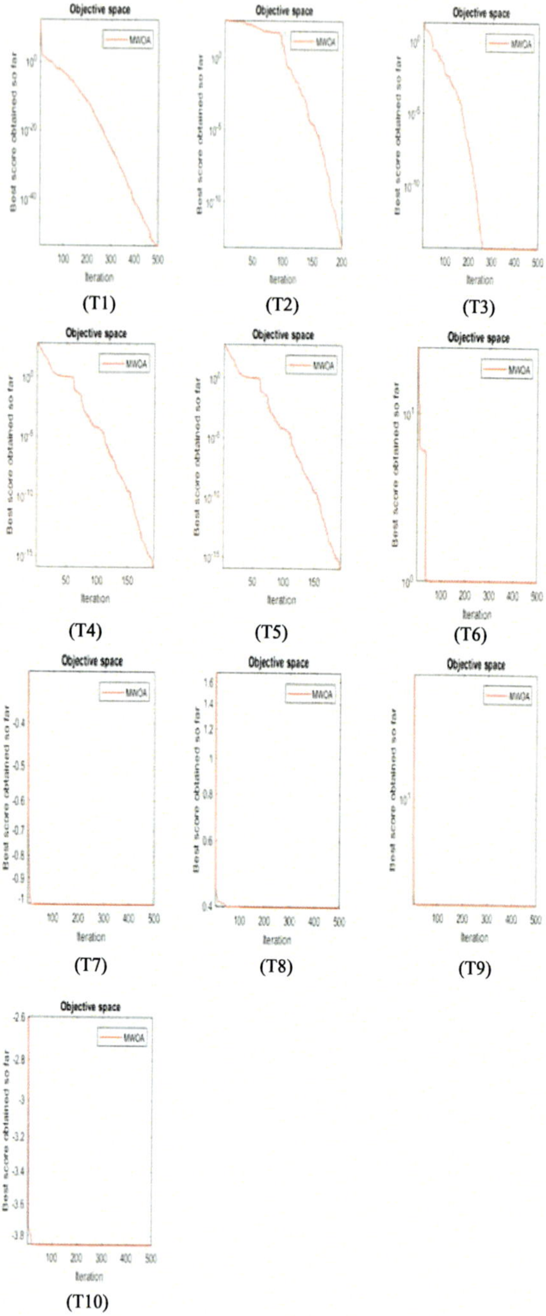

Fig. 5 Objective space of the MWOA

Acknowledgements The authors acknowledge VIT for providing VIT SEED GRANT for carrying out this research work.

References

1. Holland JH (1992) Genetic algorithms. Sci Am 267:66–72
2. Koza JR (1992) Genetic programming
3. Rechenberg I (1978) Evolutionsstrategien. Springer, Berlin, pp 83–114
4. Simon D (2008) Biogeography-based optimization. IEEE Trans Evol Comput 12:702–713
5. Webster B, Bernhard PJ (2003) A local search optimization algorithm based on natural principles of gravitation. In: Proceedings of the 2003 international conference on information and knowledge engineering (IKE'03), pp 255–61
6. Erol OK, Eksin I (2006) A new optimization method: big bang–big crunch. Adv Eng Softw 37:106–111
7. Formato RA (2007) Central force optimization A new metaheuristic with applications in applied. Prog Electromag Res 77:425–491
8. Hatamlou A (2013) Black hole. A new heuristic optimization approach for data clustering. Inf Sci 222:175–184
9. Kaveh A, Khayatazad M (2012) Ray optimization. A new meta-heuristic method. Comput Struct 112:283–294
10. Moghaddam FF, Moghaddam RF, Cheriet M (2012) Curved space optimization. A random search based on general relativity theory. In: arXiv, pp 1208–2214
11. Kennedy J, Eberhart R (1995) Particle swarm optimization. In: Proceedings of the 1995 IEEE international conference on neural networks, pp 1942–1948
12. Dorigo M, Birattari M, Stutzle T (2006) Ant colony optimization. IEEE Comput Intell 1:28–39
13. Rao RV, Savsani VJ, Vakharia DP (2012) Teaching–learning-based optimization: an optimization method for continuous non-linear large scale problems. Inf Sci 183:1–15
14. Rao RV, Savsani VJ, Vakharia DP (2011) Teaching–learning-based optimization: a novel method for constrained mechanical design optimization problems. Computer-Aided Des 43:303–315
15. Geem ZW, Kim JH, Loganathan G (2001) Harmony search. A new heuristic optimization algorithm 76:60–68
16. Fogel D (2009) Artificial intelligence through simulated evolution. Wiley-IEEE Press
17. Glover F (1989) Tabu search—Part I. ORSA J Comput 1:190–206
18. Glover F (1990) Tabu search—Part II. ORSA J Comput 2:4–32
19. He S, Wu Q, Saunders J (2006) A novel group search optimizer inspired by animal behavioural ecology. In: Proceedings of the 2006 IEEE congress on evolutionary computation, CEC 2006, pp 1272–1278
20. He S, Wu QH, Saunders J (2009) Group search optimizer. An optimization algorithm inspired by animal searching behavior. IEEE Trans Evol Comput (13):973–90
21. Tan Y, Zhu Y (2010) Fireworks algorithm for optimization. Advances in swarm intelligence. Springer, pp 355–364
22. Mirjalili S, Andrew L (2016) The whale optimization algorithm. Advances in engineering software, vol 95, pp 51–67
23. Rashedi E, Nezamabadi-Pour H, Saryazdi S (2009) A gravitational search algorithm. Inf Sci 179:2232–2248
24. Storn R, Price K (1997) Differential evolution. A simple and efficient heuristic for global optimization over continuous spaces. J Glob Optim (11):341–59
25. Yao X, Liu Y, Lin G (1999) Evolutionary programming made faster. IEEE Trans Evol Comput 3:82–102

Optimal Design of Water Distribution System Considering Water Quality and Hydraulic Criteria Using Multi-objective Harmony Search

Mun Jin Ko and Young Hwan Choi

Abstract The objective of the water distribution system (WDS) is to supply water from sources to consumers with safe water quality, stable water quantity, and appropriate water pressure under normal as well as abnormal conditions. To satisfy these objectives, there is a minimum pressure required for each WDS. In addition, to supply safe water, the World Health Organization has stipulated that residual chlorine standards should be maintained at 0.2–5.0 mg/L. However, the change in residence time, pressure, and flow rate can affect the design factor depending on the usage pattern. Even if the network size of WDS is similar, the shape of WDS affects hydraulic and water quality characteristics, such as not meeting the minimum pressure required by WDS. Therefore, in this study, branched type, hybrid type, and looped type were classified to consider the types and size of WDS. The objective functions applied the maximum system resilience and the minimum design cost. The optimal solution compared with the optimal design considering only pressure as the constraint and the optimal design considering the pressure and residual chlorine concentration simultaneously. The derived optimal design will meet the system's resilience and the economic, water quality, hydraulic aspects, and it will increase the usability of customers.

Keywords Optimal design of water distribution systems · Multi-objective harmony search · Water quality and hydraulic indices

1 Introduction

The water distribution systems (WDS) purpose to supply water from water sources to consumers within a range that satisfies stable quantity, safe water quality, and adequate water pressure. In order to prevent the sharp pressure change condition for

M. J. Ko · Y. H. Choi (✉)
Department of Civil and Infrastructure Engineering, Gyeongsang National University, Jinju 52725, South Korea
e-mail: yh.choi@gnu.ac.kr

J. H. Kim et al. (eds.), *Proceedings of 7th International Conference on Harmony Search, Soft Computing and Applications*, Lecture Notes on Data Engineering and Communications Technologies 140, https://doi.org/10.1007/978-981-19-2948-9_46

the destruction of pipes or a sudden increase in demand, the required minimum pressure for WDS was determined. To satisfy safe water quality, World Health Organization (WHO) recommends a minimum residual chlorine concentration of 0.2 mg/L and a maximum residual chlorine concentration of 5 mg/L for effective drinking water disinfection [6]. However, due to abnormal conditions, pressure changes, resulting in a problem of not satisfying the minimum pressure required by WDS or not satisfying the water quality standard. For this reason, in the field of WDS, optimal design using optimization algorithms has been carried out since the 1970s [14, 20].

Initially, research was conducted to minimize the final design cost by considering only the design cost for the optimal WDS. However, this design was vulnerable to coping with abnormal conditions and did not meet the water quality standard. Therefore, to solve this problem, the multi-objective optimization technique was developed that simultaneously considers design factors such as design cost, reliability, resilience, and robustness of WDS, and it was possible to secure the cost aspect of WDS and the safety of water supply [4, 12]. In addition, a study was conducted to compare and analyze the optimal design to apply the optimization algorithm suitable for WDS [5, 10, 21].

However, the hydraulic and water quality characteristics of WDS affect the change in demand and the type and characteristics of WDS. Therefore, research is being conducted to supply stable water in terms of hydraulic-water quality to consumers by conducting a design using a term extended period simulation (EPS) to study the analysis results over time, such as pressure at points and residual chlorine concentration, in consideration of changes in demand [13, 18, 17]. In addition, research is being conducted to analyze according to the format and characteristics of WDS. Initially, the classification of WDS was classified through a review of WDS, but additional classification criteria are needed to classify WDS with a clear format. In order to solve this problem, the method of classifying the characteristics of WDS with the average pipe diameter of WDS, skeletonizing and quantifying WDS, and classifying it was suggested.

Therefore, in this study, the maximization of system resilience and design cost minimization were set as design factors to enable system operation similar to normal conditions even in abnormal conditions, and the optimal design using multi-objective harmony search was compared and analyzed according to the type and characteristics of WDS.

2 Methodology

In this study, the multi-objective harmony search (MOHS; [3]) was used as an optimization algorithm that can consider two or more objective functions having a trade-off relationship based on the harmony search (HS) [9]. HS has a simple optimization process and has been applied in various ways, such as optimal structure design, engineering problems, and optimal design of WDS.

MOHS uses the concept of harmony search, non-dominated Sorting [8] technique, and crowding distance [7] to derive the multi-objective optimal solutions considering various objective functions. MOHS has the same parameter as the original HS such as harmony memory considering rate (HMCR), pitch adjustment rate (PAR), harmony memory size (HMS).

The non-dominated sorting technique judges the governing relationship of a solution, considers the set of non-dominated solutions as optimal solutions, and prioritizes them. If the priorities are the same, the concept of crowding distance is introduced and preserved in the large order of crowding distance values among the remaining optimal solution values except for the solution of both extremes after normalization. The minimum and maximum normalization enable effective crowding distance calculation by unifying the range of possible solutions of each objective function. The operation is to set the largest value of each objective function value to 1 and the smallest value to 0 and replace the remaining solution with a value between 0 and 1 according to the ratio. As following (1)

$$\text{Normailzation Formula}(i) = (X_i - X_{\min})/(X_{\max} - X_{\min}) \tag{1}$$

where the crowding distance equation means ith value of the optimal design after non-dominated sorting, $X_{\min}$ means the smallest value of the optimal design solution, and $X_{\max}$ means the largest value of the optimal design solution. As following (2)

$$\text{Crowding distance}(j) = |f_n(j) - f_n(k)| + |f_{n+1}(j) - f_{n+1}(k)| \tag{2}$$

where $f_n(i)$ is the value of the nth objective function in solution i, and $f_n(k)$ is the value of the nth objective function in solution k, and in this study, the crowding distance was calculated using two objective functions.

Therefore, hydraulic analysis and water quality analysis were analyzed through EPS through the EPANET 2.0 program [16] provided by the US Environmental Protection Agency for hydraulic analysis and water quality analysis of WDS. The period of consideration of the extended period simulation was set to 14 days (336 h), and the water quality analysis period was the time until the initial flow rate of the water source reached the last node and the time until the change in the residual chlorine concentration was stabilized (patterned). 168 h was considered by subtracting the initial 7 days out of 14 days.

Figure 1 is the flow chart of the MOHS. The HMS, HMCR, and PAR were determined through parametric sensitivity analysis. The initial harmony memory was set by randomly selecting each benchmark diameter candidate group. The penalty functions were given if the penalty function did not satisfy the minimum pressure required by WDS during the 336 h considered. As the water quality, the penalty function was given if the standard of residual chlorine concentration was not satisfied during 169 h. The objective functions were determined through the sum of all nodal pressure when the minimum pressure occurred during 336 h.

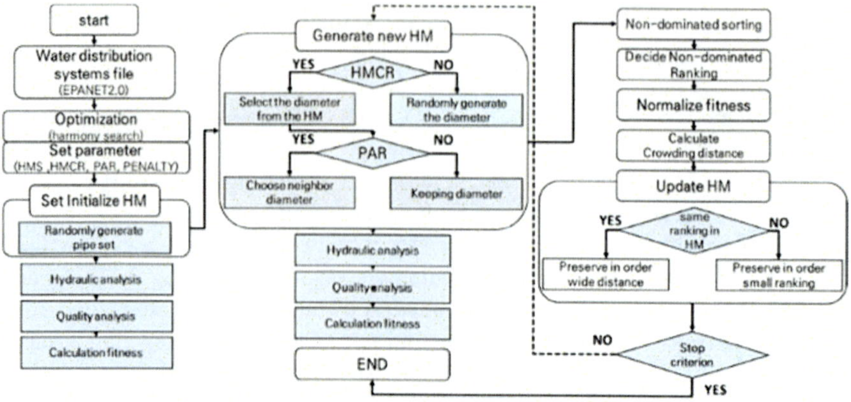

Fig. 1 Flowchart of multi-objective harmony search

2.1 Design Criteria

The first objective function is the minimum design cost. As following (3)

$$\text{Design Cost} = \sum_{i=1}^{N} C(D_i)L_i \tag{3}$$

where N is the number of pipes in WDS, $C(D_i)$ is the cost per unit length of pipe i, and L_i is the length of pipe i.

The second objective function is the maximum sum of the surface head for each node under satisfying pressure constraints, and it means the system resilience which recovers against abnormal system conditions. As following (4)

$$\text{System Resilience} = \sum_{j=1}^{n} \left(h_j - h_{\min}\right) \tag{4}$$

where n means the number of nodes excluding the water source and tank of WDS, h_j is the pressure at the jth node, and $h_{\min}$ is the minimum pressure required by WDS.

In this study, the minimum pressure or standard of residual chlorine concentration of the solution does not satisfy the penalty functions that were considered to eliminate the solution. If the minimum pressure or residual chlorine are not satisfied by the method of considering the constraints, the penalty functions are considered large so that it is eliminated as the number of iterative calculations increases. For the penalty functions, 10^{10} was used to apply a value in which the value of the relatively optimal solution among the objective functions is greater than the design cost. In addition, the absolute value was taken because the pressure may come out as a negative value due to insufficient flow rate or small pipe diameter. As following (5) and (6)

$$\text{Penalty}_{\text{Press}_j} = \text{if } h_j < h_{\min} \text{ then } \sum_{j=1}^{n}\left(\left|h_j - h_{\min}\right|\right) \times a, \text{ others } 0 \tag{5}$$

$$\text{Penalty}_{\text{Quality}_i} = \begin{matrix} \text{if } Q_{i\ \min} < Q_{s,\min} \text{ then } \sum_{i=1}^{n}\left(\left|Q_j - Q_{s,\min}\right|\right) \times b \\ \text{if } Q_{i\ \max} < Q_{s,\max} \text{ then } \sum_{i=1}^{n}\left(\left|Q_j - Q_{s,\max}\right|\right) \times b \end{matrix}, \text{ others } 0 \tag{6}$$

where n is the number of nodes excluding water sources and tanks in WDS. a and b mean the penalty constant that can be eliminated during repeated calculations, and 10^{10} was used. $Q_{i\ \min}$ and $Q_{i\ \max}$ each mean the minimum and maximum residual chlorine standard value, and Q_i means the residual chlorine concentration of ith node.

2.2 WDS Classification Approach

In this study, WDS classification approach was proposed by Hawng and Lansey [11]. WDS was classified into branched type, hybrid type, looped type. The number of non-essential nodes was reduced by simplifying the network with the node-reduction algorithms. In order to determine the non-essential node, the grade was assigned to the entire WDS, and the non-essential node was deleted using the grade. Branch Index (BI) was used to classify the network type (i.e., branched type, hybrid type, looped type) as the below equation branch index. In this study, the pipe network was classified considering the network shape rather than the average pipe diameter. Therefore, network categorization was performed using only the BI without classification based on the average diameter. As following (7)

$$\text{Branch Index} = \frac{e_b}{e_r + e_b} \tag{7}$$

where e_r is the number of edges in the reduced network and e_b is the number of edges in the existing pipe network.

2.3 Determination of the Amount for Initial Chlorine

In order to calculate the effective chlorine input concentration in the water source, the optimal design considering only the pressure was set as the initial harmony memory, and the chlorine input concentration in the water source was gradually increased and analyzed.

The initial set value was repeatedly calculated while increasing the initial chlorine input concentration in the water source until the penalty constants were not taken into

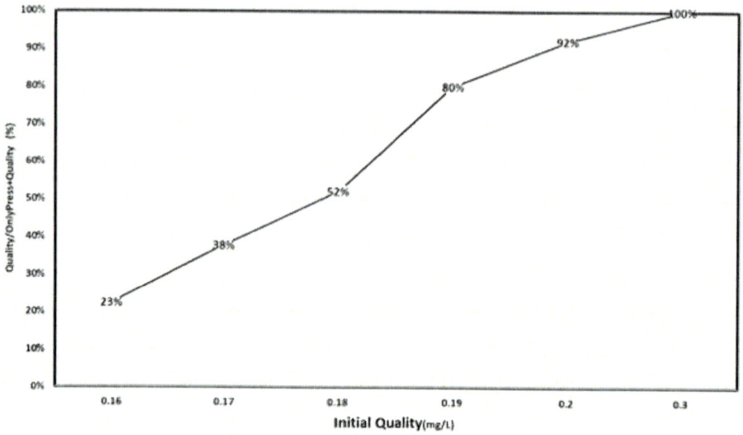

Fig. 2 Sensitivity analysis of initial quality

account in the objective functions. Figure 2 shows the results of sensitivity analysis for initial quality.

3 Applications and Results

The bulk coefficient and wall coefficient of the applied WDS is determined by the temperature and the type of pipe. Therefore, the Bulk Coefficient based on the cast iron pipe was set to 0.801, and the wall Coefficient was set to 0.0801 [1]. Similarly, the Hazen-Williams friction factor and Darcy-Weisbach roughness coefficient were set to 130 and 0.012 based on the cast iron pipe, respectively, to derive the optimal design.

In order to compare the optimal design according to the type and characteristics of WDS, the benchmark network was classified into branched type, hybrid type, looped type according to the BI. In addition, the optimal design considering only the pressure at the node as the constraint and the constraint setting the pressure and residual chlorine standard as constraints were compared and analyzed according to the type and characteristics. Therefore, this study applied three benchmark networks (Anytown network [19], Jilin network [2], and Net2 network [15]) reflected network type and characteristic. The applied network's layout as shown Fig. 3. Based on the BI index applied by Hawng and Lansey [11], the Anytown network was classified as 0.03 as looped type, the Jilin network as 0.31 as hybrid type, and the Net2 network as 0.73 as branch type.

The optimal design considering only pressure and the optimal design considering the pressure and the minimum and maximum residual chlorine concentration were compared for Anytown, Jilin, and Net 2 WDS. In the case of Anytown network, the difference between the two smallest design cost differences is 96,002.85 USD, and

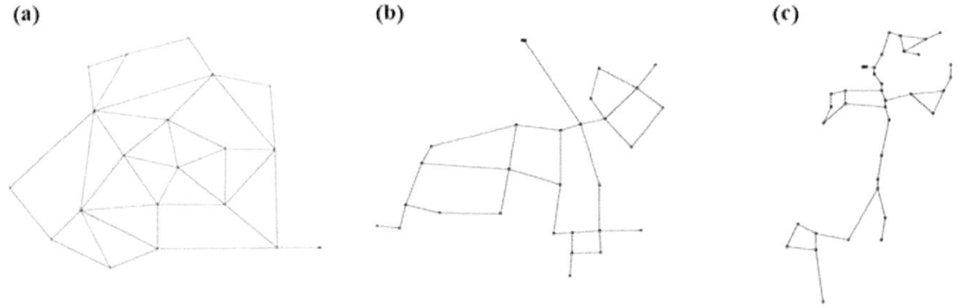

Fig. 3 The layout of benchmark networks **a** Anytown network, **b** Jilin network, **c** Net 2 network

the difference in system resilience is 0.22 psi. The difference in design cost of the Jilin network is 2341 USD, and the difference between the points with the smallest difference in system resilience is 0.51 m. In the Net2 network, the difference between the two small design cost differences is 3395 USD and the resilience of the system is 0.54 psi. However, in all optimal designs considering only pressure, the water quality standard was not satisfied when the residual chlorine concentration of the same concentration was injected from the water source.

The optimal design considering only pressure and the optimal design considering pressure and residual chlorine concentration as constraints were compared and analyzed with two main points: design cost and system resilience. Compared to the optimal design that considers pressure and residual chlorine concentration as constraints in the benchmark networks, in WDS considering only pressure as the constraint, even if there is no difference in system resilience, the standard value of residual chlorine when the same concentration of chlorine is input from the water source was not satisfied.

Similarly, even if there was no difference in design cost, the residual chlorine standard was not satisfied in cases when the same concentration of chlorine was added. It is judged that convergence is excellent in terms of design cost. It is judged that the convergence in terms of design cost increased as the area of the possible solution was reduced when the pressure and residual chlorine concentration were set as constraints compared to the optimal design considering only pressure. If the pipe diameter increases, the resilience of the system based on pressure increases, but the flow rate decreases and the residual chlorine concentration is not satisfied due to the increase of the residence time. Therefore, it was confirmed that as the chlorine input concentration in Source increased, the optimal design plan was possible, so the area increased, and in Anytown, an optimal design similar to the optimal design considering only pressure was derived.

The optimal design solutions of analysis depend on the network categorization. In the case of the Anytown network (Fig. 4), which is the looped type network, the solution satisfying the constraints condition was derived from the time the chlorine input concentration was set to 0.17 mg/L in the water source, and the slope of the graph rapidly increased as the chlorine input concentration increased. On the other hand, the Net2 network, which is the branched type, has a higher chlorine input

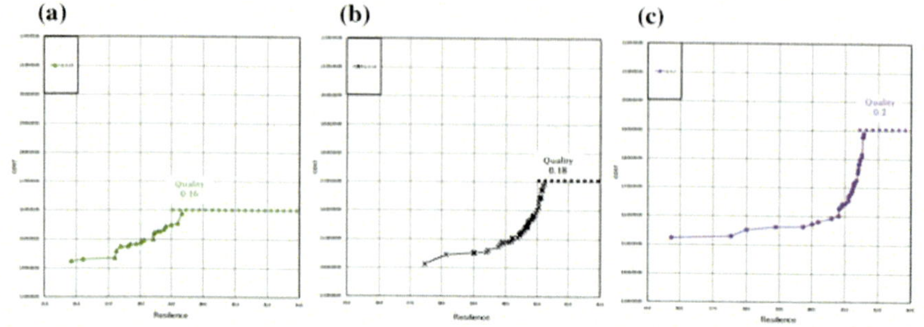

Fig. 4 Pareto optimal solutions of Anytown networks according to the initial water quality **a** initial water quality 0.16 mg/L, **b** initial water quality 0.18 mg/L, **c** initial water quality 0.2 mg/L

concentration in water source than the Anytown, which is the looped type network but is gentle unlike the looped type network, and the hybrid network, Jilin network, draws the graph between the looped type and branched type. When quantitatively evaluated, it was confirmed that the slope of the looped type of network graph was 25, and the slope of the graph of Net 2, the branched type network, was 3.7, which was 6.75 times larger. The reason was that the water coming from one node was diversified, resulting in a rapid increase in the graph slope of the minimum residual chlorine concentration as the chlorine input concentration increased.

4 Conclusions

In this study, the multi-objective optimal design of WDS was performed considering water quality and hydraulic criteria depending on the network characteristics. The optimal considering pressure and water quality can satisfy the standard of residual chlorine concentration. But the optimal solution for considering only pressure does not meet the water quality constraints Regardless of the type and characteristics of WDS. It means that to maximize of system resilience, the diameter near the water source is enlarged, but the residual chlorine concentration, which is closely related to the flow rate, slows the flow rate and increases the residence time.

In future research, quantitative evaluation should be performed with various types and sizes of benchmark networks and real-world networks. Furthermore, as an optimization aspect, the number of objective functions should be increased to reflect the real-world condition.

Acknowledgements This work was supported by the National Research Foundation of Korea (NRF) grant funded by the Korean government (MSIT). (NRF-2021R1G1A1003295).

References

1. Ahn JC, Lee SW, Rho BS, Choi YJ, Choi JH, Kim HI, Park TJ, Park CM, Park H, Koo JY (2007) Modeling residual chlorine and THMs in water distribution system. Korean Soc Environ Eng 29(6):706–714
2. Bi W & Dandy GC (2014) Optimization of water distribution systems using online retrained metamodels. J Water Resour Plan Manage 140(11):04014032
3. Choi YH, Jung D, Lee HM, Yoo DG, Kim JH (2017) Improving the quality of pareto optimal solutions in water distribution network design. J Water Resour Plan Manag 143(8):04017036
4. Choi YH, Kim JH (2019) Development of multi-objective optimal redundant design approach for multiple pipe failure in water distribution system. Water 11(3):553
5. Choi YH, Lee HM, Yoo DG, Kim JH (2015) Optimal design of water supply system using multi-objective harmony search algorithm. J Korean Soc Water Wastewater 29(3):293–303
6. Cotruvo JA (2017) 2017 WHO guidelines for drinking water quality: first addendum to the fourth edition. J Am Water Works Ass 109(7):44–51
7. Deb K, Agrawal S, Pratap A, Meyarivan T (2000) A fast elitist non-dominated sorting genetic algorithm for multi-objective optimization: NSGA-II. International conference on parallel problem solving from nature
8. Fonseca CM, Fleming PJ (1993) Genetic algorithms for multiobjective optimization: formulation discussion and generalization. Icga
9. Geem ZW, Kim JH, & Loganathan GV (2001) A new heuristic optimization algorithm: harmony search. Simul, 76(2):60–68
10. Hong AR, Lee HM, Choi YH, Choi JH, Kim JH (2016) Application and comparison of genetic algorithm and harmony search algorithm for optimal cost design of water distribution system. In: Proceedings of the Korea water resources association conference, pp 521–521
11. Hwang H, Lansey K (2017) Water distribution system classification using system characteristics and graph-theory metrics. J Water Resour Plan Manag 143(12):04017071
12. Jung D, Lansey KE, Choi YH, Kim JH (2016) Robustness-based optimal pump design and scheduling for water distribution systems. J Hydroinf 18(3):500–513
13. Kang D, Lansey K (2009) Real-time demand estimation and confidence limit analysis for water distribution systems. J Hydraul Eng 135(10):825–837
14. Quindry GE, Brill ED, Liebman JC (1981) Optimization of looped water distribution systems. J Environ Eng Div 107(4):665–679
15. Rossman LA, Clark RM, Grayman WM (1994) Modeling chlorine residuals in drinking-water distribution systems. J Environ Eng 120(4):803–820
16. Rossman LA (1994) EPANET users manual
17. Siew C, Tanyimboh TT, Seyoum AG (2016) Penalty-free multi-objective evolutionary approach to optimization of Anytown water distribution network. Water Resour Manage 30(11):3671–3688
18. Siew C, Tanyimboh TT (2010) Pressure-dependent EPANET extension: extended period simulation. In: Water distribution systems analysis 2010, pp 85–95
19. Walski TM, Brill ED Jr, Gessler J, Goulter IC, Jeppson RM, Lansey K et al (1987) Battle of the network models: epilogue. J Water Resour Plan Manag 113(2):191–203
20. Yates D, Templeman A, Boffey T (1984) The computational complexity of the problem of determining least capital cost designs for water supply networks. Eng Optim 7(2):143–155
21. Yazdi J, Choi YH, Kim JH (2017) Non-dominated sorting harmony search differential evolution (NS-HS-DE): a hybrid algorithm for multi-objective design of water distribution networks. Water 9(8):587

Car Counting Based on Road Partitioning and Blobs Analysis

Farimehr Zohari and Raziyeh Sadat Okhovvat

Abstract Traffic congestions and traffic jams are major challenges in some cities due to the substantial surge in the population and the number of cars. Thus, controlling the traffic density on roads and highways is essential. In this paper, we explore a method that can be used to direct traffic. This method is based on image processing and comparing two images with and without cars. At first, the foreground of the image is obtained. Then the road is partitioned to calculate the area occupied by a car in each part of the road. After that, the number of cars is estimated through dividing the total area of cars by that of a typical car. The camera angle and road distance from the camera are two essential factors in this method. Therefore, it is vital to use a fixed camera in this method. This proposed model was tested with the GRAM-RTM dataset and exhibited an average accuracy of 95%.

Keywords Car counting · Background subtraction · Image processing · Control traffic · Road partition

1 Introduction

With the increase in the number of cars, especially in modern cities, traffic jams have become a severe problem. Traffic has a significant impact on people's daily lives, and traffic congestion can put many lives in danger. Imagine some are caught in a fire and are waiting helplessly for the firefighters to save them. What catastrophe may occur if a fire truck gets stuck in heavy traffic? If the fire truck fails to arrive on time, the victims will have irreversible damage. Therefore, designing a traffic control system that can improve the traffic flow is vital. Such a system can ensure the use of road capacity is optimized. For instance, if there are two intersecting lines, and there is

F. Zohari (✉)
Department of Electrical Engineering, Alzahra University, Tehran, Iran
e-mail: farimehrzohari@gmail.com

R. S. Okhovvat
Department of Electrical Engineering, University of Science and Culture, Tehran, Iran
e-mail: Rokhovat@usc.ac.ir

J. H. Kim et al. (eds.), *Proceedings of 7th International Conference on Harmony Search, Soft Computing and Applications*, Lecture Notes on Data Engineering and Communications Technologies 140, https://doi.org/10.1007/978-981-19-2948-9_47

more traffic on one line than the other, the duration of the green light should not be the same for both. Otherwise, the capacity of the road will be wasted, but if the line with more traffic has the green light for a longer time, traffic in two lines will be balanced.

It is also important to know the number of vehicles on the road to better manage the roads. Applying the statistic principles to the reports on the number of cars provides us with the chance to make the transportation system smarter.

Here are some common image processing methods for car counting. In [1], authors used the frame subtraction method to detect moving cars and count them. In the frame subtraction method, if the car speed is high or the number of frames per second is low, the car would not be detected. In [2], authors used a morphological image processing technique to remove the road background and count the cars. Blobs overlapping is a problem for this method. In [3], authors used the background subtraction method to detect cars in the current frame. Because of background subtraction, this method is very dependent on light, so the sunlight affects the method's accuracy. Another drawback of this method is that blobs overlapping problem occurs if the cars are close. In [4], authors solved the first problem of background subtraction by adaptive background estimation, and we tried to provide a solution to solve the overlapping problem. In this paper, a method based on background subtraction is introduced. First, an appropriate threshold value is used to create a binary image of the acquired difference. The total number of automobiles is then calculated by partitioning the road and dividing the number of blobs pixels by the area of a car in the same region.

This paper is organized as follows: Sect. 2 presents the necessary preprocessing for this system that includes road partitioning and estimates the area of a car in each part. In Sect. 3, the different steps of our proposed system for car counting are introduced. Experimental results and analysis are in Sect. 4, and finally, in Sect. 5, our results are summarized, and we present our conclusions.

2 Preprocessing

In the preprocessing step, the road must be partitioned, and then the car area in each section is obtained. For this purpose, we need the frames from a film of the desired road where a medium-sized car has passed through it. Road partitioning is done by tracking the medium-sized car. The partitioning is done in such a way that the width of each part is equal to the length of that car. Also, to obtain the car area, it is necessary to figure out the number of pixels of that car in each part. For this purpose, we consider the car as a rectangle. The total number of pixels in an automobile is then calculated by multiplying the length and width of the rectangle (Fig. 1).

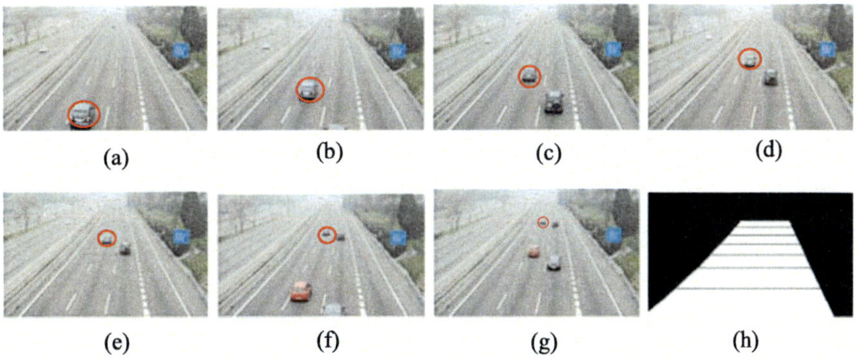

Fig. 1 **a–g** Tracking a car. **h** Road partitioning

3 Steps of Car Counting

The following block diagram describes the complete algorithm for car counting (Fig. 2).

1. Image acquisition
2. Preprocessing
3. Background subtraction
4. Selecting the region of interest
5. Morphological process
6. Car counting.

3.1 Input Image and Background Image

In the car counting process, two input images are given. One is without any cars, and the other is with cars. These photographs must be obtained with a fixed camera

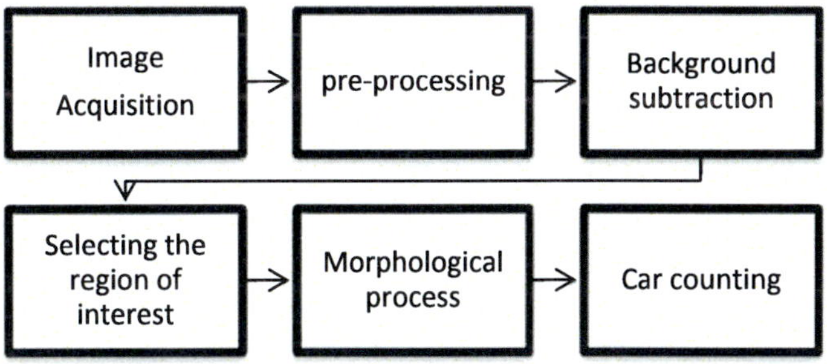

Fig. 2 Overview of the system

(a) (b)

Fig. 3 **a** Background image. **b** Image with cars on the road

Fig. 4 After background subtraction

because the images must match together in the background subtraction step. If the input images are RGB, they should convert to gray. It is done to reduce calculations [5] (Fig. 3).

3.2 Background Subtraction

The two images are compared, and the differences between pixels are found based on a threshold value. If the pixel difference is more than the threshold value, that pixel is assigned to "1", otherwise to "0" [6]. The obtained result is a binary image that blobs show the differences between two input images (Fig. 4).

3.3 Selection of the Region of Interest

We intend to count the number of cars on the road. Therefore, it is necessary to remove the blobs on the side of the road and other roads in the image. We only need

(a) (b)

Fig. 5 **a** Region of interest. **b** Selection of the region of interest

the blobs in the considered road. In this step, the region of interest for counting the cars is selected by multiplying the previous image by a proper mask [7] (Fig. 5).

3.4 *Morphological Processes*

The binary image obtained from the previous step must be filled with holes, closed, and then opened. Closing morphological operation is applied to close the space between correlated blobs, but opening morphological operation is used to remove faulty connections and unwanted objects. As a result of these operations, only the cars' blobs remain. Choosing appropriate structural elements in the morphological operation of closing and opening is critical [8].

Typically a structural element is chosen in the same size and shape as the objects are wanted to process in the output image. Because the blobs of cars look like rectangles, it is better to choose a structural element in a rectangle shape (Fig. 6).

3.5 *Car Counting*

The binary mask obtained in the preprocessing step must be labeled based on the area of a car in each part. Then, each part of the mask is multiplied by the final binary image in morphological processes step one by one. If there are some pixels in the output equal to one, the presence of cars is detected. For each part, the number of blobs pixels is divided by the area of a typical car in that part. After that, the results of the divisions are added together. Finally, the obtained number is rounded to positive infinity. Now the number of cars can be estimated (Fig. 7).

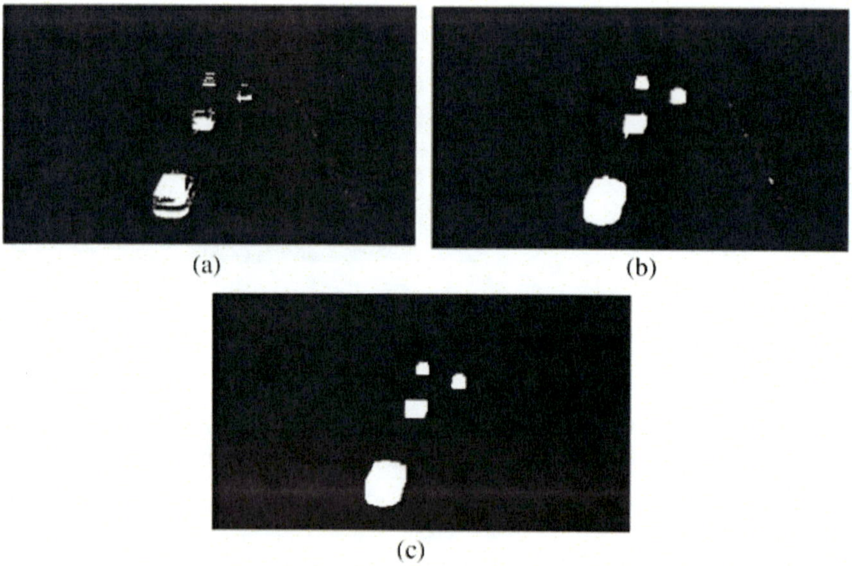

Fig. 6 **a** Filling holes. **b** Closing image. **c** Opening image

Fig. 7 Labeling of the mask obtained in the preprocessing step

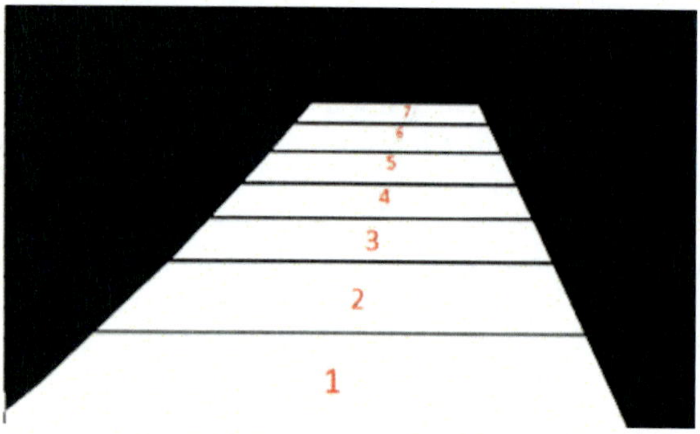

4 Experimental Results

The experiments are performed using MATLAB version R2017a on a PC with an Intel Core i5 processor with 8 GB random access memory (RAM). In our experiment, we extracted images from a real-life video. The video is captured from a camera installed on the top of the road. This video is called M-30-HD (9390 frames). It has been recorded during a cloudy day with a high-resolution camera: a Nikon DX3100 at 1200×720 @30 fps [9].

We randomly selected ten frames from this video. The total number of cars was equal to 45. There were a total of three errors in the car counting: two of them were negative errors (the number of cars counted was less than the number of available

Table 1 Testing the proposed method

Frame number	Number of cars	Number of cars counted	Positive errors	Negative errors
004232	4	4	0	0
009272	3	3	0	0
002710	4	4	0	0
002506	5	5	0	0
003220	3	3	0	0
003272	6	7	1	0
000883	5	4	0	1
000520	4	4	0	0
003856	6	5	0	1
00079	5	5	0	0

cars), while one of them was a positive error (the number of cars counted was more than the number of available cars). As a result, 43 vehicles out of 45 available cars were counted correctly (Table 1).

Therefore, according to the following formula, the method's accuracy was estimated to be 95.5%.

$$\frac{\text{total number of cars} - \text{total number of errors}}{\text{total number of cars}} \times 100$$

Compared to the accuracy of 85% in [3] or 70.31% in [10] on a cloudy day, the accuracy obtained in this method is more.

5 Conclusion

This paper designs a car counting system based on background subtraction and blobs analysis. Blob counting is a method of counting vehicles, but overlapping errors may occur in congestion conditions. The proposed method is based on counting the number of pixels corresponding to the detected blobs. Because of different areas for the car at various distances from the camera, the road must be partitioned. Then the number of pixels corresponding to the cars in each part is divided by the area of a car in the same part. In this way, the number of cars in an image is obtained. The advantages of this new method include such benefits as (1) solving the blobs overlapping problem (2) low cost (3) easy setup and relatively good accuracy and speed. Because this method has been implemented using image processing and MATLAB software, production costs are low while achieving high speed and accuracy. However, one of the most important disadvantages of this method is that errors occur in counting large-sized cars such as vans or small-sized cars.

References

1. Tourani A, Shahbahrami A (2015) Vehicle counting method based on digital image processing algorithms. In: 2015 2nd international conference on pattern recognition and image analysis (IPRIA). IEEE
2. Chowdhury PN, Ray TC, Uddin J (2018) A vehicle detection technique for traffic management using image processing. In: 2018 international conference on computer, communication, chemical, material and electronic engineering (IC4ME2). IEEE
3. Ajmal A, Hussain IM (2010) Vehicle detection using morphological image processing technique. In: 2010 international conference on multimedia computing and information technology (MCIT). IEEE
4. Srijongkon K et al (2017) SDSoC based development of vehicle counting system using adaptive background method. In: 2017 IEEE regional symposium on micro and nanoelectronics (RSM). IEEE
5. Hasan MM et al (2014) Smart traffic control system with application of image processing techniques. In: 2014 international conference on informatics, electronics & vision (ICIEV). IEEE
6. Cao Y et al (2012) A vehicle detection algorithm based on compressive sensing and background subtraction. Aasri Procedia 1:480–485
7. Niksaz P (2012) Automatic traffic estimation using image processing. Int J Signal Process Image Process Pattern Recogn 5(4):167–174
8. Javadzadeh R, Banihashemi E, Hamidzadeh J (2015) Fast vehicle detection and counting using background subtraction technique and prewitt edge detection. Int J Comput Sci Telecommun 6(10):8–12
9. Guerrero-Gómez-Olmedo Ricardo et al (2013) Vehicle tracking by simultaneous detection and viewpoint estimation. In: International work-conference on the interplay between natural and artificial Computation. Springer, Berlin
10. Lei M et al (2008) A video-based real-time vehicle counting system using adaptive background method. In: 2008 IEEE international conference on signal image technology and internet based systems. IEEE

Optimal Water Allocation in Zarrineh River Basin: PSO-WEAP Model

Sara Asadi and S. Jamshid Mousavi

Abstract A simulation–optimization framework is presented for allocating releases from Bukan Dam, the largest infrastructure in Urmia Lake Basin, to downstream uses in Zarrineh river basin in Iran. The framework links the water evaluation and planning system (WEAP) simulation module and particle swarm optimization (PSO) algorithm. WEAP by itself optimizes spatial water allocations to demand nodes in a single time period terminal subject to known demand priorities and supply preferences. However, PSO linked to WEAP determines optimal reservoir releases and downstream water allocations over a multi-period planning horizon. The PSO objective function is minimizing the sum of squared water shortages over the planning horizon considering the end-of-year reservoir storage conditions. Results of the model application in year 2012–2013 reveal different demands are met satisfactorily, whereas historical operations have led to shortages in meeting Lake Urmia's environmental water requirement.

Keywords Optimal water allocations · WEAP · PSO · Bukan reservoir

1 Introduction

The Urmia Lake (5000 km^2) which is the second largest saline lake is at risk of dying. It is located in North West of Iran, at the lowest point within the closed Urmia Lake basin as a terminal lake. Therefore, evaporation is the only outflow from the lake. Besides, direct precipitation, 17 permanent rivers, and 12 seasonal rivers mostly discharging into the south lake supply the lake.

After recording the highest water level of the lake in 1995 (1278.48 m above sea level), the area and water level of Urmia Lake have dramatically decreased and the

S. Asadi (✉) · S. J. Mousavi
Department of Civil and Environmental Engineering, Amirkabir University of Technology, Tehran, Iran
e-mail: asadi.s@aut.ac.ir

S. J. Mousavi
e-mail: jmosavi@aut.ac.ir

J. H. Kim et al. (eds.), *Proceedings of 7th International Conference on Harmony Search, Soft Computing and Applications*, Lecture Notes on Data Engineering and Communications Technologies 140, https://doi.org/10.1007/978-981-19-2948-9_48

lake salinity increased, creating serious threat to the inhabitants of the region. The area and volume of Urmia Lake are the resulting net balance of inflowing and outflowing water. The high-water consumption in different sectors, especially by the agricultural sector (93% according to Iran Ministry of Energy), upsets water resources balance in Urmia Lake basin. Bukan reservoir which is the largest reservoir of the ULB (with a capacity of 825 MCM) built on the largest river of the basin, Zarrineh River. Zarrineh River with 12,000 km^2 catchment area and 22.9% of the whole lake basin area contributes to more than 43% inflow to the lake. The downstream of the dam is the major water consumer of Zarrineh River basin. Bukan reservoir multipurpose dam is supplying water for agricultural, domestic, and industrial uses of Zarrineh River basin and part of Simineh River basin. In addition, flood control is one of the main purposes of constructing Bukan dam. In this critical situation, it is important to optimize operation of Bukan reservoir to meet water demands as well as considering the water required by the lake in order to survive. Recent researches have modeled Bukan reservoir operation optimization using different methods including real-time operation [1–3], hydro-economic model considering climate impact [4], and so forth. Simulation–optimization-based framework is widely used for purposes such as short-term deficit irrigation management for large-scale irrigation districts [5], the optimal conjunctive use of surface and groundwater [6], water management strategies for meeting environmental and agricultural water demands [7], and reliability-based inter-basin water allocation [8].

This study develops a simulation–optimization framework for optimal water allocations downstream the Bukan reservoir using WEAP simulation module and PSO algorithm. The paper is organized as follows: We introduce materials and modeling tools in Sect. 2 with two subsections explaining study area and simulation–optimization framework. Obtained results and discussion are presented in Sect. 3, and finally, conclusions are provided in Sect. 4.

2 Materials and Modeling Tools

2.1 Study Area

Urmia Lake (UL) is the second largest salt lake in the world being drying out. Irrigation as a local human intervention is the main water consumer of Urmia Lake basin (ULB), obstructing water flow from uphill sub-basins to UL. Bukan reservoir and its downstream system including Zarrineh River basin (ZRB) and part of Simineh River basin (SRB) recharging from Bukan reservoir releases are shown in Fig. 1. As Fig. 1 shows, Miandoab plain with about 85,000 ha of farming lands is the main irrigated croplands of ZRB and SRB located downstream of Bukan dam and upstream of Urmia Lake. Supplying water for agricultural, domestic, and industrial uses, as well as flood control were the main purposes of constructing dam in 1971. The reservoir initial storage capacity was 640 MCM, and it has increased to 825

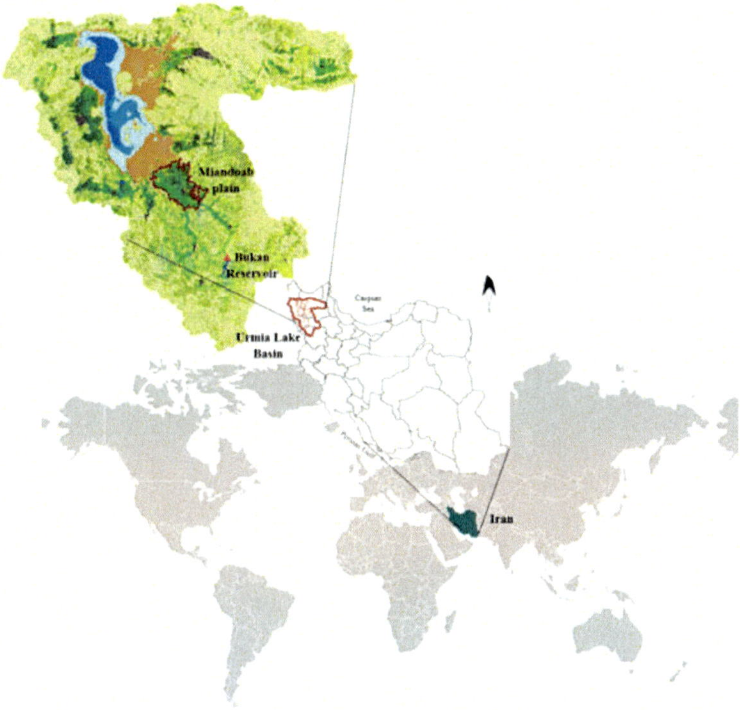

Fig. 1 Bukan reservoir and its downstream system recharging with reservoir release

MCM in 2005. The long-term average annual inflow to the reservoir is 1600 MCM, and the outflow is distributed to Miandoab farming lands by Norozlu diversion dam. In addition, Bukan reservoir outflow supplies domestic water demand of adjacent cities.

2.2 *Simulation–Optimization-Based Reservoir Operation Model*

The water evaluation and planning system (WEAP) is a simulation modeling tool benefiting from a fast, single-period linear programming approach. WEAP simulates the allocation of the available water resources among competing users considering their priorities, in an optimal way. Therefore, it is not able to identify the optimal priority level, optimal reservoir storage volume, or optimize multi-period water allocations. Multi-period optimization models are normally difficult to be solved. Moreover, other programs, programming languages, or scripts can control WEAP directly and change data values, and get results [9].

Metaheuristic optimization algorithms are useful in solving various types of complex optimization problems [10, 11]. The possibility of linking metaheuristic optimization algorithms to simulation models facilitates employment of them. Therefore, simulation–optimization framework is an appropriate alternative for solving multi-period optimization models without solving nonlinear equations. Figure 2 shows the PSO-WEAP algorithm flowchart used in this paper.

The objective function of the proposed model is to minimize the sum of squared shortages in meeting water demands for agricultural lands and environment flows. The optimization model also accounts for retaining enough water in the reservoir as the terminal storage volume, which must be equal to, or more than initial storage volume. Therefore, for final reservoir storage volumes less than the initial storage, a penalty term has been added to the objective function. It prevents the model from emptying the reservoir at the end of the water year. Table 1 reports the decision

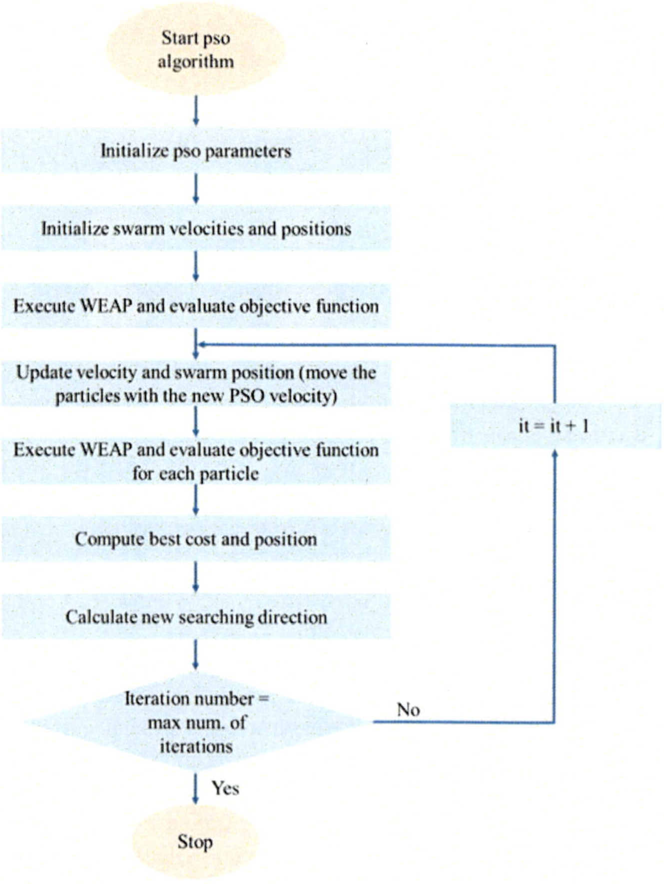

Fig. 2 Flowchart of the simulation–optimization framework

Table 1 Decision variables and their upper and lower bounds

Decision variable	Maximum	Minimum
Ordinal priority of Zarrineh River environmental water requirement (ENV_ZAR)	4	9
Ordinal priority of Simineh River environmental water requirement (ENV_SIM)	4	9
Ordinal priority of ZariEnd demand site (Bukan reservoir downstream demand site on Zarrineh River near to Urmia Lake) (ZariEnd)	4	9
Ordinal priority of ODW1 demand site (Bukan reservoir downstream demand site on Zarrineh River close to dam) (ODW1)	4	9
Ordinal priority of SIM demand site (Bukan reservoir downstream demand site on Simineh River) (SIM)	4	9
Priority number of the Bukan reservoir target storage level	4	9
Initial storage volume of Bukan reservoir	193	825

variables and their upper and lower bounds selected. Ordinal priority numbers of domestic and industrial demands have been defined equal to two. The priorities of agricultural demand sites have been set five, and these priority numbers range from four to nine. Bukan reservoir dead storage is 135.8 MCM. However, to protect the water quality, the minimum storage volume has been constrained to 193 MCM.

Monthly environmental water requirements of Zarrineh River and Simineh River have already been determined [12], which are used in WEAP as input data. In addition, agricultural demands as well as domestic and industrial water demands have been specified based on the data received from the Ministry of Energy [13]. Schematic of the Bukan reservoir and its upstream and downstream system has been demonstrated in Fig. 3.

3 Results and Discussion

The PSO-WEAP model ran over iterations. Figure 4 shows the convergence curve of the algorithm, showing that the model finds the optimal solution after 440 function evaluations. The results are presented in Table 2 where the optimal solution including decision variables, i.e., priority numbers, are reported in column 2 and the objective function term representing the penalty terms imposed by not meeting different demands are mentioned in column 3. These results imply the possibility of almost completely meeting agricultural water demands and environmental requirements through Bukan reservoir optimal releases. The reservoir's initial storage is 408 MCM, which has decreased to 306 MCM at the end of the first month. The terminal reservoir storage is the same as the initial storage.

Bukan reservoir's optimal monthly storage volumes and those of observed volumes are presented in Fig. 5. It shows that water is retained in the reservoir dam

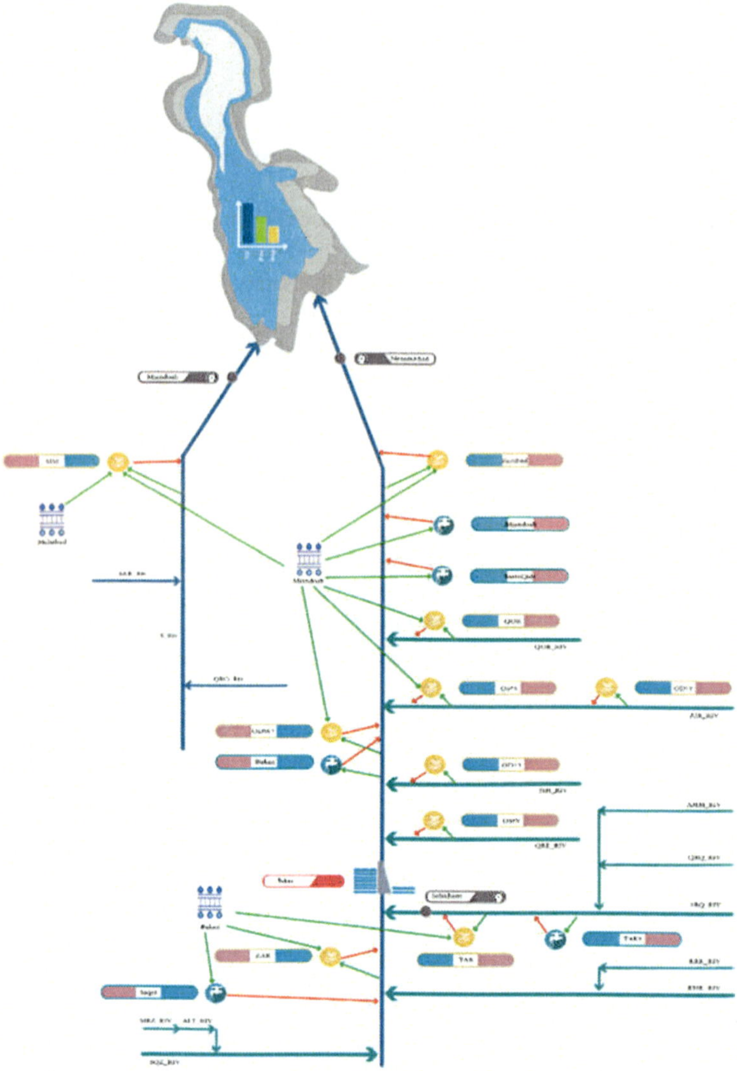

Fig. 3 Schematic representation of the Bukan reservoir and its upstream and downstream system

in winter and is released in spring. Environmental water requirements are mostly needed in winter while the main part of agricultural water demands is required in spring. The strategy of retaining water for agricultural water security in spring should not lead to not meeting the environmental water requirements in winter.

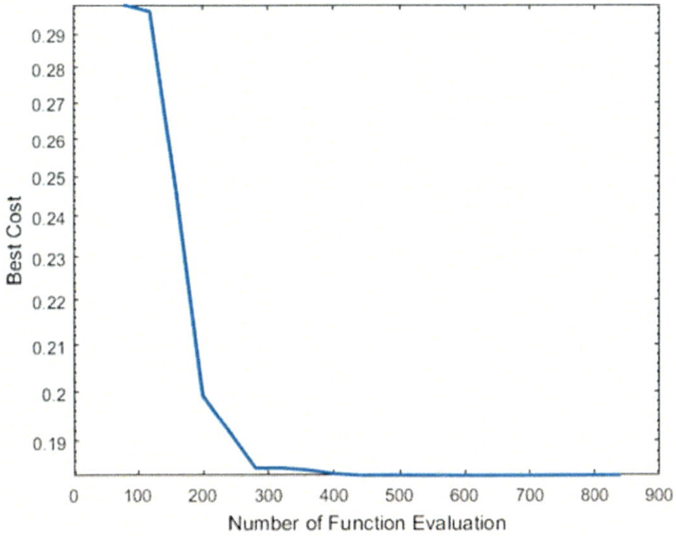

Fig. 4 Convergence of PSO-WEAP model to the optimal solution

Table 2 Optimal values of decision variables and the objective function penalty terms

Decision variable	Priority	Cost
ENV_ZAR	5	0.02
ENV_SIM	7	0.04
ZariEnd	6	3.3E−5
ODW1	7	5.5E-5
SIM	7	0.12
Bukan reservoir	8	0
Summation		0.18

4 Conclusion

In this study, the PSO-WEAP model was presented for optimal water allocations downstream the Bukan Dam reservoir in Lake Urmia basin. Optimal reservoir releases and water allocations to different agricultural and environmental flows were determined and compared to those of historical values. The model determined the terminal end-of-year storage volume left at the end of a yearly planning horizon that accounted for future water uses. The model simultaneously optimized Bukan reservoir operations and downstream water allocations including the Lake Urmia environmental using a hybrid PSO-based multi-period optimization and iterative single-period linear optimization used in WEAP. The results show the possibility of meeting demands in a normal water year almost completely if relative water supply

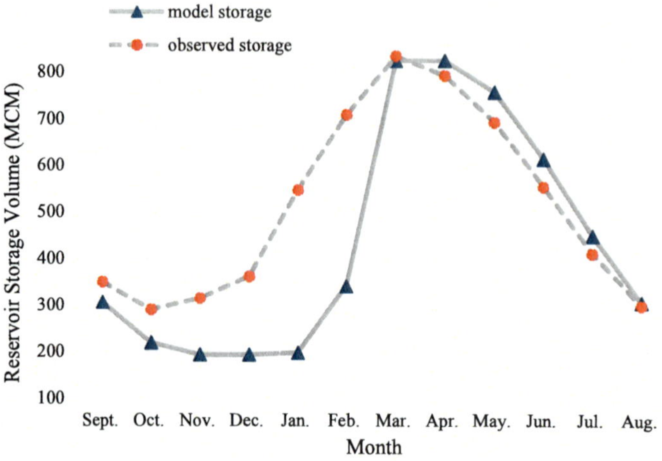

Fig. 5 Optimal storage volumes of Bukan reservoir compared to observed historical volumes in year 2012–2013

and demand priorities are appropriately set considering the requirements of Lake Urmia vital ecosystem.

References

1. Gavahi K, Mousavi J (2017) Real time operation model for optimum operation of Bukan Reservoir in Lake Urmia Basin. Long-term behaviour and environmentally friendly rehabilitation technologies of dams (LTBD 2017). https://doi.org/10.3217/978-3-85125-564-5-076
2. Gavahi K, Mousavi SJ, Ponnambalam K (2018) Comparison of two data-driven streamflow forecast approaches in an adaptive optimal reservoir operation model, vol 3, pp 755–745. https://doi.org/10.29007/wrn8
3. Gavahi K, Mousavi SJ, Ponnambalam K (2019) The role of sreamflow forecast horizon in real-time reservoir operation. Sustainable and safe dams around the world, pp 1603–1610.https://doi.org/10.1201/9780429319778-143
4. Emami F, Koch M (2018) Agro-economic water productivity-based hydro-economic modeling for optimal irrigation and crop pattern planning in the Zarrine river basin, Iran, in the wake of climate change. Sustainability 10(11):3953. https://doi.org/10.3390/su10113953
5. Alizadeh H, Mousavi SJ (2013) Coupled stochastic soil moisture simulation-optimization model of deficit irrigation. Water Resour Res 49(7):4100–4113. https://doi.org/10.1002/wrcr.202822013
6. Chakraei I, Safavi HR, Dandy GC, Golmohammadi MH (2021) Integrated simulation-optimization framework for water allocation based on sustainability of surface water and groundwater resources. J Water Resour Plan Manag 147(3):05021001. https://doi.org/10.1061/(ASCE)WR.1943-5452.0001339
7. Dehghanipour AH, Schoups G, Zahabiyoun B, Babazadeh H (2020) Meeting agricultural and environmental water demand in endorheic irrigated river basins: a simulation-optimization approach applied to the Urmia Lake basin in Iran. Agric Water Manag 241:106353. https://doi.org/10.1016/j.agwat.2020.106353

8. Mousavi SJ, Anzab NR, Asl-Rousta B, Kim JH (2017) Multi-objective optimization-simulation for reliability-based inter-basin water allocation. Water Resour Manage 31:3445–3464. https://doi.org/10.1007/s11269-017-1678-6
9. SEI (1999) WEAP: water evaluation and planning. Tellus Institute, Boston
10. Gendreau M, Potvin JY (2010) Handbook of metaheuristics, vol 2. Springer, New York
11. Maier HR et al (2014) Evolutionary algorithms and other metaheuristics in water resources: current status, research challenges and future directions. Environ Model Softw 62:271–299. https://doi.org/10.1016/j.envsoft.2014.09.013
12. East Azerbaijan Department of Environment (2016) Determining environmental water requirements of wetlands and rivers in Urmia Basin
13. MG Consulting Engineers (2013) Water master plan of Urmia Lake Basin. Preliminary water resources planning study. Tehran, Iran

A Hybrid of Artificial Electric Field Algorithm and Differential Evolution for Continuous Optimization Problems

Dikshit Chauhan and Anupam Yadav

Abstract The artificial electric field algorithm (AEFA) is a minimum and maximum value charged-based function optimization scheme that was created for continuous optimization problems in the first place. The deep-rooted AEFA is always one of the best-designed techniques for discovering optimal solutions to real-world optimization problems. Differential evolution is one of the best established optimization algorithm in the literature. This article aims to utilize the potential of AEFA and DE together to produce an efficient hybrid. Both AEFA and DE are synchronized to sum their strengths. The performance of the proposed hybrid algorithm is validated on seventeen benchmark problems including IEEE-CEC 2019 single objective unconstrained optimization problems and obtained experimental results are compared with several optimization algorithms. We used a statistical test, the Wilcoxon signed-rank test, to assess the randomness of the results produced by the suggested AEFA-DE. The experimental results suggest that the AEFA-DE has superior performance than others.

Keywords Meta-heuristic algorithms · Optimization · AEFA · Benchmark functions · Hybrid algorithm

1 Introduction

In recent decades, a large number of optimization techniques had been introduced to solve real-world optimization problems. Most of the real-world problems can be designed in terms of an optimization model that involves objective and many difficult criteria. For solving a particular group of problems such as integer program-

D. Chauhan (✉) · A. Yadav
Department of Mathematics Dr. B.R. Ambedkar National Institute of Technology Jalandhar, Jalandhar, Punjab 144011, India
e-mail: dikshitc.ma.19@nitj.ac.in

A. Yadav
e-mail: yadava@nitj.ac.in

J. H. Kim et al. (eds.), *Proceedings of 7th International Conference on Harmony Search, Soft Computing and Applications*, Lecture Notes on Data Engineering and Communications Technologies 140, https://doi.org/10.1007/978-981-19-2948-9_49

ming problems, linear programming problems, non-convex optimization problems, quadratic programming problems and many more [15] several optimization techniques are developed. Optimization is a huge area for research in which there is a different methods to solve different types of aforementioned problems. Researchers face difficulties to identify the nature of an optimization problem, and it becomes very hard to choose a suitable method for finding a good solution when the nature is unknown. From the last few decades, researchers are focusing on the appropriate method/algorithm that may solve a wide range of optimization problems. Meta-heuristic algorithms are algorithms that may solve different types of problems or a wide range of problems. Researchers have found that meta-heuristic algorithms solve large scale problems and real-world problems efficiently. Simulated annealing (SA) [22] and Tabu search [3] algorithms are some of the neighborhoods-based algorithms; ant colony optimization (ACO) [2], salp swarm algorithm (SSA) [9], etc., are some of the swarm-based algorithms; genetic algorithm (GA) [8], genetic programming (GP) [5], and differential evolution (DE) [16] are some of the evolutionary-based algorithms; artificial electric field algorithm (AEFA) [24], gravitational search algorithm (GSA) [17], charged system search algorithm (CSS) [4], etc., are some of the physics-based algorithms and etc.

Storn R. et al. introduced an evolutionary-based algorithm differential evolution (DE) in 1995 [19]. It is originally proposed for global continuous problems and contains three operators: mutation, crossover, and selection same as GA. DE has been combined with other optimization algorithms due to its excellent performance. In 2003, Zhang et al. combine DE with PSO called DEPSO [27]. Omran et al. in 2007 proposed a new hybrid PSO and DE in which DE is used to mutant for each particle and attracted particle which is defined as a weighted average of the best solution and its position [12]. A hybrid PSO and DE were proposed by Zhang et al. in 2009 in which particle update according to the operations of DE and mechanism of PSO [26]. Hybrid of DE combined with PSO proposed by Thangaraj et al. in 2008 in which proposed algorithm can preserve the strength of both PSO and DE algorithms [14]. A self-adaptive strategy differential evolution algorithm is called SaDE is proposed by Omran et al. in which both trial vector generation strategies and association control parameter values are self-adaptive by learning from their previous experiences [13]. Fuzzy adaptive differential evolution is called FADE proposed by Liu. et al. in 2005 which based on the fuzzy controllers to adopt the parameters are used [6]. Another variant of DE is success history-based parameter adaption differential evolution is called SHADE proposed by R. Tanabe et al. in 2013 which is based on the parameter adaption technique [20]. In 2014, Tanabe et al. introduced an extended version of SHADE that uses linear population size reduction, which continually decreases the population size [21]. Ali W. Mohamed et al. in 2017 proposed a new variant LSHADE-SPA which is based on the semi-parameter adaption approach [10]. Stanovov et al. proposed a new variant of the LSHADE algorithm by using a rank-based selective pressure strategy called LSHADE-RSP in 2018 [18].

Artificial electric field algorithm is a novel meta-heuristic algorithm introduced by Yadav et al. in 2019 [24]. The origin idea of this scheme is based on the electrostatic force between charged particles. Charged particles shared the information between their own best position and best location in the search space by using electrostatic force. It compared with other meta-heuristic algorithm and provided better results.

Since DE is simple to implement and has fewer parameters to be tuned and DE uses mutation, crossover, and selection operators are a remarkable advantage. Therefore, in this paper, we proposed a hybrid algorithm integrating AEFA with DE algorithm called AEFA-DE to addressed the issues in the AEFA scheme that are less memory to store previous best solution and convergence speed. To resolve the issue of storing the previous best solution of it, a global best solution component is add in the velocity update equation and to increase the convergence rate, DE algorithm is embedded. Several types of benchmark problems are solved by the proposed algorithm. The experimental results obtained by the AEFA-DE are compared with five population-based algorithms and validate with Wilcoxon signed-rank test. The experimental findings suggest that the performance of the AEFA-DE is superior to other comparative algorithms. The organization of this article is presented in the next paragraph.

A brief detail of the AEFA and DE schemes is presented in Sect. 2. The proposed algorithm is explained in Sect. 3. Section 4 introduces the experiments and discussion about the results, and the conclusion and future work are presented in Sect. 5.

2 The Brief Description of Standard AEFA and DE Algorithms

In this section, we present a brief description of standard artificial electric field algorithm (AEFA) and differential evolution (DE) algorithm.

2.1 Artificial Electric Field Algorithms (AEFA)

AEFA is a novel meta-heuristic optimization algorithm that has been proposed by Yadav et al. [24]. The basic theory of the AEFA algorithm is based on Coulomb's theory that states: every charged particle in the universe attract each other with an electrostatic force that is directly proportional to the product of their charges and inversely proportional to the square of the distance between them. AEFA can be considered as a collection of candidate solutions whose charges are the function of fitness. After iterations, all charged particle attracts each other by the electrostatic force between them and a heavier charge particle has a bigger attraction tendency. Therefore, the heavier charge which may be close to the best optimum attracts the other charge particle. The working mechanism of the AEFA is explained in Algorithm 1. The detailed theory of the AEFA is available in article [24].

Algorithm 1 Pseudo code of artificial electric field algorithm

Initialize position and velocity
Calculate the fitness values $(fit_1(t), fit_2(t), \ldots, fit_N(t))$ of agent x
Iteration $t = 0$
Calculate Coulomb's constant $K(t) = K_0 \exp\left(-\alpha \frac{t}{t_{\max}}\right)$, where α and K_0 are the two initial parameters, best fitness value $b(t) = \min(fit_i(t))$ and worst fitness value $w(t) = \max(fit_i(t))$, $i \in (1, 2, \ldots, \tilde{N})$
for $i = 1 : N$ **do**
  Calculate the fitness values $fit_i(t)$, force $F_{ijd}(t) = K(t)\frac{Q_i * Q_j(pbest_{jd}(t) - x_{id}(t))}{R_{ij}(t)+\epsilon}$ and acceleration $ac_{id}(t) = \frac{F_{id}(t)}{Ms_i(t)}$, where $Ms_i(t)$ is the mass (unit) of the ith agent
  Update the velocity and position as
  $vel_i(t+1) = \rho_2 * vel_i(t) + \tilde{ac}_i(t)$
  $x_i(t+1) = x_i(t) + vel_i(t+1)$
  Check stopping criteria is satisfied
  If yes then stop the execution otherwise repeat this process
end for

2.2 *Standard Differential Evolution*

Differential evolution is a novel evolutionary algorithm proposed by Storn and Price [19]. It is an excellent global optimization algorithm and was initially proposed for global continuous optimization problems. It contains three main strategies: mutation, crossover, and selection same as GA. The main difference between GA and DE is DE uses distance and direction information from the current population to search for solutions. In order to obtain a trial vector, the performance of DE depends on the manipulation of the target vector and difference vector. If the fitness of the trial vector is less than a predetermined vector, then the new trial vector will be accepted and compared in the following iteration. In this paper, we use two type mutation strategies as DE/rand/1/bin and DE/best/1/bin and one simple strategy as DE/best/bin; defined in the next paragraph.

Mutation

Let D is the dimension of target vector and t is the tth iteration then consider each target vector as $x_i(t)$ and trial vector as $y_i(t)$. The mutation strategies are given as follows:

1. For DE/rand/1/bin

$$y_i(t) = x_a(t) + \tilde{\beta}(x_b(t) - x_c(t))$$

2. For DE/best/1/bin

$$y_i(t) = x_{\text{best}}(t) + \tilde{\beta}(x_b(t) - x_c(t))$$

3. For DE/best/bin

$$y_i(t) = x_{\text{best}}(t) \quad (1)$$

where $a, b, c \in \{1, 2, \ldots, \tilde{N}\}$ are randomly chosen integers, $a \neq b \neq c \neq i$ and $\tilde{\beta}$ is the scaling factor that control the amplification of the differential evolution.

Crossover
In crossover, a recombination of trial vector and parent vector to produce offspring.

$$z_{j,i}(t) = \begin{cases} y_{j,i}(t), & \text{if rand} < \text{CR or } j = j_{\text{rand}} \\ x_{j,i}(t), & \text{otherwise} \end{cases} \tag{2}$$

where $j, j_{\text{rand}} \in \{1, 2, \ldots, D\}$; rand $\in [0, 1]$ are the randomly chosen index, CR is the crossover rate, and $y_{j,i}(t)$ and $z_{j,i}(t)$ are the difference vector and trial vector of jth particle in the ith dimension at tth iteration respectively.

Selection
To choose the next population between trail and target population selection operator is used

$$x_i(t+1) = \begin{cases} z_i(t), & \text{if } f(z_i(t)) < f(x_i(t)) \\ x_i(t), & \text{otherwise} \end{cases} \tag{3}$$

The more detail of DE is given in Algorithm 2

Algorithm 2 Pseudo code for differential evolution

Evaluate fitness
Use mutation operator and crossover operator and obtain a trial vector
Use selection operator, if the fitness of trial vector is less than a predetermined vector, then the new trial vector will be accepted and compared in the following iteration
$t = t + 1$; repeat the process until the stopping criteria are reached

3 Proposed Hybrid AEFA and DE Algorithms (AEFA-DE)

In this section, we present the theory of hybrid AEFA with DE that called AEFA-DE. AEFA-DE first starts with AEFA algorithm, in this algorithm, we used global best solution to update the velocity, because this algorithm is memory less to store previous best solution, as follows:

$$\text{vel}_i(t+1) = r_i * \text{vel}_i(t) + C1\tilde{a}_i(t) + C2(\text{Gbest} - x_i(t)) \tag{4}$$

Here, $\text{vel}_i(t)$, $\tilde{ac}_i(t)$, and $x_i(t)$ are the velocity, acceleration, and position of the ith agent in the tth iteration, Gbest is the global best solution found so far, $C1 = -1 + t^3/t_{\text{max}}^3$ and $C2 = t^3/t_{\text{max}}^3$ are acceleration coefficients, and r_i is the uniformly distributed number in between 0 and 1. From Eq. 4, a question may be arise, does this modification affect the exploration phase of the AEFA scheme? To balance this phase of it, we used to parameters as $C1$ and $C2$. These parameters are compliment of

Table 1 Selected benchmark functions for evaluation

Function	Functions name	Range	min F	Dim	P
F1	Sphere function	[−100, 100]	$F(0) = 0$	30	U, S
F2	De jong f4 function	[−5.12, 5.12]	$F(0) = 0$	30	U, S
F3	Ackley function	[−30, 30]	$F(0) = 0$	30	M, N
F4	Alpine function	[−10, 10]	$F(0) = 0$	30	M, S
F5	Exponential function	[−1, 1]	$F(0) = -1$	30	M, S
F6	Schwefel 222 function	[−10, 10]	$F(0) = 0$	30	U, N
F7	Sum of different power	[−1, 1]	$F(0) = 0$	30	U, S
CEC 2019 (F8-F17)		Range [1]	min $F = 1$	Dim [1]	

each others, i.e., one is increasing and other is decreasing. From the nature of them, it ensures that the exploration phase is continuously decreases and exploitation phase is reverse of it. After updating the position by AEFA scheme, we used DE algorithm to further locate the best optimal solution. In DE algorithm, trial vector is generated from Eq. 3. If this trial vector is better than the corresponding particle, then it includes in the population, otherwise, the algorithm entered in AEFA algorithm again and generate a new solution with the hope of finding a better solution. The proposed algorithm repeated iteratively till stopping criteria is reached. Since pCR is the crossover rate that affects the diversity of the population for next iteration. It is difficult to choose a good value for crossover rate because if the value of pCR is low, it is unfavorable to population diversity and losses the ability to search global solution; if the value of pCR is high, it is unfavorable to local search and speeds up the convergence rate. Hence, a good value of crossover rate is needed to choose. In this paper, the crossover rate pCR is given as follows:

$$\text{pCR} = \text{CR} + 0.2 * \frac{(\exp(\frac{10(i-1)}{(\tilde{N}-1)}) - 1)}{\exp(10) - 1}, \quad i \in \{1, 2, \ldots, \tilde{N}\} \tag{5}$$

where CR is the initial value of crossover rate which is equal to 0.1. The values of Eq. 5 may enhance speed of convergence rate as well as the diversity of the population. In proposed algorithm, the important role act one factor more that is scaling factor $\tilde{\beta} \in (0, 2)$ because it affects the differential variation between two positions. In this paper, the value of scaling factor $\tilde{\beta}$ is given in Eq. 6:

$$\tilde{\beta} = \text{unifrnd}(\tilde{\beta}_{\min}, \tilde{\beta}_{\max}, \tilde{N}); \tag{6}$$

where $\tilde{\beta}_{\min} = 0.2$ and $\tilde{\beta}_{\max} = 0.8$ are the minimum and maximum value of scaling factor. The more detail of proposed scheme is given in Algorithm 3.

Algorithm 3 Working procedure of AEFA-DE algorithm

Initialization
Randomly initialize $(x_1(t), x_2(t), \ldots, x_{\tilde{N}}(t))$ of population size $\tilde{N}$
Initialize the velocity, $\tilde{\beta}_{\min} = 0.2$, $\tilde{\beta}_{\max} = 0.8$ and CR = 0.1
set iteration $t = 0$
Reproduction and Updating
while Stopping criterion is not satisfied **do**
  Evaluate the fitness values $(fit_1(t), fit_2(t), \ldots, fit_{\tilde{N}}(t))$ of agent x and choose x_{best} position
  Calculate the global best fitness found so far
  Apply AEFA algorithm as:
  Calculate Coulomb's constant, force and acceleration from Algorithm 1
  Update velocity from Eq. 4 and position x from Algorithm 1
  for $i = 1 : \tilde{N}$ **do**
    calculate crossover rate pCR by using Eq. 5
    calculate scaling factor $\tilde{\beta}$ by using Eq. 6
    choose three randomly integers $a, b, c \in \{1, 2, \ldots, \tilde{N}\}$
    Mutation
    if $a \neq b \neq c$ **then**
      Use $DE/rand/1/bin$ as
      $y_i(t) = x_a(t) + \tilde{\beta}(x_b(t) - x_c(t))$
    else if $a \neq b$ **then**
      Use $DE/best/1/bin$
      $y_i(t) = x_{best}(t) + \tilde{\beta}(x_b(t) - x_c(t))$
    else
      $y(_it) = x_{best}(t)$
    end if
    Crossover
    for $j = 1 : D$ **do**
      if $j = j_{rand}$ or $rand < pCR$ **then**
        $z_i(t) = y_j(t)$
      else
        $z_i(t) = x_j(t)$
      end if
      Selection
      if $fit(z_i(t)) < fit(x_i(t))$ **then**
        $x_i(t+1) = z_i(t)$
      end if
    end for
  end for
end while

4 Experiments and Results

4.1 Algorithm Used and Parameter Settings

To analyze the performance of the AEFA-DE, it is compared with five state-of-the-art algorithms, these algorithms are: (i) Bat algorithm [25], (ii) Dragonfly algorithm [7], (iii) Salp swarm algorithm (SSA) [9], (iv) Gravitational search algorithm (GSA) [17], (v) Artificial electric field algorithm (AEFA) [24], (vi) the proposed algorithm. The

Table 2 Fine tuned parameter settings

$\tilde{N}$	K_0	$t_{\max}$	α
30	500	500	30

parameter settings of these algorithms are kept same as proposed by the authors of these algorithms in their original work. The tuned parameter settings for the AEFA-DE are presented in Table 2.

4.2 Evaluation Criteria

To compare the results, the mean and standard deviation (std.) of each algorithm are evaluated for every problem. The following text defines the definition of these performance measures in $\tilde{N}$ independent runs:

Definition 1 *Mean* is the average value of the obtained results by an algorithm as shown

$$\text{Mean} = \frac{\sum_{i=1}^{\tilde{N}} F(x_i^*)}{\tilde{N}} \tag{7}$$

Definition 2 *Standard deviation (std.)* is the degree of dispersion of sample points relative to its mean, from a statistics point of view. It tells how the values are spread across the sample data. If the standard deviation obtained by the optimization algorithm for a problem is very small, it means that the algorithm converges to the same solution and further improvisation need not be required. On the other hand, if it takes a large value, it means that it is near to random results and further improvisation is needed. It is measured by the variation of the sample points from the mean as

$$\text{Std} = \sqrt{\left(\frac{\sum(F(x^*) - \text{mean})^2}{\tilde{N} - 1}\right)} \tag{8}$$

4.3 Experimental Setup

All the algorithms are coded on the MATLAB (2013a) platform. The experiments are conducted on a machine with system configuration Windows 10, Processor Intel(R) Core(TM) i5-8265U CPU, and 8 Logical Processor(s). The experiments are conducted for a fixed $t_{max} \times \tilde{N}$ maximum number of function evaluations, and the best result is recorded over 20 independent runs.

Table 3 Experimental results for all benchmark problems

		DA	BA	SSA	GSA	AEFA	AEFA-DE
F1	Mean values	1.4100E+03	6.6200E−02	4.6700E−07	2.8600E−16	2.0900E+00	6.0600E−30
	Std values	2.4200E+03	1.1500E−01	2.4800E−07	1.4900E−16	5.4900E+00	1.2500E−29
	P values	8.8449E−05 (+)	8.8575E−05 (+)	8.8575E−05 (+)	8.8575E−05 (+)	8.8575E−05 (+)	− −
F2	Mean values	4.6900E+01	2.5400E−04	2.7400E−08	8.0500E−32	5.3200E+02	9.2300E−58
	Std values	5.3200E+01	1.0000E−04	8.6200E−09	1.0800E−31	1.0900E+03	2.9300E−57
	P values	8.8575E−05 (+)	8.8575E−05 (+)	8.8575E−05 (+)	8.8575E−05 (+)	8.8575E−05 (+)	− −
F3	Mean values	1.9000E+00	2.9400E+00	7.4000E−01	4.6600E−02	1.7100E+00	3.9100E−15
	Std values	9.8000E+00	1.3000E+00	2.4600E+00	2.0800E−01	1.2400E+00	3.5100E−15
	P values	8.8449E−05 (+)	8.8575E−05 (+)	8.8324E−05 (+)	8.8449E−05 (+)	8.8324E−05 (+)	− −
F3	Mean values	1.0300E+01	4.0700E+00	1.9200E+00	2.0700E−03	1.4000E+00	8.5700E−16
	Std values	1.3800E+01	2.9200E+00	3.2900E+00	2.1600E−03	1.5200E+00	2.7000E−15
	P values	8.8575E−05 (+)	8.8575E−05 (+)	8.8575E−05 (+)	8.8575E−05 (+)	8.8575E−05 (+)	− −
F5	Mean values	3.8900E−02	4.3900E−03	1.8700E−11	3.9400E−16	9.6600E−01	0.0000E+00
	Std values	7.2400E−02	1.3600E−03	1.3900E−11	1.8200E−16	1.1900E−02	0.0000E+00
	P values	8.8575E−05 (+)	8.8575E−05 (+)	8.8575E−05 (+)	8.3542E−05 (+)	8.8074E−05 (+)	− −
F6	Mean values	7.9800E+00	3.4600E+00	1.6500E+00	7.9400E−02	5.6200E+01	4.6100E−15
	Std values	1.5200E+01	5.4800E+00	1.9000E+00	1.2600E−01	2.8400E+01	5.4000E−15
	P values (+)	8.8575E−05 (+)	8.8575E−05 (+)	8.8575E−05 (+)	8.8575E−05 (+)	8.8575E−05	− −
F7	Mean values	5.8000E−13	1.1000E−37	7.9500E−31	3.9800E−43	7.3300E−02	5.5600E−23
	Std values	1.4100E−13	3.4600E−37	2.7000E−31	1.1800E−42	7.3000E−02	1.7600E−22
	P values	8.8575E−05 (+)	8.8575E−05 (+)	8.9676E−03 (+)	8.8575E−05 (+)	8.8324E−05 (+)	− −
F8	Mean values	9.4300E+10	2.2400E+11	5.5200E+09	3.9900E+12	2.3400E+12	9.2300E+04
	Std values	6.0800E+10	5.8300E+11	6.2500E+09	3.5000E+12	1.5900E+12	3.1600E−03
	P values	8.8575E−05 (+)	8.8575E−05 (+)	8.8575E−05 (+)	8.8449E−05 (+)	8.8575E−05 (+)	− −
F9	Mean values	1.2300E+02	1.7400E+01	5.8700E−02	1.6100E+04	1.3500E+04	3.0900E+02
	Std values	7.9100E+01	1.1400E−01	1.7400E+01	3.9800E+03	3.2500E+03	4.1700E+02
	P values	6.8061E−04 (+)	1.3183E−04 (+)	8.8575E−05 (+)	8.8575E−05 (+)	8.8575E−05 (+)	− −

(continued)

Table 3 (continued)

		DA	BA	SSA	GSA	AEFA	AEFA-DE
F10	Mean values	9.3700E−04	1.2700E+01	1.8300E−12	1.2700E+01	1.2700E+01	1.2700E+01
	Std values	1.2700E+01	3.4100E−04	1.2700E+01	3.6500E−15	9.8200E−04	3.4000E−04
	P values	1.000E+00 (=)	1.000E+00 (=)	1.000E+00 (=)	1.000E+00 (=)	1.000E+00 (=)	− −
F11	Mean values	7.4000E+02	4.4500E+02	1.8200E+01	2.8800E+02	1.4700E+02	3.1900E+01
	Std values	6.6500E+02	2.1800E+02	4.1400E+01	2.5600E+02	1.7300E+02	2.5900E+01
	P values	1.0335E−04 (+)	8.8449E−05 (+)	1.2583E−01 (−)	1.0335E−04 (+)	2.2039E−03 (+)	− −
F12	Mean values	4.3700E−01	1.6300E+00	1.2300E−01	1.0500E+00	1.0100E+00	1.0100E+00
	Std values	1.6800E+00	4.2100E−01	1.2100E+00	5.3100E−02	9.6500E−03	8.1000E−03
	P values	8.8575E−05 (+)	8.8199E−05 (+)	8.7699E−05 (+)	2.2883E−04 (+)	6.2134E−01 (−)	− −
F13	Mean values	1.5200E+00	7.2600E+00	1.8600E+00	1.0000E+00	1.3100E+00	4.1600E+00
	Std values	9.9100E+00	1.9000E+00	4.7300E+00	5.0300E−05	6.3100E−01	3.8900E+00
	P values	1.8901E−04 (+)	8.9676E−03 (+)	4.7809E−01 (−)	3.9063E−03 (+)	4.8828E−03 (+)	− −
F14	Mean values	2.6600E+02	6.6200E+02	2.3700E+02	3.1500E+02	2.1700E+02	2.1600E+02
	Std values	6.0900E+02	4.1400E+02	3.0700E+02	1.8100E+02	1.2800E+02	1.0500E+02
	P values	3.3845E−04 (+)	2.4950E−03 (+)	3.9053E−01 (−)	5.0159E−01 (−)	3.7026E−01 (−)	− −
F15	Mean values	7.1000E−01	5.8200E+00	6.3900E−01	5.6100E+00	4.7600E+00	4.0300E+00
	Std values	5.9400E+00	6.0000E−01	5.4300E+00	4.3900E−01	9.4300E−01	1.0900E+00
	P values	5.1613E−04 (+)	1.8901E−04 (+)	1.4013E−04 (+)	1.6286E−04 (+)	2.0611E−02 (+)	− −
F16	Mean values	2.9600E+00	2.7200E+00	2.0300E−01	4.3500E+00	3.8300E+00	3.4600E+00
	Std values	4.8100E+00	2.2000E−01	2.7000E+00	9.4400E−01	7.3900E−01	4.6000E−01
	P values	9.9899E−03 (+)	1.0320E−04 (+)	8.8324E−05 (+)	5.1072E−03 (+)	8.5924E−02 (−)	− −
F17	Mean values	1.8500E−01	2.0100E+01	1.0400E−01	2.0000E+01	1.9000E+01	1.7800E+01
	Std values	2.0300E+01	1.4100E−01	2.0100E+01	2.5000E−02	4.4700E+00	6.1500E+00
	P values	4.5119E−04 (+)	3.0762E−02 (+)	5.5176E−02 (−)	4.5313E−01 (−)	7.1484E−01 (−)	− −

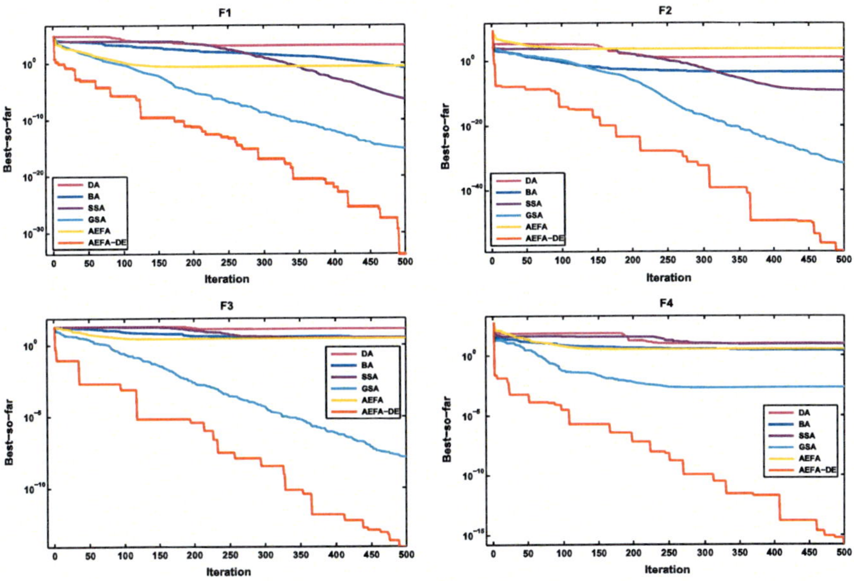

Fig. 1 Convergence plots for selected benchmark problems

4.4 Discussions

To validate the performance of the proposed algorithm, five existing algorithms are selected including the AEFA scheme over seventeen benchmark problems with 100 digit CEC-19 [1]. These benchmark problems are divided into four groups as per their properties: unimodal, multimodal, separable, and non-separable. Table 1 lists these benchmark problems where range represents the range of function, dim indicates the dimension of the problem and min F represents the optimal value of the problem, and P symbolizes the characteristics of benchmark problems where U, M, S, and N stand for unimodal, multimodal, separable, and non-separable problems, respectively. The details of these selected optimization problems are given in [1, 11]. The objective of all algorithms including the AEFA-DE is to find the optimal value of the benchmark problems. In order to decide whether the results of the proposed algorithm differ from other competitive algorithms, a non-parametric Wilcoxon signed-rank test [23] is performed for the statistical validity of the results.

Table 3 and Figs. 1 and 2 are presented the experimental results in terms of mean and standard deviation (std) values along with p-values and convergence curves of selected benchmark problems, respectively. According to the results of this table, it can be observed that the performance of the AEFA-DE is superior to others on ten problems F1–F6, F8, F14, F15, and F17. For F7, the performance of it is better than DA and AEFA but slightly worse than others. The performance of it for F9 is good as AEFA and GSA but not DA, BA, and SSA. For F12, AEFA-DE achieved better mean and std values than others except for SSA. It can be also observed that the standard

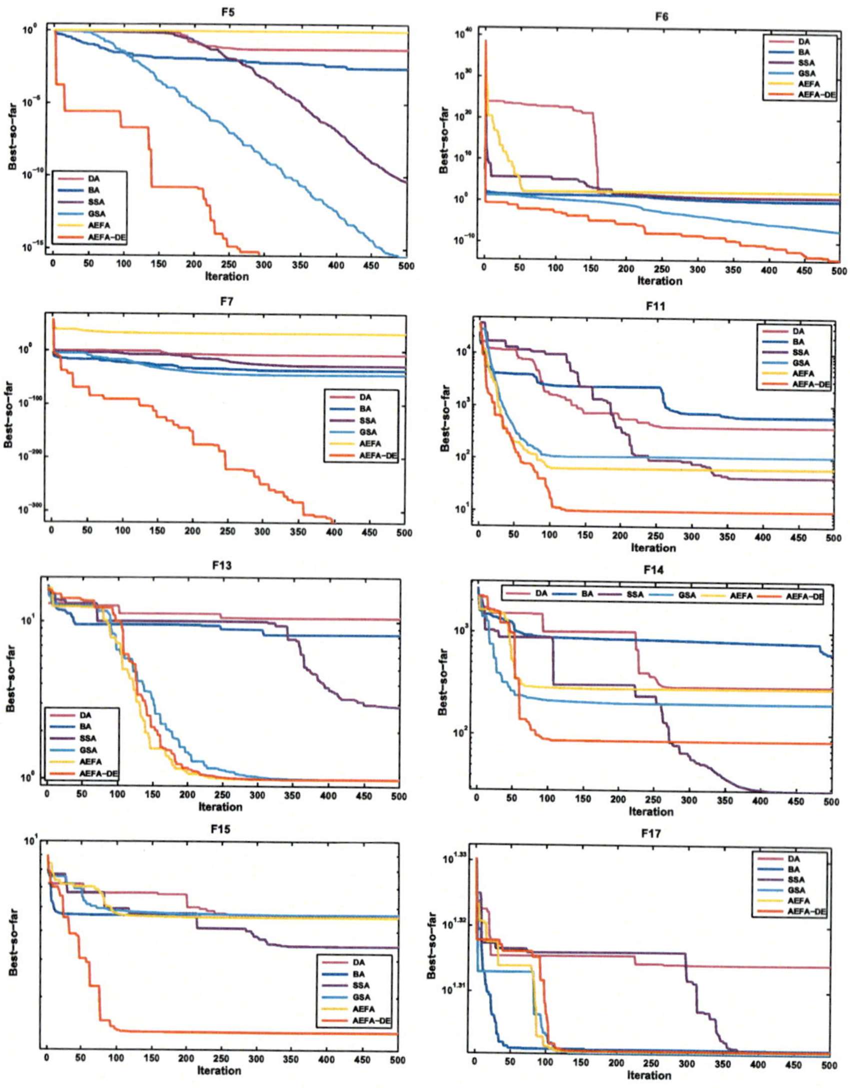

Fig. 2 Convergence plots for selected benchmark problems

deviation of the AEFA-DE is close to zero almost all problems except some this indicates that it has a better ability to achieve the stable optimal values and form the p-values, the results of it are different and better than others due to majority of the $+$ sign. Moreover, the performance of the proposed algorithm is also validated on the convergence curves from Figs. 1 and 2. From this figure, the convergence rate of it is superior to others on problems F1–F17 except for F9 and F14. All these experimental results and curves prove that the performance of the AEFA-DE is superior to other competent algorithms.

5 Conclusions

In this article, a hybrid of artificial electric field algorithm and differential evolution is proposed. To propose this hybrid, the velocity update equation of the AEFA scheme is modified by global best knowledge. The experimental results obtained by AEFA-DE is validated on five state-of-the-art algorithms by their mean and std values. All experimental findings suggested that its performance is superior to other comparative algorithms. The std values proved that the finding results by the proposed algorithm are more stable than others. The p-values or signs indicated that the optimal values of the proposed algorithm are different and excellent than other competent schemes. Hence, this hybrid version of AEFA has superior performance than AEFA and other selected algorithms. From the superiority of the AEFA-DE, in the future, it can apply to discrete, constrained, and multi-objective optimization problems. The performance of it can also validate real-life optimization problems such as image segmentation, high-order graph matching, assignment problems, hydrothermal optimization problems, etc.

References

1. Ali MZ, Suganthan PN, Price KV, Awad NH (2019) The 2019 100-digit challenge on real-parameter, single objective optimization: analysis of results
2. Dorigo M, Birattari M, Stutzle T (2006) Ant colony optimization. IEEE Comput Intell Mag 1(4):28–39
3. Glover F, Laguna M (1998) Tabu search. In: Handbook of combinatorial optimization. Springer, pp 2093–2229
4. Kaveh A, Talatahari S (2010) A novel heuristic optimization method: charged system search. Acta Mech 213(3):267–289
5. Koza JR, Koza JR (1992) Genetic programming: on the programming of computers by means of natural selection, vol 1. MIT Press
6. Liu J, Lampinen J (2005) A fuzzy adaptive differential evolution algorithm. Soft Comput 9(6):448–462
7. Mirjalili S (2016) Dragonfly algorithm: a new meta-heuristic optimization technique for solving single-objective, discrete, and multi-objective problems. Neural Comput Appl 27(4):1053–1073
8. Mirjalili S (2019) Genetic algorithm. In: Evolutionary algorithms and neural networks. Springer, pp 43–55
9. Mirjalili S, Gandomi AH, Mirjalili SZ, Saremi S, Faris H, Mirjalili SM (2017) Salp swarm algorithm: a bio-inspired optimizer for engineering design problems. Adv Eng Softw 114:163–191
10. Mohamed AW, Hadi AA, Fattouh AM, Jambi KM (2017) LSHADE with semi-parameter adaptation hybrid with CMA-ES for solving CEC 2017 benchmark problems. In: 2017 IEEE congress on evolutionary computation (CEC). IEEE, pp 145–152
11. Molga M, Smutnicki C (2005) Test functions for optimization needs, vol 101, p 48
12. Omran MG, Engelbrecht AP, Salman A (2007) Differential evolution based particle swarm optimization. In: 2007 IEEE swarm intelligence symposium. IEEE, pp 112–119
13. Omran MG, Salman A, Engelbrecht AP (2005) Self-adaptive differential evolution. In: International conference on computational and information science. Springer, pp 192–199

14. Pant M, Thangaraj R, Abraham A (2011) DE-PSO: a new hybrid meta-heuristic for solving global optimization problems. New Math Nat Comput 7(03):363–381
15. Pant M, Zaheer H, Garcia-Hernandez L, Abraham A et al (2020) Differential evolution: a review of more than two decades of research. Eng Appl Artif Intell 90:103479
16. Price KV (2013) Differential evolution. In: Handbook of optimization. Springer, pp 187–214
17. Rashedi E, Nezamabadi-Pour H, Saryazdi S (2009) GSA: a gravitational search algorithm. Inf Sci 179(13):2232–2248
18. Stanovov V, Akhmedova S, Semenkin E (2018) LSHADE algorithm with rank-based selective pressure strategy for solving CEC 2017 benchmark problems. In: 2018 IEEE congress on evolutionary computation (CEC). IEEE, pp 1–8
19. Storn R (1996) On the usage of differential evolution for function optimization. In: Proceedings of North American fuzzy information processing. IEEE, pp 519–523
20. Tanabe R, Fukunaga A (2013) Success-history based parameter adaptation for differential evolution. In: 2013 IEEE congress on evolutionary computation. IEEE, pp 71–78
21. Tanabe R, Fukunaga AS (2014) Improving the search performance of shade using linear population size reduction. In: 2014 IEEE congress on evolutionary computation (CEC). IEEE, pp 1658–1665
22. Van Laarhoven PJ, Aarts EH (1987) Simulated annealing. In: Simulated annealing: theory and applications. Springer, pp 7–15
23. Wilcoxon F (1992) Individual comparisons by ranking methods. In: Breakthroughs in statistics. Springer, pp 196–202
24. Yadav A et al (2019) AEFA: artificial electric field algorithm for global optimization. Swarm Evol Comput 48:93–108
25. Yang X-S, Gandomi AH (2012) Bat algorithm: a novel approach for global engineering optimization. Eng Comput
26. Zhang C, Ning J, Lu S, Ouyang D, Ding T (2009) A novel hybrid differential evolution and particle swarm optimization algorithm for unconstrained optimization. Oper Res Lett 37(2):117–122
27. Zhang W-J, Xie X-F (2003) DEPSO: hybrid particle swarm with differential evolution operator. In: SMC'03 conference proceedings. 2003 IEEE international conference on systems, man and cybernetics. Conference theme-system security and assurance (Cat. No. 03CH37483), vol 4. IEEE, pp 3816–3821

Author Index

J. H. Kim et al. (eds.), *Proceedings of 7th International Conference on Harmony Search, Soft Computing and Applications*, Lecture Notes on Data Engineering and Communications Technologies 140, https://doi.org/10.1007/978-981-19-2948-9